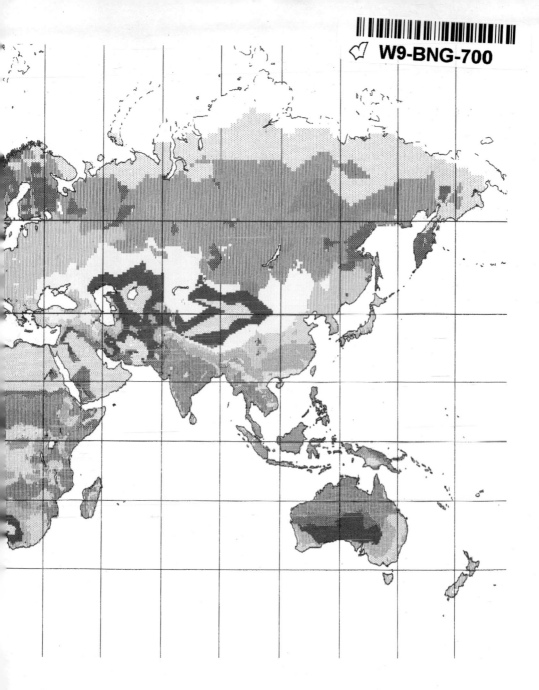

	TROPICAL WET FOREST		TROPICAL THORN WOODLAND
	SUBTROPICAL WET FOREST		TEMPERATE THORN STEPPE
	TROPICAL MOIST FOREST		COOL TEMPERATE STEPPE
	SUBTROPICAL MOIST FOREST		TEMPERATE DESERT BUSH
	TROPICAL DRY FOREST		TROPICAL DESERT BUSH
	WARM TEMPERATE FOREST		DESERT
	DRY WARM TEMPERATE FOREST		BOREAL DESERT
	COOL TEMPERATE FOREST		TUNDRA
	MOIST BOREAL FOREST		ICE
	WET BOREAL FOREST		HIGH ALTITUDE

SCOPE 29

The Greenhouse Effect, Climatic Change, and Ecosystems

Edited by

Bert Bolin

Department of Meteorology, University of Stockholm

Bo R. Döös

Department of Meteorology, University of Stockholm

Jill Jäger

Fridtjot Nansen Strasse 1, Karlsruhe

and

Richard A. Warrick

Climatic Research Unit, University of East Anglia, Norwich

Published on behalf of the
Scientific Committee on Problems of the Environment (SCOPE)
of the
International Council of Scientific Unions (ICSU)
with the support of the United Nations Environment Programme
and the World Meteorological Organisation

by
JOHN WILEY & SONS
Chichester · New York · Brisbane · Toronto · Singapore

Library of Congress Cataloging-in-Publication Data:

Main entry under title:

The Greenhouse effect, climate change, and ecosystems.

   (SCOPE; 29)
   Includes bibliographies and index.
   1. Greenhouse effect, Atmospheric.   2. Climatic
changes.   3. Atmospheric carbon dioxide—Environmental
aspects.   I. Bolin, Bert, 1925–.   II. Döös, Bo R.
III. International Council of Scientific Unions.
Scientific Committee on Problems of the Environment.
IV. Series: SCOPE (Series) (Chichester, West Sussex);
29.
QC912.3.G73   1986   551.5   86-1319
ISBN 0 471 91012 0

British Library Cataloguing in Publication Data:

The Greenhouse effect, climatic change, and
   ecosystems.—(SCOPE; 29)
   1. Atmospheric carbon dioxide
   I. Bolin, Bert   II. Döös, Bo R.
   III. International Council of Scientific Unions.
   *Scientific Committee on Problems of the
   Environment*   IV. Series
   551.6   QC879.8

   ISBN 0 471 91012 0

Printed and bound in Great Britain

International Council of Scientific Unions (ICSU)
Scientific Committee on Problems of the Environment (SCOPE)

SCOPE is one of a number of committees established by the nongovernmental group of scientific organizations, the International Council of Scientific Unions (ICSU). The membership of ICSU includes representatives from 68 National Academies of Science, 18 International Unions and 12 other bodies called Scientific Associates. To cover multidisciplinary activities which include the interests of several unions, ICSU has established 10 Scientific Committees, of which SCOPE is one. Currently representatives of 34 member countries and 15 Unions and Scientific Committees participate in the work of SCOPE, which directs particular attention to the needs of developing countries. SCOPE was established in 1969 in response to the environmental concerns emerging at that time: ICSU reconized that many of these concerns required scientific inputs spanning several disciplines and ICSU Unions. SCOPE's first task was to prepare a report on Global Environmental Monitoring (SCOPE 1, 1971) for the UN Stockholm Conference on the Human Environment.

The mandate of SCOPE is to assemble, review, and assess the information available on man-made environmental changes and the effects of these changes on man; to assess and evaluate the methodologies of measurement of environmental parameters; to provide an intelligence service on current research; and by the recruitment of the best available scientific information and constructive thinking to establish itself as a corpus of informed advice for the benefit of centres of fundamental research and of organizations and agencies operationally engaged in studies of the environment.

SCOPE is governed by a General Assembly, which meets every three years. Between such meetings its activities are directed by the Executive Committee.

R. F. Munn
Editor-in-Chief
SCOPE Publications

Executive Secretary: V. Plocq

Secretariat: 51 Bld de Montmorency
                F-75016 PARIS

Funds to meet SCOPE expenses are provided by contributions from SCOPE National Committees, an annual subvention from ICSU (and through ICSU, from UNESCO), an annual subvention from the French Ministère de l'Environment et du Cadre de Vie, contracts with UN Bodies, particularly UNEP, and grants from Foundations and industrial enterprises.

Contents

6 Empirical Climate Studies 271

Warm World Scenarios and the Detection of Climatic Change
Induced by Radiatively Active Gases

T. M. L. Wigley, P. D. Jones, and P. M. Kelly

7 Changing the Sea Level 323

Projecting the Rise in Sea Level Caused by Warming of the
Atmosphere

G. deQ. Robin

Preface

The problems of the increasing atmospheric carbon dioxide concentration and possible future climatic changes have attracted considerable attention in recent years. A number of assessments of this problem have been made by national groups, notably in the United States. The problem is clearly an international one and an assessment at the international level therefore seems desirable to serve as a basis for discussion and possibly, at some stage, for the development of an action plan. The present analysis is aimed at serving such a purpose and is the result of an agreement between the United Nations Environment Programme (UNEP), the World Meteorological Organization (WMO) and the International Council of Scientific Unions (ICSU), the organizations which jointly implement the World Climate Programme.

In the present assessment the following major questions have been considered:

— How much CO_2 has been and will be released into the atmosphere as a result of fossil-fuel combustion (Chapter 2)?

— What are the natural sources and sinks of carbon (the global carbon cycle) and what projections can be made of future atmospheric CO_2 concentrations (Chapter 3)?

— What are the expected increases of other greenhouse gases that affect the Earth's radiation budget (Chapter 4)?

— How will global and regional climates change as a result of increases in CO_2 and and other greenhouse gases (Chapter 5)?

— When and how will climatic changes be detected? Is it possible to design climate scenarios which can be used for climate impact studies (Chapter 6)?

— What changes of sea level can be expected as a result of a warming of the atmosphere (Chapter 7)?

— What are the responses of terrestrial ecosystems to direct effects of an increased atmospheric concentration of CO_2 and climatic change (Chapters 8–10)?

For each one of these questions one or a few well-recognized scientists have been asked to summarize our present knowledge and, in doing so, also to present the main uncertainties and controversial opinions that exist. These contributions have been exposed to critical reviews by panels of scientists in respective fields who have expressed their views either during panel meetings, by correspondence or in direct personal contact with the authors. The aim has been to arrive at analyses of the different aspects of the problem areas that describe current knowledge in a balanced and well-documented manner.

Chapter 1 is intended to address these more general problems based on the detailed presentations in Chapters 2–10. Such an analysis certainly can be written in many different ways depending on how our present knowledge is evaluated. The present effort is aimed at giving a balanced overview and at indicating major controversies.

With these presentations as a background the UNEP/WMO/ICSU International Conference on the Assessment of the role of carbon dioxide and of other greenhouse gases in climate variations and associated impacts at Villach, Austria, in October 1985 could be devoted primarily to discussions of

— How to deal with the major uncertainties in our factual knowledge base, which will probably not be significantly altered until a change has been observed.

— The nature of international collaboration and action that might be necessary now, on the basis of present knowledge, or later, when more information has become available.

— Research priorities to provide the best possible knowledge about the results of man's future exploitation of his natural environment and their impact on climate on Earth.

Included in the present volume is also the Conference Statement from the conference at Villach, which is aimed at serving the nations of the world and the international organizations concerned in their further attending to the problem of possible future man-induced changes of the global climate.

The assessment has been carried out at the International Meteorological Institute in Stockholm with the participation of a large number of scientists, who are listed at the beginning of each chapter. Their devoted work to complete the task is gratefully acknowledged.

We also wish to thank Ingrid Gustafson and Benita Wahlström for their secretarial assistance.

Financial support for the conduct of the present assessment has come primarily from the United Nations Environment Programme (UNEP). Major contributions have also been given by the World Meteorological Organization (WMO), the Scientific Committee on Problems of the Environment (SCOPE/ICSU), the World Resources Institute (WRI) Washington, DC (USA), the Swedish Board for Energy Research (Energiforskningsnämnden) and by the International Meteorological Institute in Stockholm.

Bert Bolin, Bo R.Döös, Jill Jäger, and Richard A. Warrick
Stockholm, November 1985

Contributors to the present SCOPE Report

M. Ya. Antonovsky USSR State Committee for Hydrometeorology and Control of the Natural Environment, Pavlika Morozova Street 12, 123376 Moscow, USSR

B. Bolin University of Stockholm, Department of Meteorology, Arrhenius Laboratory, S-106 91 Stockholm, Sweden

H.-J. Bolle Universität Innsbruck, Institut für Meteorologie und Geophysik, Schöpfstrasse 41, A-6020 Innsbruck, Austria

R. E. Dickinson National Center for Atmospheric Research, Boulder, Colorado 80307-3000, USA

B. R. Döös International Meteorological Institute in Stockholm, Arrhenius Laboratory, S-106 91 Stockholm, Sweden

R. M. Gifford CSIRO, Division of Plant Industry, GPO Box 1600, Canberra, ACT 2601, Australia

J. Jäger Fridtjof Nansen Strasse 1, D-7500 Karlsruhe 41, FRG

P. G. Jarvis University of Edinburgh, Department of Forestry and Natural Resources, Darwin Building, Mayfield Road, Edinburgh, EH9 3JU Scotland

P. D. Jones Climatic Research Unit, University of East Anglia, Norwich NR4 7TJ, England

W. Keepin The Beijer Institute of the Royal Swedish Academy of Sciences, Box 50005, S-104 05 Stockholm, Sweden

P. M. Kelly Climatic Research Unit, University of East Anglia, Norwich NR4 7TJ, England

L. Kristoferson The Beijer Institute of the Royal Swedish Academy of Sciences, 50005, S-104 05 Stockholm, Sweden

I. Mintzer The World Resources Institute, 1735 New York Avenue, N.W. Washington D.C. 20006, USA

M. L. Parry University of Birmingham, Department of Geography, PO Box 363, Birmingham B15 2TT, England

G. deQ. Robin Scott Polar Research Institute, Lensfield Road, Cambridge, CB2 1ER England

A. P. Sandford University of Edinburgh, Department of Forestry and Natural Resources, Darwin Building, Mayfield Road, Edinburgh EH9 3JU, Scotland

W. Seiler Max Planck Institut für Chemie, Otto Hahn Institut, Postfach 3060, D-6500 Mainz, FRG

H. H. Shugart Department of Environmental Sciences, Clark Hall, Charlottesville, Virginia 22903, USA

J. R. Tarrant School of Environmental Sciences, University of East Anglia, Norwich NR4 7TJ, England

C. J. Tucker Goddard Space Flight Center, National Aeronautics and Space Administration, Greenbelt, Maryland 20771, USA

R. A. Warrick Climatic Research Unit, University of East Anglia, Norwich NR4 7TJ, England

T. M. L. Wigley Climatic Research Unit, University of East Anglia, Norwich NR4 7TJ, England

Statement by the UNEP/WMO/ICSU International Conference on

THE ASSESSMENT OF THE ROLE OF CARBON DIOXIDE AND OF OTHER GREENHOUSE GASES IN CLIMATE VARIATIONS AND ASSOCIATED IMPACTS
VILLACH, AUSTRIA, 9–15 OCTOBER 1985

A joint UNEP/WMO/ICSU Conference was convened in Villach (Austria) from 9 to 15 October 1985, with scientists from twenty nine developed and developing countries, to assess the role of increased carbon dioxide and other radiatively active constituents of the atmosphere (collectively known as greenhouse gases and aerosols) on climate changes and associated impacts. The other greenhouse gases reinforce and accelerate the impact due to CO_2 alone. As a result of the increasing concentrations of greenhouse gases, it is now believed that in the first half of the next century a rise of global mean temperature could occur which is greater than any in man's history.

The Conference reached the following conclusions and recommendations:

1) Many important economic and social decisions are being made today on long-term projects—major water resource management activities such as irrigation and hydro-power, drought relief, agricultural land use, structural designs and coastal engineering projects, and energy planning—all based on the assumption that past climatic data, without modification, are a reliable guide to the future. This is no longer a good assumption since the increasing concentrations of greenhouse gases are expected to cause a significant warming of the global climate in the next century. It is a matter of urgency to refine estimates of future climate conditions to improve these decisions.

2) Climate change and sea level rises due to greenhouse gases are closely linked with other major environmental issues, such as acid deposition and threats to the Earth's ozone shield, mostly due to changes in the composition of the atmosphere by man's activities. Reduction of coal and oil use and energy conservation undertaken to reduce acid deposition will also reduce emissions of greenhouse gases, a reduction in the release of chloro-fluorocarbons (CFCs) will help protect the ozone layer and will also slow the rate of climate change.

3) While some warming of climate now appears inevitable due to past actions, the rate and degree of future warming could be profoundly affected by governmental policies on energy conservation, use of fossil fuels, and the emission of some greenhouse gases.

These conclusions are based on the following consensus of current basic scientific understanding:

- The amounts of some trace gases in the troposphere, notably carbon dioxide (CO_2), nitrous oxide (N_2O), methane (CH_4), ozone (O_3) and chlorofluorocarbons (CFCs) are increasing. These gases are esentially transparent to incoming short-wave solar radiation but they absorb and emit longwave radiation and are thus able to influence the Earth's climate.

- The role of greenhouse gases other than CO_2 in changing the climate is already about as important as that of CO_2. If present trends continue, the combined concentrations of atmospheric CO_2 and other greenhouse gases would be radiatively equivalent to a doubling of CO_2 from pre-industrial levels possibly as early as the 2030's.

- The most advanced experiments with general circulation models of the climatic system show increases of the global mean equilibrium surface temperature for a doubling of the atmospheric CO_2 concentration, or equivalent, of between 1.5 and 4.5 °C. Because of the complexity of the climatic system and the imperfections of the models, particularly with respect to ocean–atmosphere interactions and clouds, values outside this range cannot be excluded. The realization of such changes will be slowed by the inertia of the oceans, the delay in reaching the mean equilibrium temperatures corresponding to doubled greenhouse gas concentrations is expected to be a matter of decades.

- While other factors such as aerosol concentrations, changes in solar energy input, and changes in vegetation may also influence climate, the greenhouse gases are likely to be the most important cause of climate change over the next century.

- Regional scale changes in climate have not yet been modelled with confidence, However, regional differences from the global averages show that warming may be greater in high latitudes during late autumn and winter than in the tropics, annual mean runoff may increase in high latitudes, and summer dryness may become more frequent over the continents at middle latitude in the Northern Hemisphere. In tropical regions, temperature increases are expected to be smaller than the average global rise, but the effects on ecosystems and humans could have far reaching consequences. Potential evapotranspiration probably will increase throughout the tropics whereas in moist tropical regions convective rainfall could increase.

- It is estimated on the basis of observed changes since the beginning of this century, that global warming of 1.5 °C to 4.5 °C would lead to a sea-level rise of 20–140 centimeters. A sea-level rise in the upper portion of this range would have major direct effects on coastal areas and estuaries. A significant melting of the West Antarctic ice sheet leading to a much larger rise in sea level, although possible at some future date, is not expected during the next century.

- Based on analyses of observational data, the estimated increase in global mean temperature during the last one hundred years of between 0.3 and 0.7 °C is consistent with the projected temperature increase attributable to the observed increase in CO_2 and other greenhouse gases, although it cannot be ascribed in a scientifically rigorous manner to these factors alone.

- Based on evidence of effects of past climatic changes, there is little doubt that a future change in climate of the order of magnitude obtained from climate models for a doubling of the atmospheric CO_2 concentration would have profound effects on global ecosystems, agriculture, water resources and sea ice.

RECOMMENDED ACTIONS

1. Governments and regional inter-governmental organizations should take into account the results of this assessment (Villach 1985) in their policies on social and economic development, environmental programmes, and control of emissions of radiatively active gases.

2. Public information efforts should be increased by international agencies and governments on the issues of greenhouse gases, climate change and sea level, including wide distribution of the documents of this conference (WMO, 1986).

3. Major uncertainties remain in predictions of changes in global and regional precipitation and temperature patterns. Ecosystem responses are also imperfectly known. Nevertheless, the understanding of the greenhouse question is sufficiently developed that scientists and policy-makers should begin an active collaboration to explore the effectiveness of alternative policies and adjustments. Efforts should be made to design methods necessary for such collaboration.

   (i) Governments and funding agencies should increase research support and focus efforts on crucial unsolved problems related to greenhouse gases and climate change. Priority should be given to national scientific programme initiatives such as (a) the World Climate Research Programme (WMO-ICSU), (b) present and proposed efforts on biogeo-chemical cycling and tropospheric chemistry in the framework of the Global Change Programme proposed by ICSU, (c) National Climatic Research Programmes. Special emphasis should be placed on improved modelling of the ocean, cloud-radiation interactions, and land surface processes.

(ii) Support for the analysis of policy and economic options should be in-creased by governments and funding agencies. In these assessments the widest possible range of social responses aimed at preventing or adapt-ing to climate change should be identified, analyzed and evaluated. These assessments should be initiated immediately and should employ a variety of available methods. Some of these analyses should be under-taken in a regional contest to link available knowledge with economic decision-making and to characterize regional vulnerability and adapt-ability to climate change. Candidate regions may include the Amazon Basin, the Indian subcontinent, Europe and Arctic, the Zambezi Basin, and the North American Great Lakes.

4.   Governments and funding institutions should strongly support the fol-lowing:

(i) Long-term monitoring and interpretation with state-of-the-art models of:

    (a) radiatively important atmospheric constituents in addition to CO_2, including aerosols,

    (b) solar irradiance, and

    (c) sea level.

(ii) Study and interpretation of the past history of climate and environ-ment, specially regarding interactions among the atmosphere, oceans and ecosystems.

(iii) Studies of the effects of atmospheric compostion and of changing cli-mate and climatic extremes on sub-tropical and tropical ecosystems, boreal forests, and on water regimes.

(iv) Investigations of the sensitivity of the global agricultural resource base with respect to:

    (a) direct effects of increases in atmospheric CO_2 and other greenhouse gases,

    (b) effects of changes in climate, and

    (c) probable combinations of these.

(v) Evaluation of social and economic impacts of sea-level rises.

(vi) Analysis of policy-making procedures under the kinds of risks implied by a significant greenhouse warming.

5.   UNEP, WMO and ICSU should establish a small task force on greenhouse gases, or take other measures, to:

 (i) Help ensure that appropriate agencies and bodies follow up the recommendations of Villach 1985.

(ii) Ensure periodic assessments are undertaken of the state of scientific understanding and its practical implications.

(iii) Provide advice on further mechanisms and actions required at the national or international levels.

(iv) Encourga research in developing countries to improve energy efficiency and conservation.

(v) Initiate, if deemed necessary, consideration of a global convention.

REFERENCE

World Meteorological Organisation (WMO) (1986) *Report of the International Conference on the assessment of the role of carbon dioxide and of other greenhouse gases in climate variations and associated impacts*, Villach, Austria, 9–15 October 1985, WMO No. 661.

Executive Summary

The amounts of some trace gases in the atmosphere, notably carbon dioxide (CO_2), nitrous oxide (N_2O), methane (CH_4), chlorofluorocarbons and tropospheric ozone, have been increasing. All of these gases are transparent to incoming short-wave radiation, but they absorb and emit long-wave radiation and are thus able to influence the Earth's climate. They are referred to in this report as greenhouse gases.

Increased concentrations of CO_2 and other greenhouse gases lead to a warming of the Earth's surface and the lower atmosphere. The resulting changes in climate and their impacts (e.g. on sea level, agriculture and forestry) can be estimated without associating the origin of the warming to any one of these gases specifically. It is, however, necessary to study the effects of these greenhouse gases separately in order to estimate their relative contributions to the warming at any given time and, consequently, to develop strategies for reducing their possible harmful effects.

A review of previous assessments of the CO_2 problem shows that there are agreements on some basic issues. The net emissions of CO_2 from the biota (due to deforestation and land use changes) in themselves will be insufficient to cause a significant change of climate, while fossil fuel reserves are large enough for climatic changes to occur if these reserves continue to be exploited at a high rate in the future.

Generally it has also been agreed that regional patterns of climatic change cannot yet be predicted. Thus, the ways in which higher CO_2 concentrations and given changes in climate would affect ecosystems and human activities cannot be predicted either. This is presumably one of the main reasons why there has been substantial disagreement among previous studies regarding recommendations for future action.

Emission of CO_2

- The observations that began in 1958 have clearly shown that the atmospheric CO_2 concentration has increased from about 315 ppmv in 1958 to about 343 ppmv in 1984. We know the amount of CO_2 that has been emitted into the atmosphere by fossil fuel combustion and changing land use and can relate the observed increase of atmospheric CO_2 to these human activities.

- An evaluation of the 'pre-industrial level' of atmospheric CO_2 (concentrations occurring in the middle of the last century), based on direct measurements and analyses of air trapped in glacier ice, yields a value of 275 $\pm$ 10 ppmv.

- Combustion of fossil fuels—primarily oil, gas and coal—currently meets about 80% of the global energy demand. Future emissions of CO_2 will depend on how this global demand changes and what role fossil fuels play in the future supply. However, even short-term (a few decades) projections of energy use are very uncertain.

- The present assessment places an upper bound on possible CO_2 emissions of about 20 Gt C/year in the year 2050, i.e. about a fourfold increase of the present emission ($\approx$ 5 Gt C/year). Higher values seem unlikely in view of environmental, social and logistic constraints.

- The lower bound on CO_2 emissions is placed at 2 Gt C/year in 2050. This value could possibly be achieved by sustained global efforts to limit the future use of fossil-fuel energy by decreasing energy demands and by increasing the use of non-fossil energy sources.

- Deforestation and land-exploitation cannot be major future sources of atmospheric CO_2. Much less carbon is stored as wood and in soils compared with fossil fuel deposits.

- Policy decisions concerning fossil-fuel use should take into account the negative effects of CO_2 emissions with regard to changes in climate while simultaneously considering their other environmental effects (e.g. air pollution, increasing ozone concentrations, acid rain).

Increases of atmospheric CO_2

- Understanding of the global carbon cycle has improved in recent years. Despite these advances, it is still not possible to balance the global carbon budget completely. However, the remaining uncertainties do not seriously influence the conclusions regarding the future levels of atmospheric CO_2 concentrations.

- Constant or very slowly increasing (0.5% per year) emissions of CO_2 during the next four decades, with slowly increasing emissions thereafter, would give an atmospheric CO_2 concentration of less than 440 ppmv at the end of the next century (i.e. less than 60% above the pre-industrial level).

- If the present increase of CO_2 emission (an average of 1–2% per year since 1973) continues over the next four decades with a slackening of the rate of increase thereafter, a doubling of the pre-industrial CO_2 concentration would be reached towards the end of the next century.

- The upper bound scenario implies that the CO_2 concentration might double by the middle of the next century, while the lower bound scenario implies that doubling of CO_2 concentration will not be reached until after 2100.

Other greenhouse gases and aerosols

- The equilibrium temperature change due to the increasing concentrations of other greenhouse gases (in particular, nitrous oxide, methane, tropospheric ozone and chlorofluorocarbons) up to the present is estimated to be about half of the temperature change attributed to the increase of atmospheric CO_2 alone. The concentrations of some of these gases, and hence their relative importance in changing the climate, are increasing more rapidly than that of CO_2.

- If present trends continue, the role of non-CO_2 greenhouse gases in changing the climate will be about as important as that of CO_2 during coming decades. The combined concentrations of atmospheric CO_2 and other greenhouse gases would be equivalent to a doubling of CO_2 possibly as early as the third decade of the next century.

- Chlorofluorocarbons may, within decades, become the greenhouse gas that next to CO_2, is increasing its importance for changing the radiative properties of the atmosphere most rapidly if no preventive measures are taken. On the other hand, their regulation would be easier to achieve than the limitation or reduction of CO_2 emissions.

- Our knowledge of the global biogeochemical cycles that determine atmospheric concentrations of methane, nitrous oxide and ozone is still inadequate as a basis for policy decisions on how to reduce or limit the future growth of their concentrations.

- Although changes in global climate due to increasing concentrations of aerosols in the atmosphere have probably not been significant in the past, the possibility that they may become of importance in the future, particularly regionally, cannot be excluded. Future changes in aerosol concentrations cannot be projected with any certainty.

Changes in climate

- An evaluation of results from climate models leads to the conclusion that the increase in global mean equilibrium surface temperature due to increases of CO_2 and other greenhouse gases equivalent to a doubling of the atmospheric CO_2 concentration is likely to be in the range of 1.5–5.5 °C. Although no value within this wide range of uncertainty can be excluded, it is plausible that the increase may be found in the lower half of this range.

- The observed increase in mean temperature during the last 100 years (0.3–0.7 °C) cannot be ascribed in a statistically rigorous manner to the increasing concentration of CO_2 and other greenhouse gases, although the magnitude is within the range of predictions (0.3–1.1 °C).

- The expected change of the global mean temperature due to a doubling of CO_2 is of about the same magnitude as the change of global temperature from the last glacial period to the present interglacial.

- Continental or regional scale changes in climate cannot yet be modelled confidently, except that there are indications that warming will be enhanced in high latitudes and that summer dryness may become more frequent over the continents at middle latitudes in the Northern Hemisphere.

Changes in sea level

- The global average sea level has risen 12 ± 5 cm during the twentieth century.

- It is estimated empirically, on the basis of observed changes since the beginning of this century, that a global warming of 1.5 °C to 5.5 °C would lead to a sea-level rise of 20 to 165 cm. The major contributing factor to such a rise would be the thermal expansion of ocean water.

- A disintegration of the West Antarctic ice sheet is not judged to be imminent and would take a century or more if it were to occur. Many glaciologists consider, however, that further research is necessary before a reliable assessment of this possibility can be made.

Assessing the impact on ecosystems

- Based on evidence from climatic changes of the distant past there is little doubt that a future change in climate of the order of magnitude obtained from climate models for a doubling of the atmospheric CO_2 concentration could have profound effects on global ecosystems.

- Prediction of the future impacts on ecosystems is precluded by the lack of reliable estimates of climatic changes at regional scales, and by the lack of knowledge concerning the interactive effects of CO_2 and climate variables on vegetation. Despite the inability to make predictions, sensitivity analyses can produce useful information for judging the possible direction and magnitude of effects for given changes in CO_2 levels or climate variables, and thus for identifying specific regions and environmental changes which may warrant policy attention in the future.

Consequences for agriculture

- In general, the direct effects of enhanced CO_2 concentrations on crop yields are beneficial. It is estimated from laboratory experiments on individual plants that, in the absence of climatic change, a doubling of the CO_2 concentration would cause a 0–10% increase in growth and yield of C_4 crops (e.g. maize, sorghum, sugar cane) and a 10–50% increase for C_3 crops (e.g. wheat, soybean, rice), depending on the specific crop and growing conditions.

- In analysing the sensitivity of crop yields to possible changes in climate without including the direct effects due to higher CO_2 concentrations, most research has focused on average yields of cereal grains in core crop regions of the temperate latitudes. Less attention has been paid to the tropics and subtropics, to the climate-sensitive margins of production and to possible changes in year-to-year climatic extremes.

- Crop-impact analyses show consistently that warmer average temperatures are detrimental to both wheat and maize yields in the mid-latitude core-crop regions of North America and Western Europe. Given current technology and crop varieties, a warming of 2 °C with no change in precipitation might reduce average yields by 10 ± 7%. Increases in precipitation could partially offset these effects, while drier conditions could exacerbate them. Changes in the length of the growing season or in the frequencies of extreme climatic events could also have important effects.

- At the margins of crop areas, spatial shifts in cropping patterns might occur as a result of changes in climate. A limited number of marginal-spatial analyses suggest that, in the mid- to high-latitude cereal growing regions, horizontal shifts of several hundred kilometres per °C change are possible, assuming unchanged technology and economic constraints. In North America, these are comparable in magnitude with shifts in crop patterns that have taken place over this century. At the cool, high-altitude limits of production, altitudinal shifts of more than 100 metres per °C may be possible.

- Models of agricultural production and trade suggest that numerous feedback mechanisms exist in many regions through which agriculture can adjust and adapt to environmental change. Over the long term, food production in such areas appears more sensitive to technology, price or policy changes than to climatic changes, and these factors are largely controllable, whereas climate is not.

• However, for some regions, particularly the lands marginal for food production in the developing world, agriculture may be acutely sensitive to climatic change, as evidenced by the tolls taken by year-to-year variations in climate. If these regions can adopt measures to reduce further the ill-effects of current, short-term climatic variability it is likely that they will be better prepared to adapt to some adverse effects of future changes in climate, should they occur.

Consequences for forests

• At present firm conclusions cannot be reached regarding the direct effects of elevated CO_2 concentrations on the productivity, species competition or size and areal extent of the world's forests. This is because of the paucity of experimental evidence for relevant tree species, particularly over one or more growing cycles, and the large uncertainties involved in 'scaling up' from the short-term responses of individual leaves or plants to complex forest systems.

• If, indeed, elevated CO_2 concentrations do result in increased growth of individual trees over the long term, increases in productivity would be most likely to occur in commercial forest plantations, and would be less likely to occur in mature natural forests, although biomass turnover rates would increase.

• The sensitivity of forests to climatic change has been analysed using forest simulation models. These studies suggest that temperature-increases of the size indicated by current climate models for a doubling of atmospheric CO_2 are potentially sufficient to produce substantial intermediate and long-term responses in the composition, size and location of forest ecosystems.

• These climate models predict the largest warming to occur at high latitudes as a result of increased concentrations of greenhouse gases, with smaller rises in temperature in the lower latitudes. The natural forests of the high latitudes in general, and the boreal forests in particular, may be most sensitive to temperature changes. Warmer conditions could thus possibly lead to large reductions in the areal extent of boreal forests and a poleward shift in their boundaries.

• The forests of the tropical and sub-tropical zones probably would be more sensitive to changes in precipitation than in temperature. However, because of the high uncertainty regarding future changes in precipitation in the tropics, and because of the lack of models that can be used to simulate the effects on tropical ecosystems caused by changes in climate variables, it is virtually impossible at present to make informed prediction of the responses of tropical forests to future climatic changes.

The possible problem of a change in climate due to the emissions of greenhouse gases should be considered as one of today's most important long-term environmental problems. It should be considered in the context of other ongoing changes of our environment also caused by human activities, such as air pollution, acid rain and deforestation. Only in this way can we achieve a realistic integrated view of the interplay between the environment as a whole and the global society that is required for thoughtful consideration of options and policies for avoiding long-term adverse consequences.

CHAPTER 1

The Greenhouse Effect, Climatic Change, and Ecosystems

A Synthesis of Present Knowledge

B. BOLIN, J. JÄGER, AND B. R. DÖÖS

1.1 INTRODUCTION

Man's expanding activities have reached a level at which their effects are global in nature. The natural systems, i.e. the atmosphere, land and sea as well as life on this planet, are clearly being disturbed. We know that some natural trace gases in the atmosphere, such as carbon dioxide (CO_2), nitrous oxide (N_2O), methane (CH_4) and tropospheric ozone (O_3), have been increasing during the last century. In addition, other gases are being emitted that are not naturally part of the global ecosystem, notably chlorofluorocarbons. These trace gases absorb and emit radiation and are thus able to influence the Earth's climate. They will be referred to collectively as greenhouse gases.

It is also clear that the major natural terrestrial biomes are changing and it is generally accepted that the area covered by tropical forests is decreasing, although there is still considerable debate about the rate of ongoing changes. Significant changes of the marine system on a global scale are less well documented, but it has been shown clearly that man-made pollutants are invading even the deep sea. Air and water pollution are generally increasing. It is, of course, hardly surprising that the presence of almost 5000 million people on Earth is altering the natural systems significantly. Some such changes must be accepted in order to permit the exploitation of the natural resources on which man is dependent. We must ask, however, if and to what extent the speed of recent development represents a threat to the renewable resources on Earth. What do the changes of the terrestrial and marine ecosystems and other ongoing changes mean to man in a long-term perspective? It is important to keep in mind throughout the following discussion that *the CO_2 problem, or rather the problem of a possibly changing climate due to emissions of greenhouse gases into the atmosphere, cannot be considered in isolation. It is one of many important environmental prob-*

1

lems that must be addressed but in a long-term perspective probably the most important one.

The realization that the climate might change as a result of emissions of carbon dioxide into the atmosphere is not new. Arrhenius (1896) pointed out that the burning of fossil fuels might cause an increase of atmospheric CO_2 and thereby change the radiation balance of the Earth. During the 1930s Callendar (1938) for the first time convincingly showed that the atmospheric CO_2 concentration was increasing. Earlier attempts had not been successful primarily due to non-representative sampling. The problem was revived again by C. G. Rossby in the 1950s, who was the driving force behind the initiation of CO_2 measurements in Scandinavia, and by R. Revelle, who was instrumental in getting C. D. Keeling engaged in the observational programmes on Mauna Loa, Hawaii, and at the South Pole in 1957–58. At about the same time Revelle and Suess (1957) presented the first more careful assessment of the likely future CO_2 increases due to fossil fuel combustion. This was followed by a more elaborate analysis by Bolin and Eriksson (1959).

The observations begun in 1958 have clearly shown that the concentration of carbon dioxide (CO_2) in the atmosphere has increased from about 315 ppmv then to about 343 ppmv in 1984. We know today approximately the amounts of CO_2 that have been emitted into the atmosphere by fossil fuel combustion and changing land use (deforestation and expanding agriculture) and can relate the observed increase of atmospheric CO_2 to these human activities. Since a continued increase of the atmospheric CO_2 concentration might lead to changes of the global climate, it is essential to be able to project the likely future concentrations that may occur due to various possible rates of CO_2 emission.

The reason for concern about climatic effects is the so-called 'greenhouse effect' of CO_2. While CO_2 is transparent to incoming short wave radiation from the Sun, it absorbs outgoing long wave radiation and re-emits this energy in all directions. Therefore, an increase of the atmospheric CO_2 concentration leads to a warming of the Earth's surface and lower atmosphere. In addition, it is becoming increasingly clear that a number of other greenhouse gases in the atmosphere similarly affect the radiation budget. Their concentrations are also changing as a result of natural and human causes. Since increased concentrations of CO_2 as well as of these other greenhouse gases all lead to a warming of the Earth's surface and lower atmosphere, the estimated climatic effects and further impacts (e.g. on sea level and agriculture) must be considered as a result of a combined effect of these potential origins of the warming. However, in order to be able to make estimates of their relative contributions to the warming and associated climatic changes at any given time, their effects are studied separately.

A major characteristic of the CO_2 problem is the uncertainty that is encountered when considering each of the aspects identified above. This uncertainty has often been ignored or underplayed in the past. Here it will be considered more specifically in the final sections of this chapter, together with other principal questions that confront us. However, the results from earlier assessments and the most important findings of the present one will be summarized first.

1.2 PREVIOUS ASSESSMENTS OF THE CO_2 PROBLEM

Of the numerous assessments that have been made of the CO_2 problem we shall limit ourselves here to considering those that have been conducted by national bodies and similar groups and those that have dealt most comprehensively with the problem. Table 1.1 compares some of the results of these previous assessments with those of the present study.

The first international assessment of the CO_2 issue organized by *UNEP, WMO and ICSU* resulted from an expert meeting held in Villach, Austria, in November 1980 (World Climate Programme, 1981). The projection of future fossil fuel use made at that meeting was largely based on a scenario developed at the Institute for Energy Analysis, Oak Ridge (USA) (Rotty and Marland, 1980). The net emissions of carbon dioxide from the biota were estimated on the basis of a review of available studies. The projection of the atmospheric carbon dioxide concentration in 2025 was made by assuming that 40–55% of the total emissions would remain in the atmosphere (the so-called airborne fraction). The globally averaged surface temperature response to a doubling of the atmospheric CO_2 concentration was estimated, as in all the studies referred to in Table 1.1, by examining the results of numerical models of the climate system. The WCP (1981) report concludes that CO_2-induced climatic change is a major environmental issue but that, because of existing uncertainties, the development of a management plan for control of CO_2 levels in the atmosphere and for preventing detrimental impacts on society would be premature. It was felt that research to place decision-making with respect to CO_2 on a firm scientific basis merits high priority. The meeting further emphasized that the CO_2 problem affects both developing and developed nations and calls for a special partnership of effort.

The report of the *Carbon Dioxide Assessment Committee* (*CDAC, 1983*) of the U.S. National Research Council gives a detailed assessment of the various aspects of the CO_2 problem. To estimate future emissions of CO_2 from fossil fuels, Nordhaus and his co-authors conducted a review of previous global long-range energy studies and developed a model of the global economy and carbon dioxide emissions. The model used a range of paths and uncertainties for major economic, energy and carbon dioxide variables, which allowed a 'best guess' of the future path of carbon dioxide emissions and a reasonable range of possible outcomes given present knowledge. The

Table 1.1 A comparison of results of recent assessments of the CO_2 problem

STUDY	Projection of future fossil fuel use in GtC/year*	Net emissions from the biosphere in GtC/year*	Future atmospheric CO_2 concentration in ppmv**	Globally averaged surface temperature response for CO_2 doubling
WCP (1981)	13.6 in 2025	present release 0–4 past total = 75–175 GtC	410–90 in 2025 (most likely 450)	1.5–3.5 °C
CDAC (1983)	10 in 2025 ('best guess')	present release 1.8–4.7 past total = 180 GtC	428 in 2025 ('best guess')	1.5–4.5 °C
EPA (1983)	10 in 2025 ('midrange baseline')	—	440 in 2025	1.5–4.5 °C
Clark et al. (1982)	growth rate of 2%/year likely until 2030	present release 2 past total 160 GtC	371–657 in 2030 (based on literature review)	2–3 °C
Jülich (1983)	1–16 in 2030	present release 0.5–4 (probable 1)	370–500 in 2030 (probable = 400)	1–3 °C
Present assessment	2–20 in 2050	present release 1.6±0.8 past total 150 ± 50 GtC	380–470 in 2025 (Figure 3.23)	1.5–5.5 °C

* 1 GtC/year = 10^9 ton carbon/year

** ppmv = parts per million volume

estimate of the net emissions of CO_2 from the biota was made on the basis of information presented in the report and available in the scientific literature. The possible future atmospheric CO_2 concentrations were calculated using an estimate of 0.60 $\pm$ 0.10 as the likely future airborne fraction of the projected emissions due to fossil fuel combustion (the value referred to here is higher than given in the CDAC report, but is a correction communicated by Nordhaus; see further Chapter 3 of the present report). The effects of CO_2-induced climatic changes on agricultural, social and economic systems were also assessed with emphasis on the United States. It was concluded that the longer-term agricultural effects are uncertain and depend strongly on the outcome of future research, development, and new technology in agriculture. The CDAC report reached a general conclusion similar to that of the WCP (1981) report, i.e. that the evidence at hand about CO_2-induced climatic change does not support steps to change current fuel-use patterns away from fossil fuels, although such steps may be necessary or desirable at some time in the future. The report pointed out that steps to control climatic change should start possibly with reductions of the emissions of other greenhouse gases, since their control may be more easily achieved. Further, the CDAC report suggested that the CO_2 problem might serve as a stimulus for increasingly effective cooperative treatment of world issues.

The study of the *US Environmental Protection Agency, EPA (Seidel and Keyes, 1983)* took a different approach to the problem by examining whether specific policies aimed at limiting the use of fossil fuels would prove effective in delaying temperature increases over the next 120 years. In contrast to the other assessments discussed here, the EPA study based its conclusions on the results of a set of three particular models, as opposed to a review of earlier model results. The projections of future energy demand and supply were made using the world energy model of the Institute for Energy Analysis (Edmonds and Reilly, 1983). A global carbon model developed at Oak Ridge National Laboratory (ORNL) (Emanuel *et al.*, 1981) was used to estimate the atmospheric CO_2 concentration. The changes of the atmospheric temperature were evaluated using a simplification of a one-dimensional radiative–convective model developed at the Goddard Institute for Space Studies (Hansen *et al.*, 1981). The EPA analysis concluded that only a ban on coal use instituted before 2000 would effectively slow down the rate of a global temperature change and delay a 2 °C increase until 2055. It was concluded further that major uncertainties include the increasing concentrations of other greenhouse gases and the atmospheric temperature response and that alternative energy futures produced only minor shifts in the calculated date of a 2 °C warming. Although the results suggested that bans on coal and shale oil are most effective in reducing temperature increases in 2100, the EPA study concluded that a ban on coal is probably economically and politically infeasible. It was suggested that action is required in the following three areas: accelerating and expanding research on improving our ability to adapt to a warmer climate; narrowing uncertainties about the future effects of greenhouse gases other than CO_2; and reducing uncertainty in our knowledge about the temperature response of the atmosphere.

In a study by the *Nuclear Centre in Jülich, F.R. Germany, completed for the Federal German parliament (Volz, 1983)* the future emissions of CO_2 due to fossil fuel use were estimated on the basis of scenarios produced by other groups (IIASA, 1981; Colombo and Bernadini, 1979; Lovins et al., 1981). The atmospheric CO_2 concentration in 2030 was calculated using a model of the global carbon cycle. The study concluded that, although our knowledge is presently not good enough to be able to describe quantitatively and unambiguously the relationship between CO_2 emissions and specific climatic changes in the individual world regions, the world-wide and long-ranging potential threat demands fast and specific action. In fact, the Jülich study concluded that if a 'threatening climatic catastrophe' is to be avoided with certainty, the necessary steps must be taken immediately.

In contrast to the latter view, *Clark* et al. (*1982*) concluded, on the basis of a detailed review of each of the aspects of the CO_2 problem, that there is no justification for immediate advocacy of a zero-growth policy for fossil fuel use, although it is possible that increased understanding, coupled with a return to high economic growth rates, could justify serious efforts to constrain future growth of fossil fuel use beginning as early as 1990. They suggested that during this 'grace period' two courses of action are rational: develop, maintain, and evaluate options for satisfying energy demand with less reliance on CO_2-emitting fossil fuels and promote further monitoring and research on the effects of increased atmospheric CO_2.

A Committee of the *Health Council of the Netherlands* made an assessment of the CO_2 problem in *1983 (CHCN, 1983)*. The energy scenarios upon which the assessment was based were taken from the IIASA (1981) study, with CO_2 emissions from fossil fuels in 2030 ranging between 8.9 GtC/year and 16.2 GtC/year. The changes in the atmospheric carbon dioxide concentration were calculated using a model of the carbon cycle. The concentration in 2030 was estimated to be 431 ppmv and 482 ppmv for the IIASA 'low' and 'high' scenarios respectively. Considerable emissions from the further exploitation of the terrestrial ecosystems were assumed but also more effective uptake by undisturbed forest systems and by soils due to charcoal formation in the process of burning during deforestation. The likely future temperature-change corresponding to the atmospheric CO_2 increase was calculated using a simple climate model. The global average surface temperature increase from 1980 values was calculated to be between 0.86 and 1.15 °C in 2030 for the two IIASA scenarios. The Netherlands study concluded by pointing out that CO_2 is an international problem and that measures taken by individual countries will hardly be effective. Further, the

Netherlands assessment found that since the CO_2 issue is so complex, the only recommendation it could make was the creation of an authority to deal with the CO_2 issue in all its aspects and to keep the Government informed.

A working group at an international conference on energy/climate interactions (Bach et al., 1980) in 1980 concluded that in view of present uncertainties prudence dictates a cautious and flexible energy strategy. The group recommended a 'low climate-risk energy policy', which would promote the more efficient end use of energy, secure the expeditious development of energy sources that add little or no CO_2 to the atmosphere, and keep global fossil fuel use, and hence CO_2 emission, at the present level. A further working group at the same conference noted that decisions that will have to be made in the decades ahead to prepare for or avert a carbon-dioxide climatic change will have to be made before all the answers have been obtained. It was concluded that the assessment of the impacts of climatic changes will have to be made now despite the uncertainties.

It is clear from this review that there is agreement on some basic issues. The net emissions from the biota (due to deforestation and land use changes) in themselves will be insufficient to bring about a significant change of climate, while fossil fuel reserves are sufficiently large that environmental disturbance would occur if the reserves are exploited at an increasing rate in the future. Although there are differences in the estimated globally averaged surface temperature response to a doubling of the atmospheric CO_2 concentration, these differences are not large, but the uncertainties of these estimates are considerable. It is also generally agreed that regional differences of climatic change cannot be predicted at present, and similarly the general way a given change of climate would influence people and nations around the world cannot be predicted. This is presumably the reason why there is substantial disagreement regarding the recommendations for future action. While some assessments conclude that there is not sufficient evidence to support changes of current fossil fuel use patterns, other assessments conclude that immediate action is necessary. The implications of the existing uncertainties have been evaluated differently in the various studies. We shall return to these questions later in this introductory chapter.

1.3 MAJOR FINDINGS OF THE PRESENT STUDY

The analysis of the 'CO_2 Problem' presented in this report has been pursued along the lines usually adopted by scientists. In the individual chapters, the different aspects of the problem are treated in the logical sequence: emissions—projections of atmospheric CO_2 and other greenhouse gas concentrations—climatic change—environmental and ecosystem effects. However, when an assessment is to be made of the problem as a whole, it is not sufficient to proceed step by step through this sequence. There will always be uncertainties in the successive analyses and it is difficult to assess what can be said with some degree of reliability with regard to the problem as a whole. It is, therefore, necessary to single out findings that are significant and upon which there is general agreement. At the same time, the nature of the uncertainties must be examined. It should be borne in mind that policy will ultimately be based on a judgment of the problem as a whole in relation to other societal problems, some caused by other environmental problems, some of a totally different nature. In this process the urgency of the problem is of decisive importance in the establishment of a policy for action. Long-term problems, such as the CO_2 issue, are therefore often deferred for later consideration. Both scientific progress and the adoption of a policy for action depend very much on the availability of factual information about ongoing changes. In the former case the information is essential for verification of model projections of future changes, in the latter case the evidence of observations usually is necessary to convince the general public and make action politically possible. We shall return to a discussion of the overall problem after having summarized the most important findings that are now available. The following sub-sections bring together the conclusions of the detailed analyses of the different problem areas given in Chapters 2–10. For a closer study of the major issues reference is made to these chapters and the extensive scientific literature.

1.3.1 Likely Emissions of Carbon Dioxide into the Atmosphere Due to Future Energy Demands (cf Chapter 2)

Combustion of fossil fuels—primarily oil, gas and coal—today meets about 80% of the total global energy demand. Future emissions of carbon dioxide will depend on how this global energy demand will change and what role fossil fuels will play in the future global energy supply system. The characteristics of the energy system ('the energy mix') change only slowly because of the large capital investments that are required to establish the units that supply the energy needed, and because the lifetime of existing installations is long and accordingly the time required to develop new supply systems is long. Even so, projections of future energy demands made during the last decade or two have not been very successful, a principal reason being the rather dramatic events that have taken place in the oil market since the early 1970s. A considerable decrease of the energy use per production unit has occurred and a further decrease is technically possible. Whether this will take place or not depends primarily on the relative costs for investments in new energy supply systems on one hand and those for conservation and energy savings on the other. Many other factors, however, also play an important role in this context.

It has been pointed out in a few recent analyses of the problem and has become increasingly clear in the present analysis that the projections of energy use beyond the early decades of the next century are very uncertain. This also implies that the options for choices increase in the long-term perspective. It is therefore not very meaningful to attempt precise projections beyond 30–40 years, but instead likely upper and lower bounds for future energy demands are estimated. The degree to which these may be met by fossil fuels or other energy supply systems is assessed in Chapter 2.

The estimated upper bound of possible scenarios implies an emission of about 20 GtC/year in the year 2050, i.e. about a fourfold increase of present emissions during the next 65 years. It is interesting to note that the last four-fold increase of emissions has occurred since the late 1930s, i.e. during the last 45–50 years. Although it is recognized that some studies have projected an even more rapid future increase, it is argued that this seems quite unlikely because of a number of environmental, social and logistic constraints. On the other hand it is not likely that the CO_2 emissions can be reduced to a value below about 2 GtC/year in 2050, and this value could only be achieved by sustained global efforts to limit the future energy demand and, particularly, the use of fossil fuels. Even lower values have been projected but are judged to be unrealistic on economic grounds. Low CO_2 emissions would only be achieved if there were some global acceptance of a much more considerate attitude towards the natural environment, implying decreasing use of fossil fuels. Although some people already consider this as fundamental for the continued well-being of human society, a general acceptance will take a long time. The range of possible future CO_2 emissions into the atmosphere as a function of time is shown in Figure 1.1 and discussed in Chapter 2.

At present about 5 GtC/year are emitted into the atmosphere through the combustion of fossil fuels. Although biotic emissions of carbon dioxide as a result of deforestation and land use changes have also contributed to the rise of the atmospheric carbon dioxide concentration in the past, it is clear that in the future the emissions from the biota will be comparatively small, since there is a physical limit to the extent of deforestation. If there is going to be a significant increase of the atmospheric carbon dioxide concentration during the next hundred years, it will come from the emissions of CO_2 by fossil fuel use, and, in a longer term perspective, particularly from the use of coal.

The range of likely carbon emissions in 2050 is not an uncertainty of the kind that natural scientists assign when they add an uncertainty range to an observation or to a prediction by a model. The future CO_2 emissions due to the burning of fossil fuels as well as changing land use will depend on human decisions. In any evaluation of the problem, economic considerations will play an important role, but other environmental concerns are also being expressed, as shown by the increasing use of concepts such as 'quality of life'. Therefore, policy with regard to CO_2 emissions will depend on how well we understand the negative effects of continued and possibly increased use of fossil fuels for energy supply, not only with regard to CO_2 emissions and possible climatic changes, but also with regard to other environmental effects such as air pollution, acid rain, etc.

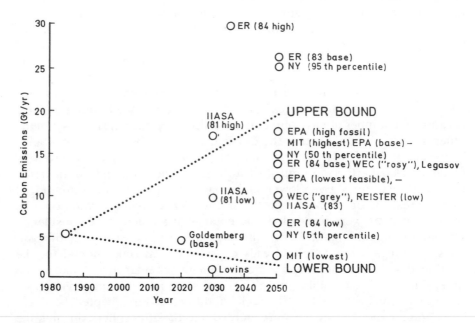

Figure 1.1 Projections of carbon emissions in 2050. Sources: EPA, Environmental Protection Agency (1983); ER, Edmunds and Reilly *et al.* (1984); Goldemberg *et al.* (1984); IIASA, International Institute for Applied Systems Analysis (1981, 1983); Legasov *et al.* (1984); Lovins *et al.* (1981); MIT, Massachusetts Institute of Technology (1983); NY, Nordhaus and Yohe (1983); Reister (1984); and WEC, World Energy Conference (1983)

1.3.2 Projections of Future Atmospheric CO_2 Concentrations (cf Chapter 3)

Future atmospheric CO_2 concentrations resulting from a given scenario of CO_2 emissions depend on the transfer processes, whereby CO_2 is partitioned between the major carbon reservoirs in nature. It is well established that the crucial questions are:

— How rapidly is the huge storage capacity of the world ocean (including the role of marine biota and sediments) becoming available as a sink for the CO_2 emitted into the atmosphere?

— In which way do the terrestrial ecosystems modulate the increase of atmospheric CO_2 and what are the implications of man's exploitation of the terrestrial biosphere, particularly by deforestation and changing land use?

The assessment shows an improved understanding of the global carbon cycle in recent years. This is particularly due to the development of more realistic models of the carbon cycle, the simultaneous consideration of the observed changes in the global distribution of all three carbon isotopes, ^{12}C, ^{13}C and ^{14}C, as well as other global data for validation of carbon cycle models. Progress in documenting past changes of the carbon cycle, particularly by analysis of the air trapped in glacial ice, has also been of great significance. It is still not clear how to balance the global carbon budget, partly because of the poor knowledge of land use changes and partly because of the uncertainty about pre-industrial atmospheric CO_2 concentrations. However, the remaining uncertainties about the global carbon cycle do not seriously influence the conclusions about future levels of atmospheric CO_2. The main source of uncertainty is, rather, the projections of future CO_2 emissions. The general conclusions illustrated in Figure 1.2 and discussed in Chapter 3 can be summarized as follows:

— A low CO_2 emission scenario, i.e. constant or only very slowly increasing emissions during the next 4 decades and slowly decreasing emissions thereafter, would give atmospheric CO_2 concentrations below about 440 ppmv, i.e. less than about 60% above the pre-industrial level. It is accordingly quite possible that a doubling of CO_2 will be reached only after year 2100.

— A modest increase (1–2%/year) of CO_2 emissions during the next 4 decades (and a decline thereafter) will lead to about a doubling of the CO_2 concentration (i.e. 550 ppmv) towards the end of the next century.

— The upper bound scenario implies CO_2 doubling by about 2050 and it is accordingly rather unlikely that the CO_2 concentration in reality will double before the middle of the next century.

These conclusions suggest a somewhat slower development than that discussed in the first WMO/ICSU/UNEP assessment (World Climate Programme, 1981) and in the U.S. National Academy of Sciences study (CDAC, 1983). It depends on the conclusion reached in Chapter 3 that the airborne fraction of the CO_2 emissions is probably 45% $\pm$ 10% rather than about 50%, as has usually been assumed previously. It is also important to note that the more slowly the emissions increase, the smaller is the fraction that remains in the atmosphere. A warmer climate induced by increasing CO_2 concentrations could, however, diminish the transfer of CO_2 to the deep sea and thereby increase the future airborne fraction.

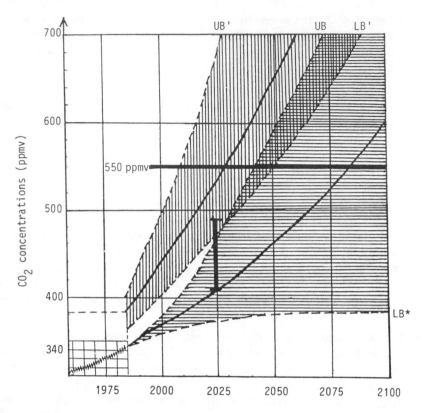

Figure 1.2 The lower set of curves indicates scenarios for future atmospheric CO_2 concentrations. The upper and lower bounds (UB and LB*) values are given in Figure 1.1. The value 550 ppmv corresponds to a doubling of the pre-industrial concentration of CO_2 (275 ppmv). The range, 410 to 490 ppmv for the year 2025, is the projection presented in the previous UNEP/WMO/ICSU CO_2 assessment (1981). The upper set of curves shows the corresponding scenarios if account also is taken of the other greenhouse gases (see Chapter 4)

1.3.3 Expected Increases of Other Greenhouse Gases that May Affect the Earth's Radiation Budget (cf Chapter 4)

Carbon dioxide is not the only atmospheric constituent that is of importance for the heat budget of the atmosphere and, thus, the global temperature. Moreover, the concentrations of some of these other constituents are also observed to be changing. Water vapour is the major radiatively active constituent of the atmosphere, but systematic and significant changes of atmospheric water vapour will primarily occur in association with changes of the climate and have been implicitly included in numerical models simulating climatic change (see sub-section 1.3.4 and Chapter 5).

It has been recognized for a long time that ozone is important in the Earth's radiation budget and, since changes of atmospheric ozone concentrations can be induced by human activity, attention should be paid to its importance in comparison with CO_2. Some slight decrease of stratospheric ozone (less than 1%) seems to have been detected and this decrease might continue in the future. Ozone in the Northern Hemisphere troposphere has probably on average increased by more than 10% presumably because of human activities. An additional increase by 10% during the remainder of this century and the first decades of the next may well occur.

Methane (CH_4, present average global concentration 1.65 ppmv) is increasing at a rate of about 1.2% per year, presumably due to more extensive use of paddy fields in cultivating rice, increasing numbers of domestic ruminants, biomass burning and to leakage of natural gas when exploiting gas fields. Analyses of air trapped in glacier ice indicate pre-industrial concentrations probably only about 40% of those observed today. Until we know better the reasons for this increase it is difficult to project likely future increases with any degree of certainty, but a 20–50% increase during the next 50 years seems plausible.

Nitrous oxide (N_2O, present average concentration 0.30 ppmv) is increasing by about 0.3% per year. Pre-industrial concentrations may have been 5–10% below the values observed today. The increase is probably due to the use of nitrogen fertilizers in agriculture and forestry and also to combustion processes and is expected to reach 0.35–0.40 ppmv towards the middle of the next century.

Man is also producing a large number of gases that were not present naturally in the atmosphere or only in small and insignificant amounts. The analysis presented in Chapter 4 shows that we may expect that also the two most common chlorofluorocarbons (CFCs), i.e. F11 and F12, will become of increasing significance for the radiation budget of the Earth, while a series of other compounds probably will be of less importance combined than any of the gases considered separately above. An annual increase of F11 and F12 emissions by 4% does not seem unlikely, if no preventive measures are taken.

It is fairly straightforward to compute the direct radiative effects of these greenhouse gases. Such computations can be made with adequate accuracy. They can also be included in General Circulation Models (GCMs). The uncertainties about the climatic changes that these greenhouse gases might cause are similar to those for CO_2. Therefore, their role can be expressed approximately in terms of an equivalent amount of CO_2.

It is concluded that the temperature change due to the changing concentrations of these greenhouse gases up to the present is about one half of the change calculated for the increase of the atmospheric CO_2 alone (about 70 ppmv). The effect of these other greenhouse gases is equivalent to an additional increase of CO_2 by 40–50 ppmv.

The concentrations of several of these gases are increasing more rapidly than that of CO_2. If the rates of increase as given above are applied during the next 50 years, we find that such a scenario would be equivalent to a doubling of atmospheric CO_2 concentrations well before the middle of the next century (see Figure 1.3). Chlorofluorocarbons would then become the most important gases in addition to CO_2, if no preventive measures are taken. On the other hand, their regulation would be easier to achieve than the limitation or reduction of CO_2 emissions.

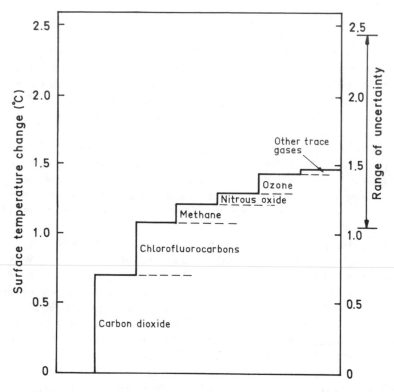

Figure 1.3 Cumulative equilibrium surface temperature warming due to increase in carbon dioxide and other trace gases from *AD* 1980 to 2030 as computed by a one-dimentional model. (After Ramanathan *et al.*, 1985.) Due to feedback mechanisms as revealed by general circulation models (cf. sub-section 4.4.3 and Chapter 5) expected changes are 0.8–2.6 times the values given in this figure

1.3.4 Modelling of Future Climate (cf Chapter 5)

There are many factors that are known to cause changes of global climate but our understanding of the cause-and-effect linkages is limited. Global climate can be affected by changes of solar energy output, changes of the Earth's orbit around the Sun, volcanic eruptions, changes of atmospheric composition, changes of cloudiness, the Earth's albedo and atmosphere–land–ocean interactions. These factors can act individually or together. In spite of recent theoretical advances and quite detailed understanding of many processes, it has not been possible to establish unequivocally the causes of documented past climatic fluctuations. It seems likely, however, that changes of the incident solar radiation caused by the slow variation of the characteristics of the Earth's orbit around the Sun have played a significant role on the glacial/interglacial time scale.

Estimates of the effects of changes in atmospheric composition on climate are made using models of the climate system. Since the climate system is very complex, various approximations and simplifications have been made in the development of such models. Different modelling approaches have been adopted, giving rise to a range of climate models from simple zero-order models, simulating global mean temperature changes, to comprehensive three-dimensional general circulation models of the atmosphere–land–ocean system. Only the latter are capable of providing adequate information for evaluating the characteristic features of future climatic change, that is, not merely average changes for the Earth as a whole, but also some details regarding regional climatic change and changes in surface hydrological processes. Even though significant advances have been made in modelling the climate system, present models are not yet able to simulate reliably the many processes that govern the regional climate. However, comparison of model computations with observed features of the general circulation of the atmosphere, particularly the model capability to reproduce seasonal variations of weather and climate, has given us some confidence in available results. Figure 1.4 shows a comparison of the estimated global equilibrium temperature change due to a CO_2 doubling as deduced in Chapter 5 with estimates made in previous assessments.

An evaluation of the large number of results from climate models leads to the conclusion that the global equilibrium temperature change expected from increases of CO_2 and other greenhouse gases equivalent to a doubling of the atmospheric CO_2 concentration is likely to be in the range 1.5–5.5 °C. The largest sources of uncertainty in modelling global average temperature change appear to be the levels of feedback from clouds, ice-albedo and, possibly, lapse-rate and water-vapour changes. Moreover, storage of heat in the oceans delays the warming expected for an equilibrium response to carbon-dioxide-induced warming and may also significantly modify the geographical distribution of climatic change. The role of the oceans is important but also uncertain. Model results suggest that the global warming resulting from increases in greenhouse gases to date is in the range 0.3 °C to 1.1 °C. Continental-scale or regional-scale climatic change cannot yet be mod-

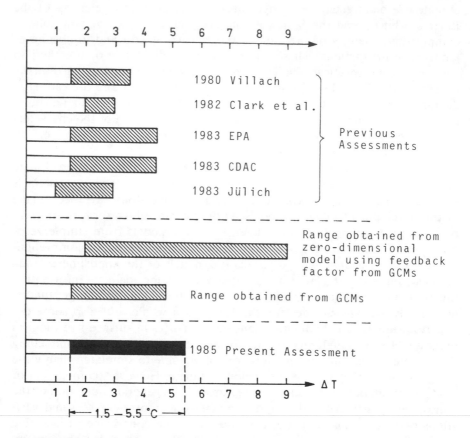

Figure 1.4 The estimated global equilibrium annual temperature change for doubling of the CO_2 concentration compared with estimates from previous assessments, and values obtained using climate models (see Chapter 5)

elled realistically but a few tentative conclusions can be drawn. All three-dimensional model results suggest that the largest temperature increases would occur in the high latitudes in the fall and winter seasons, and there is unanimous agreement that the stratosphere will cool. There is also some evidence for mid-latitude, mid-continental drying during the summer.

1.3.5 The Detection of a CO_2-induced Climatic Change (cf Chapters 5 and 6)

An analysis of surface temperature data since the middle of the last century shows that both hemispheres experienced a general warming from the late 19th century until about 1940 and a cooling until the mid-1960s. Since then the globe as a whole appears to have warmed, with a delay in this warming trend in the Northern Hemisphere. The observed global temperature record from 1850 is shown in Figure 1.5 and discussed further in Chapter 6.

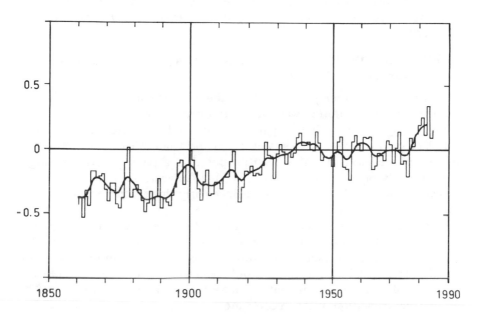

Figure 1.5 Variation with time of the global mean annual surface temperature obtained from land and marine temperature records The filtered curve has been obtained by suppressing variations on time scales less than 10 years (see chapter 6)

Detection of a CO_2-induced warming in the observational record has become a high priority issue. The detection problem can be viewed in terms of the concept of the signal-to-noise ratio, i.e. one can claim to have detected a change in climate once the signal (e.g. the increasing temperature) has risen appreciably above the background noise level (e.g. natural variability of temperature). Past records show that climate has varied naturally on timescales from a few years to centuries, millennia and longer. Chapter 6 concludes that the observed global mean temperature increase (about 0.5 °C) during the last 100 years cannot be ascribed in a statistically rigorous manner to the changing CO_2 concentration, although the magnitude is within the range of predictions. Considering also the warming due to the increase of other greenhouse gases, it seems that the observed temperature change during the last 100 years is in the lower part of the projected range. It is not possible to conclude whether this might be due to a more modest role of the positive feedback mechanisms as described in the climate models or to a delay of a warming due to the inertia of the oceans. A major problem in detecting the climatic effects of CO_2 increases is to explain the medium to longer time-scale (decadal or greater) fluctuations in the observed temperature record. Until the comparatively rapid global warming of 1920 to 1940 and the cooling between 1940 and the mid-1960s, in particular, have been adequately explained, claims regarding the detection of CO_2 effects can be easily criticized.

1.3.6 Projecting the Rise in Sea Level Caused by a Warming of the Atmosphere (cf Chapter 7)

Changes of global temperature affect the various components of the hydrological cycle in different ways and with different response times. For example, changes in precipitation patterns over land affect runoff from rivers and glaciers into the sea. Ocean waters expand when warmed. Catastrophic collapse of ice sheets has also been proposed as a potential risk causing a comparatively rapid rise of sea level, i.e. 1–2 cm/year.

Measurements of sea level changes since early this century show an average rate of rise for global sea level of 14 ± 5 cm per century. The changes of global sea level are illustrated in Figure 1.6. If the observed (modest but significant) correlation between sea level and air temperature during this time is assumed to be valid in the future, it is estimated empirically that a global warming of 1.5 °C to 5.5 °C would lead to a sea level rise of 25 to 165 cm. Probably the major contributing factor to such a sea level rise would be the thermal expansion of the ocean water.

The small glaciers would probably decrease in extent and might contribute to a rise of sea level by 20 ± 10 cm in the course of a century for a warming of 3.5 ± 1.0 °C. It seems likely that the Greenland ice sheet would decrease, but probably enhanced precipitation (snow) in the Antarctic would increase the accumulation and balance approximately the net flux of ice and water from Greenland to the sea.

Better oceanographic knowledge is required to assess the influence of any climate warming on the stability of the West Antarctic ice sheet. One possibility is that global warming could warm ocean waters and change ocean circulation sufficiently to cause a catastrophic collapse of the West Antarctic ice sheet following a global warming of 3–4 °C. Another possibility is that increased precipitation over the ice sheet may outweigh any effect of increased ocean temperatures, which may be confined mainly to the top 100 m, in which case the volume of West Antarctic ice may slowly increase in a stable manner. Even if a collapse started, it is likely to take several hundred years to raise sea level by, say, 5 m, corresponding to the possible discharge of the West Antarctic ice volume into the ocean.

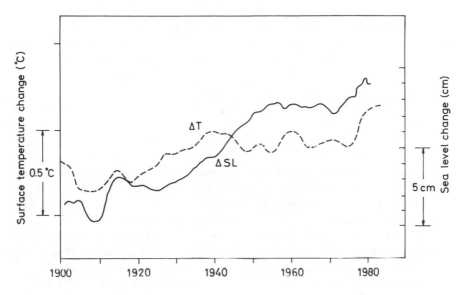

Figure 1.6 Comparison of the global mean sea level (Gornitz *et al.*, 1982) with the global mean surface temperature given in figure 1.5

1.3.7 Global Issues in Assessing the Effects on Terrestrial Ecosystems (cf Chapter 8)

Global vegetation can be affected in two general ways; by the direct effects of higher CO_2 concentrations on plant growth, and by changes in climate. At the scale of individual plants, higher CO_2 concentrations have been shown consistently in short-term laboratory experiments to stimulate growth and yield of both C_3 and C_4 plants. This results principally from enhanced net photosynthesis, and from increased water use efficiency through a reduction of the stomatal aperture in the leaf and, hence, transpiration. However, there are large uncertainties involved in extrapolating these experimental results to longer time scales and larger area scales.

A comparison of global vegetation and climatic changes of the distant past leaves little doubt that a climatic change of the order of magnitude indicated by climate models for a doubling of atmospheric CO_2 could, potentially, have profound effects on global ecosystems. However, the prediction of the direction, magnitude and rate of changes in ecosystems requires reliable estimates of climatic change at the regional scale, and such estimates are currently unavailable.

Despite the inability to make predictions, sensitivity analyses can produce useful information for judging the possible direction and magnitude of effects for given changes in CO_2 levels or climate variables, and thus for identifying specific regions and environmental changes which may warrant policy attention in the future. There is ample opportunity to test the sensitivity of global ecosystems to changes in climate variables by using scenarios of regional climatic change derived from GCM results or the instrumental record, or simply arbitrary climatic changes. In this volume, the emphasis is placed on agriculture and forests.

From a global perspective, geographic differences in agricultural regions have important implications for assessing the effects of increased CO_2 and climatic change. For example, rainfall is the principal constraint on agriculture in the tropics and sub-tropics, whereas in the temperate and higher latitudes, temperature has a relatively greater influence. In many developing countries of the lower latitudes, advances in food production have been achieved in large part through the expanded use of marginal lands, a trend which may be increasing the sensitivity of agriculture in these regions to climatic change or variability. In other areas, notably most of the major grain-producing countries of temperate latitudes, food production has risen principally through intensification and, hence, increases in average yields within the core crop regions. An unresolved issue here is whether the technological applications, which have made these long-term yield gains possible, have increased or decreased the sensitivity of crops to short-term climatic variations.

Another important trend has been the expanding volume of global grain trade during the past several decades, a trend which has increased inter-regional reliance for food production and distribution. Any adverse climatic change in the core crop regions of the temperate zones, where both major centres of supply and demand are located, could have large socio-economic impacts on the developing countries of the lower latitudes, whose purchasing power cannot always compete during times of scarcity. In this respect, the world has become increasingly interconnected in the face of increasing CO_2 and climatic change which, itself, is global in nature.

For the forest ecosystems of the world, three major issues need to be addressed, according to scale. First, at the microscale, the issue is how changes in processes such as photosynthesis, stomatal conductance or development may modify plant growth and, ultimately, forest productivity. The effects of changes in CO_2 concentrations or climate variables (like temperature) have to be analysed with respect to factors that currently limit productivity (like nutrients or moisture supply). Second, at the mesoscale, the issue is how the dynamic interaction and competition between ecosystem species would be affected, with consequences for local forest composition. Third, at the macroscale, the issue is how the size and areal extent of the world's forests may be altered. The largest spatial response could be expected from changes in temperature and precipitation at the cold or semi-arid margins of forest extent. At this scale, the lag time between climatic change and the response of forests could range from decades to hundreds, or even thousands, of years, based on evidence from the last glaciation.

In short, for both agriculture and forests the basic questions regarding increased CO_2 and climatic change are similar. How would crop yields or forest productivity be affected? How would crop types or forest composition change, particularly at the margins of production or ecological transition zones? How would the global patterns of forest ecosystems or food production be altered? These issues are addressed in Chapters 9 and 10.

1.3.8 The Response of Global Agriculture to Increasing CO_2 and Climatic Change (cf Chapter 9)

In the context of agriculture, four broad approaches to assessing the impacts of increasing CO_2 and climatic change address the major issues noted above: (a) crop impact analysis; (b) marginal–spatial analysis; (c) agricultural sector analysis; and (d) historical case studies.

Crop impact analyses are concerned specifically with the effects on plant growth and crop yields. Considering the direct effects of higher CO_2 concentrations in the absence of climatic change, it is estimated from laboratory experiments on individual plants that a doubling of CO_2 from 340 to 680 ppmv could result in a 0–10% increase in the growth and yield of C_4 crops (e.g. maize, sorghum, sugarcane) and a 10–50% increase for C_3 crops (e.g. wheat, soybean, rice), depending on the specific crop and growing conditions. These positive effects are obtained under most environmentally stressful as well as non-stressful conditions (see Table 1.2), and would therefore benefit both the environmental margins and the core of crop regions. Greater yield benefits would be expected to accrue to those regions of the world where C_3, rather than C_4, crops predominate.

Considering the sensitivity of crop yields to climatic change including the direct CO_2 effect, crop impact analyses have focused largely on grain yields in temperate and higher latitudes, to the neglect of the tropics and sub-tropics. From studies using various types of crop-climate models and climatic change scenarios, it is estimated that, with no precipitation change, a warming of 2 °C might reduce average yields of wheat and maize in the mid-latitudes of North America and Western Europe by 10 ± 7% assuming instantaneous warming and no change in cultivars, technology, or management. These yield reductions would be offset by wetter conditions and exacerbated by drier conditions. These estimates pertain to core regions; in contrast, average yields at the cool margins of cereal production, for example, might well benefit from a lengthening of the growing season and a reduction of damaging frosts.

Marginal–spatial analyses examine the margins of production where conversion to other crops (or genotypes) or land uses is most likely to take place. A very limited number of studies in the mid- and high-latitudes suggest po-

Table 1.2   Relative effects of increased CO_2 on growth
and yield: a tentative compilation[1]

	C_3	C_4
Under non-stressed conditions	+ +	0 to +
Under environmental stress:		
Water (deficiency)	+ +	+
Light intensity (low)	+	+
Temperature (high)	+ +	0 to +
Temperature (low)	+	?
Mineral nutrients:	0 to +	0 to +
Nitrogen (deficiency)	+	+
Phosphorus (deficiency)	0?	0?
Potassium (deficiency)	?	?
Sodium (excess)	?	+

[1] Sign of change relative to control CO_2 under similar
   environmental constraints.

+ + = strongly positive

+ = positive

0 = no effect

? = not known or uncertain

tential shifts in the boundaries of cereal regions of the order of magnitude
of several hundred kilometres per °C change (assuming that existing crop
regions are largely climatically determined and optimally located). Other
studies of high altitude locations with steep environmental gradients suggest
potential altitudinal shifts of more than a hundred metres per °C change.
While these estimates are highly uncertain, any such shifts at the margins
would certainly modulate the effects of climatic change on regional crop
yields and production.

Agricultural sector analyses and historical case studies examine the range
of environmental, agricultural and socio-economic impacts and explore the
ways in which agricultural systems adjust to climatic change and variability. There are many feedback mechanisms that can enhance or diminish the
potential impacts of environmental changes on crop yields and food production, as illustrated in Figure 1.7. A limited number of studies, using linked
regional models or global agriculture and trade models, suggest that, in many
regions of the world, agriculture would readjust crop yields and food production to changing climate. Over the long term, yields and production in
such regions may be more sensitive to technology, price or policy changes
than to climatic changes, and these factors are largely manipulatable whereas
climate is not.

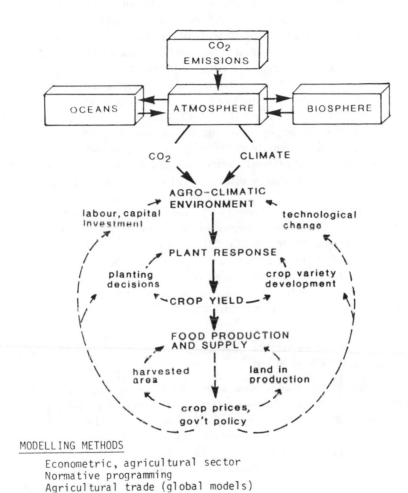

MODELLING METHODS

    Econometric, agricultural sector
    Normative programming
    Agricultural trade (global models)

Figure 1.7 A generalized diagram of some feedbacks influencing crop-yields and production over time. The *agricultural systems approach* to impact assessment examines the dynamics of agriculture and the mechanisms which can diminish or accentuate the primary yield effects of increased CO_2 and climatic change

In past studies the pragmatic approach has been to consider the impacts of climatic change separately from primary responses of plants to increased CO_2. But it is clear that the effects are interactive and non-linear, not simply additive, and should be studied accordingly. In any case, there exist large uncertainties in extrapolating laboratory results to field conditions.

Furthermore, most studies of agricultural impacts have focused on average climatic changes. However, it has been demonstrated in several instances that small changes in average climate can result in relatively large shifts in the frequencies of climatic extremes like droughts. These shifts may be equally, if not more, important to farmers than long-term changes in mean climate, especially in the marginal lands of many developing countries where climatic extremes take heavy economic and human tolls. The degree to which agriculture in such regions can be assisted in developing adjustments to buffer the ill-effects of present climatic variability, the better prepared they will be to adapt to any adverse effects of future climatic change, should they occur.

In general, given the uncertainties in regional scale estimates of climatic change and the numerous deficiencies in methodologies of impact assessment, there is presently no firm evidence for believing that the net effects of higher CO_2 and climatic change on agriculture in any specific region of the world will be adverse rather than beneficial. But it is certain that some will gain and others will lose, although we know neither where they will be found nor the magnitude of the impacts.

1.3.9 The Response of Global Forests to Increasing CO_2 and Climatic Change (cf Chapter 10)

The forests of the Earth constitute a complex system with many possible responses, both to the direct effects of an increase in atmospheric CO_2 concentration and to the possible changes in climate. These responses may originate from phenomena that operate on very different scales of time and space. In general, formidable difficulties are encountered in the 'scaling-up' of the short-term physiological and biochemical response of leaves and individual plants to estimate the intermediate and long-term responses of forests. The difficulties arise from the large uncertainties involved in the methods of extrapolation and from the complex interactions that occur at larger scales. The two uncertainties are presently large enough to preclude meaningful estimates of the effects on forests of higher CO_2 concentrations and climatic change except in the most general way.

With respect to the direct effects of CO_2, these problems of scaling-up are compounded by the lack of experimental evidence for relevant forest species, particularly for plants that have been allowed to acclimate to enhanced CO_2 concentrations over one or more growing cycles. Although higher concen-

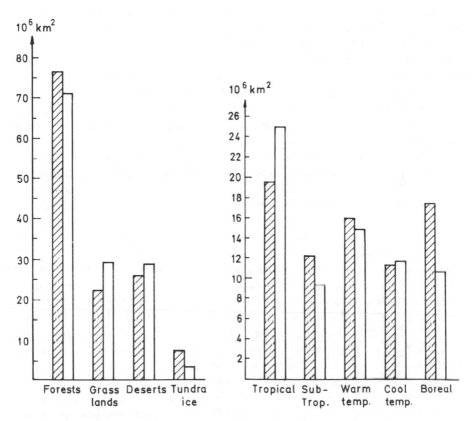

Figure 1.8 Estimation of the change in life zone extents for a CO_2-induced warming according to Emanuel *et al.* (1985) from present climate (hatched columns) to that deduced by Manabe and Stouffer (1980) in their climate simulation experiment for a doubling of the atmospheric concentration of CO_2 (open columns). It should be stressed that in calculating these changes in potential vegetation the changes in precipitation were not taken into account. Note also that potential vegetation corresponds to extension of ecosystems unaffected by man's direct impact

trations of CO_2 have been shown to increase the growth rates of individual trees in controlled conditions over the short term, it is highly uncertain whether such effects would be sustained and would lead to increased productivity in actual forest environments over the long term. In uncontrolled environments, the direct CO_2 effects are complicated by micrometeorological differences in the degree of coupling between forests and atmosphere (within as well as between forest systems), and by species competition and interaction. If, indeed, elevated CO_2 concentrations do result in long-term growth enhancement, increases in productivity would be more likely to occur in commercial forests than in mature forests in which the capacities for increased carbon storage are more limited. Direct experimentation at this scale, however, is largely impracticable. In order, therefore, to assess the responses of forest systems to both higher CO_2 concentrations and changes in climate, experimental studies must be augmented by empirical observation and simulation modelling.

With respect to the effects of climatic change, empirical climate-vegetation models and forest simulation models have been used to assess the responses of forests at scales ranging from a single point in a forest to an entire continental system. In general, the results of a limited number of such studies suggest that climatic changes of the order of magnitude predicted by climate models for a doubling of atmospheric CO_2 are potentially sufficient to produce substantial intermediate and long-term changes in the composition, size, and location of the forests of the world. The natural forests of the high latitudes in general and the boreal forests in particular, appear sensitive to predicted temperature changes, and it is at these latitudes that climate models predict the largest warming to occur as a result of increased concentrations of greenhouse gases (see Figure 1.8). Warmer conditions could possibly lead to large reductions in the areal extent of boreal forests and a poleward shift in their boundaries. The forests of the tropical and subtropical zones, on the other hand, would probably be more sensitive to changes in precipitation than temperature. Because of the high uncertainty regarding future changes in precipitation in the tropics, and because of the present lack of models that can be used to simulate the effects of tropical ecosystems to changes in climate variables, our knowledge of the responses of tropical forests to future changes in climate is meagre.

1.3.10 Concluding Remarks

The later chapters analyse the individual aspects of the 'greenhouse gas' problem in detail.

Clear priorities can be set for future research in carbon-cycle modelling, better estimation of the effects of other trace substances, sea-level projections and climate modelling. In the modelling of agricultural impacts, model validation and cross-model comparisons are required. The assessment of impacts on forest ecosystems indicates an essential lack of direct observations of whole system performance of forests under altered environmental conditions. Finally, it is clear that there needs to be a better integration of available methods and approaches in impact assessment.

1.4 WHERE DO WE GO FROM HERE?

It is clear from the above review of the following chapters and of previous assessments that there are uncertainties with regard to each aspect of the 'greenhouse gas problem'. Some are due to incomplete knowledge about the natural systems with which we are concerned, including uncertainties about the way these systems can be described in models. Others arise because it is not certain how global society will develop and respond to changes. We are not able to predict factors, such as future energy demand, use and management of renewable and non-renewable resources, very far into the future. It has been pointed out in previous studies and must be emphasized here that some of the uncertainties will still exist in the future, despite intensive research on the individual topics. However, *prevailing uncertainty does not mean that the problem can or should be dismissed.* Instead, it is necessary to examine the characteristics of these uncertainties and assess what can be said about future changes and to consider if and when some actions are needed in view of such possible changes.

1.4.1 Climatic Change

As far as the expected climatic change is concerned, this assessment concludes that a doubling of the CO_2 concentration would lead to an increase of the globally averaged surface temperature by 1.5–5.5 °C. The uncertainty is considerable, but *there is almost unanimous agreement that a substantial warming would occur.*

Some global mean surface warming most likely has occurred but may well be partly obscured by natural climatic fluctuations. Model estimates as well as observed changes are subject to considerable uncertainty. Further, the observed warming could be attributable to other causal factors. It is not possible to state unequivocally that a CO_2 or greenhouse gas signal has been detected. *The observed general increase of mean global surface temperature during the last hundred years is, however, in general accord with model results.*

Our knowledge about past changes of climate as well as model computations of future changes indicate that marked regional differences can be expected. This implies, obviously, that some parts of the globe will experience significantly larger changes than indicated by the average global change, in other regions they will be less. *The spatial patterns of future changes are, however, not known. Some of the regional climatic anomalies observed today and usually attributed to the natural variability of the climate, might be due to some extent to an ongoing man-induced change.*

1.4.2 Future Changes of Global Society, Emission Scenarios and Atmospheric Concentrations of Greenhouse Gases

Although there are gaps in our understanding of the response of the climate system to emissions of greenhouse gases into the atmosphere, it seems clear that the major uncertainty with regard to future environmental changes is due to our inability to predict man's future behaviour. We cannot predict the future of society in the way that natural phenomena can be predicted using the fundamental laws of nature. Some limited success in foreseeing changes of global society has been achieved by recognizing its inertia and by using econometric models. These methods are not applicable for long-term changes. In the present case, in particular, we do not know to what extent a wider awareness about general environmental problems might influence the behaviour of individuals, groups or even nations during the next 100 years. It is interesting in this context to note the shift of opinion about the impact of air pollution and acid deposition that has taken place in Europe, since an impact on forests has been detected.

The uncertainties of projections of future atmospheric CO_2 concentrations from a given CO_2 emission scenario are considerably less than those of the emission scenarios themselves. The major uncertainty is thus related to the difficulty of projecting the future use of fossil fuels. It is also difficult to foresee how mankind will respond if there is conclusive evidence that the increasing emission of other greenhouse gases into the atmosphere is an equally important factor in causing climatic change.

We conclude that the main aim of studies of future changes of atmospheric greenhouse gases is not only to predict their concentrations but also to analyse to what extent and how future increases can be limited and/or the impact on our environment can best be managed.

1.4.3 The Holistic View of Man and His Environment

Different kinds of emissions contribute to the same environmental problem (i.e. the greenhouse effect), while, on the other hand, one single human activity may cause different environmental problems (e.g. fossil fuel combustion emits CO_2, sulphur dioxide and nitrogen oxides into the atmosphere leading to air pollution, acid deposition, etc.).

It is becoming increasingly clear that many of these key problems are closely related both because physical, chemical and biological processes interact and also because one single human activity can contribute to several processes. It is obvious that environmental policy must be developed with such a holistic view of man and his environment.

It is emphasized in Chapter 2, that even if physical, economic and environmental variables were fixed and immutable, future energy and greenhouse gas emission policies would still be subject to considerable uncertainties. The same is true for other environmental problems. On the other hand *the full spectrum of policy choices has usually not been emphasized*. Such choices are for example the improvement of energy use efficiency, the introduction of conservation measures, accelerated development of specific technologies, the introduction of stricter environmental regulations and development of strategies of how to adapt to a changing climate. There is a wide range of possible energy futures because such choices can be made. It would be of great interest to know under which circumstances a fuller awareness of potential or real threats to our environment, in particular due to climatic change and impacts on agriculture, might lead to policy decisions to limit further emissions.

1.4.4 Man's Response to Environmental Change

With regard to the impact of climatic change on natural terrestrial ecosystems, agriculture, forestry, etc., the lack of projections of regional climatic change due to increasing concentrations of greenhouse gases in the atmosphere means that more precise assessments of likely future changes on this scale cannot be made. For quite some time, research will be restricted to examining the sensitivity of these ecosystems to given climatic change scenarios. This is a serious shortcoming, since it will be difficult to tell more specifically what the implications of a change might be for particular groups of people (e.g. farmers). Those engaged in national planning will also usually be hampered because of lack of regionally specific information.

In view of these difficulties it is important to adopt a strategy for action that would yield useful information in any case of future changes of climate. Rather than predicting likely future impacts *we should study the sensitivity of the agricultural system and explore the question of what adaptation to climatic change would imply*, e.g. the development of new varieties to safeguard crop diversity under a wide range of climatic conditions. Obviously such a measure would be of value, since it could increase efficiency in the agricultural system even if there is no change of climate.

Agricultural systems analysis looks at the readjustment that would be needed on the local, national and international scale in response to climatic changes. It is clear that these would be quite different depending on the spatial scale being considered. A shift of a climatic zone might not change the total production of a particular crop in a large and well developed country, because of possible shifts within the country of the regions where production takes place and the introduction of new varieties is possible. Such alternatives are not usually available to small countries, and considerable difficulties may be encountered at the national level. The individual farmer will also be affected very differently depending on whether or not cultivation of other crops than the traditional ones is feasible and on whether or not migration to regions with improved climate is possible. The seriousness of the problem further depends on the rate of the expected changes, which also warrants study.

Although a careful analysis of such impacts requires knowledge about the likely changes of all key variables and their variability in the region concerned, the assessment of the most sensitive parameters might still be possible. The agricultural impacts in the tropics and sub-tropics have not been studied in detail, although the impacts could be felt the most in these regions. Clearly, this imbalance should be redressed in future studies.

We obviously encounter considerable difficulties in trying to assess the more specific consequences of a possible change of the global climate. It does not seem likely that even intensified research will make it possible to foresee the geographical patterns of a climatic change in more detail, before clear signs of an ongoing change are available. In view of the far-reaching consequences that might be expected for some parts of the globe, it might be desirable to take some preventive steps before much more clear evidence is available than today. We note that the emissions of greenhouse gases other than CO_2, particularly chlorofluorocarbons, could be more easily controlled than CO_2 emissions. It is also clear that there are technically and economically viable strategies for long-term global energy development that are compatible with a high level of concern about the effects of a CO_2 increase. Actions that simultaneously contribute to solving other environmental and societal problems should then of course be given high priority. However, because of the uncertainties and the difficulty of arriving at and implementing a global policy the problem of adaptation of society to a climatic change will also become important.

1.5 REFERENCES

Arrhenius, S. (1896) On the influence of carbonic acid in the air upon the temperature of the ground, *Phil. Mag.*, **41**, 237.

Bach, W., Pankrath, J., and Williams J. (eds) (1980) *Interactions of energy and climate*, Dordrecht, D. Reidel.

Bolin, B., and Eriksson, E. (1959) Changes in the carbon content of the atmosphere and the sea due to fossil fuel combustion, in Bolin, B. (ed.) *The atmosphere and the sea in motion*, Rossby Memorial Volume, New York, The Rockefeller Institute Press, 130–142.

Callendar, G. S. (1938) The artificial production of carbon dioxide and its influence on temperature, *Q. J. R. Meteorol. Soc.*, **64**, 223.

CDAC (1983) *Changing climate,* Report of the Carbon Dioxide Assessment Committee, Washington, D.C., National Academy Press.

CHCN (1983) Part one: Report on CO_2 Problem, Committee of the Health Council of the Netherlands, The Hague, Netherlands.

Clark, W. C. *et al.* (1982) The carbon dioxide question: perspectives for 1982, in Clark, W. C. (ed.) *Carbon Dioxide Review:1982,* Oxford, Clarendon Press.

Colombo, U., and Bernadini, O. (1979) *A low energy growth 2030 scenario and the perspectives for Western Europe.* Report prepared for the Commission of the European Communities Panel on Low Energy Growth.

Edmonds, J., and Reilly, J. (1983) A longterm, global energy economic model of carbon dioxide release from fossil fuel use, *Energy Econ.,* **5,** 74.

Edmonds, J. A., Reilly, J., Trabalka, J. R., and Reichle, D. E. (1984) *An analysis of possible future atmospheric retention of fossil fuel CO_2.* Report No. DOE/OR/21400-1, Washington, D.C., Department of Energy.

Emanuel, W. R., Killough, C. G., and Olson, J. S (1981) Modelling the circulation of carbon in the world's terrestrial ecosystems, in Bolin, B. (ed.) *Carbon cycle modelling,* SCOPE 16, Chichester, Wiley.

Emanuel, W. R., Shugart, H. H., and Stevenson, M. P. (1985) Climate change and the broad scale distribution of terrestrial ecosystem complexes, *Clim. Change,* 7, 29–43.

EPA (1983), see Seidel and Keyes (1983).

Goldemberg, J., Johansson, T. B., Reddy, A. K. N., and Williams, R. H. (1984) *Energy for a sustainable world* (forthcoming).

Gornitz, V., Lebedeff, L., and Hansen, J. (1982) Global sea level trend in the past century, *Science,* **215** (4540), 1611–1614.

Hansen, J, Johnson, D., Lacis, A., Lebedeff, S., Lee, P., Rind, D., and Russel, G. (1981) Climate impacts of increasing atmospheric CO_2, *Science,* **213,** 957.

IIASA (1981) *Energy in a finite world,* Cambridge, Mass., Ballinger Publishing Co

IIASA (1983) *IIASA '83 Scenario of Energy Development: Summary.* Rogner, H. H. (ed.) Laxenburg, Austria, IIASA.

Jülich (1983), see Volz (1983).

Legasov, V. A., Kuzmin, I. I., and Chernoplyokov, A. I. (1984) The influence of energetics on climate, USSR Academy of Sciences, *Fizika Atmospheri i Okeana* 11:1089–1103.

Lovins, A. B., Lovins, L. H., Krause, F., and Bach, W. (1981) *Energy strategy for low climatic risks.* Report for the German Federal Environmental Agency, R & D No. 10402513, June 1981.

Manabe, S., and Stouffer, R.J. (1980) Sensitivity of a global climate model to an increase of CO_2 concentration in the atmosphere, *J. Geophys. Res.,* **85,** 5529–5554.

MIT (1983), see Rose *et al.* (1983).

Nordhaus, W. D., and Yohe, G. (1983) Future paths of energy and carbon dioxide emissions, in *Changing Climate,* Washington, DC, National Academy Press.

Ramanathan, V., Cicerone, R. J., Singh, H. B., and Kiehl, J. T. (1985) Trace gas trends and their potential role in climate change, *J. Geophys. Res.,* **90,** D3, 5547–5566.

Reister, D. B. (1984) *An assessment of the contribution of gas to the global emissions of carbon dioxide.* Final Report GRI-84/003, Chicago, IL, Gas Research Institute.

Revelle, R., and Suess, H. E. (1957) Carbon dioxide exchange between atmosphere and ocean and the question of an increase of atmospheric CO_2 during the past decades, *Tellus,* **9,** 18.

Rose, D. J., Miller, M. M., and Agnew, C. (1983) *Global energy futures and CO₂-induced climate change*, MITEL 83-015, Cambridge, MA, MIT Energy Laboratory.

Rotty, R. M., and Marland, G. (1980) Constraints on fossil fuel use, in Bach, N., Pankrath, J., and Williams, J. (eds) *Interactions of Energy and Climate*, Dordrecht, D. Reidel, 191–212.

Seidel, S., and Keyes, D. (1983) *Can we delay a greenhouse warming?* Washington, D.C., Environmental Protection Agency.

Volz, A., (1983) *Studie über die Auswirkungen von Kohlendioxidemissionen auf das Klima*, Kernforschungsanlage Jülich, F. R. Germany.

WCP (1981), see World Climate Programme (1981).

WEC (1983) Energy 2000–2020: *World prospects and regional stresses*. Frisch, J. R. (ed.) World Energy Conference Conservation Commission; and *Oil substitution: world outlook to 2020*, London, Graham and Trotman and Oxford University Press.

World Climate Programme (1981) *On the assessment of the role of CO₂ on climate variations and their impact*. Report of a WMO/UNEP/ICSU meeting of experts in Villach, Austria, November 1980, Geneva, WMO.

PART A

How is Man Changing the Composition of the Atmosphere?

CHAPTER 2

Emission of CO_2 into the Atmosphere

The Rate of Release of CO_2 as a Function of Future Energy Developments

W. KEEPIN, I. MINTZER, AND L. KRISTOFERSON

2.1 INTRODUCTION

Numerous gases present in low concentrations in the Earth's atmosphere are transparent to incoming solar radiation but absorb the infrared radiation emitted by the Earth's surface and by certain atmospheric constituents. An increase of the concentration of these 'greenhouse' gases in the Earth's atmosphere could lead to considerable global temperature increases and other climatic changes. One of these gases is carbon dioxide (CO_2), which is emitted to the atmosphere through both human activities and natural processes.

The main anthropogenic source of CO_2 emissions is the combustion of carbon-based fuels. In 1983, approximately five gigatons of carbon per year were emitted globally from fossil fuels (Rotty and Masters, 1984). Other net CO_2 emissions from terrestrial biota, including the clearing and burning of forests, have contributed significantly to the present atmospheric CO_2 concentration. However, these sources are expected to contribute relatively less in the future compared to the combustion of fossil fuels. The role of terrestrial biota in the carbon cycle is discussed more fully in Chapter 3.

Besides CO_2 in the atmosphere, the concentrations of other radiatively active trace substances (greenhouse gases) are observed to be increasing. The accumulation of these other gases may eventually affect global temperature as much as or more than CO_2 does. For this reason, these non-CO_2 gases have received much recent attention, and concern about global warming has increased. This new dimension of the atmospheric greenhouse problem is discussed more fully in Chapter 4.

The purpose of this chapter is to explore the feasible range of future CO_2 emissions from energy-related activities. Numerous recently published scenarios for the future global energy system are analysed to determine how each will affect the rate and cumulative level of CO_2 emissions to the atmosphere. Particular attention is paid to the underlying assumptions and uncertainties in projections of future energy developments and CO_2 emissions.

In Section 2.2, the discussion focuses first on the basic links between energy uses, economic activity, and CO_2, and then on the historical pattern of CO_2 emissions. Section 2.3 presents a detailed discussion of the uncertainties that affect projections of future emissions of CO_2. Section 2.4 provides an overview of methodological and other problems that arise in forecasting future CO_2 emissions, together with a critical review of several recent projections of CO_2 emissions to the year 2050. In Section 2.5, feasible upper and lower bounds (or 'extreme scenarios') for CO_2 emissions between now and 2050 are suggested, based on the review in the preceding sections. Section 2.6 presents key findings and conclusions.

2.2 THE LINK BETWEEN ENERGY USE, ECONOMIC ACTIVITY, AND CO_2 EMISSIONS

Modern economic growth is generally attributed to the spread of industrialization. Historically, growth has depended on the availability of energy and on the combustion of fossil fuels. For numerous countries, energy and economic data for the years between 1860 and 1973 suggest that economic growth and energy use increased together. Yet, the data do not establish that one caused the other. (See Section 2.3.)

Since 1973, the ratio of energy per unit of economic output has declined in many countries belonging to the Organization for Economic Cooperation and Development with attendant increases in the real price of energy. Of course, this ratio also tended to decrease even when the real price of energy was declining, as a result of factors not related solely to energy prices. A recent study of energy use in U.S. business has shown that the ratio of primary energy inputs to the economic value of outputs has declined steadily since 1920 (Schurr, 1984).

Increased energy efficiency has resulted from technological improvements in industrial processes, transportation systems, and building design. In addition, some industrialized countries have shifted economic activity away from the production of basic materials and toward fabrication and finishing activities. At the same time, the relative economic role of the services sector has increased (Marley, 1984).

The minimal level of per capita energy use needed to support modern medical, educational, and leisure services varies somewhat in different cultures and regions. In many developing countries, considerable growth in per capita consumption of high-quality fuels is essential to bring the population above the subsistence level. However, it now appears likely that continued economic growth and an attractive standard of living do not depend on ever greater rates of per capita energy consumption. Williams *et al.* (1984) have argued that, depending on the energy-using technologies employed, demand of as little as 1 kW per capita is theoretically enough to support the whole world at the standard of living enjoyed by Western Europe in the 1970s. To date, the social and economic transition process for reaching this level of demand has not been addressed in detail. Because developing countries must, in many cases, build up the basic components of industrial infrastructure already existing in industrialized countries, some analysts have argued that these societies will temporarily have to experience levels of energy consumption in the transition period that are significantly in excess of an average level of 1 kW per capita (Sh'pillrain, 1985) even if they were later to reduce energy demand to lower levels. However, referring to analyses of energy use patterns during the 'infrastructure-building' phase of industrialized countries, other analysts suggest that this may not be necessary (Goldemberg *et al.*, 1985).

For many countries, the price of energy in the short term will heavily influence the rate of economic growth. Over the 70-year time horizon considered here, most economies will adjust to higher energy prices by substituting labour, information, or other inputs and perhaps by shifting the balance between production of goods and provision of services. Different energy strategies, e.g. using different fractions of coal or nuclear fuel in the energy system, will also directly influence CO_2 emission rates. In short, the weak correlation between economic growth and energy consumption means that steady economic growth need not entail steadily increasing CO_2 emissions. (This will be discussed in later sections.)

2.2.1 Energy Consumption and CO₂ Emissions

No one seeks energy for its own sake. Industrialized nations use energy primarily for industry, transportation, and space heating and cooling in buildings (see Table 2.1). Many recent studies show that current usage in industrialized countries could be reduced by using more energy-efficient technologies. Mintzer and Miller (1984) showed that if the United States—the world's largest emitter of CO_2—switched to the most energy-efficient technologies now available to produce key commodities and services, its CO_2 emissions would fall approximately 25 percent (0.32 GT/year), with output remaining constant. These changes would reduce global carbon emissions by 7 percent. Similar, but smaller, reductions are possible in other industrialized countries. (A small but growing number of studies assess the economic cost of some of these technological possibilities, e.g. Goldemberg *et al.*, 1984.)

Further reductions occur when the composition of economic activity shifts as the result of more efficient methods of providing goods and services. In studies to assess the global potential for improving energy productivity or displacing energy demand (Goldemberg *et al.*, 1984; Rose *et al.*, 1984; Lovins

Table 2.1   Primary energy use by sector in selected Western European countries. Reproduced from Jäger (1984)

	% of total primary energy				
	Buildings		Transport	Industry	Other
	Domestic	Commercial			
Belgium	29		12	53	6
Denmark	23	9	17	31	20
France	22	13	19	31	15
West Germany	25	11	22	34	8
Greece	20	10	27	43	—
Ireland	33	12	22	33	—
Italy	25	10	18	42	4
The Netherlands	28		13	41	18
UK	25	13	22	35	5

et al., 1981), a wide range of technologies has been considered—such as improvements in building insulation, automobile design, and industrial processes. Overall, these studies show that energy efficiency is currently far from optimal and technological improvements could significantly reduce demand for fossil fuels. At the same time, however, some of the technical fixes being developed and applied to existing industrial processes to reduce negative environmental impacts other than CO_2 emissions—the use of flue-gas desulphurization techniques in coal-electric power plants to reduce acid precipitation, for example—may increase overall fuel consumption and, consequently, emissions. (See Sections 2.4 and 2.5.)

2.2.2 Energy Supply, Fuel Mix, and CO_2 Emissions

Useful energy is produced from several primary energy forms: fossil, nuclear, and renewable (solar, biomass, hydro, and wind). The approximate total world production from these sources is shown in Table 2.2, together with estimated reserves and resources. Future levels of CO_2 emissions will depend on both the quantity and the mix of fuels burned. These two factors, in turn, will depend as much on the future structure of national economies as on the monetary value of total outputs. Globally, the quantities and mix of fuel consumed are determined partly by national trends in industrialization, urbanization, and consumer preferences.

Current global shares of primary energy supply are illustrated in Figure 2.1. For each form of primary energy, several technologies contribute to commercial and non-commercial energy supplies. Most commercial energy comes from burning fossil fuels, which are reviewed below with other supply technologies.

Table 2.2　Global primary energy

Source	Present		Prospects	
	TW	%	Reserves[a]	Total resources[a]
Oil	4.24	41	125 TWyr　88.3 btce	365 TWyr　257.2 btce
Gas	1.61	17	76 TWyr　64×10^{12} m^3	333 TWyr　280×10^{12} m^3
Coal	2.59	24	591 TWyr　636 btce	9405 TWyr　10.127 btce
Tar sands/oil shale	0.0	0	99 TWyr　70 btce[b]	
Nuclear fission	0.22	2		Virtually unlimited (with breeders)
Nuclear fusion	0.0	0		Virtually unlimited (if realized)
Hydropower	0.19	2		1.1 TW[b]
Solar PV	0.01	<1		Very large
Wind	0.01	<1		1.2 TW[c]
Biomass	1.49	14		4.6 TW[c]
Geothermal	0.01	<1		Very large (if 'Hot Dry Rock' realized)
Total	10.3			

[a] Based on IIASA (1981), converted to energy-equivalent units

[b] WEC (1983)

[c] Rose et al. (1984)

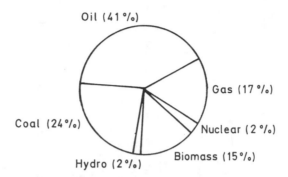

Figure 2.1 Global primary energy shares by source. Redrawn from Williams *et al.* (1984)

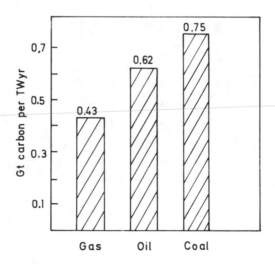

Figure 2.2 Comparisons of the release of carbon (in gigatonnes) for the production of 1 TWyr of energy from gas, oil, and coal (from Marland, 1982)

Fossil Fuels

In the combustion of any fossil fuel, carbon is oxidized and CO_2 is released to the atmosphere. When burned, different fuels produce varying amounts of CO_2 for a given release of thermal energy (see Figure 2.2). Excluding oil shale, coal is the most 'CO_2-intensive' fossil fuel, followed by oil and gas.[1]

Recent estimates for proven reserves and total recoverable resources [2]

Resources represent the total quantity of ore in the ground whose presence can be inferred from the results of direct exploration and may be recoverable in the future.—all highly uncertain—are shown in Table 2.2 for oil, coal, and natural gas. Estimated coal resources are considered in more detail in Table 2.3, which indicates that estimates of both reserves and resources have been continually revised upwards. In just six years, coal reserves 'increased' by 40 percent and total resources of coal went 'up' by 29 percent, largely because of rising prices. These revisions indicate the uncertainty surrounding estimates of reserves and resources. Moreover, not all known deposits have been extensively explored. Vast areas—including much of Africa, Central America, eastern Siberia, and northern China—have seen comparatively little fossil-fuel exploration (Wood, 1983). Indeed, large deposits have recently been found in Colombia, Sumatra in Indonesia, and Botswana (Rose, 1984a). Such finds cast further doubt on the figures shown in Table 2.3 and suggest that the future of CO_2 emissions may not remain within the control of just a handful of nations.

Besides conventional fossil resources, unconventional forms—notably tar sands and shale oil—should be taken into account. Although the future de-

Table 2.3 Global coal resource estimates (billion tonnes coal equivalent, tce). From Rose *et al.* (1983)

	Geological resources	Reserves[a]	Ratio of:	
			Reserves/ resources (%)	Reserves production (yrs)
1974 (WEC)	8 603	473	5.5	189
1976 (WEC)	9 045	560	6.2	207
1978 (WEC)	10 127	637	6.3	230
1980 WOCOL estimate	10 750	663	6.2	239
1980 (WEC)	11 060	687	6.2	248

[a] Technically and economically recoverable reserves

---

[1] Refining and burning of oil shale produces up to 30% more CO_2 per exajoule released than does combustion of coal (depending on the process of refining used). Most of the excess comes from emissions in the refining process.

[2] Reserves represent the quantity of ores known to exist as a result of direct exploration, which can be extracted using available technology and economically exploited at current price levels.

velopment of these resources is uncertain, Canada has been exploiting the Athabasca tar sands since the 1960s, Venezuela is expected to develop the huge oil sands deposits in the Orinoco Basin, and the Soviet Union is beginning to develop its oil shale resources. Taken together, global exploitable resources of shale oil and tar sands equal an estimated 70 billion tons of oil (WEC, 1983), as Table 2.2 shows.

Conventional oil and gas resources are limited, but will not be depleted by 2050. If the production of deep or geopressurized gas becomes economical, then gas reserves will increase considerably. Known reserves of coal are more than sufficient to meet expected energy demand for solid fuels well beyond 2050. Tar sands and oil shale represent supplies substantially larger than global resources of conventional oil and gas.

Non-fossil Supplies of Primary Energy

Nuclear energy, solar heat and photovoltaics (PV), hydropower, wind energy, and biomass can power numerous end-use applications of energy without net emissions of CO_2. If any of these sources (or geopressurized 'deep' gas deposits) were deployed widely and rapidly, future global demand for coal would be significantly affected. However, for various economic, technological, and environmental reasons, they provide a much smaller share of global energy today than conventional fossil fuels (see Figure 2.1). Both in industrialized and developing countries, most electricity is still produced by burning fossil fuels (see Figure 2.3).

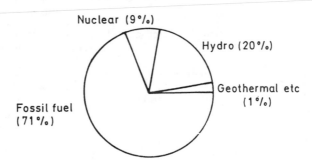

Figure 2.3 Global electricity supply by source. Redrawn from Williams *et al.* (1984)

The nuclear technologies that are economically competitive with fossil fuel systems employ nuclear fission in light-water and heavy-water reactors. Both consume uranium fuel in a 'once-through' conversion cycle. Commercial light-water reactors use artificially 'enriched' uranium while heavy-water reactors can use either natural or man-made isotopes as fuel. The most promising candidates for the future are fission breeder reactors (which

can produce fissile plutonium from uranium fuel elements) and fusion reactors (which use natural or man-made isotopes of 'heavy' hydrogen for
fuel). Environmental and safety problems in connection with nuclear power
have been hotly debated for many years, including environmental contamination during routine disposal or storage of waste material, unauthorized
diversion of fuel and the risks of catastrophic accidents. In addition, nuclear
power and the proliferation of nuclear weapons are linked—though how
strongly is also hotly debated. These issues add to the uncertainty about
how widespread nuclear power is to become as a future source of electricity, though nuclear programmes are well under way in the United States,
the Soviet Union, France, Great Britain, Japan, Korea, Taiwan, and several
other countries. Because nuclear energy would substitute primarily for coal
combustion in electrical generation, the problem of future CO_2 emissions
to the atmosphere will of course be substantially reduced if nuclear energy
becomes the world's main source of electricity.

Although the share of global electricity generated by fission reactors rose
rapidly during the 1970s, the growth rate has tapered off recently. In many
industrial countries, rising construction and running costs, long lead-times,
unanticipated problems during routine operations, and safety-related environmental protection requirements have combined to make the rapid expansion of fission electric power systems seem less likely today than a decade
ago. As for the longer term, how and how much commercial nuclear power
will develop is difficult to anticipate.

Geothermal energy is another alternative to fossil fuel combustion in some
applications. Because high-quality geothermal resources are limited to specific geographic areas, this energy form currently contributes less than one
percent to the global energy supply. However, it does provide an alternative in a few locations. Dry steam plants are currently economical and
being developed in, e.g., Italy, New Zealand, Mexico, and the United States.
Geothermal resources are also being developed in the Soviet Union, Australia, Hawaii, Africa, and Europe. Economic and environmental considerations make it unlikely that, by the year 2000, geothermal energy will be
as widely used as commercial nuclear power is now, even if so-called 'hot
dry rock' resources prove to be economically attractive. Yet, this variant
of geothermal technology, now being tested in prototype units, would allow the heat of deep strata in the bedrock to be recovered, thus making an
essentially unlimited source of heat available.

Renewable energy has contributed to global energy supplies for millennia,
and a number of those technologies may continue to make significant contributions also in the future. These include hydropower, wind, biomass and
solar energy.

Hydropower resources are exploited throughout the world. Recently,
world hydroelectric generation has increased: by 1982, it accounted for ap-

proximately 14 percent of electricity generation (about equal to the nuclear contribution) (see Figure 2.3). Today, large dams are being built in, e.g., the Soviet Union, South America, and Canada, and thousands of small-scale hydropower plants (250 kW or less) are under development in other parts of the world. Considerable resources remain to be tapped worldwide, particularly in developing countries. Several industrial countries are now reaching their limits for environmental and economic exploitation of hydropower whereas resources are still plentiful in many developing regions, and could play a very significant role in the development process. The total realizable global resource of available hydropower sites has been estimated to be up to 1.5 TW (IIASA, 1981; Kristoferson, 1977; WEC, 1983).

Wind electric systems, both small scale and large scale, contribute only marginally to the energy supply today. The fastest growing markets for these systems are in California, Australia, and Scandinavia. Although development is still limited by economics and by the resource's geographic distribution, the global potential for recoverable wind energy has been estimated to be of the order of 1 TW (Rose *et al.*, 1984; IIASA, 1981).

Biomass has historically contributed most to global energy supplies. Although seldom recorded in commercial transactions or in national income accounts in many developing countries, biomass use (primarily fuelwood and charcoal) accounts for more than 80 percent of annual energy consumption in many developing countries (O'Keefe *et al.*, 1984). It also plays an important role in developed countries. For example, the use of wood for home and office heating is the fastest growing energy supply technology in the United States; in 1981, more than 15 million households used wood to meet some part of home heating demand. The combustion of biomass for industrial process heat and the cogeneration of steam and electricity is also expanding, for example in the United States, Brazil, Sweden, Jamaica, and the Philippines. Today, large quantities of biomass are converted to liquid fuel for transportation in Brazil. In the future, other countries may also enter into ethanol or methanol production on a large scale. Biomass's total potential as a sustainable energy source has been estimated to be about 5 TW (Rose *et al.*, 1984; IIASA, 1981).

Whether biomass combustion adds to the global atmospheric CO_2 concentration depends on how the fuel cycle is managed. If feedstock production equals consumption, the carbon is merely being cycled between terrestrial biota and the atmosphere. However, if forests or other ecosystems are cleared or harvested more rapidly than they can be regrown, the standing biomass is, in effect, being 'mined', and CO_2 emissions from combustion would increase the carbon content of the atmosphere. Current and past rates of increase are matters of controversy. (See Chapter 3.)

Solar energy: although the direct conversion of solar energy to heat or electricity currently contributes negligible amounts to commercial energy consumption on a global scale, it shows promise for large scale applications for the future if economic and other problems can be overcome. In many regions, solar energy cannot currently compete economically with fossil fuels, and its major contribution is such decentralized uses as home heating. However, according to some analysts recent technological developments in photovoltaic power systems show promise for delivering bulk electricity at costs comparable with those of fossil and nuclear fuels by the turn of the century (see Maycock and Stirewalt, 1982; Green, 1983). Also by that time R & D efforts may have brought photovoltaic power systems to such a stage where they may be economically coupled to electrolyzers to produce hydrogen as a fuel for transportation. The timing or success of these systems cannot be predicted and is also subject to debate, but if successful, either could seriously challenge fossil fuel applications (Rose *et al.*, 1983). In part because their environmental impacts are less severe than those of the technologies that they might replace, new or emerging renewable technologies could successfully displace fossil fuel use if their production costs continue to fall. Also a few other renewable energy technologies, such as ocean thermal energy conversion and wave or tidal power, show the theoretical potential of delivering significant amounts of energy in the future. Future contributions are however very hard to estimate, but are likely to be limited.

Over the long term, any one of several advanced technologies (e.g., deep gas, fusion, fission breeders, hot dry rock, or photovoltaics) may become economically attractive as a substitute for fossil-fired electricity production. In addition, improvements in the efficiency of industrial processes, residential and commercial building design, and transportation systems may greatly reduce the amount of fossil fuel consumed per unit of energy service. The 70 or more years time-scale applied in this study is enough to allow substantial technological innovations and changes in behaviour to occur. Thus, it is also necessary to be prepared for technological breakthroughs and substantial improvements, and policy-makers may have more room to manoeuvre than current practices and historical energy trends alone would suggest.

2.2.3 Historical Development of CO_2 Emissions

Since 1860, industrialization has progressed with an increase in the use of fuels—especially fossil fuels—and a corresponding increase in CO_2 emissions. Keeling (1973) identified a striking increase in global CO_2 emissions from commercial sources of energy between 1860 and 1953 (Figure 2.4). Rotty and Masters (1984) reported that those results cannot be substantially improved. As Figure 2.5 shows, growth has been steady and exponential at 4.2 percent per year, interrupted only by the two World Wars and the Great Crash of 1929. According to Rotty and Masters, this growth 'should not be taken as indicative of the future' since it 'cannot continue indefinitely' (p. 6).

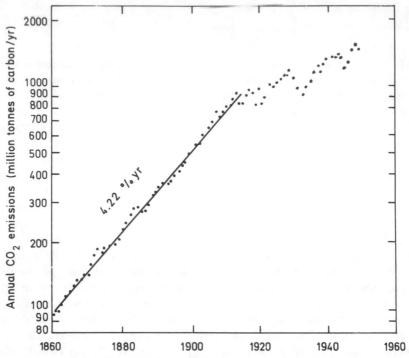

Figure 2.4 Annual carbon dioxide emissions resulting from fossil fuel combustion, 1860–1949. Redrawn from Rotty and Masters (1984)

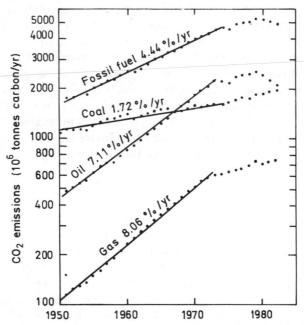

Figure 2.5 Annual carbon dioxide emissions resulting from fossil fuel combustion, 1950–1982. Redrawn from Rotty and Masters (1984)

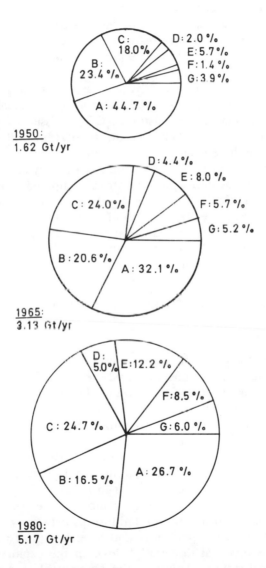

Figure 2.6 The changing patterns of global CO_2 emissions. Redrawn from Rotty and Masters (1984). A: North America; B: Western Europe; C: USSR, C.P. Europe; D: Japan, Australia; E: Developing; F: Centrally Planned Asia, USSR; G: Other

The detailed data collected and published each year since 1950 in the United Nations' *Energy Statistics Yearbook* show growth at about 4.4 percent annually until 1973, when the average growth rate dropped significantly. In the last decade, emission rates grew more slowly and then actually declined (see Figure 2.5). The global economic recession between 1979 and 1982 accounts largely for the decline in CO_2 emissions observed during this period. Because industrialized nations have become more energy-efficient and have turned in part to non-fossil fuel sources, CO_2 emissions have declined since their historic peak in 1979 (approximately 5.3 GT/year; Marland and Rotty, 1984).

On the other hand, developing nations have cut their growth rate in energy consumption only slightly. Thus, their share of global CO_2 emissions has increased. As Figure 2.6 illustrates, CO_2 emissions from Western industrialized countries have declined from 68 percent in 1950 to 43 percent in 1980, while the share from developing countries has risen sharply. From 1960 to 1973, the rate of growth in commercial energy consumption was 7.3 percent per year in the developing countries, compared with 4.8 percent per year in the industrialized countries, and 4.3 percent per year in the centrally planned economies. Since 1973, growth of energy use has fallen dramatically to 0.5 percent per year in the industrialized countries (Darmstadter, 1984), with smaller reductions in the developing countries (6.2 percent per year) and centrally planned economies (4.0 percent per year).

2.3 UNCERTAINTIES IN FUTURE ENERGY USE AND CO_2 EMISSIONS

2.3.1 Introduction

The only way to project future CO_2 emissions is to estimate the annual amount of primary energy consumed and the mix of fuels used. In turn, this requires estimating economic growth and the dependence of energy consumption on economic activity. Fundamental and irresolvable uncertainty surrounds both the future events affecting energy use and the simulation of future energy systems. Scientists' understanding of the complex interaction of technological, cultural, and political forces that determine the development of national energy economies is limited and whether today's forces and relationships will persist largely unchanged for 50 to 75 years is unknown.

Often, investigations of uncertainty in modelling experiments and policy analysis are confined to structural relationships between variables perceived to influence the system significantly and to estimates of these key parameters at various times, as discussed below. In the exploration of possible outcomes in real systems, however, the element of ignorance itself must be considered. In the case of future estimates of CO_2 emissions, the structures of national economies and energy systems in the distant future are unknown and largely unpredictable. Too often, analysts' ignorance of these factors remains implicit and unacknowledged when short-term engineering and economic models are extrapolated into the more distant future, only to erupt as an unwelcome surprise when the real system is modelled.

As often noted, no existing model could have predicted the development of the global energy system in the 1970s, using data from the 1950s and 1960s. Even the best models could not predict the oil price shocks of 1973 and 1979. Indeed, even if modellers had been given advance notice of these events, the models probably would not have anticipated such other key systemic shifts as the slowdown in the development of commercial nuclear power at the very time it was expected to advance rapidly and edge out increasingly costly fossil fuels.

The presence of uncertainty about economic and political issues does not cripple the analysis of energy systems. Any single-value estimate or scenario of future energy use and CO_2 emissions is of limited value as a foundation for policy formulations. The prudent analyst must instead consider a range of feasible energy futures that might logically result from alternative policy choices.

With this caveat voiced, what are the uncertainties in demographic, economic, technological, and institutional factors that will affect the actual level of future energy demand, the mix of energy supplies consumed, and the associated rates of CO_2 emissions? In the following discussion, these first-order uncertainties are explored along with those characteristic of simulation models and projections of energy economic systems.

2.3.2 Factors Contributing to Uncertainty about Future Energy Use and CO_2 Emissions

Demographic Uncertainties

Today, most of the world scientific community agrees on rates of world population increase over the next century and generally accepts the United Nations' mid-range estimates of six-billion people by the year 2000 and ten-billion by the year 2100. However, Ausubel and Nordhaus (1983) still found estimates for rates of global population growth for the period 1975 to 2025 ranging from 1.2 to 1.7 percent per year, compared to the estimates of 1.6 percent between 1975 and 2000, and 0.5 percent between 2000 and 2100 used in the mid-range estimates of the United Nations.

Many global population estimates do not reflect the effects of regional social, and cultural issues and national population policies. For example, the current campaign for 'one family, one child' in the People's Republic of China has substantially lowered the birthrate in that nation, where nearly one-quarter of humanity lives. The age structure of a population also affects its patterns of energy consumption and its productivity. In countries where much of the population is under 15 or over 65, economic output per capita may be lower than in a region with an equal level of investment but a higher percentage of its population in the more productive 20-to-60 age group.

Urbanization also plays a role in per capita energy demand. In developing countries, urbanization can triple or quadruple per capita demand for primary energy, as charcoal is substituted for fuelwood (Goodman, 1983). However, when kerosene, liquid petroleum gas, electricity, or other high-quality fuels are substituted for traditional fuels, urban households able to afford these fuels can use as little as 30 percent of the energy required for cooking (the largest single household demand) as do their rural counterparts (Williams *et al.*, 1984). Thus, while urbanization helps determine energy demand in developing countries, the direction and magnitude of the effect on energy demand depend not only on the technologies and fuels employed but also on the level of economic development and household buying power.

Global population trends are important for predicting future energy use, but they are not enough. Energy demand cannot be predicted simply by assuming that there will be mit X percent more people who will each use Y amount more energy. The rate of population increase in any period interacts with, and may correlate with, the rate of GDP growth, but not necessarily with the rate of GDP growth per capita (Radetski, 1984). As discussed below, other factors contribute to uncertainty in estimates of economic growth.

Uncertainties in Future Economic Variables

The major analyses of future energy use and CO_2 emissions incorporate differing estimates of future rates of economic growth. Even when looking forward for only a decade and at only one or two factor prices, economic experts offer no consensus. To analyse changes in global climate, one must look at 50 to 100 years—an impossibly long time frame for economic forecasts. When compounded over half a century, even a small difference in the estimated rate of economic growth has a significant effect on the global GDP projected for the final year of the period.

Manne and Schrattenholzer (1984a) have surveyed estimates for future global GDP and oil prices. Looking only 25 years ahead, they requested estimates from 70 governmental and international agencies, corporations, individuals, research institutes, and universities. Estimates of world GDP varied by 50 percent for the year 1990, by a factor of two for the year 2000, and by nearly a factor of four for 2010. Projections of the world price of crude petroleum (in constant economic units) varied by a factor of three for 1990, four in the year 2000, and five in 2010 (Manne and Schrattenholzer, 1984b). Similar differences have been obtained in other reviews (see Ausubel and Nordhaus, 1983).

Many factors, some not easily quantifiable, contribute to the uncertainty of these estimates. These include:

1. Natural resource endowments.
2. International relations and stability, particularly East–West relations and North–South dynamics.
3. Consumer values and expectations.
4. Technological developments.
5. Annual investments and the efficiency of national and international investment allocation.
6. The availability of infrastructure to coordinate and distribute goods and services.
7. Economic policies to provide counter-cyclical stimuli to growth, promote infrastructure development, and encourage specialization through international trade.

Cleveland *et al.* (1984) argue that increasing the thermodynamic quality of energy used in production would spur productivity and increase the GDP. International economic policies to promote stability, free trade and peace may be the most important determinant of the pattern of long-term energy-use and CO$_2$ emissions over the next 50 to 100 years. Few, if any, of these factors can be easily incorporated into computer models.

Technological Factors Affecting Future CO$_2$ Emissions

The magnitude and composition of future energy demand will be influenced by four technological factors: the difficulty of substituting fossil for non-fossil fuels and vice versa, the rate of general productivity growth, the ease of substitution between energy and non-energy inputs to production (Nordhaus and Yohe, 1983), and trends in the real cost of producing fuel. In addition, technical breakthroughs could well occur between now and 2050 that would change the energy economy.

These four factors are the product of interactions involving four other factors: (1) incremental change in both energy supply technologies and the efficiency of energy use; (2) climatic change and other environmental constraints on the maximum mobilization of various energy technologies; (3) the market for new technologies; and (4) changes in the composition of energy demand.

Predicting the opportunities and timing of technological advances is a gambler's art. Edmonds and Reilly (1983a, see also Edmonds *et al.*, 1984) estimate the aggregate rate of technological change in terms of a single parameter with values between zero and 1 percent per year. Using the same model, Rose *et al.* (1984) found higher rates of technological innovation to be plausible and consistent with other analyses. Such aggregated rates do not provide a precise description of the dynamics of change, but they are one way of representing the effects.

Environmental constraints on mobilizing energy technologies may take many forms. In some cases, energy activities will deplete limited local resources of high-quality ore, clean water, or the ability to absorb wastes and effluents. Major prolonged damage to lakes and forests from acid deposition, for example, could lead to public protest against a technology that seems risky.

Most analyses of future energy demand ignore the fact that climatic changes caused by a CO_2 buildup can alter energy supply-and-demand patterns. Analysing the results of two general circulation models for five urban sites representative of much of western Europe, Jäger (1984) found that doubling atmospheric CO_2 levels could cut the heating season by one to three months and increase the cooling season significantly. This suggests a decrease of up to 10 percent in the demand for space heating and a reduction of regional energy demand by as much as 2 to 3 percent.

Changes in regional climate can also affect energy supply. Quirk (1981) has shown that droughts can substantially reduce hydroelectric output. Jäger (1984) observed that climatic changes can also significantly affect supplies of both conventional and renewable energy.

An array of economic, institutional, and physical factors may affect the rate of market penetration of new energy technologies. Laurmann (1984), drawing on the work of Marchetti (Marchetti, 1980; Marchetti and Naki-cenovic, 1979), states that past evidence on shifts in global primary energy supply implies a characteristic time of 50 years for new technologies to capture half of the market. Conditions for market penetration, however, differ among cultural and economic systems, local institutional and regulatory settings, and technological characteristics. Thus, use of a single-value estimate for all societies and time periods is inappropriate.

Another potential technological development that could affect the rate of CO_2 buildup is the introduction of a process to remove (or 'scrub') CO_2 from other exhaust gases after combustion. Once such a process were developed and its economics became attractive, there would still be a problem of where to dispose of the solid effluent (for example, as $CaCO_3$). To date, no economically efficient method has been identified and none is within immediate sight. In fact, most approaches use more energy than the original fuel releases. Some recent studies however suggest that the cost of such techniques may be coming down (see Steinberg *et al.*, 1984). If technological breakthroughs occur, they could seriously alter the atmospheric impacts of fossil-fuel use, although the problems with other greenhouse gases than CO_2 would remain to be solved.

Social, Political, and Institutional Factors

Changes in cultural values, lifestyles, political structures, and institutional commitments could substantially affect energy use (and CO$_2$ emissions) over long periods of time. One of the factors to be taken into account is social resistance to changes in lifestyles and familiar technologies, as well as institutional, professional, and political resistance to innovations. In addition, inherent instability in natural and social systems could limit the range of predictability of future energy needs.

Policy choices regarding efficiency and conservation measures, accelerated development of specific technologies, strict environmental regulations (or absence thereof), or lack of such policies will determine which of many possible energy futures will actually be realized.

However, a theoretically feasible policy may be impossible to implement. Technologies with long lead times could be stopped by the considerable inertia in the global energy/economic system. In addition, a technology (e.g., nuclear power or hydropower sited in a sensitive area) that may look good to scientists, engineers, and economists may prove to be socially unacceptable.

On the other hand, infrastructural and institutional changes (or lack thereof) can enhance or inhibit rapid changes and shifts of emphasis. For example, a national policy to subsidize local biomass production, nuclear power and waste disposal or coal exploitation can make an economically marginal technology much more attractive in a country where it does not yet exist or is under development. Reasons for adopting such national policies may include job creation, national security or environmental considerations. Energy policies are thus often decided by factors completely outside the energy sector.

In addition to the uncertainty about future policy choices is that surrounding implementation—an uncertainty that will vary with each policy and the evolution of institutional structures. Because policy implementation does help determine the future, uncertainty as to which future policies bearing on the development of energy technology will be implemented or what constraints on the permissible level of CO$_2$ emissions will be imposed matters greatly. Moreover, unlike the date of a future technological breakthrough, policy is one determinant of the future that people can control—a self-evident fact that is too often ignored in modelling efforts.

Uncertainties in Models and Projections

The models built to simulate the evolution of the global energy system are themselves subject to uncertainty. Most studies acknowledge uncertainty as a difficulty. But only recently has it been featured prominently in global forecasting. In long-range forecasting, uncertainty grows with time. However, in energy- and economic-modelling, disagreement about the near future is frequently greater than about the distant future (Nordhaus and Yohe, 1983). Indeed, many of the most common assumptions about the future are the least analysed—one reason why long-range uncertainties are so frequently underestimated. Authors can influence one another greatly, and when taken together, their 'agreement' creates an unjustified sense of certainty. In several cases, all forecasts turned out to be uniformly incorrect, especially in the energy field (see Greenberger *et al.*, 1983). Moreover, projections that downplay uncertainty frequently carry a false sense of accuracy or inevitability. A model with lots of data and equations can appear to be analytically rigorous, even though most of the data and equations may reflect assumptions with little empirical or theoretical basis.

In forecasting, especially with the use of computer modelling, little attention is usually given to the importance of surprises or atypical events, even though the global energy system is likely to respond to these unpredictable episodes. It matters greatly, for instance, if prices triple in the first year of a twenty-year planning period (due to, say, a national disaster or a war), instead of tripling gradually over 20 years. When looking forward 50 to 75 years, it is worthwhile to return to 1910 and consider how well 1980 could have been forecast then, given the intervening world wars, the Great Depression, and the discovery of nuclear fission or computers among other things. Thus, although the evolution of social and institutional factors cannot be predicted, it must be acknowledged.

Parametric and Structural Uncertainty

In modelling, the two general categories of uncertainty are parametric and structural. Parametric uncertainty relates to such questions as what the GDP or energy demand will be in the future and refers to doubts about a variable's precise value. Structural uncertainty is more fundamental and subtle, involving relationships between variables and the way they affect one another. For example, what is the relationship between GDP and energy demand? Will the relationship hold for the next 50 years? Uncertainties of this kind relate to unknown or poorly understood behavioural and causal relationships between variables.

The concepts of parametric and structural uncertainty are well illustrated by a familiar example of energy economics: the energy/GDP coupling. For many years, aggregate energy consumption and GDP seemed to march in lock-step, and this was empirically well represented by the income elasticity of demand. But since 1973, the GDP grew but energy use did not. This 'breaking of the coefficient' or 'decoupling' has shifted attention from parametric uncertainties to the more fundamental structural issues, which are not well understood.

Handling Uncertainty

Methods for handling uncertainty in models usually focus on parametric uncertainties, including sensitivity analyses, statistics, stochastic processes, and such probabilistic techniques as 'Monte Carlo' simulations. These methods have proved to be powerful tools when carefully applied in energy and economic modelling. But few methods for handling structural uncertainty exist. Most models have one basic structure, which makes exploring structural uncertainty difficult. One solution is to build different models with different structures in order to test the impact of alternative mathematical constructs on the robustness of simulation results.

The growing emphasis on sensitivity analyses and probabilistic treatments is welcome, but not without pitfalls. In sensitivity analyses, the usual approach is to search for output variables that are either greatly affected, or largely unaffected, by varying input data. Such variables, if found, are commonly considered as representative of highly sensitive policy 'levers' or indicative of fundamental changes in structural relationships. But such conclusions should be drawn with caution since sensitivities of the model itself are one thing and the sensitivities of the natural system being modelled are another.

In building a complex dynamic simulation model incorporating many specified formal relationships, the model's overall behaviour—how the most important relationships and parameters interact to influence the outcome of simulation experiments—may not be well understood. Sensitivity analysis should reduce this uncertainty, but even a robust model can be a poor simulation if the correspondence between the behavioural characteristics of the model and real world characteristics is not verified.

Which historical data are the most relevant to test model validity? The analyst's choice may reflect only implicit beliefs and prejudices about the driving forces already built into the central structural relationships in the model. In such cases, validation becomes self-referential and circular, and more fundamental uncertainties about the real systems are concealed.

Often, the implicit assumption in modelling exercises is that uncertainty in analyses arises only from inaccurate measurement of the objective system that exists 'out there' in reality. Yet, the far more subtle, misleading, and unadmitted kind of uncertainty results from ignorance about the dynamics of the real system and from the analyst's undetected prejudices and implicit preconceptions. The impact of these types of uncertainty goes beyond those introduced by the choice of specific parameter values. In short, they introduce a 'hidden', unmanageable element of analytical brittleness.

The increasingly wide use of probabilistic techniques and of extensive sensitivity testing helps uncover and combat built-in biases, but surprises will inevitably occur. Thus, prudent public policy should be formulated on the basis of a range of feasible outcomes and developed using a variety of modelling tools and simulations that identify underlying assumptions, uncertainties, and biases.

2.3.3 Developing Countries—A Special Case

Some 75 percent of the world's population live in developing countries and earn about 20 percent of the global income (Goodman, 1983). Less than 10 percent of the world's industry is located in these regions. In most global energy studies, developing countries are considered 50 to 100 years 'behind' the industrialized countries. Some analysts also believe that the various aid programs operated by the industrialized countries are accelerating developing countries' efforts to 'catch up' and mimic the economies of industrialized states. These and other assumptions are often built into energy and economic models and forecasts, despite mounting evidence that existing aid programs are not always helping to build balanced, self-sustaining economies in developing countries (Cole, 1983). Leaving aside the question of whether developing countries should be measured with an industrialized countries' yardstick, the measurements and projections themselves have often been inaccurate. The economies of many developing countries are deteriorating, and their economic growth falls short of that assumed in early global energy studies.

In addition, in comparisons of energy systems in developing countries with those in the industrialized states, it is not always noted that in many developing countries the largest fractions of primary energy consumption are in the non-commercial sectors (fuelwood, crop residues, and dung, for example). Informal economic activities, or 'shadow economies', often contribute significantly to GNP. Continuing problems of hunger, disease, and inadequate shelter undermine economic growth. The need to address these issues simultaneously, and in an integrated fashion in the process of economic development, makes linear extrapolations of past trends in energy consumption of little value. New systematic methodologies must be developed to simulate the changing conditions in developing countries and to project appropriate patterns of energy use.

Variations in development potential among countries must also be recognized. Brazil, for example, is more likely to develop an economy that resembles that of the U.S. than that of Belgium. In 70 years, Ethiopia may look more like Spain or Portugal than like Britain or the Netherlands. No single model of economic development or energy consumption is well-matched to the range of circumstances facing developing countries.

2.3.4 Conclusions

A host of complex, interrelated factors combine to make it impossible to predict long-term future global energy and CO_2 developments precisely. Thus, if energy development policies are to take environmental and other long-term impacts into account, the full range of feasible outcomes must be explored and the impacts of specific policy choices evaluated in the light of the uncertainties inherent in the processes of formulating long-term projections of energy use and CO_2 emissions. Keeping in mind the dimensions of uncertainty discussed above, the next section presents a review and analysis of a range of recently published energy/economic projections and explores their implications for future CO_2 emissions.

2.4 REVIEW OF FUTURE ENERGY AND CO₂ PROJECTIONS

2.4.1 Introduction

This section provides a critical review of several global energy and CO_2 forecasts. Other reviews have appeared recently, notably those by Ausubel and Nordhaus (1983), Perry (1982), and Darmstadter (1984). Rather than duplicate these efforts, a complementary discussion and analysis is presented here. Earlier reviews have concentrated on final results from various studies and comparisons of these, giving only brief summaries of how the results were obtained. Here, the focus is on the methods employed to obtain results, with brief summaries of the results.

Many studies covered in previous reviews are omitted. Attention is focused on major and/or recent studies. To obtain a general picture of results in the field of energy/CO_2 projections, the reader is also urged to consult the earlier reviews (particularly Ausubel and Nordhaus, 1983).

2.4.2 Modelling and Forecasts

Any forecast is the result of some kind of model, even if only a mental model. In the last decade, formal models, particularly computer models, have been increasingly utilized for investigating the effects of alternative national and international policies. Models vary in complexity from simple accounting schemes to sophisticated systems of behavioural functions involving intricate feedback mechanisms. The process of building, testing, applying, and interpreting models needs to be clarified and demystified in order to make the strengths and limitations of the results more apparent and open for discussion and the models thus more useful as planning and scientific tools.

A model is a representation of a structural relationship or other underlying aspect of reality. In principle, the major advantages of mathematical models are that they require formal representation of the relationships between factors and they demand that assumptions and inputs be made explicit. This helps to control bias or at least make them more transparent. In addition, models permit investigation of different policy options and simulation of unforeseen developments. Thus, they can provide counterintuitive insights that might otherwise go unnoticed.

Characteristics of Models

All models have certain elements in common. Every model is a simplification of the real system it is designed to reflect. Values and prejudices are rarely explicit in the description of the model, but the world view and biases of the analyst often shapes the model structure, the assumptions made about underlying processes, and the range of 'plausible results'. Politics, for example, are a crucial element of energy systems the world over. But they are often ignored or only implicitly incorporated into formal models. Koreisha and Stobaugh (1979) point out that models 'are often modified by personal judgments to make the results correspond more closely to the specialist's understanding of the real world'.

Such modifications can take the form of selection of input data and parameter values in order to steer a model toward a desired result. In other cases, it may be accomplished by adjusting the internal workings of the model so that the input data are transformed in ways more believable to the analyst. This is sometimes regarded as the application of informed judgment in order to fertilize the sterile workings of the mathematical model. Provided that adjustments to the model structure are clearly and fully documented, this is not an improper practice.

An important aspect of modelling is that models are built for different purposes. In global energy forecasting, one researcher may be interested in the future role of nuclear power while another may focus on the interactions between the industrialized countries and the Third World. Both researchers will produce global energy models, but the resulting models will be different. This makes direct comparison difficult, if not meaningless.

There are a number of simplifying assumptions that characterize all models, and these are discussed briefly below.

1. *Omissions*: Regardless of the degree of sophistication or detail, a model can treat only a rather limited number of relationships. The implicit assumption made by the modeller is that the influence of excluded factors is unimportant. In fact, a model can sometimes mislead its users by inadvertently focusing attention on unimportant aspects of the real system while ignoring critical variables.

2. *Approximations*: Even a well understood relationship can often only be

approximated in a model. This introduces an immediate source of error, the size of which is not always known.

3. *Assumptions*: When a relationship is not well understood, or is uncertain, or not easily quantified, then either it is omitted, or an assumption must be made to represent it. Such assumptions are often arbitrary and unverifiable.

4. *Relevant range of variables*: Data used to develop equations and forecasts from historical relationships are typically derived from experiences with one range of values. If the underlying behaviour and circumstances undergo little change, the relationships will probably remain constant. But if conditions change rapidly or radically (as happened to oil prices in 1973), the historic trends and variable ranges may have no relevance. Most analytic forecasting methods are applicable for very short time horizons at best (e.g., five years), and the validity of extending the time horizon to several decades is very questionable.

5. *Reversibility*: Many models assume that fundamental economic relationships are reversible. They assume that, for example, consumer behaviour patterns typical of a period of declining energy prices will continue in a period of rising prices. But there is no evidence that the relationships will remain true under different conditions.

Model Size

An important issue in model building is how much detail should be included in the structure. Large, disaggregated models offer considerable detail in their pictures of the energy/CO$_2$ system of the future; they are capable of incorporating a variety of technological, economic, and demographic features. Such projections can highlight the differences between technologies and the effects of changes in lifestyles and values. Regional detail can reflect differences in circumstances, policies, or attitudes.

Models of this type often require much detailed data and many assumptions regarding lifestyles and values. These data are often incomplete or unobtainable, and thus assumptions are often fabricated in an *ad hoc* manner. Systematic treatment of uncertainty is difficult with such models and accurate documentation is a mammoth task.

Small models offer complementary advantages and drawbacks. They are easier to understand, modify, and document; and they facilitate systematic treatment of uncertainty. Computational simplicity can lend transparency to the analysis. However, small models do not include much detail: important aspects of the real system may be lost in the aggregation, e.g., political and socioeconomic contrasts between developing and developed countries. Also, it can prove difficult to establish realistic interpretations for the data or variables.

Types of Future Projections

The simplest types of future projections are forecasts based on the extrapolation of historical data. The key mathematical relationships in these models are usually either linear functions or exponential growth curves. The relationships that produced the historical data are assumed to continue uninterrupted. Such forecasts cannot easily incorporate radical changes. Many of the projections of energy demand made before 1973 were forecasts of this type, and they were unable to anticipate the results of events in the last decade.

Some recent studies have used the scenario approach, a picture of an unfolding future resulting from specific assumptions. Scenarios usually incorporate both technological and economic elements, and may also address such issues as social values, capital accumulation, and infrastructure development. They offer a richness of description, but not necessarily a forecast.

Backcasts begin with descriptions of a desirable future and work backwards to identify specific actions or limits to action that might make that future possible. One example is an attempt to identify energy futures that hold global atmospheric CO_2 concentrations below some specified level (see Perry, 1982). Backcasts can be used, however, to provide important guidance in formulating policy on issues such as CO_2 buildup in the atmosphere which have uncertain long-term consequences.

2.4.3 Future Energy and CO_2 Projections

Several specific energy/CO_2 projections are now considered in detail. The principal results are summarized in Tables 2.4 and 2.5. The first table presents global primary energy projections for the year 2050; the second table gives the corresponding projections for CO_2 emissions. These results are illustrated graphically in this section's concluding discussion.

Edmonds and Reilly (1983a, b, 1984)

The Edmonds and Reilly (ER) model is a detailed partial equilibrium model, specifically designed to investigate long-term alternative energy policies and their implications for future CO_2 emissions. The world is disaggregated into nine geopolitical regions, as shown in Figure 2.7. A total of nine primary and four secondary energy categories are considered. The major inputs are assumptions about population and economic growth, supply and demand schedules for each type of fuel, and initial conditions. Interregional trade is incorporated into the market clearing calculations for all fuels except electricity. Demand for energy services is driven by population and GNP, the latter being calculated from the labour force and labour productivity.

Table 2.4   Projected primary energy in 2050 (in TWyr/yr)

Report	'Base case'	Range TWyr/yr	Page ref. in source
MIT (1983)	—	16.3–32.3	p.56
EPA (1983)	30.0[a]	Not given	p.4–4
ER (1983)	52.2	52.2	p.31
ER (1984)	28.4	17–62.5	p.49
Goldemberg *et al.* (1984)	11.2(2020)	11.2–14.8	b
Lovins *et al.* (1981)	4.56	4.56	c
IIASA (1981)	—	22.4–35.7(2030)	p.522
IIASA (1983)	26.3	26.3	p.2[d]
Legasov *et al.* (1984)	42	42	e
NY (1983)	—	9.7–109.3	f
Reister (1984)	29.9	24.0–41.3	p.21
WEC (1983)	29.8	18.9–29.8	g

[a] Estimated from Figure 4.1, p. 4–4.

[b] From Tables 9 and 14.

[c] Based on linear interpolation between 2030 and 2080; Krause (1983), Table 2.

[d] Based on linear extrapolation from 2030 to 2050.

[e] Estimated from Figure 3a, p. 1092

[f] Estimated from Figure 2.15, p. 136

[g] Calculated from Table 1, p. 252

The GNP is also affected by price via a constant elasticity feedback mechanism. Demand for energy services is converted into actual energy demand by a process involving interfuel substitution elasticities, and efficiency indices (reflecting technological change) which determine the coupling between energy demand and GNP.

The nine sources of primary energy supply are divided into three categories: constrained nonrenewable resources (conventional oil and gas), constrained renewable sources (hydroelectricity and biomass), and unconstrained sources (coal, unconventional oil and gas, nuclear, and solar). The cumulative extraction for resources in the first category is modelled by a logistic function of time (except for the Middle East, which is determined by OPEC policy, specified exogenously). Thus, the extraction rates for these resources are prescribed, and the total supply is offered without regard to market conditions. In the second category, hydroelectric power is also price insensitive, and is phased in by an exogenous logistic function (and exogenous price). Biomass is a price-sensitive resource. Finally, the five supply sources in the unconstrained category (called backstop technologies) are price sensitive, and contribute if production costs exceed exogenously spec-

Table 2.5   Projected CO_2 emissions in 2050 (present level is 5.0 Gt/yr)

Report	'Base case' emissions	Range (Gt/yr)	Page ref. in source	Approximate carbon intensity (Gt/TWyr)
MIT (1983)	15[a]	2.7–15	p.57	0.17–0.46
EPA (1983)	15	10–18	4–25	—
ER (1983)	26.3	15.7–26.3	p. 41	0.50
ER (1984)	14.5	6.8–47.4	p.37	0.40–0.76
Goldemberg *et al.* (1984)	4.6(2020)	4.6–5.9	b	0.40–0.43
Lovins *et al.* (1981)	<1	<1(2030)	c	0.15
IIASA (1981)	—	10–17(2030)	d	0.45–0.48
IIASA (1983)	9.4	9.4	e	0.43
Legasov *et al.* (1984)	13	13	f	0.31
NY (1983)	15	5–26[g]	p.94	—
Reister (1984)	10.7	9.7–27.1	p.30	0.29–0.66
WEC (1983)	14.4	10.0–14.4	h	0.48–0.53

[a] This assumes no CO_2 abatement: it is not a best guess (Rose, 1984b).

[b] For 2020, calculated from Tables 6 and 11 and Note (11) in Williams *et al.* (1984).

[c] From Krause (1983) Figure 9, p.28.

[d] p.586.

[e] Approximate value shown in Rotty (1984), p.34.

[f] Estimated from Figure 3b, p.1092.

[g] These are the 5th and 95th percentile scenarios.

[h] Calculated from data on p.252, using carbon conversion factors in Gt/yr of 0.78, 0.63, and 0.48 for solid mineral fuels (SMF), oil, and gas, respectively.

ified 'breakthrough' prices. The total CO_2 emissions are calculated from the fossil supply projections by summing contributions from various source points in the conversion process.

The ER model has certain weaknesses. Reister (1984) reports that the supply function for coal is highly inelastic, thus influencing the equilibrium between supply and demand. His conclusion is that the model is not completely specified and that it needs a more elastic supply function for the backstop technologies. This suggests that the analytic behaviour of the model may be highly sensitive to the arbitrary specification of exogenous supply parameters, which is characteristic of many large-scale energy models. Other criticisms are that conventional oil and gas are not price sensitive, and that capital formation and depreciation are not incorporated. In addition, the ER model is inappropriate for detailed end-use analysis, because end-use efficiency is aggregated into a few parameters. Also, the major focus

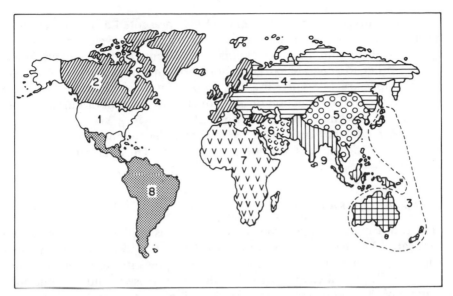

Figure 2.7 Typical geopolitical disaggregation in large energy/CO_2 models. Redrawn from Edmonds and Reilly (1983a). 1: USA, 2: OECD West, 3: OECD Asia, 4: Centrally planned Europe, 5: Centrally planned Asia, 6: Middle East, 7: Africa, 8: Latin Americ, 9: South and East Asia

is on commercial energy technologies of highly developed countries. This makes proper analysis of the energy sectors in many developing countries difficult, since non-commercial energy (fuelwood and charcoal) accounts for a large part of energy consumption (up to 80 percent). Finally, because supply and demand are not determined by price, the appropriateness of using an equilibrium model for analysing centrally planned economies or other economies where governments intervene actively in energy markets can be questioned.

Despite these limitations, the ER model is the only globally disaggregated model sufficiently well documented and tested to be useful. Considerable sensitivity testing has been performed and most researchers report no major vulnerabilities.

Edmonds and Reilly (1983a) developed a 'base case' scenario of the global energy future from 1980 to 2050. This is not a forecast, but a kind of 'best guess' scenario, incorporating what are thought to be likely developments. By the year 2050, global CO_2 emissions reach 26.3 Gt carbon per year (Gt/yr) as shown in Table 2.5. In several CO_2 taxation scenarios, a range of 15.7 to 26.3 Gt/yr is obtained. The conclusion is that the U.S. acting alone can do little to ameliorate the worldwide buildup of CO_2 and even a stringent global tax policy would delay a doubling of atmospheric CO_2 by only a decade or so.

More recently, Edmonds and Reilly (1984) presented a new set of three scenarios: a base case and two extreme cases that were 'developed by varying key parameters within the range of what are currently felt to be likely future values in order to generate extremes in carbon emissions'. In the base case, CO_2 emissions are 14.5 Gt/yr in 2050, 45 percent lower than the 1983 base case. This substantial reduction results from the adjustment of unrealistically low coal prices in the earlier runs (Rose, 1984a).

The ER model has been used by several independent research groups to investigate the global CO_2 question in detail. These efforts are briefly reviewed below.

EPA (Seidel and Keyes, 1983)

The U.S. Environmental Protection Agency has applied the ER model to investigate 13 global scenarios to the year 2100. A 'mid-range baseline' scenario was presented that was 'believed to be representative of likely future conditions'. It presumed an annual 0.6 percent increase in end-use efficiency for non-OECD countries, and a corresponding increase of 1 percent/yr in the industrial sector of the OECD countries. Five other baseline and seven policy-specific scenarios are also investigated. The study focused on the effectiveness of reduced fossil fuel consumption in delaying global warming. Thus, one 'high fossil' scenario is included. The other scenarios considered fossil fuel consumption and CO_2 emissions below the midrange baseline scenario (see resulting carbon emissions in Table 2.5).

The baseline scenarios include, besides the midrange case, increased renewable and solar energy, nuclear energy, and reduced energy demand scenarios. For these cases, 'the variation in CO_2 emissions ... appears small', ranging between 10 and 18 Gt/yr by 2050 (Table 2.5).These results are reported to be quite robust to moderate changes in GNP growth rates and income elasticity of demand.

The EPA study also included seven scenarios that simulated specific policies. Three scenarios include taxes on fuel use (resulting in CO_2 emissions between 12 and 14 Gt/yr by 2050); two involved bans on shale oil and synfuels. These five scenarios resulted in CO_2 emissions between about 10.9 and 14 Gt/yr. Two coal ban scenarios were considered (with CO_2 emissions of about 4.7 and 9 Gt/yr in 2050). But these were judged economically and politically infeasible.

In conclusion, the 11 feasible scenarios investigated by the EPA, using the ER model, resulted in CO_2 emissions ranging from about 10 to 18 Gt/yr by the middle of the next century. This is between two and four times the current level of about 5 Gt/yr. The EPA analysis concluded that the timing of a two degree global warming is rather insensitive to feasible future energy policies, based on two factors: the feasible scenarios entailed at least a 100 percent increase (over today's values) in CO_2 emissions by 2050 and the assumption that other greenhouse gases will significantly contribute to global warming and that the size of this contribution is not sensitive to any of the energy policies discussed above.

Rose et al., *(1983, 1984)*

A group of researchers at the Massachusetts Institute of Technology (MIT) applied the ER model to study available energy options relevant to ameliorating the future buildup of CO_2 Results were reported in a detailed report (1983) and a summary journal article (1984). Accounting for technological, economic, and environmental opportunities and constraints, the study explored the energy options for holding down CO_2 emissions. Eleven scenarios were investigated, extending to the year 2050 incorporating increased end-use efficiency and other possibilities, such as increased fossil fuel prices, reduced resource bases, increased nuclear costs, reduced photovoltaic costs, a nuclear moratorium, higher oil prices, a cut-off of Middle East oil, and combinations of these possibilities. Assessments of major energy supply technologies were also provided, and analyses of electricity demand fluctuations and energy storage were provided. All scenarios investigated were judged to be feasible (no fuel bans were considered).

The MIT study differed from other studies that have employed the ER model. First, detailed calculations of major materials requirements are provided and the study investigated realistic possibilities for improved energy efficiency, and found that 'we are far from the limits of what can be achieved by more efficient energy usage'. Substantial evidence was provided to support the specific conclusions that:

'The effectiveness of energy use on a global scale can be increased by about 1 percent per year for decades without any social strain. This seemingly small figure leads to a halving of energy use by the year 2050 and a 50 percent reduction in CO_2 emissions. This result is quite independent of the effect on CO_2 of any shifts to non-fossil sources for primary energy supplies.'It is not yet clear whether the prevailing societal, institutional, and economic circumstances will favour this course of development.

By 2050, the range of CO_2 emissions in the 11 scenarios is from less than 3 to about 15 Gt/yr. Thus, 'the bounds appear to be fairly wide, with a spread of a factor of five in annual carbon emissions by the middle of the next century' (1983). The conclusion is that 'an option space exists in which the CO_2/climate problem is much ameliorated'. Note the discrepancy between this conclusion and the corresponding conclusion reached in the EPA study, and the fact that consideration of the effect of greenhouse gases has not been incorporated explicitly into the MIT analysis.

Reister (1984)

Another study using the ER model was that of Reister, in which two detailed scenarios were developed: a base case and a 'high carbon' case that focus on the contribution of gas (both natural and synthetic) to future CO_2 emissions. Four different gas options were considered, all of which increased gas demand and reduced electricity demand. However, these gas options were found to have minor effects on future CO_2 emissions (the combined effect of all four options is, at most, a 10% reduction in emissions). In addition, sensitivity analysis to variations in GNP in developing countries is included. Resulting CO_2 emissions in 2050 are 10.7 Gt/yr for the base case and 27.1 Gt/yr for the high carbon case.

Comparison of Studies Using the Edmonds–Reilly Model

The results from the ER, EPA, MIT, and Reister studies appear to differ significantly. All four employed the ER model, and so their results may be directly compared. (Since the time horizons differ, results are compared here for the most distant year common to all, 2050.) The results are shown in Table 2.5. The basis for these differing results is not the model or its detailed analytic structure, but rather the informal judgments that determined what the model inputs would be.

As an example, both the EPA and MIT studies look into possible ways to reduce CO_2 emissions. There is a difference of a factor of 3 in the effective lower bound projected for CO_2 emissions into the middle of the next century. Since today's emissions are approximately 5 Gt/yr, MIT finds it possible to postpone the timing of an effective doubling of atmospheric CO_2 emissions by 40% over the next 65 years, while EPA, using the same model, finds it infeasible to delay the doubling for any significant period. Thus, it becomes apparent that the subjective selection of input data sometimes plays a more important role in determining model results than does the internal analytical structure.

To investigate this a bit more thoroughly, it is of interest to examine some of the key input data used in the different studies that employed the ER model. Table 2.6 shows the annual growth rate assumptions for population and GNP (in %/yr). It is clear that in all cases, the same input assumptions were employed for the base case. Moreover, only one study investigated sensitivity to variations in these assumptions. Note in particular that the differences between the MIT and EPA studies are not explained by differences in population or GNP assumptions. To help explain these differences, Table 2.7 displays the range of input assumptions for technological efficiency improvements. The most interesting observation from this table is the wide variation in efficiency improvements in non-OECD countries. This is part

Table 2.6 Growth rate assumptions for population and GNP

	Population (%/yr, 1975–2050)		GNP (%/yr, 1975–2050)	
	Base	Range	Base	Range
MIT (1983)	1.0	—	2.6	—
EPA (1983)	1.0[a]	—	2.6[a]	—
ER (1984)	1.0[b]	0.7–1.3[b]	2.6[b]	2.1–3.4[b]

[a] Based on data from Seidel and Keyes (1983).

[b] Calculated from Table 2.11 in Edmonds *et al.* (1984)

Table 2.7   Assumed range of technical efficiency improvements (%/yr, 1975–2050)

	OECD countries			Non-OECD countries
	Res.comm.	Transport	Industrial	
MIT (1983)	0.0–1.0	0.0–1.0	1.0	0.4–4.0
EPA (1983)	0.0–1.0	0.0–1.0	1.5	0.6–1.0
ER (1984)	0.0–0.5	0.0–0.5	0.0–1.5	0.00–1.0

Sources: MIT (1983), EPA (1983), Edmond *et al.* (1984).

of the reason for the different results obtained in the EPA and MIT studies (another principal reason is that the price of coal was much lower in the EPA study than in the MIT study).

The values chosen for key input data such as those shown in Tables 2.6 and 2.7 are crucial in determining model results, and yet this is not always clear in reports. The data in these tables reflect a host of behavioural and institutional assumptions that are represented by a few simple but key parameters.The different conclusions reached in different studies often boil down to basic differences in assumptions such as these that are in fact quite arbitrary.

Most energy/CO₂ studies provide little justification for the selection of input data. In this example, the principal differences between the EPA and MIT inputs relate to energy efficiency improvements: only the latter study presented analysis to support its choice of input data.

IIASA (1981, 1983)

The Energy Systems Program at the International Institute for Applied Systems Analysis (IIASA) conducted probably the most ambitious global energy study ever undertaken (Perry, 1982), which is described in *Energy in a Finite World* (IIASA, 1981). There were at least three distinct energy and/or CO_2 projection efforts within the study: the focus here is on the most prominent, the IIASA global energy scenarios, which have been a benchmark in energy and CO_2 projections. Other modelling efforts within the IIASA program were reviewed in Ausubel and Nordhaus (1983).

The stated purpose of the IIASA study was to provide an objective analysis of the facts and conditions for any energy policy (Häfele, 1980). Two detailed global energy scenarios were presented, labelled 'high' and 'low', to span the conceivable evolutions of the global energy system from 1980 to 2030. The world was disaggregated into seven geopolitical regions similar to the ER regions (Figure 2.7).

The detailed documentation of the IIASA scenarios described an iterative procedure involving three computer models. The iteration began with population and economic growth projections and lifestyle parameters, in an energy demand model. The resulting projections of energy consumption (from 1980 to 2030) were then fed into an energy supply model that computed the least expensive supply strategy that met the specified consumption levels. This was fed into a model that calculated economic and environmental impacts. These variables were fed back to modify the original economic projections (closing the iterative model loop). The entire procedure was repeated until internal consistency was achieved (interfaces between models are not formalized). The IIASA models have been widely considered to be 'the closest existing approach to an appropriate disaggregated technique for forecasting CO_2 emissions' (Ausubel and Nordhaus, 1983). The CO_2 emissions in 2030 were about 10 Gt/yr in the low scenario and about 17 Gt/yr in the high scenario.

The IIASA scenarios have recently been criticized by Keepin and Wynne (1984) who argued that the iterative modelling procedure was never achieved in practice, and because the models were primarily an accounting framework for displaying the subjective projections of the analysts. Moreover, key results were found to depend strongly on uncertain cost and resource estimates. Sensitivity analysis showed that the models' analytic structure was unstable. Thus, although the models included many detailed aspects of the energy system, the variability of future energy costs—an issue of overriding importance—was omitted. These findings bring into question the widely publicized policy recommendations drawn from the scenarios.

Recently, IIASA has produced a new scenario (1983), using an improved energy supply model, which foresees an expanded role for natural gas, and is otherwise similar to the 1981 low scenario. The resulting CO_2 emissions would be about 9.4 Gt/yr in 2050 (see Table 2.5).

Nordhaus and Yohe (1983)

Nordhaus and Yohe (NY) employed a compact global model to analyse CO_2 emissions from 1975 to 2100. The work was a principal component of a recent U.S. National Academy of Sciences study, *Changing Climate*. The principal focus was the systematic treatment of uncertainties in the overall energy/economic/CO_2 nexus. The model employed a rather general-ized Cobb–Douglas production function, in which global GNP is basically a product of labour productivity, population, and energy consumption. There was no regional disaggregation. Energy was divided into two categories: fos-sil and non-fossil, which facilitated the calculation of CO_2 emissions and highlighted the substitution effects between the two. Demand for fossil and non-fossil energy was based on price. Calculations of CO_2 emissions were fed into a simple carbon cycle model to obtain atmospheric CO_2 concentration.

The unique feature of this work was the systematic treatment of uncertain-ties in the overall system. For this purpose, NY selected ten variables likely to greatly influence the outcome, but whose values were not well known. Each of the variables was assigned a low, medium, and high value, reflecting historical uncertainties or spreads in the published literature. This gave 3^{10} (or 59,049) possible scenarios. Results were then presented by randomly se-lecting 1000 different cases. This procedure was called probabilistic scenario analysis.

The results from the NY analysis were not presented simply as forecasts, but in percentiles. An example is given in Figure 2.8, showing the 5th, 25th, 50th, 75th, and 95th percentile scenarios of carbon emissions. The 50th percentile for carbon emissions in 2050 is 15 Gt/yr. The authors drew the following major conclusions:

> 'Given current knowledge, we find that the odds are even whether the doubling of carbon dioxide will occur in the period 2050–2100 or outside that period. We further find that it is a 1-in-4 possibility that CO_2 doubling will occur before 2050, and a 1-in-20 possibility that doubling will occur before 2035.'

These conclusions were derived from the statistical nature of their anal-ysis. However, since all parameter variations were judgmental, this result is also judgmental. Moreover, a difficulty with probabalistic scenario analysis is that the global energy future is not the same as tossing a coin. Choosing among energy policies can significantly alter the probabilities of occurrence for particular energy futures. Thus, due to policy choices, reality is closer to

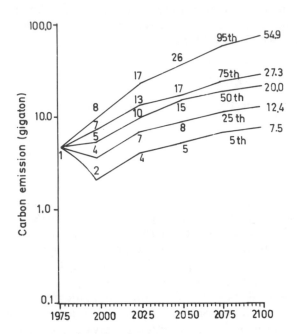

Figure 2.8 Carbon dioxide emissions (gigatons of carbon) from a sample of 100 randomly chosen runs. The 5th, 25th, 50th, 75th and 95th percentile runs for yearly emissions, with emissions for the years 2000, 2025, 2050, and 2100 indicated. Redrawn from Nordhaus and Yohe (1983)

having the opportunity to load the coin before tossing it, e.g., if we go all out for nuclear power, then a nuclear future is likely. Assigning percentiles to various scenarios implicitly assigns low probabilities to certain policy options which the world might choose to pursue vigorously. For example, nine feasible scenarios in the MIT study (Rose *et al.*, 1983) yielded CO_2 emissions, in 2050, ranging from 3 to 14 Gt/yr (corresponding to a CO_2 doubling time of up to 'several centuries'). Meanwhile, the NY analysis implies that the odds are 50 percent that emissions would be less than 15 Gt/yr (see Figure 2.8). However, if the world chooses full implementation of the MIT scenarios (particularly, the lower CO_2 emission cases) the chances are certainly greater than 50 percent that CO_2 emissions in 2050 would be less than 15 Gt/yr, assuming that the MIT study was not greatly in error. In fact, such policy decisions would shift all percentiles indicated in Figure 2.8, downward. Other policy decisions could shift the percentiles upward. In either case, this would alter the conclusions, quoted in the preceding paragraph, regarding the doubling time for atmospheric CO_2 concentration.

The ten uncertain variables selected for detailed consideration in the NY work were tested for relative importance via sensitivity analysis. They were then ranked from one to ten, according to importance. A parameter representing the ease of substitution between fossil and non-fossil fuels turned out to be the most important uncertain variable. This was described as a surprising finding, because sensitivity to assumptions about substitution had not been previously noted. Such 'sensitivity ranking' is useful but it should be regarded with caution. Such conclusions implicitly require that uncertainties in all variables which have been omitted would not significantly alter the ranking. Furthermore, the ranking itself depends on the analytical structure of the model. As an example, two of the most sensitive parameters in the ER model were effectively excluded or constant in the NY model (see discussion in Section 2.5). It is possible that the sensitivity ranking in the NY analysis would have been significantly altered if these or other uncertain variables had been incorporated into the NY analysis.

Nordhaus and Yohe included a set of experiments involving five scenarios in which taxes on CO_2 were imposed (having five different dynamic variations as a function of time). These were compared with the 50th percentile scenario and the tentative conclusion was that forceful taxation policies would be required to significantly reduce future CO_2 emissions. This conclusion is consistent with results obtained by both Edmonds and Reilly (1983a) and the EPA study.

In conclusion, the NY work represented an overdue step in energy/economic forecasting because it was the first systematic attempt to handle parametric uncertainties and address model validation issues. The fundamental message is that the range of possible futures is wide.

World Energy Conference (WEC, 1983)

The World Energy Conference (WEC) is an international organization of energy economists, planners, manufacturers, scientists, and politicians. Eighty countries are members and have played a significant role in shaping global energy policy and planning. A number of forecasts have been produced; the most recent (WEC, 1983) is considered here.

The objective of the study was to forecast primary energy balances in 2000 and 2020 for the globe, disaggregated into ten regions. These regions are roughly similar to those of other globally disaggregated studies (see Figure 2.7). A primary motivation for the study was to offer a decentralized alternative to the 'centralized forecasting models which have almost exclusively dominated the minds of researchers'. There was a separate regional working team (RWT) for each of the regions. Each RWT was composed of experts living in that region, and the RWTs were allowed freedom in formulating the regional forecasts. Fifty experts participated and a central team coordinated the effort.

The major advantage of this approach was that a single analytical framework, developed by researchers from a single cultural background, was not imposed on the entire globe. (This is especially advantageous for regions consisting largely of developing countries, which have traditionally been modelled using a framework appropriate for industrialized nations.) The challenge was to obtain results from all regions that were reasonably consistent and meaningful. To facilitate this, the central team provided each RWT with a homogeneous historical reference base for the years 1960 and 1978, a common system of units, and two general forecasting scenarios that defined a general framework. The two scenarios were characterized as optimistic ('rosy') and pessimistic ('grey') clearly reflecting the values and prejudices of the WEC authors. They were based in spirit on scenarios B and C from the Interfutures study (OECD, 1979). Each RWT was supplied with fixed input forecasts for population and economic growth. The principal results generated by each RWT included forecasts for primary energy demand, energy supply (from eight sources), total fossil imports required, and fossil exports available. Forecasts were cross-sections for the years 2000 and 2020. Only primary supplies were analysed. No analyses of useful energy, demand sectors, or secondary conversion were included. The global sum of imports and exports turned out to be roughly in balance without any need for iteration.

CO_2 emissions are not included as part of the study. But they can be calculated from the primary fossil fuel supply data in Table 2.4. The results are 14.4 Gt/yr for the 'rosy' and 10.0 Gt/yr in 2020 for the 'grey' scenarios.

The greatest strength of the study was probably also its greatest weakness: the knitting together of ten regional forecasts, prepared by different experts, using various methods. Although care was taken to ensure overall consistency, it is likely that significant inconsistencies remain embedded.

Goldemberg et al. *(1984)*

This study was a global end-use analysis, based on extrapolation from four national case studies: India, Sweden, Brazil, and the U.S. These nations have about one-fourth of the world's population, and represent a wide range of cultures, economies, politics, and geography. The study did not include forecasts, but presented a kind of 'existence proof' that outlined how population and economy could grow, while consuming less energy than historical trends would suggest.

With conservation and improvements in energy efficiency, the per capita demand for primary energy in the industrialized nations could be halved. Meanwhile, the per capita demand in developing countries would grow by only about 10 percent during the next 40 years, and the economic activity levels could approach that of Western Europe and Japan today. This surprising result was not presented as a target strategy but rather as a demonstration that 'dramatic improvements in living standards can be achieved without increasing per capita energy use'. A range of technologies were considered. Assumptions about technological performance, economic cost, and feasibility were conservative, rather than optimistic.

A base case was analysed to the year 2020, and various sensitivity analyses were performed. Primary energy consumption remains essentially constant at 10.3 TW in the base case (growing to 14.8 TW in a high demand case). The authors considered this outcome to be readily feasible. The problems ahead were not seen to be primarily technological or financial because all of the technologies now exist. The challenge would be largely that of information transfer and encouraging appropriate investments in capital goods.

The supply side incorporated a range of technologies and resources. In both scenarios, coal consumption declined about 20 percent from today's level, and natural gas consumption increased by 80 percent. Oil consumption declined by 30 percent in the base case (with CO_2 emissions of 4.6 Gt/yr) and grew by 17 percent in the high demand case (with CO_2 emissions of 5.9 Gt/yr; see Table 2.5). Nuclear power, which expanded almost fourfold in both scenarios, was regarded by Goldemberg *et al.* as a last resort because of the danger of proliferation of nuclear weapons, via recovery of plutonium or other weapons-usable material. In particular, low-cost uranium is sufficiently abundant that advanced technologies requiring fuel reprocessing (such as the breeder reactor) were not needed. Nevertheless, the share of electricity grows with time. Hydropower plays an increasingly important role, and wind and solar PV have a share of 5 percent by 2020.

The difficulty with this study is that it extrapolated from a few detailed case studies to a broader domain. The Soviet Union and Eastern Europe are treated as 'Sweden/U.S.-like' countries, in the sense that opportunities for efficiency improvements are equally large there (or perhaps even larger). Nevertheless, the four countries in the scenarios do span a broad range of socioeconomic and political differences, and the authors have diverse backgrounds representing four countries. Most other globally disaggregated models use a fixed framework for analysing all regions which is equivalent to extrapolation.

This study (along with that of Rose *et al.*) demonstrates that economic growth can be decoupled from growing energy consumption. In addition, the global energy future was analysed in the context of agriculture and food shortages, nuclear weapons proliferation, and the environment. As such, the study is unique among those reviewed here.

Lovins et al. (1981)

The starting point for this study was the population and economic growth projections for the seven world regions of IIASA's 'low' global scenario.

Rather than attempting to forecast the future, the objective was to describe a future that minimizes the total energy consumption required to meet IIASA's population and economic growth projections. To implement the scenario outlined would require a vigorous shift of emphasis for global energy policy, from expanded supply capabilities toward technological improvements in efficiency and increased conservation. For this reason, the scenario is controversial. The authors stated that their analysis projects the least-cost energy future, and resolves the problem of CO_2 emissions by largely eliminating them.

The scenario is based on extrapolation from a case study for the Federal Republic of Germany (FRG), which is examined briefly (Krause, 1983). The core of the analysis was the application of 120 measures to improve energy efficiency and supply in 15 sectors of the West German economy. All technologies discussed exist today.The required investments were compared with the marginal costs of expanding supply within the present energy system. The conclusion was that the alternative scenario could supply the same services at lower cost. The measures would involve replacement, expansion and repair of capital stock, but not accelerated retirement.

In the household sector, space heating requirements for new buildings would be reduced with superinsulation and improved heating systems, including passive solar design. Existing buildings would be retrofitted with superinsulation. The newest electrical appliances and light bulbs would be introduced (up to three or four times more efficient than older ones).

In the commercial sector, office buildings would use existing state-of-the-art technologies. There are Canadian and northern European buildings now which use only 20 to 30 percent of the energy consumed in conventional buildings.

In the transport sector, superefficient automobiles and trucks would be introduced. Trains and airliners would be replaced and retrofitted with more efficient technology.

Major gains in efficiency would be achieved in the steel, chemicals, cement, and aluminium industries. Electric drive technology would be improved, waste heat utilized, and materials used more efficiently.

Taken together, these measures would replace about 70 percent of end-use energy in all sectors combined, with a shift toward greater shares for electricity and liquid fuels. Total energy consumption is reduced to 30 percent of the 1973 level for the FRG.

These measures would be introduced gradually, cutting global energy projections for 2030 in the IIASA low scenario by approximately 50 percent. This would be further reduced by assumed economic structural changes, resulting in 5.23 TW primary energy consumption in 2030 (about 25 percent of IIASA's projection of 24 TW), with CO_2 emissions and acid rain greatly reduced as a consequence.

The validity of such an extrapolation from a single nation to the entire globe is problematical, as discussed above for Goldemberg *et al.* (1984). In particular, there is little basis for assuming that the economies of the developing countries will evolve to look like the FRG. Furthermore, details on capital and labour requirements were not included. Finally, many institutional changes not considered in the scenario would be required.

The scenario was not presented as a forecast, but as another 'existence proof'. In view of the last 70 years, the potential for tremendous change over a long time horizon should not be underestimated. Even if it were possible to only partially achieve the scenarios described, it indicates a future global energy strategy that would permit continued economic growth along with reduced environmental strain.

Legasov et al. *(1984)*

This study is the latest of several Soviet analyses of the energy/climate nexus. In an earlier study, Legasov and Kuzmin (1981) analysed two long-term global scenarios to study the effects of waste heat discharge. Annual per capita energy consumption was represented by a logistic function of time, whose parameters were estimated, based on historical data.

More recently, Kuzmin *et al.* (1984) have investigated various CO_2 abatement strategies, concluding that a whole complex of feedback relations must be accounted for in technological risk assessments.

Legasov *et al.* present two scenarios in which global energy consumption reaches 6 kW/capita and 20 kW/capita, respectively, by 2100. The lower of these scenarios (called the 'minimal variant') is used for the detailed analysis of CO_2 emissions, and thus we focus on this case here.

Assuming a global population of 10 billion by 2100, the minimal variant results in global energy consumption of 60 TW by 2100. This scenario is elaborated in Stirikovich *et al.* (1981), and leads to approximately 42 TW by 2050. Fossil fuel consumption peaks at less than 20 TW before 2050, and slowly tapers off thereafter (Legasov *et al.*, 1984). Resulting CO_2 emissions follow a bell-shaped curve, reaching a peak of approximately 13.3 Gt/yr in 2050 (and decreasing thereafter), as shown in Table 2.5. The study proceeds to analyse a variety of broader climatic implications (including effects of atmospheric electroconductivity changes due to krypton emissions from nuclear fuel cycle plants).

Discussion

It is clear after reviewing these projections that the global energy/CO_2 future is highly uncertain (for a more detailed review, see Keepin, 1984b). It is important to note that even if future GDP or energy consumption were known with certainty, future CO_2 emissions would still exhibit a range of possible values. To illustrate this, rough calculations of 'carbon intensity' are given for most of the studies in the last column of Table 2.5. These numbers were obtained by computing the ratio of CO_2 emissions to primary energy consumption. Thus, a low value means a relatively 'clean' energy supply system in regard to CO_2; a higher value means a more fossil-based supply system. These ratios are only intended to illustrate that CO_2 emissions can be decoupled from energy consumption. (They are not appropriate for judging the merits of individual projections.)

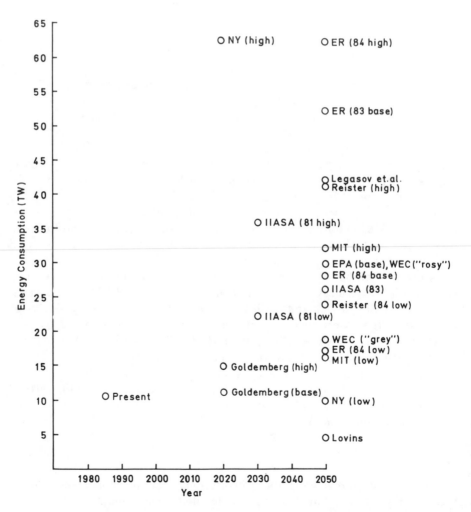

Figure 2.9 Projections of primary energy consumption in 2050

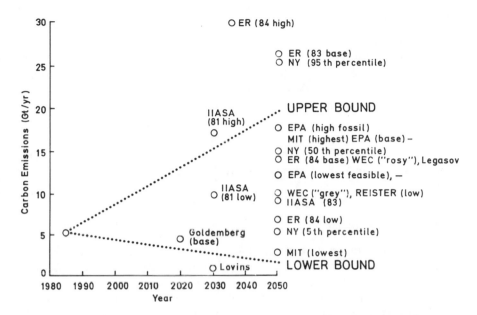

Figure 2.10 Projections of carbon emissions in 2050. Sources: EPA: Environmental Protection Agency (1983), ER: Edmonds and Reilly (1984), Goldemberg (1984), IIASA: International Institute for Applied Systems Analysis (1981, 1983), Legasov *et al.* (1984), Lovins *et al.* (1981), MIT: Massachusetts Institute of Technology (1983), NY: Nordhaus and Yohe (1983), Reister (1984), and WEC: World Energy Conference (1983)

Results from the various projections are illustrated in Figures 2.9 and 2.10, showing projections for global primary energy and CO_2 emissions. These figures present various point estimates only, omitting specific trajectories. This unusual format was adopted because the uncertainties in primary energy and CO_2 emissions for the year 2050 are much greater than uncertainties associated with specific trajectories (see Section 2.5 for further discussion).

The uncertainty exhibited in Figures 2.9 and 2.10 is more fundamental than simple lack of knowledge of future parameter values. Indeed, as discussed in Section 2.3, the large variation in future projections is due to a general ignorance about future cultural, political, social, and technological developments. These factors are the most important determinants of the future, but they cannot be anticipated by analytic forecasting methods. Thus specific projections can only focus on second- and third-order effects, and it is unavoidable ignorance of the first-order effects that gives rise to the great uncertainty observed in Figures 2.9 and 2.10.

2.5 BOUNDS ON FUTURE CO_2 EMISSIONS

2.5.1 Introduction

Future anthropogenic CO_2 emission rates are highly uncertain, primarily because they result from a complex set of interactions among a number of uncertain sometimes unknown factors. Thus, it is natural to question whether it is possible to identify upper and lower bounds on likely future CO_2 emissions. A word of caution is needed regarding the term 'bounds'. It is clearly not possible to specify actual upper and lower bounds within which future CO_2 emissions will lie. The term 'bound' is used here for lack of a better word, and it is used informally to suggest an interval or range of projections, outside of which conditions and scenarios are regarded as extreme.

2.5.2 General Approach

No attempt is made here to produce new projections. In fact, the many uncertainties in existing estimates are of greater significance than the individual projections themselves. Rather than make additions to the existing array of forecasts, we attempt to establish extremes for the range of uncertainties in future carbon emissions. The approach will be to consider the past projections and identify tentative upper and lower 'bounds' on future CO_2 emissions. This does not mean that two scenarios will be singled out and labelled 'upper' and 'lower' limits. On the contrary, each bound corresponds to more than one scenario, because there are a multitude of ways to achieve a given level of CO_2 emissions. Moreover, we do not specify a quantitative measure of confidence that future CO_2 emissions will indeed fall within the bounds identified here. To do so would suggest that greater precision inheres in this process than can be justified.

2.5.3 Comparative Assessment of Future CO_2 Projections

To help explain the choice of what seem to be reasonable bounds offered in this section, a comparative assessment and discussion of the projections and forecasts will be reviewed. The purpose is to illustrate why the results of various studies differ and why we should (or should not) believe the results.

Nordhaus and Yohe (1983); Edmonds and Reilly (1983b, 1984)

A significant difference in results between the NY and ER models concerns the sensitivity to the ease of substitution between fossil and nonfossil fuels. Each model has a parameter which represents this factor, the most sensitive parameter in the NY model. The authors reported the significance of this factor to be important and unexpected. Since the NY model estimated the CO_2 concentration with a simple atmospheric accumulation model, this finding implies that the CO_2 emissions themselves are sensitive to variations in the substitution parameter. In the ER study, however, 'carbon emissions are very insensitive to changes in the interfuel substitution parameter' (Reilly and Edmonds, 1984b). Thus, opposite conclusions regarding the importance of this parameter are reached, and it is natural to ask why. Nordhaus and Yohe observed sensitivity to the substitution parameter because it entered their model as an exponent that heavily influences the relative demand for fossil and nonfossil fuels. In the ER model, 'fossil and nonfossil fuel prices are relatively close together and therefore the substitution parameter is less important' (Reilly and Edmonds, 1984b). Thus, one model is sensitive to variations in a particular parameter, while the other is not. Either structural representation is plausible. The point here is not to decide which is 'correct', but to emphasize the importance of inherent structural uncertainties that characterize all energy and CO_2 models. These differences in sensitivity suggest the importance of model validation exercises that attempt to establish a correspondence between model and reality. Behavioural patterns suggested in model runs may either provide genuine insights concerning the underlying systems, or they may be simply artifacts of the model (Keepin, 1984b).

Given the observed lack of sensitivity in the ER model to the most sensitive parameter in the NY model, the question can be reversed: what is the most sensitive parameter in the ER model and what is its effect in the NY model? The two most sensitive parameters in the ER model are income elasticity of demand and the growth rate of non-price-induced efficiency improvements. However, the former of these parameters is explicitly represented in the NY model. Thus, there is no way to assess the importance of these parameters in the NY analysis. This illustrates another problem: one model effectively holds constant a variable whose variation is critical in another model.

When the same conclusion is reached independently, by different researchers, employing different analytic approaches or methods, such a conclusion deserves careful scrutiny. This is the case with the ER and NY models on the issue of fuel tax policies for reducing CO_2 emissions. Both models indicate that such policies would have to be extreme in order to have a significant effect.

IIASA (1981) and Lovins et al. *(1981)*

Various independent researchers have compared the IIASA and Lovins *et al.* projections (Rose *et al.*, 1983; Meinl *et al.*, 1984; Wynne, 1984). Robinson

(1982b) finds that the two studies began from quite dissimilar conceptual frameworks, and that the conclusions of each study are rooted in fundamental presuppositions that do not result from the quantitative analysis, but rather guide it. Thus, the 'driving force is either an explicit or implicit view of what constitutes a desirable energy future' (Rose *et al.*, 1983). Lovins *et al.* are explicit about this, whereas IIASA claims to be objective, particularly because of its use of sophisticated computer models. Robinson (1982b) reported that each study labels the other's scenario as highly undesirable and unaffordable, and that policymakers have tended to ignore these mutual condemnations, while the policies themselves have slowly shifted toward conservation and more efficient end uses.

Goldemberg et al. *(1984) and MIT (Rose* et al.*, 1983)*

These two studies reach similar conclusions. As Darmstadter (1984) has observed, 'few pairs of energy analyses are ever totally free of incest'. However, the analysts were unaware of the others' work. They employ different analytic and philosophical approaches although their authors shared the view that the possible consequences of conservation on the greenhouse problem were worthy of systematic exploration. Goldemberg *et al.* provided a detailed end-use analysis, focusing on low energy futures that would incorporate a variety of energy-saving measures, and technological improvements. The analysis focused on the demand side, included detailed treatment of developing countries, and embedded the energy system in a global context that included major agricultural, environmental, and political factors. The MIT study employed a classical economic equilibrium model that focused on the supply side (commercial technologies), is best suited to developed market economies, and treated the energy sector in isolation. Thus, the two studies may be considered 'analytically orthogonal'. That is, although using very different methods and assumptions, the two studies reach very similar conclusions about the feasibility and implications of a low-demand end-state.

An important conclusion common to both Goldemberg *et al.* and MIT is that over the next several decades, CO_2 emissions can be kept at or even substantially below today's levels (see Figure 2.10). Both studies find considerable room for improvement in overall energy efficiency, particularly in the developed countries, and recommend that policies be formulated to that end.

Bounds on Future CO$_2$ Emissions

For the purpose of specifying bounds on the range of future CO_2 emissions, the focus here is on the end point of the time horizon considered in this study: the year 2050. A discussion of emission trajectories to the year 2100 is also included.

In the attempt to identify bounds on future CO_2 emissions, it is necessary to consider the feasibility of different futures. The notion of feasibility occurs frequently in analyses that assess future possibilities, opportunities, and consequences of policy options. Although 'feasibility' appears to be a simple concept, it actually encompasses a broad range of technical, economic, environmental, sociopolitical, and even cultural factors. Thus a truly feasible future is one that is feasible simultaneously in each of several dimensions.

In this work the focus is largely on technical, economic, and environmental feasibility, which can be analysed quantitatively to some extent. However, consideration is also given to subjective and human dimensions that are not quantifiable but can limit what is technically, economically, and environmentally feasible.

To estimate an upper bound for CO_2 emissions in 2050, several different studies are relied upon. High fossil fuel projections are first considered, especially coal projections, since coal is the most 'CO_2-intensive' of the conventional fossil fuels and also, by far, the most abundant. A typical projection is the IIASA (1981) high scenario, which projected a five-fold expansion in global coal extraction by 2030. Another example is the world coal study (WOCOL, 1980) which produced coal projections to the year 2000 roughly equivalent to IIASA. Although the IIASA and WOCOL scenarios do not extend to the year 2050, they are of particular interest because they are among the few that have been studied by independent researchers to determine overall feasibility. Alcamo (1984a,b) has carried out a detailed analysis of the coal projections in the IIASA high scenario. He found that for the U.S. and the USSR (which have some of the largest known coal reserves), the projected four- or five-fold expansion is probably not feasible, due to insufficient coal reserves and lack of water. Chadwick and Lindman (1982) provided a detailed investigation of the environmental implications of expanded coal utilization, based on the IIASA coal projections. Perry (1982) investigated the CO_2 implications of the IIASA/WOCOL scenarios and found emissions to be extremely high, due to the huge coal projections. Prior (1982) performed a 'bottom-up' analysis of future coal consumption in Western Europe by considering each country's actual and likely performance in converting industrial and power plants to coal firing. He concluded that demand for coal will fall considerably below the WOCOL and IIASA levels. Long (1982) investigated constraints on future coal demand relating to the development of port facilities and shipping terminals, comparative economics of competing coal-exporting regions, and resource distribution issues. His analysis supported the conclusions of Prior, that future coal production will be below the IIASA and WOCOL levels.

More recently, Fischer (1984) has found that observed increases in coal demand have been only 40 percent of that predicted in many forecasts. He suggests that this is due to unrealistic assumptions about price differentials between oil and coal, and the ease of substitution (e.g., failure to appreciate the effects of conservation and the slowdown in the world economy since 1973; failure to appreciate the strength of the 'antisocial factor' associated with coal).

Since coal is the most 'CO_2-intensive' fossil fuel (and the IIASA high scenario projects major expansions for conventional oil, shale oil, and natural gas), these findings are interpreted to mean that the IIASA high scenario (and others equivalent to it) was unrealistic. Consequently, the upper bound for carbon emissions in this study will be below these levels. (In Table 2.5, the IIASA high scenario shows carbon emissions of approximately 22.3 Gt/year by 2050, based on a linear extrapolation from 2030.)

For equivalent levels of consumption, high coal scenarios would mean greater carbon emission levels. Several studies have explored such base case scenarios which represent the extrapolation of current trends, without policies to restrict carbon emissions. The results are shown in the final column of Table 2.5 and in Figure 2.10. Note that the projected emissions are at or below 15 Gt/yr in all base cases except one (Edmonds and Reilly, 1983b). (This last case has recently been replaced by the authors with a new base case scenario (Edmonds et al., 1984) that yielded CO_2 emissions of 14.5 Gt/yr by 2050.) Thus, 15 Gt/yr appears to be a plausible upper bound. EPA (1983) explored a 'high fossil' scenario that resulted in approximately 18 Gt/yr by 2050.

Based on these considerations, an upper bound for carbon emissions in 2050 of approximately 20 Gt/yr was selected. Although emissions could exceed this value, it appears to be unlikely. A few studies do include estimates above 20 Gt/yr, but none of these have analysed the realism of such projections. For example, Edmonds *et al.* (1984) included an unusual scenario in which CO_2 emissions reach 47.4 Gt/yr by 2050, increasing to 91.1 Gt/yr by 2075. This scenario incorporated forecasts of future coal consumption that are higher than the IIASA high scenario (judged to be unrealistic), in addition to a shale oil program of comparable magnitude. While this scenario is perhaps technically feasible, no analysis of the other implications is provided. The scenario appears unrealistic in view of the studies cited above (Long, 1982; Prior, 1982; Alcamo, 1984a,b; Perry, 1982; Chadwick and Lindman, 1982). In fact, the upper bound identified here corresponds more closely to the Edmonds and Reilly (1984) base case.

In the statistical study by Nordhaus and Yohe (1983), the 75th and 95th percentile scenarios yielded 17 and 26 Gt/yr by 2050. The NY study was primarily an aggregated statistical analysis rather than an investigation of feasibility. For the latter purpose, a more disaggregated approach would be more appropriate. In terms of the IIASA scenarios, the upper bound chosen here lies between the high and low scenarios (1981), and well above the 1983 scenario.

To estimate a lower bound for CO_2 emissions in 2050, the work of Goldemberg *et al.* (1984), MIT (Rose *et al.*, 1983), and Lovins *et al.* (1981) is used. The first two studies reached similar conclusions regarding the extent to which CO_2 emissions can be limited. This consistency is particularly significant because they utilize such different analytic approaches and tools. The lowest figure obtained by MIT is just under 3 Gt/yr for 2050. This is less than half the lowest value, of 6.8 Gt/yr, obtained by Edmonds and Reilly (1984), using the same model. However, MIT developed three different scenarios using the ER model that resulted in lower CO_2 emissions than the Edmonds and Reilly (1984) low case, and also included additional analysis to illustrate overall feasibility. Thus, the lower bound obtained by MIT will be used here. This scenario is characterized by a combination of efficiency improvements and large non-fossil energy shares (nuclear 29%, hydro 19%, and solar 12%). This results in CO_2 emissions of 2.7 Gt/yr in 2050. Meanwhile, Goldemberg *et al.* obtained about 4.3 Gt/yr in 2020 for their base case. (The trajectory between 1980 and 2020 is flat. Thus, it is assumed that 4.3 Gt/yr is maintained through 2050.)

Lovins *et al.* (1981) found that carbon emissions could be less than 1 Gt/yr by 2030. Recall that this scenario was offered as an 'existence proof' rather than a forecast. While this scenario may be technically feasible, it seems unlikely that global energy/economic policy will be sufficiently stringent and well-coordinated for this scenario to actually be realized. But it suggests that concerted effort could greatly reduce carbon emissions. Based on these considerations, a lower bound for 2050 of approximately 2 Gt/yr was selected.

The upper and lower bounds specified here yield an estimated uncertainty factor of ten for carbon emissions by the middle of the next century. If we include the effects of biological CO_2 sources (which could add another 1 to 2 Gt/yr), the lower bound could increase to 3 Gt/year and the upper bound extend to 22 Gt/yr. The higher emission levels are associated with relatively high fossil fuel consumption. The lower levels are associated with a broader range of possibilities that include displacement of energy demand (conservation, efficiency improvements, etc.) and/or vigorous development of CO_2-benign supply sources (renewable, nuclear, etc.).

The upper bound corresponds to an annual growth rate for carbon emissions of 2.3 percent/yr, which is considerably greater than the average of 1.46 percent/yr since 1973. The corresponding growth rate for the lower bound is −1.4 percent/yr. However, these figures presume an exponential model (either growth or decay) which is not realistic for CO_2 emissions, because it suggests that emissions could increase to infinity or decrease to zero, making the model highly sensitive to the value of the growth rate parameter.

In choosing upper and lower bounds, it is not meant to suggest that a 'best guess' for the future lies halfway between. Choosing a median or best guess scenario is intentionally avoided here. A stringent policy to reduce CO_2 emissions would yield a lower outcome for future carbon emissions than a policy that ignores the greenhouse effect. Since it is not possible to predict policy, it is impossible to make an authoritative 'best guess' for future CO_2 emissions.The prerequisites for arriving at different end-values for CO_2 emissions within the upper and lower bounds will be briefly discussed in the last section of this chapter.

What is actually needed for analysis of future global warming is not a best guess of future CO_2 emissions, but the resulting concentration of CO_2 in the atmosphere. To estimate the latter requires analysis of the global carbon cycle, which is considered in Chapter 3. Using the carbon cycle model developed by Siegenthaler (1983), an idea of the bounds on atmospheric CO_2 concentration that would result from our selected upper and lower bounds on CO_2 emissions has been produced. The resulting atmospheric CO_2 concentrations in 2050 are 531 and 367 ppmv for the upper and lower bounds, respectively (Svenningsson, 1985). These reflect increases over the pre-industrial level (270 ppmv) by factors of 2.0 and 1.4. These results assume a contribution of CO_2 from the terrestrial biota of 1.2 Gt/yr in 1980, decreasing linearly to zero in 2055.

When assessing projections, it is prudent to try to bound the greatest uncertainties first. Thus, while there is more uncertainty related to different emission trajectories, it is of secondary importance compared with the greater uncertainty as to the actual level of CO_2 emissions in the middle of the next century. For this reason, bounding the latter, within a range that seems reasonable, has been the primary focus of the efforts described in this study. Moreover, there is probably no justification for placing confidence in any particular shape of trajectories of long-term energy consumption or carbon emissions. For this reason, the projections reviewed here are represented with points only in Figures 2.9 and 2.10. The issue is considered again in Chapter 3 in connection with limiting the atmospheric concentration of CO_2.

Conclusions

Future emission rates of CO_2 resulting from fossil fuel combustion are highly uncertain, especially when looking into the distant future. Based on a critical examination of the literature and consultations with experts, tentative upper and lower 'bounds' on these emission rates for the year 2050 were selected. These bounds are not rigorous limits, but rather scenarios that are judged to be extreme. The upper bound is 20 Gt/yr, which was arrived at by considering several recent studies of constraints on future coal consumption. The lower bound is 2 Gt/yr, which was obtained from consideration of various recent analyses of the future opportunities for efficiency improvements and non-fossil supply technologies. Both the upper and lower bounds reflect extreme cases that are judged to be very difficult to achieve in practice. In addition, biotic sources of CO_2 could contribute up to 1 or 2 Gt/yr, resulting in a range of 2–22 Gt/yr for total carbon emissions by 2050. It is possible that the actual emissions will fall outside this range, but this is judged to be unlikely.

The 'envelope' of uncertainty identified here is illustrated graphically with dotted lines in Figure 2.10. This wide range of uncertainty is not likely to be significantly narrowed by further research, because this envelope signifies a basic ignorance about future cultural evolution. Thus any particular trajectory implies some combination of political, technological, and social developments that cannot be predicted.

2.6 CONCLUSIONS

Several conclusions emerge from this analysis. First, future emissions of carbon dioxide (CO_2) cannot be predicted with precision. Attempts to forecast energy and CO_2 developments into the distant future are hindered by great uncertainties. The precision and elegance of analytic forecasting methods are overwhelmed by the arbitrary and imprecise nature of the structural assumptions and parameters. For this reason, recent forecasts have generally included a wide range of possible outcomes. It is estimated that CO_2 emissions from fossil fuel combustion in the year 2050 will be between 2 and 20 Gt/yr (compared with today's level of 5 Gt/yr). Coupled with an additional contribution of up to 1 or 2 Gt/yr from terrestrial biota and other sources, the resulting uncertainty factor in CO_2 emissions by the middle of the next century exceeds an order of magnitude.

This does not imply that nothing meaningful can be said about the future, however. One reason for uncertainty is that the energy/CO_2 future depends, in significant measure, on present and future policies. Recent studies have shown that future economic growth can be largely decoupled from growing energy consumption, at least in the industrialized countries, and that future CO_2 emissions can be decoupled in industrialized countries from both economic growth and energy consumption. Thus, the causal relationship between energy and climate is weaker than has hitherto been believed.

Perhaps the most important conclusion in the more recent published studies is that global energy productivity and efficiency is far from optimal. Those studies which have not explicitly considered the 'efficiency factor' are without practical interest. Numerous opportunities exist for improving the efficiency of energy use, and are economically attractive. Global energy efficiency probably can be increased at a sustained rate of at least 1 percent/year for several decades with no adverse social effects. By 2050, this would cut energy consumption and CO_2 emissions in half from what they would otherwise be. In addition, the global energy supply mix may well move away from fossil fuels (especially coal) and toward CO_2-benign technologies. To the extent that energy demand is reduced via increased efficiency (e.g. along the lines suggested in the low energy scenarios discussed), this shift to a cleaner supply system is made easier.

The effects of global warming are potentially serious. Estimates show that substantial warming may occur in the future due to other greenhouse gases as well as CO_2. If the global warming expected to result from a doubling of CO_2 is the upper limit of 'acceptable' climatic changes, then because of the effects of other greenhouse gases, CO_2 accumulation must be limited to approximately 1.5 to 1.7 times the pre-industrial level. According to several studies, there are, however, technically and economically viable strategies for global energy development that are compatible with a high level of concern about the climatic impacts of a continued CO_2 build-up. Some of these strategies allow for a considerable continued growth in the industrialized countries and for bringing developing countries to the economic level of Western European countries in the 1970s.

For global society to reach the 'lower bound' CO_2 emission scenario it would undoubtedly require a cooperative effort based on a common view of the seriousness of the greenhouse and climate issue. In order to reach the 'upper bound', a series of concurrent policy choices based on a disregard of the greenhouse and climate issue would likewise have to be implemented on the global scale. A much more likely outcome of the myriad of energy policy decisions to be taken in the future is that CO_2 emission rates will therefore be in the middle range of the interval.

Actual future emission rates of CO_2 (and other greenhouse gases) will be influenced by many factors. These include policy feedback from scientific studies about the greenhouse and climate issue, economic growth rates and the penetration rates of existing and new technologies for increased energy efficiency and non-fossil energy production. Given the current situation and trends, it seems likely that the net result of all these influences will be to push global CO_2 emissions down, towards the lower parts of the interval between the bounds.

However, monumental uncertainties surrounding all long-term energy projections must not overshadow the fact that it is within our power to control the future concentration of CO_2 through the implementation of systematic policy choices, although these choices must be made in the face of substantial scientific uncertainty.

NOTE ON AUTHORSHIP AND ACKNOWLEDGEMENTS

Although the three authors provided critical discussion and commentary on one another's material, principal responsibility for researching and writing the Sections of this Chapter is as follows: 2.1: all authors, 2.2 and 2.3: I. Mintzer, 2.4 and 2.5: W. Keepin, and 2.6: all authors.

L. Kristoferson co-ordinated, critiqued and supervised the production of the Draft Report. Dr. A. Miller contributed most helpfully to Sections 2.2 and 2.3.

The authors are indebted to Professor Thomas B. Johansson for his helpful commentary on the Draft; also to Academician M. A. Stirikovich for personal discussions and valuable commentary on the Draft. We also wish to thank the following for many helpful suggestions and advice on the content of an earlier draft: B. Bolin; B. R. Döös; P. Gleick; G. T. Goodman; J. Harte; J. Mathews; A. Natkin; R. Repetto; G. Speth; P. Thacher; S. Wennerberg; R. Williams; and, especially, D. Rose.

Additionally, by arrangement with the USSR Academy of Sciences (Moscow), a small Workshop was convened in Tbilisi, USSR (5–7 February, 1985) under the auspices of the Academy of Sciences of the Soviet Socialist Republic of Georgia. We are particularly pleased to acknowledge the support of these two Academies of Science for arranging such a successful Meeting and to the following participants for all their useful proposals, which have now been incorporated in this Final Draft: M. Ja. Antonovsky; B. Bolin; E. Buettner; M. Chadwick; P. Godoy; V. Gomelauri; G. Goodman; J. Hollander; V. Kakabadze; I. Kuzmin; T. Lagidze; M. Lönnroth; and B. Wynne.

We are also thankful for the comments and advice given by the WMO/ICSU/UNEP Review Panel, including W. Böhme, R.E. Dickinson, J. Edmonds, H.W. Ellsaesser, H.L. Ferguson, R.M. Gifford, F.K. Hare, P.M.R. Kiangi, H.E. Landsberg, J.A. Laurmann, W.J. Maunder, A.S. Monin, O. Preining, R.M. Rotty, G.E. Suarez, A. Weinberg, and M.M. Yoshino.

We hope that we have been able to successfully coordinate and as far as possible incorporate all the many valuable suggestions made by all the persons listed above. The Chapter has benefited greatly from their advice. However, any errors that remain are entirely our own.

We also wish to acknowledge the Beijer Institute (Royal Swedish Academy of Sciences) of Stockholm, the Swedish Energy Research Commission (Stockholm) and the World Resources Institute (Washington D.C.) for financial support.

2.7 REFERENCES

Alcamo, J. (1984a) A regional perspective of global coal scenarios, *Energy Policy*, **12** (2), 213–15.

Alcamo, J. (1984b) Fire and water: Water needs of future coal development in the Soviet Union and the United States, *Resources and Energy*, **6** (2).

Ausubel, J., and Nordhaus, W. D. (1983) A review of estimates of future carbon dioxide emissions, in *Changing Climate*, Washington, D. C., National Academy Press.

Caputo, R. (1984) Worlds in collision: Is a rational energy policy possible for Western Europe? *Futures*, **16** (3), 233–59.

Chadwick, M. J., and Lindman, N. (eds) (1982) *Environmental Implications of Expanded Coal Utilization.* A study by the Beijer Institute, UNEP, and the USSR Academy of Sciences, Oxford, Pergamon.

Cleveland, C. J., Costanza, R., Hall, C. A., and Kaufmann, R. (1984) Energy and the U.S. Economy: A Biophysical Perspective, *Science*, **225** (4665), 890–897.

Cole, S. (1983) After Cancun: The magic of the market, *Futures*, **15** (1), 2–12.

Darmstadter, J. (1984) Energy Patterns, draft, (Washington, D. C., Resources for the Future.

Edmonds, J., and Reilly, J. (1983a) A long-term global energy-economic model of carbon dioxide release from fossil fuel use, *Energy Econ.*, **5** (2), 74–88.

Edmonds, J., and Reilly, J. (1983b) Global energy and CO_2 to the year 2050, *Energy Journal*, **4** (3), 21–47.

Edmonds, J. A., Reilly, J., Trabalka, J. R., and Reichle, D. E. (1984) *An Analysis of Possible Future Atmospheric Retention of Fossil Fuel CO_2*, Report No.DOE/OR/21400-1, Washington, D. C., Department of Energy

EPA (1983), see Seidel and Keyes (1983).

Fischer, W. (1984) *Coal Trade Statistics*, Vol.II, Llandudno, UK, Robertson Research.

Goldemberg, J., Johansson, T. B., Reddy, A. K. N., and Williams, R. H. (1984) *Energy for a Sustainable World* (forthcoming).

Goldemberg, J., Johansson, T. B., Reddy, A. K. N., and Williams, R. H. (1985) Basic needs and much more, with 1 kW per capita (to be published).

Goodman, G. T. (1983) Societal problems in meeting current and future energy needs, *Ambio*, **12** (2), 97–101.

Green, M. A. (1983) *Solar Cells: Operating Principles, Technology, and Systems Applications*, Englewood Cliffs, NJ, Prentice-Hall.

Greenberger, M., *et al.* (1983) *Caught Unawares: The Energy Decade in Retrospect*, Cambridge, MA, Ballinger.

Häfele, W. (1980) A global and long range picture of energy developments, *Science*, **209**, 174–182.

Hammond, P. B. (1984) The energy model muddle, *Policy Sci.*, **16**, 227–243.

IIASA (1981) *Energy in a Finite World*. Ed. W. Häfele, Cambridge, MA, Ballinger.

IIASA (1983) *IIASA '83 Scenario of Energy Development: Summary*. Ed. H.-H. Rogner, Laxenburg, Austria, IIASA.

Jäger, J. (1984) Impacts on the energy sector, in Meinl, H. *et al.* (1984), Comparison of two energy scenarios. *Socioeconomic Impacts of Climatic Changes due to a Doubling of Atmospheric CO_2 Content*, Brussels, EEC, 396–418.

Keeling, C. D. (1973) Industrial production of carbon dioxide from fossil fuels and limestone, *Tellus*, **28**, 174–198.

Keepin, B. (1984a) A technical appraisal of the IIASA energy scenarios, *Policy Sci.*, **17** (4), Special issue.

Keepin, B. (1984b) *Review of Global Energy Forecasts*, Stockholm, Beijer Institute, Royal Swedish Academy of Sciences.

Keepin, B., and Wynne, B. (1984) Technical analysis of IIASA energy scenarios, *Nature*, **312**, 691–695.

Koreisha, S., and Stobaugh, R. (1979) Limits to models, in Stobaugh, R., and Yergin, D. (eds) *Energy Future*, New York, Vintage, 309–342.

Krause, F. (1983) A low cost/low risk global energy future. Paper submitted to *Annu. Rev. Energy*.

Kristoferson, L. (1977) Waterpower, *Ambio,* **6**, 44–45.
Kuzmin, I. I., Romanov, S. V., and Chernoplyokov, A. I. (1984) *On a Quantitative Approach to the Safety Assessment,* Moscow, Kurchatov Institute of Atomic Energy.
Laurmann, J. A. (1984) *Market Penetration as an Impediment to Replacement of Fossil Fuels in the CO₂ Environmental Problem.* Paper prepared for the H. H. Humphrey Institute of Public Affairs Symposium on Greenhouse Problem Policy Options, Minneapolis, MN, May 29–31, 1984 (Proceedings in preparation).
Legasov, V. A., and Kuzmin, I. I. (1981) The problem of energy production, *Prioroda,* **2**, 8–23.
Legasov, V. A., Kuzmin, I. I., and Chernoplyokov, A. I. (1984) The influence of energetics on climate, *Fizika Atmospheri i Okeana,* **11**, 1089–1103, USSR Academy of Sciences.
Long, R. (1982) *Constraints on International Trade in Coal,* London, IEA Coal Research.
Lovins, A. B., Lovins, L. H., Krause, F., and Bach, W. (1981) *Least Cost Energy: Solving the CO₂ Problem,* Andover, MA, Brick House Press.
Manne, A., and Schrattenholzer, L. (1984a) *International Energy Workshop Part II: Individual Poll Responses,* Laxenburg, Austria, IIASA.
Manne, A., and Schrattenholzer, L. (1984b) International Energy Workshop: A summary of the 1983 poll responses, *Energy Journal,* **5**, (1), 45–65.
Marchetti, C. (1980) *On Energy Systems in Historical Perspective,* Working Paper WP-80, Laxenburg, Austria, IIASA.
Marchetti, C., and Nakicenovic, N. (1979) *The Dynamics of Energy Systems and the Logistic Substitution Model,* Research Report RR-79-13, Laxenburg, Austria, IIASA.
Marland, G. (1982) The impact of synthetic fuels on global CO₂ emission, in W. Clark (ed.) *CO₂ Review 1982,* Oxford, Clarendon Press, 406–410.
Marland, G., and Rotty, R. M. (1984) Carbon dioxide emissions from fossil fuels: A procedure for estimation and results for 1950–82, *Tellus* (forthcoming).
Marley, R. C. (1984) Trends in industrial use of energy, *Science* **226**, 1277–1283.
Maycock, P., and Stirewalt, E. (1982) *Photovoltaics: Sunlight to Electricity in One Step,* Andover, MA, Brick House Press.
Meinl, H., *et al.* (1984) *Comparison of Two Energy Scenarios. Socioeconomic Impacts of Climatic Changes due to a Doubling of Atmospheric CO₂ Content,* Brussels, EEC, 25–30.
Mintzer, I., and Miller, A. (1984) *Impacts of Energy Conservation Measures and Renewable Energy Technologies on Global Emissions of Carbon Dioxide,* Washington, D. C., World Resources Institute.
MIT (1983), see Rose *et al.* (1983).
Nader *et al.* (1979) *Impacts of Consumption, Location, and Occupational Patterns on Energy Demand:* A Report to the Demand Panel of National Academy of Sciences Committee on Nuclear and Alternative Energy Systems, Washington, DC, National Research Council.
Nordhaus, W. D., and Yohe, G. (1983) Future paths of energy and carbon dioxide emissions, in *Changing Climate,* Washington, D.C., National Academy Press.
OECD (1979) OECD *Interfutures Project 'Facing the Future',* Paris, OECD.
O'Keefe, P., Raskin, P., and Bernow, S. (eds) (1984) Energy and development in Kenya: opportunities and constraints. *Energy, Environment and Development in Africa,* Vol. 1, Stockholm, Beijer Institute and Scandinavian Institute of African Studies.

Peng, T. H., *et al.* (1983) A deconvolution of the tree ring based on the ^{14}C record, *J. Geophys. Res.*, **88**, 3609–3620.

Perry, A. N. (1982) CO_2 production scenarios: An assessment of alternative futures, in Clark, W.C. (ed.) *Carbon Dioxide Review:1982*, New York, Oxford University Press.

Prior, M. (1982) *Steam Coal and Energy Needs in Western Europe Beyond 1985*, EIU Special Report No.134, London, Economist Intelligence Unit.

Quirk, W. J. (1981) Climate and Energy Emergencies, *Bull. Am. Meteorol. Soc.*, **62**, 623–632.

Radetski, M. (1984) *The Art of Forecasting Long Run Economic Growth.* Internal Memorandum, Beijer Institute, Royal Swedish Academy of Sciences.

Reilly, J., and Edmonds, J. (1984a) *Time and Uncertainty: Analytic Paradigms and Policy Requirements*, Oak Ridge, TN, Institute for Energy Analysis.

Reilly, J., and Edmonds, J. (1984b) 'Changing climate' and energy modeling: A review, *Energy Journal* (forthcoming).

Reister, D. B. (1984) *An Assessment of the Contribution of Gas to the Global Emissions of Carbon Dioxide.* Final Report GRI-84/003, Chicago, IL, Gas Research Institute.

Robinson, J. B. (1982a) Backing into the future: On the methodological and institutional biases embedded in energy supply and demand forecasting, *Technological Forecasting and Social Change*, **21**, 229–240.

Robinson, J. B. (1982b) Apples and horned toads: On the framework-determined nature of the energy debate, *Policy Sci.*, **15**, 23–45.

Rose, D. J. (1984a) Private communication.

Rose, D. J. (1984b) *Interpreting the Rose–Miller–Agnew Global Energy Scenarios* (unpublished memorandum).

Rose, D. J., Miller, M. M., and Agnew, C. (1983) *Global Energy Futures and CO_2-Induced Climate Change.* MITEL 83-015, Cambridge MA, MIT Energy Laboratory.

Rose, D. J., Miller, M. M., and Agnew, C. (1984) Reducing the problem of global warming, *Technol. Rev.*, May/June, 49–58.

Rotty, R. M., and Masters, C. (1984) *Past and Future Releases of CO_2 from Fossil Fuel Combustion*, Oak Ridge, TN, Institute for Energy Analysis (forthcoming).

Schurr, S. H. (1984) Energy use, technological change, and productive efficiency: An economic-historial interpretation, *Annu. Rev. Energy*, **9**, 409–425.

Seidel, S., and Keyes, D. (1983) *Can We Delay a Greenhouse Warming?* Washington, D.C., Environmental Protection Agency.

Sh'pillrain, E. (1985) Personal communication.

Siegenthaler, U. (1983) Uptake of excess CO_2 by an outcrop-diffusion model of the ocean, *J. Geophys. Res.*, **88**, 3598–608.

Skalkin, F. V., Kanaev, A. A., and Kopp, I. Z. (1981) *Energetics and the Environment*, Leningrad.

Steinberg *et al.* (1984) *A Systems Study for the Removal, Recovery, and Disposal of Carbon Dioxide from Fossil Fuel Power Plants in the US*, Report No.DOE/CH/00016-2, Upton, NY, Brookhaven National Laboatory.

Stirikovich, M. A., Sinyak, Y. V., and Chernyavskii, S. Y. (1981) Long term global energy developments—achievements and perspectives, *Energitika*, **14**, Toplivo, No. 3.

Svenningsson P. (1985) *Energy Strategies and Atmospheric CO_2 Concentrations.* Report, Stockholm, Beijer Institute, Royal Swedish Academy of Sciences.

Thompson, M. (1984) Among the energy tribes: A cultural framework for the analysis and design of policy, *Policy Sci.*, **17** (4) Special issue.

WEC (1983) *Energy 2000–2020: World Prospects and Regional Stresses,* Ed. J. R. Frisch, World Energy Conference Conservation Commission; and *Oil Substitution: World Outlook to 2020,* London, Graham and Trotman and Oxford University Press.

Williams, R. H., Goldemberg, J., Johansson, T. B., Reddy, A. K. N., and Larson, E. (1984) *Overview of An End-Use Oriented Global Energy Strategy.* Paper presented at the Greenhouse Problem: Policy Option Symposium, University of Minnesota, May 1984.

WOCOL (1980) *Coal—Bridge to the Future*: Report of the World Coal Study, Ed. C.Wilson, Cambridge, MA, Ballinger.

Wood, G. V. (1983) Private communication to D. J. Rose.

Wynne, B. (1983) Redefining the issues of risk and social acceptance. *Futures,* **20,** 1–21.

Wynne, B. (1984) The institutional context of science, models and policy: The IIASA energy study, *Policy Sci.,* **17** (4) Special issue.

CHAPTER 3

How Much CO_2 Will Remain in the Atmosphere?

The Carbon Cycle and Projections for the Future

B. BOLIN

3.1 INTRODUCTION

The atmospheric concentration of carbon dioxide is rising and is now 25% higher than during the first half of the last century, i.e. before the rapid expansion of agriculture and industry began. Fossil fuel combustion and changing land use (deforestation and expanding agriculture) have caused large emissions of CO_2 into the atmosphere. Since the continued rise of atmospheric CO_2 concentrations might lead to changes of the global climate, it is essential to be able to project future concentrations which may occur due to alternative future emission scenarios. Atmospheric CO_2 is exchanging with terrestrial vegetation and soils and with the oceans. For this reason the observed increase during the last hundred years is less than half of the total emissions during this same period. This chapter on the carbon cycle will serve as a basis for assessing likely future levels of atmospheric CO_2 concentrations.

During the last few decades considerable progress has been made in improving our understanding of the key features of the global carbon cycle. Rather detailed quantitative models have been developed. Many overviews of the carbon cycle have been published, e.g. Keeling (1973), Revelle and Munk (1977), Bolin *et al.* (1979), Broecker *et al.* (1980), Peng *et al.* (1983). It is not the purpose of this chapter to present an extensive synthesis of past research, but we shall restrict ourselves to those aspects of the global carbon cycle that are of importance for the determination of likely future atmospheric CO_2 concentrations.

3.2 CARBON IN NATURE

3.2.1 Time Scales and Scope of the Treatise

Although many elements are essential to living matter, carbon is the key element of life on Earth. The carbon atom's ability to form long covalent chains and rings is the foundation of organic chemistry. The biogeochemical cycle of carbon is necessarily very complex, since it includes all life forms on Earth as well as the inorganic transfers between and within the different carbon reservoirs. Being the most complex element cycle in nature, it has been most extensively studied and is better understood than other element cycles.

Figure 3.1 is a schematic diagram of the global carbon cycle, showing the major carbon reservoirs with which we will be concerned, i.e. the atmosphere, the terrestrial biosphere including soils, the hydrosphere, including marine life, and the lithosphere. The transfers between these reservoirs as well as some processes in their interior are also shown schematically. They will be analysed in more detail in the following.

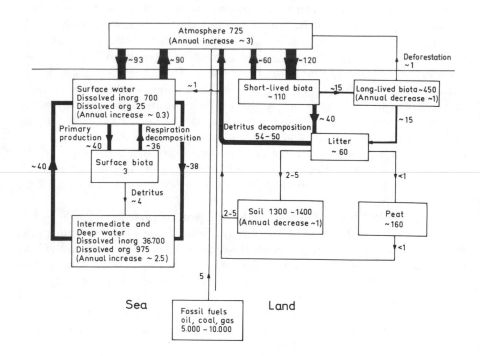

Figure 3.1 Schematic diagram of the global carbon cycle. Approximate reservoir sizes are given in units of 10^{15} g C and fluxes between reservoirs in units of 10^{15} g C yr^{-1}, valid as of the early 1980s. Data from compilation by Bolin (1983) modified in accordance with most recent data (cf also Table 3.2). Although single numbers usually are given, there is considerable uncertainty about both reservoir sizes and fluxes. A complete balance is not achieved as later discussed in Section 3.7

We note first that the characteristic time scales which govern the carbon cycle range from millions of years for processes controlled by the movements in the Earth's crust, to days or even seconds for processes related to air–sea exchange and photosynthesis. We will limit the following discussion to time scales less than 20,000–30,000 years and in this time perspective we may disregard lithospheric processes except those of erosion, chemical denudation and sedimentation. The former two are controlled by the partial pressure of CO$_2$ in soil water, they release carbon into the atmosphere–biosphere–hydrosphere system and are balanced by sedimentation of carbon at the bottom of the sea (Broecker, 1973). The pre-industrial transfer of this kind was probably quite small, less than $0.5 \cdot 10^{15}$ g C yr^{-1}, but may in spite of this be important in the analysis of changes from glacial to interglacial periods. Relative to the more rapid changes which have occurred during the last two centuries, however, these transfer processes are of secondary importance and accordingly we may largely limit our analysis to atmosphere–biosphere–hydrosphere interactions including the change of carbon content of soil due to changing land use.

3.2.2 Key Chemical Compounds and Reactions

There are more than a million known carbon compounds, thousands of which are vital to biological processes. Carbon atoms have nine possible oxidation states ranging from $+IV$ to $-IV$. The most common state is that of complete oxidation, i.e. $+IV$, in CO$_2$ and CO$_3^{-2}$ (carbonate). The former constitutes more than 99% of the carbon in the atmosphere. About 97% of the carbon in the oceans is in the form of dissolved carbonate ($H_2CO_{3(aq)}$, $HCO_{3(aq)}^{-1}$ and $CO_{3(aq)}^{-2}$), and in the lithosphere carbonate exists as minerals ($CaCO_{3(s)}$, $CaMg(CO_3)_2$, $FeCO_3$). In oxidation state $+II$, we find CO, a trace gas in the atmosphere, which is rather quickly oxidized to CO$_2$. Neutral carbon is present in the lithosphere in small quantities as graphite and diamond and in soil as charcoal. Assimilation of carbon by photosynthesis creates reduced carbon ($C_nH_{2n}O_n$) which is present as biota, as dead organic matter in the soil and in the top layers of the sediments, as coal, oil and gas reserves at greater depths, and as highly dispersed reduced carbon in the lithosphere. Some gaseous compounds that contain reduced carbon (CH_x, particularly CH_4, i.e. methane) find their way into the atmosphere by further reduction due to anaerobic processes. Although there are thus several different gaseous carbon compounds formed by bacterial decomposition, we need only consider CO$_2$ in the present context, since the oxidation of these other gases proceeds rather quickly. Since methane is another greenhouse gas, the concentration of which also is changing, we need to analyse its budget in the atmosphere, which is done in Chapter 4. In the oceans there is a considerable amount of carbon in the form of dissolved organic carbon. The processes of oxidation of these compounds into carbonate ions are poorly known.

3.2.3 Carbon Isotopes

There are three carbon isotopes (from a total of seven) that are of importance in nature. Two are stable, ^{12}C and ^{13}C, one is radioactive, ^{14}C, with a half-life of 5730 years.

The amount of ^{13}C is expressed by the permille deviation ($\delta^{13}C$) of the ratio $^{13}C/^{12}C$ for the sample from a universally accepted standard (cf Craig, 1957b; Stuiver and Polach, 1977).

The abundance of ^{14}C ($\delta^{14}C$) is expressed in terms of the deviation (in ‰) of the activity of the sample from that of the standard (National Bureau of Standards, normalized to $\delta^{13}C = -19‰$), and corrected for radioactive decay since 1950. In geosciences the concept $\Delta^{14}C$ is used and defined as follows. The ^{14}C-atoms are subject to fractionation in processes of chemical reaction and transfer, due to the fact that a molecule containing ^{14}C is heavier than one containing ^{13}C or ^{12}C. Since the weight difference between ^{14}C and ^{12}C is twice that between ^{13}C and ^{12}C the fractionation of ^{14}C is approximately twice that of ^{13}C. By introducing a correction to the measured $\delta^{14}C$ value, which is twice the ^{13}C fractionation, the ^{14}C and ^{12}C fields in nature to a first approximation can be compared without the need for considering the different effects of fractionation of ^{14}C and ^{12}C. Concentrations are therefore usually expressed in terms of $\Delta^{14}C$ (see Stuiver and Polach, 1977; Wigley and Mueller, 1981).

The importance of the different carbon isotopes as observed in nature is related to the fact that the rates of transfer of carbon compounds and the equilibria of chemical reactions are dependent on the carbon isotope they contain. Accordingly, the distributions of the stable isotopes ^{12}C and ^{13}C in nature differ. The distribution of ^{14}C in addition depends on the formation of ^{14}C due to nuclear reactions between neutrons (formed by cosmic radiation) and nitrogen atoms in the atmosphere on one hand, and radioactive decay on the other. The different distributions of these three isotopes provide important information on key characteristics of the carbon cycle.

3.3 CARBON IN THE ATMOSPHERE

3.3.1 Atmospheric CO_2

Accurate measurements of atmospheric carbon dioxide were begun by Keeling at Mauna Loa in 1957 (cf Bacastow and Keeling, 1981). This record as well as that from the South Pole are invaluable pieces of geophysical information that reveal several important features of the carbon cycle (Figure 3.2). The average atmospheric CO_2 concentration in 1984 was 343 ± 1 ppmv. Several other series of regular measurements have been begun since then. They confirm in principle the annual cycle of the atmospheric concentrations and the gradual increase with time as shown in Figure 3.2 but provide in addition information on the global variations of these features (cf Pearman and Hyson, 1980). We conclude from analysis of these records that the annual variations are primarily due to the regular variations of terrestrial photosynthesis, to a minor degree to the annual variations of sea surface temperature, which influence the solubility of CO_2 in sea water and thirdly to annual variations of photosynthesis in the sea. The steady increase of atmospheric CO_2 in the atmosphere is slightly larger in the Northern Hemisphere than in the Southern one, presumably due to the fact that the net emissions caused by human activities are located primarily in the Northern Hemisphere (see Keeling and Heimann, 1985). In addition there are small interannual variations, which are associated with variations of the general circulation of the atmosphere, e.g. the Southern Oscillation and the related variations of the equatorial ocean currents in the Pacific (El Niño). These variations of atmospheric CO_2 reflect features of the air–sea exchange of CO_2 which are of interest in the analysis of the global carbon cycle. Their quantitative interpretation is, however, uncertain. In the validation of models for the global carbon cycle the regular increase of atmospheric CO_2 as observed during the last 25 years is of fundamental importance.

Earlier measurements of atmospheric CO_2 (since the middle of the last century) were usually rather inaccurate and samples were not collected with necessary care and adequate consideration of representativeness. Wigley (1983) has concluded that the most likely value about 1870 was 270 ± 15 ppmv, primarily based on data from the Southern Hemisphere. Siegenthaler (1984) on the other hand points out that a careful selection of Northern Hemisphere data is more trustworthy and considers the most likely value in the 1880s to have been 285–290 ppmv. The uncertainty of these values is, however, still considerable.

Measurements of atmospheric concentrations of CO_2 have been extended about 30,000 years back in time by the analysis of air trapped in glacier ice (Delmas *et al.*, 1980; Neftel *et al.*, 1982). The mean concentration during the Holocene was about 270 ppmv, but values as low as about 200 ppmv have been recorded towards the end of the last glaciation. It has recently been possible to cover also the period 1750–1960 with such measurements (Neftel *et al.*, 1985). Figure 3.3 shows that the values determined for samples from the 1950s agree well with the Mauna Loa data (cf Figure 3.2), while concentrations during 1750–1800 seem to have been about 280 ppmv, starting to rise slowly soon thereafter. Similar analyses by Raynaud and Barnola

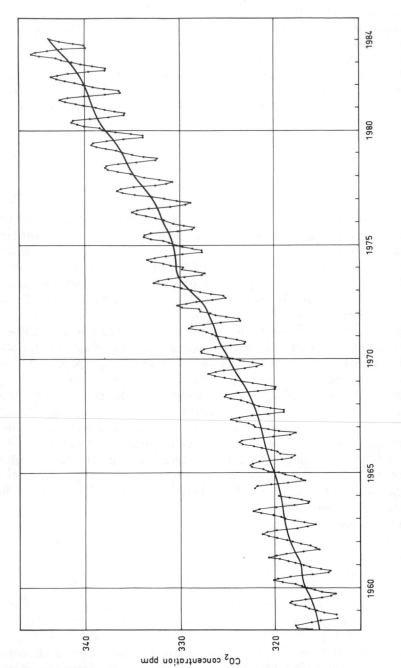

Figure 3.2 Concentration of atmospheric CO_2 at Mauna Loa Observatory, Hawaii. Dots indicate monthly averages determined from continuous measurements. Based on data reported by Bacastow and Keeling (1981), supplemented by data from recent years supplied by personal communication

(1985) show systematically somewhat lower values and also slowly decreasing concentration from late medieval times until an increase due to man began about 250 years ago. For a more detailed discussion of previous work on this problem reference is made to WMO (1983). We shall here adopt the most likely value for the atmospheric CO_2 concentration to have been 275 ± 10 ppmv before the expansion of agricultural activities and the industrial revolution during the 19th century.

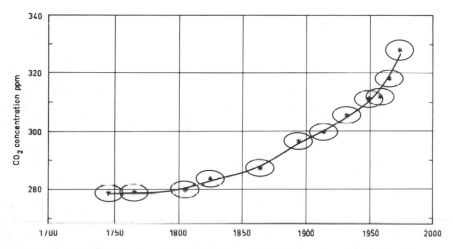

Figure 3.3 Atmospheric CO_2 concentrations measured in glacier ice formed during the last 200 years (Neftel *et al.*, 1985)

3.3.2 The ^{13}C Content of Atmospheric CO_2

The first measurements of the ^{13}C-isotope in the atmosphere were made by Keeling in 1956 and similar measurements were repeated in 1978 (cf Keeling *et al.*, 1979). The ^{13}C abundance in atmospheric CO_2, δ^{13}C, was −7.0‰ in 1956 and −7.65‰ in 1978. The change during this period is shown by a dashed–dotted line in Figure 3.4.

Variations of δ^{13}C of atmospheric CO_2 in the past can be inferred from measurements of δ^{13}C in wood from tree rings, which have been dated by using dendrochronology (see Figure 3.4 and Freyer, 1979; Freyer and Belacy, 1983; Stuiver *et al.*, 1984). Isotopic fractionation occurs in the process of photosynthesis and the further transformation of the primary carbon compounds into cellulose (cf Francey and Farquhar, 1982; Farquhar *et al.*, 1982).

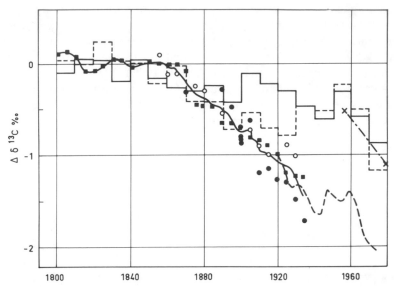

Figure 3.4 Changes of $\delta^{13}C$ in tree rings and of atmospheric CO_2 relative to pre-industrial conditions (1800–1850). The data points and squares and the continuous and dashed curves show measurements by Freyer (1979) using tree samples (see also Peng *et al.*, 1983). The histograms are measurements on trees according to Stuiver *et al.* (1985), where the one drawn by solid lines shows results from trees on open sites. The dash–dotted line gives the change in the atmosphere 1956–1978 according to Keeling *et al.* (1979)

The ^{13}C fractionation in the sequence of events is about 19‰ (for C_3 plants) resulting in $\delta^{13}C$ values for wood of about −26‰. If we assume that the fractionation is independent of the $\delta^{13}C$ value of CO_2 in the ambient air from which carbon is extracted, the observed change as a function of time can be taken as a measure of $\delta^{13}C$ changes of atmospheric CO_2. The data are, however, quite uncertain because of the influence of other environmental factors on the fractionation process, e.g. stress (Francey and Farquhar, 1982). By careful selection of samples we may, however, assume that the change of $\delta^{13}C$ in wood during the last few hundred years reflects a corresponding change of $\delta^{13}C$ in atmospheric CO_2. Recently $\delta^{13}C$ measurements of CO_2 trapped in glacier ice have also been reported (Friedli *et al.*, 1984). These data show smaller changes during the last 200 years than most of the data from tree ring analyses (cf Figure 3.4). On the basis of the data available today the decrease of $\delta^{13}C$ in atmospheric CO_2 during the last 200 years seems to have been 1.0–1.5‰.

The observed changes are primarily due to low $\delta^{13}C$ values (−24 to −27‰) for the emissions of CO_2 into the atmosphere by deforestation, changing land use and by fossil fuel combustion but may also reflect natural variations of the carbon cycle. Fossil fuels have on the average the same

[13]C concentrations as wood, since they were formed from organic material buried in the ground. Although the measured δ^{13}C values for atmospheric CO_2 are uncertain, they are of importance for the calibration of global carbon cycle models as will be discussed in Section 3.6.4.

3.3.3 The ^{14}C Content of Atmospheric CO_2

The amount of the ^{14}C isotope on Earth depends on an approximate balance between the formation of ^{14}C by cosmic radiation and radioactive decay. The quasi-steady global carbon cycle before the agricultural and industrial revolutions presumably maintained an approximately constant partitioning of ^{14}C between the different carbon reservoirs in nature. The amount of ^{14}C in the atmosphere at that time, as well as changes since then, can be used for analysis of the mechanisms that govern the global carbon cycle.

In the scale adopted for ^{14}C (see Section 3.2.3) the value for atmospheric CO_2 in 1954 was close to 0‰, i.e. before any significant changes had been caused by the injections due to nuclear bomb testing, which had begun in 1952. Variations of ^{14}C before that time can be determined by analysis of wood samples from tree rings. Careful measurements have been presented by Stuiver and Quay (1981) as shown in Figure 3.5. A marked decrease has

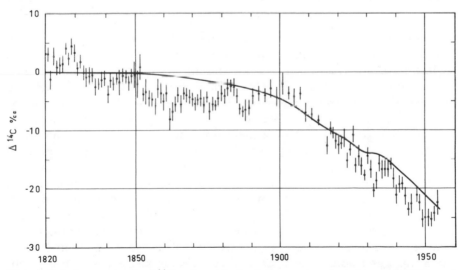

Figure 3.5 Atmospheric Δ^{14}C values derived from tree rings between 1820 and 1954. Single-year determinations are given, except for the 1890–1915 interval. The vertical bars denote one standard deviation. Δ^{14}C levels during the period 1895–1915 probably give an upper limit only. For details see Stuiver and Quay (1981). (The declining curve shows the model calculations by Peng *et al.*(1983); see further Section 3.6.4)

apparently occurred since early last century amounting to approximately 25‰. It must be due primarily to the emission of ^{14}C-free fossil CO_2, which has diluted the pre-industrial ^{14}C content in atmospheric CO_2. Significant variations, of the order of 10‰, are also present in the record shown in Figure 3.5 and similarly during previous centuries. We do not understand adequately the reasons for their occurrence. In the later analysis we shall only make use of the general decrease observed since early last century.

Since the first nuclear bomb tests in 1952 and 1954 very marked changes of the $\Delta^{14}C$ values for atmospheric CO_2 have been observed. Measurements primarily by Nydal and Lövseth (1983) are displayed in Figure 3.6 and reveal the large injections of ^{14}C in 1958 due to US Pacific tests and in 1961–1962 from Soviet tests. Since then very limited injections have been made. Initially most of the radioactive debris is transferred into the stratosphere. Since the exchange time between the stratosphere and the troposphere is several years, the declining concentrations in the troposphere from 1965 due to transfer from the atmosphere into the terrestrial biosphere by photosynthesis and into the oceans due to air–sea exchange are delayed by inflow from above. Obviously the observations shown in Figure 3.6 represent important data for testing models of the carbon cycle.

3.3.4 Mixing in the Atmosphere

The mixing of tropospheric air is rather rapid. The westerlies at middle latitudes in the two hemispheres circle the Earth on the average in about a month, vertical overturning between ground level and the tropopause (at 12 to 16 km elevation) also takes about a month, north–south mixing within a hemisphere is accomplished in about three months and effective exchange between the two hemispheres takes about a year. In the present context we are concerned with processes of change over periods of several years, decades and centuries. We may therefore assume that the troposphere at any time is well mixed. The approximate validity of this assumption is borne out by the fact that the annual latitudinal averages of the atmospheric CO_2 concentration differ by merely 1.5–2.0 ppmv between high northerly and high southerly latitudes. The concentration in the Northern Hemisphere is larger than that of the Southern Hemisphere. Since this difference has been enhanced during the last few decades with increasing fossil fuel combustion, it seems likely that it is primarily due to the fact that about 90% of the present emissions from fossil fuel combustion take place in the Northern Hemisphere. The observed north-south gradient is in general accord with the rates of horizontal mixing of tropospheric air as deduced by other means. Although it may be necessary to consider the seasonal variations of the atmospheric CO_2 concentrations to account in more detail for the air–sea exchange, terrestrial photosynthesis and soil respiration, their inclusion

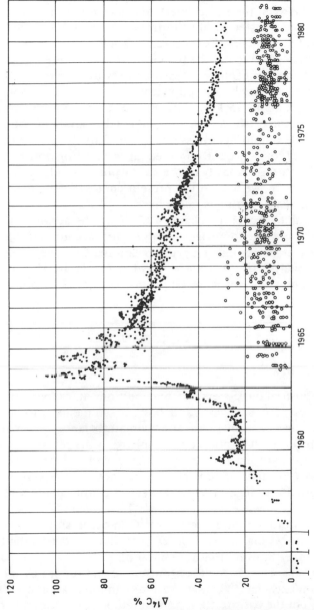

Figure 3.6 Direct measurements of atmospheric $\Delta^{14}C$ (surface data) during the period 1954 to 1981 (upper set). Data 1954–1962 as compiled by Tans (1981) and 1962–1981 according to Nydal and Lövseth (1983). Data from the ocean surface water (lower set) are from a few sites only and not representative for the oceans as a whole, see further Figures 3.1(–3.11

into a global carbon cycle model is not crucial for our understanding of its long-term behaviour and for projecting likely future atmospheric CO_2 concentrations as a result of given emission scenarios.

The exchange between the stratosphere and the troposphere is considerably slower than the transfer within the troposphere. Accordingly the seasonal variations of atmospheric CO_2 decrease rapidly above the tropopause. The upward trend as observed in the troposphere (see Figure 3.2) is considerably delayed in the stratosphere. Measurements by Bischof *et al.* (1980) reveal CO_2 concentrations about 7 ppmv lower at 36 km elevation than at tropopause level (at about 15 km). This is equivalent to a mixing time of 5–8 years between these two levels. In a detailed analysis of the increase of atmospheric CO_2 due to net emissions at the Earth's surface this delay should be accounted for. Since the observed lag is changing only slowly with time and since the storage capacity of the stratosphere is only about 15% of that of the troposphere, we may in the present context safely disregard this delayed transfer of CO_2 into the stratosphere.

When attempting to model the changes of ^{14}C in the atmosphere during and after the injections due to nuclear bomb testing, account must be taken, however, of the fact that a major part was emitted into the stratosphere. A more detailed analysis of this transfer has been given by Machta (1972). In the present context stratosphere–troposphere transfer is not of prime concern.

3.4 AIR–SEA EXCHANGE

3.4.1 Rate of Transfer

During pre-industrial quasi-steady state conditions more than 90% of ^{14}C on Earth was present in sea water and bottom sediments (the latter containing merely a few per cent). An approximate balance was maintained between the transfer of ^{14}C from the atmosphere to the sea and radioactive decay within the sea. By measuring the difference of ^{14}C between atmospheric CO_2 and CO_2 dissolved in the surface layer of the sea, the global average of the gross exchange of CO_2 between the atmosphere and the sea can be determined (cf review by Bolin *et al.*, 1981). The decrease of atmospheric ^{14}C concentrations and increase of ^{14}C in ocean surface water since the time of nuclear testing offers another possibility to determine the rate of gaseous exchange (Stuiver, 1980). Peng *et al.* (1979) have estimated the rate of gaseous exchange between the atmosphere and the sea by measuring the disequilibrium between ^{226}Ra and ^{222}Rn due to the evasion of ^{222}Rn into the atmosphere. On the basis of these three methods the mean rate of gross exchange of CO_2 between the atmosphere and the sea at an atmospheric CO_2 concentration of 300 ppmv has been determined to be 18 ± 5 moles

m^{-2} yr^{-1}. This implies a mean residence time for CO_2 in the atmosphere before transfer into the sea of 8.5 ± 2 years.

Gas exchange across the air–sea interface varies across the ocean surface particularly depending on winds and waves. Since we do not know adequately the relative role of the physical processes of importance for the exchange, it has not been possible to determine the magnitude of these variations. However, the global CO_2 exchange between the atmosphere and the sea can be deduced reasonably well with global carbon cycle models calibrated against the global changes that have taken place in the past.

3.4.2 Chemical Buffering

When CO_2 dissolves in sea water the molecule undergoes hydration reactions forming $H_2CO_{3(aq)}$, which in turn dissociates into H^+, HCO_3^- and CO_3^{2-}. For a detailed account of the chemical reactions and equilibria that establish themselves reference is made to Keeling (1973), Takahashi *et al.* (1980) and Baes (1982). The carbonate system is defined by total carbon or dissolved inorganic carbon ($\Sigma C = DIC$); total borate (ΣB); alkalinity (Alk $= A$), which describes the acid neutralizing capacity; acidity, pH; the partial pressure, P_{CO_2}, which in equilibrium with the atmosphere is equal to the atmospheric CO_2 pressure. For our present purpose we may consider ΣB = const, the magnitude of which is well known. We can then deduce any two of the quantities that characterize the carbonate system, if the other two are known. The two usually chosen to describe the system are DIC and A. For the following discussion it is important to keep in mind that, if CO_2 is transferred to sea water, alkalinity remains unchanged, while on the other hand, the formation and decomposition of inorganic and organic compounds change both DIC and alkalinity, A.

In the present context we note two features of this carbonate system

(i) The solubility of CO_2 in sea water, and accordingly total carbon in equilibrium with a given atmospheric CO_2 concentration, is temperature dependent. The functional relationship is given in Figure 3.7.
(ii) The CO_2 exchange between gas phase and solution is buffered, the 'buffer' factor, ξ, also called the 'Revelle' factor (Revelle and Suess, 1957), being given by

$$\xi = \frac{\Delta P_{CO_2}/P_{CO_2}}{\Delta(DIC)/(DIC)}$$

where Δ indicates a finite change of the variable concerned. The variations of ξ in the range of the key variables DIC and alkalinity observed in the ocean are shown in Figure 3.8.

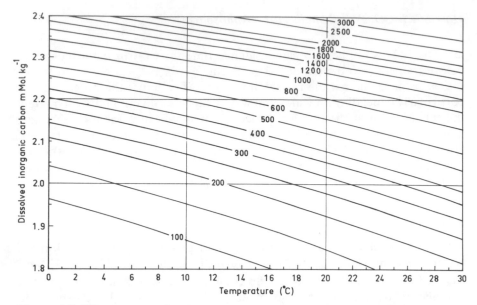

Figure 3.7 CO_2 concentration in sea water (ppmv) as function of dissolved inorganic carbon (DIC) and temperature (assuming 35‰ salinity and 2.35 meq kg^{-1} alkalinity (Holmén, 1985)

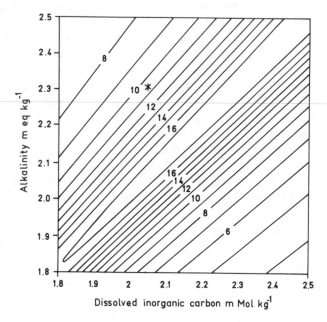

Figure 3.8 The buffer factor of sea water (at temperature of +15 °C and salinity 35‰) as function of dissolved inorganic carbon (DIC) and alkalinity (A). For present concentrations in surface water, DIC = 2.05 mmol kg^{-1}, A = 2.30 meq kg^{-1}, ξ = 10.6 (Holmén, 1985)

We note that the solubility and the buffer factor increase with decreasing temperature. Since the atmospheric CO_2 partial pressure does not change much from pole to equator, in the mean CO_2 is transferred from the atmosphere to the sea at high latitudes during winter and in the opposite direction at low latitudes, although there are significant deviations from this pattern due to the fact that upwelling also brings CO_2-rich water from deeper layers to the surface. The buffer factor is of the order of 10 and increases with increasing values for *DIC*. This implies that P_{CO_2} is sensitive to quite small changes of the amount of *DIC* in the water. The change of the atmospheric CO_2 concentration by about 25%, which has occurred during the last 100 years, has been associated with a change of *DIC* in the surface waters by merely 2–2.5% to maintain equilibrium. The storage capacity of the ocean for the excess atmospheric CO_2 is reduced by a factor $\xi \approx 10$, as compared with what might be expected at first sight when comparing reservoir sizes as shown in Figure 3.1. Exchange equilibrium between the atmosphere and the sea is established about ten times more quickly for total carbon than for the ^{13}C and ^{14}C isotopes.

3.5 CARBON IN THE SEA

3.5.1 Total Carbon and Alkalinity

As was shown in Figure 3.1, the oceans contain more than 50 times as much carbon in the form of *DIC* ($38\,000 \cdot 10^{15}$ g) as found in the atmosphere in the form of CO_2 ($727 \cdot 10^{15}$ g $= 343$ ppmv) in 1983. In addition significant amounts of dissolved organic carbon are present in the sea (cf below). The vertical distribution of *DIC* is not homogeneous, but concentrations are higher in the deep sea than in the surface layers (Figure 3.9). There is also a systematic change from rather low deep sea values in the Arctic Sea, to higher concentrations in the Atlantic Ocean, still higher values in the Antarctic Ocean and in the Indian Ocean and maximum values in the Pacific Ocean.

The vertical distribution of alkalinity is rather similar to that of *DIC* (Takahashi *et al.*, 1981). The range of variations is, however, considerably smaller (about 30% of that of *DIC*) (see Figure 3.9).

It is interesting to note that the surface concentrations of DIC would be about 15% higher if the oceans were well mixed, which in turn would imply an atmospheric CO_2 concentration of about 700 ppmv. The maintenance of the vertical gradients of *DIC* (as well as alkalinity) in the oceans is crucial for the present atmospheric CO_2 concentrations, see further Section 3.5.2.

3.5.2 Photosynthesis, Decomposition and Dissolution of Biogenic Material

Marine life is almost exclusively restricted to the surface layers with in-

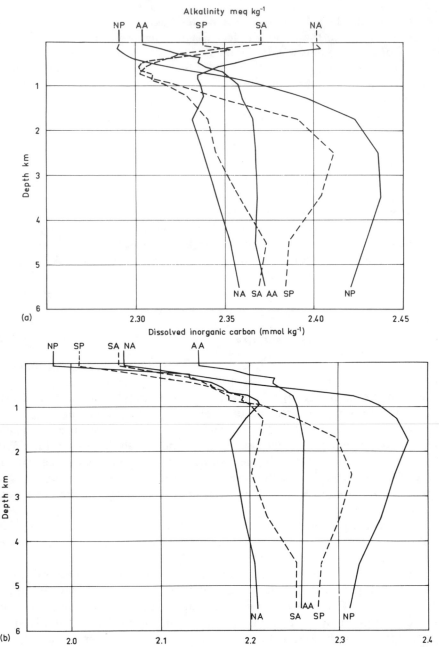

Figure 3.9 The mean vertical distribution of (a) alkalinity and (b) total dissolved inorganic carbon (DIC) in five regions of the world oceans. NA = North Atlantic, SA = South Atlantic, AA = Antarctic region (south of 45°S), SP = South Pacific Ocean and NP = North Pacific Ocean. From Takahashi *et al.* (1981)

tense photosynthesis in the photic zone and bacterial decomposition also taking place primarily in the top 100 m of the sea. Some dead organic matter largely in the form of faecal pellets, and so called macroflocs (Honjo, 1980), probably only about 10% of the primary production, reach deeper layers and perhaps 1% is deposited on the ocean floor. The total primary production in the sea is about $40 \cdot 10^{15}$ g C yr^{-1} (de Vooys, 1979) but the rate of photosynthesis per unit area has large variations from more than 0.5 g C m^{-2} day^{-1} in areas of intense up-welling to less than 10% of this value in the desert regions of the ocean which are characterized by down-welling and lack of nutrients.

Photosynthesis is dependent on nutrients being available. Wherever light is adequate the nutrients are quickly used up. In particular, lack of nitrogen and phosphorus often limits the rate of primary production. At high latitudes, however, particularly in the Antarctic Ocean, the presence of rather high concentrations of both nitrogen and phosphorus in the surface waters clearly indicates that some other environmental factor, probably light, limits primary production (cf Knox and McElroy, 1984; Sarmiento and Toggweiler, 1984).

Primary production, both the formation of inorganic and organic compounds, obviously reduces the amount of *DIC* by direct uptake. The alkalinity on the other hand is affected differently. For each μmol of carbon that is used in the formation of organic tissue, alkalinity is increased by about 0.16 μeq, while it is decreased by 2 μeq if the carbon is used for $CaCO_3$ formation. The differences between the spatial distributions of *DIC* and alkalinity therefore implicitly contain information on the relative magnitudes of the formation and decomposition or dissolution of organic and inorganic particulate matter in the sea. The distribution of *DIC* as shown in Figure 3.9 may be interpreted to a first approximation as a quasi-steady state, but undoubtedly the increasing concentration of atmospheric CO_2 has induced a net CO_2 flux into the oceans, which in turn must have modified the pre-industrial distribution of *DIC* at upper levels in the sea. We shall later determine by indirect means how much the distribution of *DIC* in the oceans may have changed during the last 100 years.

The total amount of dissolved organic carbon (*DOC*) in the sea has been deduced from rather few measurements, and a value of about 1000 $\cdot 10^{15}$ g C is usually quoted (Williams, 1975). An unknown portion of this carbon pool may be of terrestrial origin. Gorshkov (1982) has proposed that a considerable amount of dissolved organic carbon is produced as excretion from zoo-plankton and that this process is enhanced when the CO_2 partial pressure in sea water increases. No observations to support this hypothesis have been presented and it is unlikely to occur.

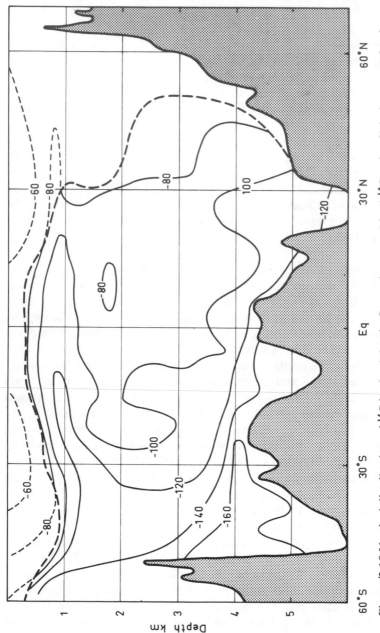

Figure 3.10 Vertical distribution of ^{14}C in the Atlantic Ocean (expressed in $\Delta^{14}C$ units, ‰) along a section in the western basin. The values shown are for natural conditions, i.e. prior to influx of bomb-produced $\Delta^{14}C$. The thick dashed line shows the depth to which significant amounts of bomb ^{14}C penetrated at the time of the GEOSECS survey in 1973. The analysis above this line is due to Broecker and Peng (1982) (data from Broecker *et al.*, 1960 and Stuiver and Östlund, 1980)

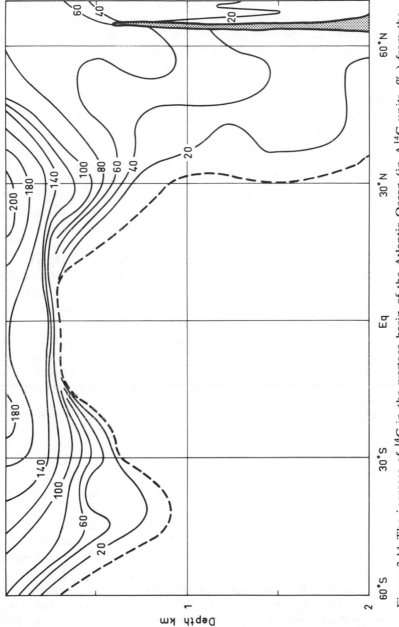

Figure 3.11 The increase of ^{14}C in the western basin of the Atlantic Ocean (in $\Delta^{14}C$ units, ‰) from the time of the first nuclear bomb tests in the middle of the 1950s until the time of the GEOSECS survey in 1973 (based on data from Broecker *et al.*, 1960 and Stuiver and Östlund, 1980)

3.5.3 ^{14}C in the Sea

The distribution of Δ^{14}C in dissolved inorganic carbon has been measured in all oceans during the GEOSECS expeditions 1972–1978. Figure 3.10 is an attempt to reconstruct the most probable pre-bomb distribution in the western basin of the North Atlantic. The GEOSECS data (Stuiver and Östlund, 1980) have been used at the depths which bomb-produced tritium had not yet reached at that time (cf Figure 3.12), while the distribution at upper layers is based on measurements in 1957 by Broecker *et al.* (1960), cf Broecker and Peng (1982). The penetration of bomb-produced ^{14}C during the period 1957–1973 can be determined as shown in Figure 3.11.

The time course of Δ^{14}C in the surface layers of the oceans based on direct measurements can be deduced approximately from the data shown in the lower part of the graph in Figure 3.6. Measurements have also been made by using carbonate deposition in corals, which can be accurately dated (Druffel and Suess, 1983). Maximum Δ^{14}C values seem to have been reached early in the 1970s. These variations should also be reproduced in efforts to model the carbon cycle and its change.

A limited number of analyses of Δ^{14}C in dissolved organic carbon (*DOC*) in the sea, mostly from deep-water samples (Williams, 1975), are available. Considerably lower Δ^{14}C values are found, -300 to $-350‰$, equivalent to an age of 3000–4000 years, which is the basis for the view that *DOC* primarily consists of refractory material. The easily degradable compounds (e.g. sugars, proteins) are quickly consumed in the surface layers by zooplankton and serve as an important energy source. We estimate that the turnover (oxidation) of *DOC* below the surface mixed layer is 0.2–0.4 · 10^{15} g C yr^{-1}, which is about 5% of the total amount of reduced carbon (particulate and dissolved organic carbon) which is oxidized in these deeper layers of the oceans.

3.5.4 Ocean Sediments

About 0.5 · 10^{15} g C yr^{-1} is deposited on the ocean floor (Broecker and Takahashi, 1977), some of which is organic carbon and some $CaCO_3$. Organic carbon is the main energy source for organisms at the sea floor and only a small part is buried in organic form in the sediments, except in the coastal zone or on the continental shelves. In some limited regions (e.g. parts of the Baltic Sea) the oxygen content of the bottom water may be depleted, the rate of oxidation decreased and significant amounts of organic carbon buried. Regions of anoxic conditions have increased due to coastal pollution and the amount of organic matter withdrawn from the rapid turnover within the sea has probably increased in recent years. Estimates of the magnitude of these changes are not yet reliable, although some attempts to assess this sink have been made (Walsh *et al.*, 1981).

Ocean water is supersaturated with respect to $CaCO_3$ above the surface of saturation, the lysocline, which is found at about 4000 m depth in the Atlantic Ocean and at merely about 1000 m in the Pacific Ocean. No appreciable dissolution of $CaCO_3$ takes place above the lysocline while dissolution reduces the net accumulation at greater depths and prevents it totally below the carbonate compensation depth. Since stirring of the sediments by bottom-living organisms (bioturbation) extends about 10 cm into sediment, an appreciable amount of carbon ($\approx 5,000 \cdot 10^{15}$ g C) in the form of $CaCO_3$ is in slow exchange with inorganic carbon in sea water, primarily at the depth of the lysocline (Broecker and Takahashi, 1977).

There are some measurements of ^{14}C in ocean sediments (cf Broecker and Takahashi, 1977) showing a rather rapid decrease with depth, which fact represents a valuable measure of accumulation rate. This rate has changed significantly since the last glaciation, which fact should be considered in the analysis of possible changes of the carbon cycle that may have occurred in association with climatic variations. We also note that the total ^{14}C inventory of the sediments is small in comparison with the atmosphere, the biosphere and the oceans, which permits us to restrict our theoretical analysis of the carbon cycle to these latter reservoirs.

3.5.5 Transfer Processes Within the Sea

Due to the CO_2-buffering (see Section 3.4.2) the change of *DIC* in sea water by CO_2 uptake, required to maintain equilibrium with an increasing atmospheric CO_2 concentration, is markedly reduced and a quasi-equilibrium between atmospheric CO_2 and sea water is established rapidly. The rate of turnover of the ocean primarily determines its role in the global carbon cycle.

The ocean surface layers are rather well mixed down to the upper boundary of the thermocline layer, i.e. down to a depth of about 75 m between about 45° N and 45° S (cf Bathen, 1972). Further polewards wintertime cooling creates mixing to considerably greater depths, which in limited areas and during short periods of time extends all the way to the bottom as is the case in the Greenland Sea and in the Weddel Sea. Furthermore, quasi-horizontal large scale mixing brings cold surface water equatorwards into the thermocline layer at depths between about 100 and 1000 m along the density (isopycnic) surfaces from the regions of the major ocean currents at latitudes 45–55°, the Gulf Stream in the North Atlantic, the Kuroshio in the North Pacific and the circumpolar Antarctic current. Vertical mixing across the isopycnic surfaces in the thermocline layer also takes place, but is restricted because of the prevailing stable stratification. Both processes are presumably of importance for the transfer of carbon in the sea.

The rate with which ocean surface water is mixed into deeper layers of the ocean was poorly known until both radioactive and stable tracers could be used for more precise estimates. The penetration of bomb-produced tritium into the thermocline region of the Atlantic and Indian Oceans has been analysed by Broecker *et al.* (1980) using estimates of tritium transfer from the atmosphere to the sea by Weiss and Roether (1980). From 1958–1962, when intense nuclear bomb testing took place, until 1972–1974, when the (GEOSECS) geochemical survey was undertaken, large amounts of tritium invaded the oceans. During this period of 10–15 years the average penetration depth (the total tritium inventory per unit area divided by the surface concentration) between 45° N and 45° S was 350–550 m, except in the equatorial upwelling region, where it was merely about 200 m. In polar regions, where mixing to great depth takes place, the invasion of tritium is much larger and gives important information on the rates of transfer as shown in Figure 3.12 (Östlund *et al.*, 1976). The Transient Tracer Ocean Program (TTO) in 1982 reveals a further substantial advance of tritium-rich water downwards and equatorwards. Similar information on the characteristics of deep water formation in the North Atlantic has been obtained by analysis of chlorofluorocarbons in sea water (Bullister and Weiss, 1982). Since their gradual increase in the atmosphere is about the same over the entire globe, the source function for transfer from the atmosphere to the sea is much better known than that for tritium. Still, a quantitative interpretation of what these data mean with regard to the penetration of excess CO_2 into the deeper layer of the ocean is not yet possible because of inadequate data coverage.

Another more direct way of determining the CO_2 penetration into the sea has been proposed by Brewer (1978). Water at the sea surface equilibrates with atmospheric CO_2 within about a year (disregarding seasonal variations). The amount of *DIC* established in the surface waters in this way is carried along as this water moves to deeper layers, but usually also increases in course of time because of decomposition and dissolution of detritus settling from the ocean surface layer. This latter increase can be computed, however, by considering the simultaneous increase of nutrients and alkalinity. No accurate determinations of the likely atmospheric CO_2 concentrations at the time of the deep water formation can be made in this way, but Brewer (1978) estimates that the *DIC* concentrations in the deep sea correspond to P_{CO_2} values at the time of deep water formation between 240 and 280 ppmv. A series of similar computations was also presented by Chen (1982). All these results are, however, uncertain. It has not been possible to estimate the total CO_2 uptake by the sea since pre-industrial times.

As has been previously emphasized, the quasi-steady state distribution of *DIC* in the oceans represents an approximate balance between downward transfer due to the detritus flux and upward transfer by mixing and upwelling from deeper layers with greater *DIC* concentrations. In the discussion above we have considered the transfer of excess CO_2 into the deep sea being brought about by a decrease of the transfer upwards by mixing

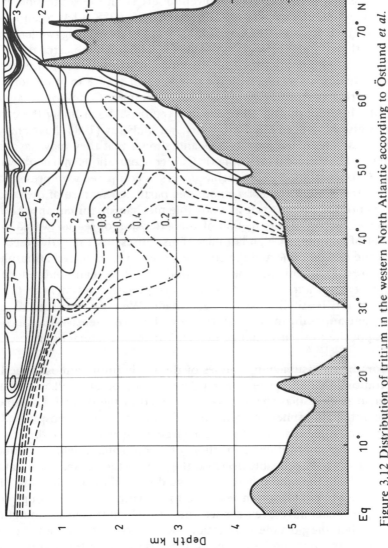

Figure 3.12 Distribution of tritium in the western North Atlantic according to Östlund *et al.* (1976), in tritium units (defined as $(T/H)^{-18}$). Prebomb concentrations were below one tritium unit in surface layers and below 0.1 in the deep sea

because of the CO_2-enriched ocean surface layers, but have simultaneously assumed that the detritus flux downward remains unchanged. The justification has been the fact that the primary production in the ocean surface layer is usually limited by lack of nutrients. It should be noted, however, that this is not necessarily so in the major upwelling areas on the south side of the Antarctic Circumpolar current in a latitude belt 55–60°S as has recently been emphasized by Knox and McElroy (1984), Sarmiento and Toggweiler (1984) and Siegenthaler and Wenk (1984). This circumstance indicates that there are other limiting factors for phytoplankton growth at these latitudes, e.g. incident radiation associated with the extension of sea ice to quite northerly latitudes during the austral spring and early summer. Conditions may have been quite different during other climatic regimes. Accordingly the carbon cycle may also have been different. Knox and McElroy (1984) have analysed some of these alternative possibilities and shown that they may have been characterized by different atmospheric CO_2 concentrations in balance with a lower *DIC* concentration in the surface layers as compared with present conditions. In the present context we note this particular feature of the oceanic component of the carbon cycle as a possible mechanism for enhanced downward flux of carbon, if a reduction of the extension of sea ice would occur as a result of warming at high latitudes. This represents a negative feedback between the carbon cycle and the climate system, i.e. increased atmospheric temperatures would enhance the CO_2 uptake by the oceans and reduce the rate of increase of atmospheric CO_2. A more precise assessment of the significance of this possibility has not been made. We note also that the lower CO_2 concentrations during the last ice age with more extensive sea ice coverage are opposite to what would be expected from the interactions described above.

It is commonly assumed in analyses of the likely future atmospheric CO_2 concentrations that the general circulation of the oceans will not change. However, it is clear that changes have occurred in the past. If a significant warming of the atmosphere were induced by the enhanced atmospheric CO_2 concentration, some change of the ocean circulation would most likely occur. Particularly the intermittent formation of deep water might be reduced, which in turn would probably decrease the role of the oceans for uptake of excess CO_2 from the atmosphere. Since we do not know much about likely changes of ocean circulation in response to climatic change, although some model experiments have been carried out, it does not seem meaningful to analyse further the possible implications with regard to the global carbon cycle. They may, however, be significant (see also Bolin, 1981, and Broecker, 1981).

A modification of the carbon circulation in the sea might also occur, if the total amount of nutrients in the sea would increase. Assuming that photosynthesis still would be limited primarily by nutrients in the surface layers their concentrations would still be low and accordingly the vertical nutrient gradient enhanced between the surface layers and the deep sea. The vertical mixing of the sea would then transfer more nutrients to the sea surface, an increased photosynthesis would be maintained and associated with increased detritus flux to deeper layers. The vertical gradient of DIC would also be larger, the surface concentrations accordingly decreased and similarly the CO_2 partial pressure.

Broecker (1981) has analysed possible mechanisms of this kind (especially emphasizing the role of phosphorus) that may have been of importance during the transition from glacial to interglacial conditions. Their possible occurrence might explain the rather low atmospheric CO_2 concentrations that seem to have prevailed towards the end of the glacial epoch and the high concentrations during the warm period around 8000 BP (cf Section 3.3.1). Although the possible series of events proposed offers an interesting hypothesis, further analyses of data are necessary to show that the observed changes of atmospheric CO_2 during the last 20,000 years can be explained in this way. Nevertheless, the works by Broecker (1981), Knox and McElroy (1984), Sarmiento and Toggweiler (1984) and Siegenthaler and Wenk (1984) show that complex secondary effects may result, which in turn call for care when we later estimate likely changes 100 years into the future as a result of man-made emissions of CO_2 (see Section 3.8.4).

Both carbon and phosphorus are carried to the sea by rivers (cf Figure 3.1). Schlesinger and Melack (1981) and others have shown that the carbon flux is about $0.5 \cdot 10^{15}$ g C^{-1} but may be increasing due to intensified agriculture and forestry. Furthermore, in the light of the possible interdependence of the carbon and phosphorus cycles we estimate below the importance of man's increasing use of phosphorus as fertilizers in agriculture and forestry and as detergents in households and industry. The annual mining of phosphorus was about $13 \cdot 10^{12}$ g P in 1972 (Pierrou, 1976) and has presumably increased since then. At most 50% reaches the aquatic systems (lakes, rivers and the sea), probably significantly less, since part of the phosphorus used to fertilize farm land and forests is immobilized in the soils. It has been well established, however, that eutrophication of lakes and the coastal zones of the oceans has occurred because of nutrient release. Changes in the open sea cannot be seen because of the great variability of marine primary production. For an approximate calculation of the possible increased primary production in the aquatic systems we assume that 20–50% of the phosphorus input by man is used in the primary production of natural systems and that the organic matter produced in this way becomes part of the carbon cycle in the sea or is buried in the sediments. If we accept a Redfield mass-ratio C : P of 40 (Redfield, 1958), a withdrawal of $0.1–0.3 \cdot 10^{15}$ g C yr^{-1} from the atmosphere and the surface layers of the aquatic systems could result. This corresponds to 2–6% of the annual release of carbon to the atmosphere due to fossil fuel combustion at the time of the phosphorus mining estimate. This is a comparatively small but not insignificant sink that may have to be considered. Independent verification of its role is, however, desirable. Some of this carbon withdrawal might be accomplished by transfer of detrital matter to the deep sea along the continental slope as proposed by Walsh *et al.* (1981).

3.5.6 Modelling the Role of the Oceans in the Carbon Cycle

We obviously need a quantitative model to describe adequately how the distribution of carbon in the sea changes, when the atmospheric CO_2 changes due to anthropogenic emissions of CO_2 into the atmosphere. It is important that we are able to explain in this way past changes to develop confidence in projections for the future. Dynamical general circulation models of the oceans with proper inclusion of biological and chemical processes will in a long-term perspective represent the most satisfactory approach to such an analysis of the role of the oceans in the carbon cycle. So far few results of this kind are available. We shall therefore have to employ simple geochemical models which, however, are most useful in providing some important constraints on the rates of exchange and turnover of carbon in the sea.

We shall limit ourselves to a brief account of the historical development and primarily be concerned with the present status of the work in this field. Reference is made to Keeling (1973), Oeschger *et al.* (1975), Björkström (1979), Bolin *et al.* (1981), Bacastow and Börkström (1981), Hoffert *et al.* (1981), Killough and Emanuel (1981), Oeschger and Heimann (1983), Siegenthaler (1983), Peng *et al.* (1983), Emanuel *et al.* (1984) for more detailed presentations of the models developed.

The simplest model advanced is the two-box model consisting of a well-mixed surface layer and a well-mixed deep sea reservoir. Photosynthesis, detritus formation and dissolution are all assumed not to change and can therefore be ignored in the analysis. Such a model was first proposed by Craig (1957a), further developed by Bolin and Eriksson (1959) and analysed in detail by Keeling (1973). Although it still may be useful for analyses of the characteristic role of the sea in the carbon cycle, it is by now clear that this simple model cannot be used for quantitative projections (Bacastow and Björkström, 1981; Bolin *et al.*, 1983). It is for example necessary to prescribe the thickness of the surface layer to be 400–600 m, as compared with an observed value of about 75 m, to obtain response characteristics of this model similar to those of more detailed models. It is thus not possible to identify model parameters with observed features of the real ocean.

Oeschger *et al.* (1975) proposed a box-diffusion model, which describes the world ocean in terms of a well-mixed surface layer (about 75 m deep) and a deep sea reservoir, within which vertical transfer is accomplished by eddy diffusion (see Figure 3.13a). The rate of air–sea exchange is given in accordance with measurements (Section 3.4.1) and the eddy diffusivity is assigned a value of $1.3 \text{ cm}^2 \text{ sec}^{-1}$, which yields a vertical distribution of $\Delta^{14}C$ in approximate agreement with the observed global average (cf Section 3.5.3). Peng *et al.* (1983) have assigned a larger value, $1.6 \text{ cm}^2 \text{ sec}^{-1}$, for the eddy diffusivity in the layer between 75 and 700 m (i.e. in the thermocline region) to reproduce the transfer in this layer as revealed by the penetration of tritium during the period 1957–1973. A reduction to $0.5 \text{ cm}^2 \text{ sec}^{-1}$ below 700 m is then necessary to maintain balance, since also deep water formation (with a rate of $50 \cdot 10^6 \text{ m}^3 \text{ s}^{-1} = 1.6 \cdot 10^{15} \text{ g yr}^{-1}$) is

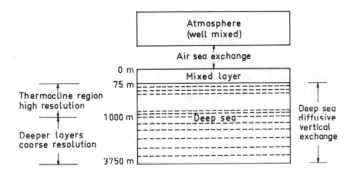

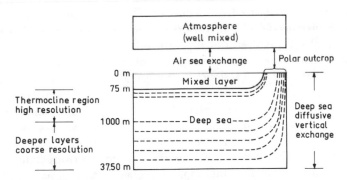

Figure 3.13 Schematic diagram showing the basic characteristics of (a) the box-diffusion model with an upper well-mixed layer and the thermocline layer and the deep sea into which transfer exclusively takes place by vertical diffusion (Oeschger *et al.*, 1975); (b) the box-diffusion model with polar outcrop in which transfer into the thermocline layer and the deep sea also is accomplished by quasi-horizontal (isopycnic) exchange from the regions of cold surface water in polar regions (Siegenthaler, 1983)

considered in approximate accordance with observed conditions (see Section 3.5.5). This version of the model has been used by Peng *et al.* (1983) in their analysis of the global carbon cycle. A similar model has also been used by Goudriaan and Ketner (1984). We shall return to a more detailed analysis of its characteristics in Sections 3.7.1 and 3.7.2. Similar work has also been presented by Soviet scientists (cf Buettner and Zacharova, 1983).

An extension of this model to include the role of intermediate and deep water formation at high latitudes has been made by Siegenthaler (1983). The deep sea is described with the aid of a series of isopycnic layers that reach the ocean surface in polar regions within an area that constitutes about 10% of the total sea surface (cf Figure 3.13b). Both cross-isopycnic transfer as described in the box-diffusion model with the aid of vertical eddy diffusivity and quasi-horizontal transfer along the isopycnic surface between the sea surface at high latitudes and deeper layers of the ocean further equatorwards are thereby accounted for. As previously, the model is calibrated by prescribing appropriate values for eddy diffusivity and air–sea exchange to account for the steady distribution of ^{14}C. To maximize the role of isopycnic exchange, horizontal mixing is assumed to be instantaneous whereby the ventilation of the deep sea becomes limited by the size of the area of deep water formation at high latitudes and the rate of air–sea exchange. Vertical (cross-isopycnic) turbulent transfer is accordingly decreased, the eddy diffusivity becoming $0.7 \text{ cm}^2 \text{ sec}^{-1}$. The model is still not able to reproduce the rather rapid penetration of bomb-produced ^{14}C (Figure 3.11) into the thermocline region (cf Section 3.5.5), nor the observed increase of atmospheric CO_2 in the atmosphere as a result of known fossil fuel emissions (cf Section 3.2). Good agreement with reality is obtained by increasing the vertical eddy diffusivity to $1.7\text{--}2.5 \text{ cm}^2 \text{ sec}^{-1}$, as obtained from a calibration using bomb-produced ^{14}C. The observed vertical steady state distribution of ^{14}C at greater depth is then not reproduced.

The models described above are quasi-linear. Their response to emissions depends on model characteristics (rates of air–sea exchange, vertical diffusion etc.) and the initial conditions assumed at the beginning of a time integration. The latter are, however, the result of past emissions. Because of the linearity of the system we may consider future changes as the sum of the response of the system at equilibrium to a given emission scenario and the decline of the initial state towards the new equilibrium state that would prevail without any future emissions (Oeschger and Heimann, 1983). The latter adjustment has been called 'the declining base line'.

We first compute the model response to exponentially increasing emissions. A reasonable scenario is then one with an e-folding time of 22.5 years (cf Chapter 2). For the two-box model we assume the well-mixed surface layer to be 75 m (cf Section 3.5.5), the $\Delta^{14}C$ value for the mixed layer to be -50% (cf Figure 3.10) and for the deep sea -160% (the approximate average for the deep sea in the Atlantic and Pacific Oceans) to determine the rate of water exchange between the mixed layer and the deep sea. We use the model parameters for the box-diffusion model as described above. Let us define the airborne fraction (α) of the emissions into the atmosphere to be the ratio between the increase in the atmosphere and the emissions (cf Section 3.7.3) and compute its asymptotic value when the models are perturbed by exponentially increasing emissions. The following values for α are obtained (cf Bacastow and Björkström, 1981; Siegenthaler, 1983).

Two-Box model	0.80
Box-diffusion model, original version	0.67
Box-diffusion model, with polar outcrop	0.61
Box-diffusion model, with polar outcrop and enhanced vertical eddy diffusivity	0.53

The differences noted are very significant, but depend markedly on the e-folding time chosen. Assuming rather an e-folding time of 50 years reduces α for the two-box model to 0.75 and similarly for the other models. These response characteristics are brought out by some recent computations by Svenningsson (1985), although his results also depend on the fact that the initial conditions imposed do not correspond to a steady state. The box-diffusion model due to Siegenthaler (1983) is used with a vertical diffusivity necessary to account for the uptake of bomb-produced ^{14}C. The model was integrated from a pre-industrial state until 1980 using estimates of past emissions due to fossil fuel combustion (Chapter 2) to provide initial conditions for integration into the future. Upper and lower bounds for future emissions as specified in Chapter 2 were prescribed and also an intermediate scenario. Although these emission scenarios were not exponential and the airborne fraction therefore changed slowly, the mean changes during the period 1980–2050 illustrate in principle the dependence of the airborne fraction on the rate of emissions, Table 3.1. The results are also compared with a similar computation by Siegenthaler (1983) using a scenario of rapidly increasing emissions.

It is obvious that the airborne fraction is markedly dependent on the rate with which the global carbon cycle is being disturbed. Although the model is simple, the general characteristics of the response as shown in Table 3.1 probably are valid.

The box-diffusion model of the oceans is undoubtedly useful for general analysis of the role of the oceans in the global carbon cycle. By consideration of the changes of the ^{13}C isotopes as observed during the last few hundred years quantitative validation of the choice of model parameters can be done (cf Section 3.8.2). Nevertheless, it is desirable to make better use of the rich data base that describes the state and circulation of the oceans, i.e.

Table 3.1   Airborne fraction of emissions when forcing the global carbon cycle using an exponential CO_2
emission, Siegenthaler (1983), (1), and with selected
emission scenarios during the period 1980 to 2050
(Svenningsson, 1985), (2)–(4)

	Emissions (10^{15} g C yr^{-1})		Average annual increase of emissions (%)	Average airborne fraction (%)
	1980	2050		
(1)	–	–	4.4	53
(2)	6.2	20	1.7	45
(3)	6.2	10	0.7	38
(4)	6.2	2	−1.6	21

measurements of temperature, salinity and currents. The increasing amount
of data on the distribution of biochemical tracers in the sea (cf Broecker and
Peng, 1982) should of course also be used systematically. Two approaches
are being pursued to develop more detailed models, which can be carefully
compared with available data.

A general circulation model of the ocean developed on the basis of dynamical principles (Hasselmann, 1982) has been used for a first simulation
of the transfer of carbon and ^{14}C (Maier–Reimer, 1984). More development
is, however, necessary to ascertain that the ocean circulation models describe
the real ocean adequately and to include biological processes.

Bolin *et al.* (1983) have approached the problem of modelling the ocean
within the context of the global carbon cycle by asking the inverse question: What steady state patterns of ocean circulation, turbulent transfer,
primary productivity, decomposition of organic detrital matter and dissolution of biogenic carbonates are required to explain the observed quasi-steady
distributions of temperature, salinity, total carbon (*DIC*), alkalinity, $\Delta^{14}C$,
oxygen and phosphorus? In a first attempt a 12-box model of the oceans was
employed, and realistic patterns of the transfer parameters were obtained.
Verification against the transient changes that have occurred in recent years
showed, however, that the spatial resolution still was inadequate to describe
these satisfactorily. An analysis of the response characteristics of the model
by using exponentially increasing emissions (e-folding time 22.5 years) as
discussed above yielded a value for the airborne fraction $\alpha = 0.74$. This
result further emphasizes that simple box models are inadequate for a more
precise analysis of the role of the oceans in the carbon cycle.

Finally, it is of interest to determine the sensitivity of the equilibrium
atmospheric CO_2 concentrations to variations of the rate of photosynthesis in the sea (cf Section 3.5.1). Viecelli (1984) has analysed this problem
and found that a decrease of the rate of photosynthesis by one per cent

would lead to an increase of the equilibrium atmospheric CO_2 concentration by 0.5–2.5%, i.e. 2–7 ppmv, and that the adjustment would lag the change of photosynthesis by merely a few years. The rather wide range of uncertainty could probably be narrowed by a more careful validation of the model against different isotopic data.

3.6 CARBON IN TERRESTRIAL BIOTA AND SOILS

3.6.1 Carbon in Biota and the Rate of Primary Production

During the last two decades many attempts have been made to determine the amount of carbon in terrestrial vegetation and the annual turnover in the form of gross primary production, respiration and detritus formation. The early survey by Bazilevich *et al.* (1970) presented estimates of the potential vegetation and primary production. Since some undisturbed grassland and forestland (about 10% of the total land surface) have been converted into farm land, particularly at middle latitudes in the Northern Hemisphere, the biomass given by Bazilevich *et al.* (1970), $1080 \cdot 10^{15}$ g C, represents a considerable overestimate of present conditions. Whittaker and Likens (1975) in a later estimate considered the state of the terrestrial biosphere in about 1950 and did not include standing dead wood. They also used a different classification. Later studies based on more data indicate, however, that also their estimate of carbon in living matter, $827 \cdot 10^{15}$ g C, most likely is an overestimate. Two studies by Duvigneaud (see Ajtay *et al.*, 1979) and Olson *et al.* (1983) give more detailed consideration of the patchiness of existing biomes particularly within tropical ecosystems. They both derive a value of $560 \cdot 10^{15}$ g C for the present (1970) carbon reservoir in the form of living terrestrial phytomass. An even lower value $445 \pm 25 \cdot 10^{15}$ g, has been obtained by Brown and Lugo (1982, 1984). They base their classification on the Holdridge (1967) life zone concept, whereby a different method is used to generalize available measurements to estimates of the biomass for whole biomes (see also Section 3.6.2).

It is difficult to compare these estimates because of the different classification systems that have been used. Table 3.2 shows an attempt to interpret the compilations by Ajtay *et al.* (1979) and Olson *et al.* (1983) in terms of the classification employed by Whittaker and Likens (1975). Although this kind of projection is approximate, it illustrates very well the difficulties encountered in any analysis of the global distribution of ecosystems.

The biomass of the forest systems has probably been markedly overestimated by Whittaker and Likens (1975), particularly so for tropical forests. It now seems clear that secondary forests contain considerably less carbon than do the natural tropical forests and their area is larger than previously assumed. Many regions which have been considered to be covered by closed

Table 3.2 Area coverage, plant carbon and net primary production for major terrestrial ecosystems according to Whittaker and Likens (1975); (2) Ajtay et al. (1979); (3) Olson et al. (1983). The amount of carbon in soil is also shown following the classification by (2) Ajtay et al. (1979), and according to (4) Schlesinger (1977), based on the classification by Whittaker and Likens (1975)

		Area (10^{12} m^2)			Plant carbon (10^{15} g C)			Primary prod. (10^{15} g C yr^{-1})			Detritus, soil (10^{15} g C)	
		(1.4)	(2)	(3)	(1)	(2)	(3)	(1)	(2)	(3)	(2)	(4)
(1)	Tropical rain forest	17.0	10.3	12.0	344	193	164	16.8	10.5	9.3	82	⎫ 288
(2)	Tropical seasonal forest	7.5	4.5	6.0	117	51	38	5.4	3.2	3.3	41	⎭
(3)	Temperate forest	12.0	7.0	8.2	174	88	65	6.7	4.6	4.9	72	161
(4)	Boreal forest	12.0	9.5	11.7	108	96	127	4.3	3.6	5.7	135	247
(5)	Woodland, shrubland interrupted woods	8.5	4.5	12.8	23	24	57	2.7	2.2	4.6	72	59
(6)	Savannah	15.0	22.5	24.6	27	66	49	6.1	17.7	10.7	264	63
(7)	Temperate grassland	9.0	12.5	6.7	6	9	11	2.4	4.4	2.6	295	170
(8)	Tundra, alpine	8.0	9.5	13.6	2	6	13	0.5	0.9	1.8	121	163
(9)	Desert, semidesert	18.0	21.0	13.0	6	7	5	0.7	1.3	0.9	168	104
(10)	Extreme desert	24.0	24.5	20.4	0	1	0	0.0	0.1	0.5	23	4
(11)	Cultivated land	14.0	16.0	15.9	6	3	22	4.1	6.8	12.1	128	111
(12)	Swamps, marshes and coastal land	2.0	2.0	2.5	14	12	7	2.7	3.3	3.6	⎫ 225	145
(13)	Bogs and peatland	–	1.5	0.4	–	3	1	–	0.7	0.2	⎭	–
(14)	Lakes and streams	2.0	2.0	3.2	0	0	1	0.4	0.4	0.4	0	0
(15)	Human areas	–	2.0	–	–	1	–	–	0.2	–	10	–
	Total	149.0	149.3	151.1	827	560	560	52.8	59.9	60.6	1636	1515

Data from Whittaker and Likens (1975) have been directly reproduced. The data from Ajtay et al. (1979) 'Mangrove forests' have been included as tropical humid (rain) forest and 'forest plantation' as temperate and boreal forests. The classification used by Olson et al. (1983) deviates from the one used in the table. Their 'interrupted woods' have here been included in 'tropical, temperate or boreal forests' if classified as 'second woods and field mosaic' and 'Tropical savannah and woodland' has been included in 'Savannah' in the present table.

forests should rather be classified as partly closed. In summary, we can conclude that the amount of carbon in biota in tropical forests is 170 ± 70 · 10^{15} g C, compared with 460 · 10^{15} g C given by Whittaker and Likens (1975).

The total amount of plant carbon and the net primary production are almost identical in the compilations of Ajtay *et al.* (1979) and Olson *et al.* (1983), 560 · 10^{15} g C and 60 · 10^{15} g C yr^{-1} respectively. Although the areas covered by forests and woodland (items (1) to (5)) differ considerably, both with regard to the distribution between biomes and total size, these two surveys yield almost the same total amounts of plant carbon (450 · 10^{15} g C) and total net primary production (24 · 10^{15} g C yr^{-1} and 28 · 10^{15} g C yr^{-1}), i.e. about 80% and 40–50% respectively of the global values. The average residence time for carbon in forest systems is 16–20 years, but the average age of trees is at least twice as large, since less than half of the net primary production is transformed into cellulose. The average residence time for carbon in plant material outside the forest systems is only about 3 years.

3.6.2 Carbon in the Soil

An attempt to analyse in some detail the amount of soil carbon distributed according to the ecosystem classification introduced by Whittaker and Likens (1975) has been given by Schlesinger (1977, 1984) and yields a total inventory of 1515 · 10^{15} g C. The result is shown in the last column of Table 3.2. A similar analysis has also been given by Ajtay *et al.* (1979) yielding quite a different distribution between biomes although the total amount of soil carbon, 1636 · 10^{15} g C, is only marginally larger. Post *et al.* (1983) have based their estimates on Holdridge's (1967) classification scheme and the integration to obtain a global inventory was obtained with the values shown in Figure 3.14. The result, 1395 · 10^{15} also agrees rather well with the other estimates. Buringh (1984) has approached the problem differently and used a classification scheme based on soil types. The value for the total amount deduced is, however, almost the same: 1477 · 10^{15} g C. The estimates by Bazilevich (1974) 1392 · 10^{15} g C (excluding peat), and particularly by Baes *et al.* (1976), 1080 · 10^{15} g C, are, on the other hand, somewhat lower. The agreement between these different total estimates is, however, adequate for our present purposes. The main uncertainty is due to poor assessments of the peatlands of the world.

Both the assessments given in Table 3.2, and particularly those by Ajtay *et al.* (1979), indicate slower rates of soil decay in cold climates reflected in larger accumulation of soil carbon (per unit area) in boreal forest and temperate grassland than in tropical ecosystems. It should be emphasized, however, that only a small amount (a few per cent or less) of the detritus received annually by the soil reservoir remains there for any appreciable

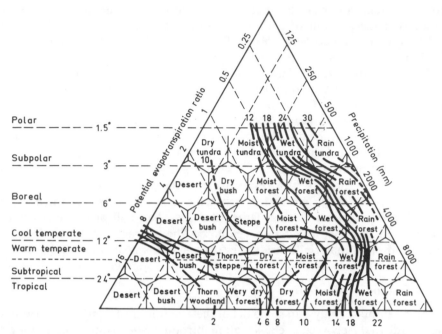

Figure 3.14 Contours of soil carbon density plotted in Holdridge's (1967) scheme for world life zone classification. The temperature and precipitation uniquely determine a life zone and associated vegetation (from Post *et al.*, 1983). With knowledge about the area covered by the different vegetation classes the global inventory can be deduced

period of time. Most of the dead organic matter is oxidized into CO_2 within a few years. Schlesinger (1977) has emphasized that in chernozem grassland soil 98% of the litter has a turnover time of merely 5 months, but that the remaining 2% on the average stays in the soil 500–1000 years. This characteristic feature of the process of soil formation is also reflected in the fact that the ^{14}C-age of soils at middle latitudes is a few hundred years to more than a thousand years.

The rate of decomposition of organic matter when virgin land is claimed for agriculture is, however, quite different. Vitousek (1983) is of the opinion that as much as 50% of the organic carbon in agricultural soil in North America may have been lost by oxidation, since it was claimed before or at the turn of the last century. A more detailed assessment of this problem will be given in the following section.

3.6.3 Changes of the Amount of Carbon in Terrestrial Ecosystems

Considerable changes of the terrestrial ecosystems have occurred during the last 200 years due to man's rapidly expanding activities. When forests or grassland were converted into farmland, organic matter, i.e living matter in plants and dead organic matter in humus and soils, was oxidized and emitted as CO_2 into the atmosphere. Some elemental carbon may also have been buried in the soil in the form of charcoal (as residues from forest burning) and in this way withdrawn from the rapid circulation in the carbon cycle (Seiler and Crutzen, 1980).

The most detailed assessment of past and ongoing changes of the terrestrial ecosystems has been made by Moore *et al.* (1981), Houghton *et al.* (1983) and Houghton (1984). By using historical records on changing land use a dynamic book-keeping model has been formulated accounting for changes that follow on disturbance, i.e. changes of vegetation and soil and estimates of products removed from the field or the forest. The amounts of carbon in these different components change because of regrowth and decay, which vary with geographical region and vegetation type. The 14 ecosystems on land (cf Whittaker and Likens, 1975, Table 3.2) are considered separately and also their distributions between 10 geographical regions. In total 69 region-specific ecosystems are defined in this way. Six different kinds of disturbances are considered:

(1) harvest followed by regrowth
(2) clearing of natural ecosystems for agriculture
(3) clearing of natural ecosystems for pasture
(4) abandonment of agricultural land
(5) abandonment of pasture
(6) afforestation

Characteristic response curves for carbon in vegetation and in soil for each of these disturbances and for all regions have been determined on the basis of available data regarding ecosystem behaviour. Figure 3.15 shows two examples of changes in response to clearing for agriculture. The model keeps track of area, age and carbon content as a function of time for each disturbed ecosystem based on historical records for the regions concerned. The use of fuel wood is also accounted for. Although computations were begun in 1700, the period until 1860 has been considered as a time of adjustment of the model and has not been used in the following analysis.

The results of this assessment of change are obviously dependent on the initial assumptions of the size of the areas covered by different ecosystems and upon the amounts of carbon per unit area that are assumed to characterize them, the rates and type of disturbances and their change over time following disturbance. The data compiled by Whittaker and Likens (1975) have been used and as we have seen in Section 3.6.1 these probably are overestimates, particularly for the tropics. The results are also sensitive to assumptions of whether virgin or secondary forests are being cut in defor-

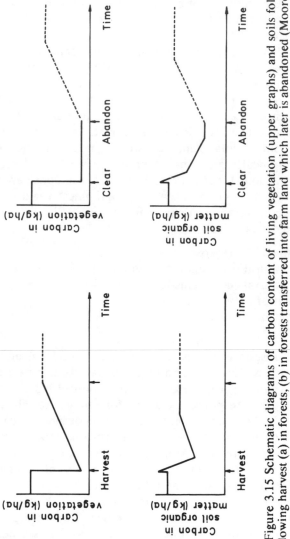

Figure 3.15 Schematic diagrams of carbon content of living vegetation (upper graphs) and soils following harvest (a) in forests, (b) in forests transferred into farm land which later is abandoned (Moore *et al.*, 1981)

estation. The estimates (Houghton *et al.*, 1983) now seem too large (cf Detwiler *et al.*, 1984). Until recently Soviet forests have been considered to be a net carbon sink. It seems now clear, however, that the opposite may rather be true. Forests at mid-latitudes elsewhere in the Northern Hemisphere, on the other hand, most likely contain more carbon now than earlier this century. A similar analysis has also been carried out by Richards *et al.* (1981), but was limited to CO_2 release due to conversion of land to agricultural uses. Their value for release of carbon since 1860, $62 \cdot 10^{15}$ g C, therefore probably is a considerable underestimate.

Seiler and Crutzen (1980) have proposed that terrestrial carbon is withdrawn from rapid circulation in the global carbon cycle by the formation of charcoal in the burning of grassland and forests ($0.8 \pm 0.5 \cdot 10^{15}$ g C yr^{-1}). This figure probably is a considerable overestimate because of the use of Whittaker and Likens's (1975) estimates of the carbon content in tropical ecosystems. It also seems possible that the amounts of charcoal formed in forest burning are overestimates, since the accumulation in regions of shifting cultivation would have to be considerably larger than observed, if their estimates were correct (cf Schlesinger, 1977).

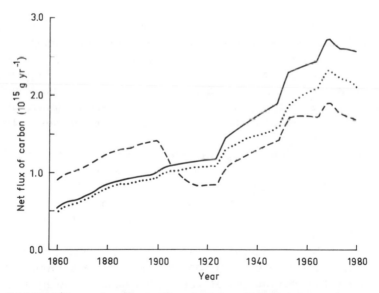

Figure 3.16 Emissions of carbon to the atmosphere due to deforestation and changing land-use according to three different procedures for analysis of past data (from Houghton, 1984; Bolin *et al.*, 1985) The solid line is based on biomass data from Whittaker and Likens 1975); the dotted line is based on tropical biomass derived from timber volumes reported by FAO (1981); the dashed line depends on another scenario for the transfer of forest land and grassland into agriculture

The historical records of changing land use are too uncertain to give a reliable picture of the change during the period for our consideration. Attempts have been made to relate land use changes to population changes (Revelle and Munk, 1977; Houghton *et al.* 1983). Considerable uncertainty still remains. Figure 3.16 shows some patterns of change as deduced by various means. We conclude on the basis of the studies referred to above that the net CO_2 release to the atmosphere during the period 1860 to 1980 totalled $(150 \pm 50) \, 10^{15}$ g C and that the release in 1980 was in the range 1.6 $\pm$ 0.8 10^{15} g C yr^{-1} (cf Bolin *et al.*, 1985). These direct assessments will be compared later with results obtained with carbon cycle models.

It should be emphasized that the possible response of photosynthesis and decay in terrestrial ecosystems to increasing atmospheric CO_2 concentrations and emission of pollutants such as SO_2 and NO_x, possibly both fertilizing and acidifying the soils, has not been considered in this context. It is likely that such secondary changes are significant but their magnitudes are poorly known (see further Section 3.6.4 and Chapters 9 and 10).

3.6.4 Modelling Changes of Carbon Storage and Isotope Composition in Terrestrial Ecosystems

In order to analyse past changes of atmospheric CO_2 concentrations and to project future changes we need to formulate a dynamic model of the terrestrial ecosystems to be combined with a model for the role of the oceans in the carbon cycle. We wish to consider the anthropogenic impacts as described in the previous section, and we should include the possible enhanced growth due to the increasing CO_2 concentration in the atmosphere. We must also be able to account for the changes of isotopic composition that have occurred.

The change of carbon storage in terrestrial ecosystems as a response to environmental changes, particularly to increasing atmospheric CO_2 concentration, is poorly understood (cf Chapter 10) and only very simple and crude models have been advanced in attempting to determine its importance for past changes of the carbon cycle as observed. Keeling (1973) and later Revelle and Munk (1977) have proposed that the change of the rate of photosynthesis might be put proportional to a power, β, of the percentage increase of the atmospheric CO_2 concentration and also to the amount of synthesizing matter. Further, the rate of plant matter death is usually assumed to be directly proportional to the amount of living matter, the proportionality coefficients being different for photosynthesizing matter (leaves, grass, etc.) and structural matter (wood). They are usually determined from crude estimates of turnover times for these components of the biomass. An increase of the net primary production is then gradually balanced by enhanced plant matter death and similarly decomposition of detrital matter. The total carbon storage increases slowly when atmospheric CO_2 concentrations increase. This approach to model the ecosystem response is not complete, since it ignores the role of important ecosystem characteristics. Although the rate of photosynthesis probably increases when atmospheric CO_2 increases, it does not necessarily follow that carbon storage in the form of biomass would also increase. The latter might to a considerable degree be determined by climate and by soil properties, particularly availability of nutrients. The turnover rate might primarily increase, but less so the biomass. Until more appropriate models of ecosystems have been validated, the inclusion of the CO_2 fertilization effect is hardly justified except in attempting to assess the general implications of such effects (see further Chapters 9 and 10).

Peterson and Melillo (1985) have attempted to estimate what the increase of CO_2 uptake might be as induced by CO_2 and nutrient fertilization. They conclude that at most 5%, i.e. 0.3–0.4 • 10^{15} g C yr^{-1}, of the net annual release to the atmosphere is being transferred into the terrestrial ecosystems by increased photosynthesis. It is not known if this will lead to increased storage in the long term or not.

As has been emphasized before the distributions of ^{13}C and ^{14}C and their changes due to anthropogenic disturbances give valuable information about the dynamic behaviour of the global carbon cycle. For their interpretation global models are required in which the terrestrial ecosystems are properly described. An early attempt of this kind was made by Machta (1972). He assumed that terrestrial ecosystems could be described by merely two compartments, 'long-term biosphere', containing 1000 • 10^{15} g C and having a characteristic adjustment time of 40 years, and 'short-term biosphere' containing about 60 • 10^{15} g C with a turnover time of 2 years. In the light of our present knowledge this simple model is not adequate. Revelle and Munk (1977) include the biospheric growth factor due to increase of atmospheric CO_2, but retain otherwise the main features of Machta's model.

Peng *et al.* (1983) use four reservoirs in direct exchange with the atmosphere as shown in Figure 3.17a. The exchange is formulated by first order processes and the mass (M) and characteristic exchange times (τ) for the different terrestrial compartments are shown in the figure. As seen the total amount of plant carbon (640 • 10^{15} g C) is somewhat larger than the recent estimates given in Table 3.2, while the turnover times agree well. The soil carbon is considerably less than the assessments by Schlesinger (1977) and the turnover rate of soils does not depict the characteristic features described in Section 3.6.2. The penetration of ^{13}C and ^{14}C into the soil is too rapid in the model with the parameter values proposed.

A dynamically more correct model of the terrestrial ecosystems has been proposed by Emanuel *et al.* (1984), cf Figure 3.17b. Detritus and soil carbon are formed from plant material and a distinction is made between rapidly

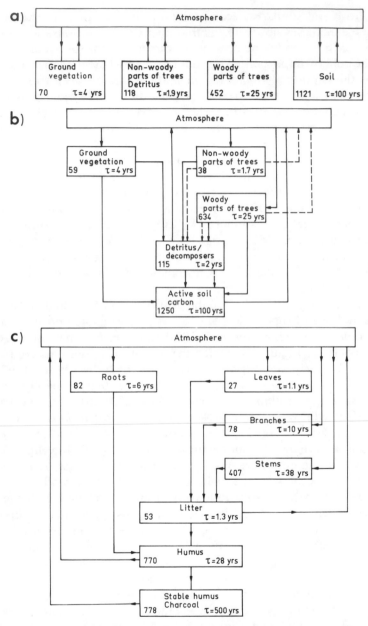

Figure 3.17 Schematic diagrams showing simple models for the terrestrial biosphere according to (a) Peng *et al.* (1983), (b) Emanuel *et al.* (1984) and (c) Goudriaan and Ketner (1984). The biomass (in 10^{15} g C) and average turnover time for carbon (in years) are shown for the different reservoirs. In the last model average characteristics for the six ecosystems when considered simultaneously are shown

overturning detritus and a slow soil reservoir. The sizes and response characteristics of the three plant reservoirs as assumed by Emanuel *et al.* (1984) are almost identical to those used by Peng *et al.* (1983), while the model of Emanuel *et al.* (1984) permits a more realistic response of the uptake of ^{13}C and ^{14}C by the soil. The values chosen in these model experiments imply, however, rather small differences between their general characteristics.

A considerably more elaborate model has been developed by Goudriaan and Ketner (1984); cf Figure 3.17c. Their assumptions regarding the main ecosystem structure are similar to those of Emanuel *et al.* (1984), but are more detailed. They distinguish between six different ecosystems (tropical forest, temperate forest, grassland, agricultural area, human area and tundra/semidesert, each one of them modelled according to the scheme shown in Figure 3.17c). The total amounts of litter, humus and stable humus/charcoal as assumed are quite different from the estimates given in the previous section. The assumption of 28 years average turnover time for humus furthermore represents a quicker response of the soils to external disturbances. Goudriaan and Ketner (1984) finally assume a very significantly enhanced impact on plant growth due to the increasing atmospheric CO_2 concentrations (the integration presented yields a biota uptake that is $2.0 \cdot 10^{15}$ g C yr^{-1} larger in 1980 than without CO_2 enhancement) and similarly a considerable withdrawal of carbon by charcoal formation ($0.9 \cdot 10^{15}$ g C yr^{-1} in 1980) which together almost completely prevent an increase of CO_2 in the atmosphere due to biomass burning and changing land use. Data are not available that validate the model in these regards.

In applying models for terrestrial ecosystems such as those outlined above to changes in ^{13}C and ^{14}C isotope distributions it is essential that the fractionation in the assimilation process is dealt with properly. It is different depending on whether diffusion through the stomata or carboxylation is of prime importance in the rate determining process. This in turn is affected by climatic conditions, site characteristics, and varies in the course of the growing season. Peng *et al.* (1983) discuss in some detail the ^{13}C records which are used in validating the model and select trees to obtain a homogeneous set. Stuiver *et al.* (1984) also emphasize the difficulty in obtaining a reliable data set. In modelling the process it is assumed that the fractionation in the assimilation process is consistent with the 19–20‰ difference observed between ^{13}C for atmospheric CO_2 and that for wood. Fractionation of ^{14}C is assumed to be twice that of ^{13}C (cf also Section 3.2.3).

Finally, it should be emphasized that the models that have been described briefly above have not been tested against independent ecosystem data. As a matter of fact it is difficult to see how this could be done. The approach is rather to explore if global models, including a description of the terrestrial ecosystems as outlined, are consistent with data available for validation as described in early sections of this chapter.

3.7 GLOBAL CARBON CYCLE MODELLING

3.7.1 Model Features

Most early attempts to model the global carbon cycle are inadequate because data on the sizes and response characteristics of the various reservoirs were not determined satisfactorily. Also data on changes of ^{13}C and ^{14}C were rarely used as additional checks on the validity of the treatment of the transfer of total carbon.

. Machta (1972) adopted a 2-box model for the oceans and similarly a 2-box model for the terrestrial biota, thereby distinguishing between short-lived and long-lived biota. Only by assuming the surface mixed layer to be 300 m deep was reasonable agreement achieved with the observed increase of atmospheric CO_2 between 1957 and 1970.

Siegenthaler *et al.* (1978) combined the box-diffusion model (see Section 3.5.6) with a simple one-compartment model for the terrestrial biota. In their analysis they also analysed model predictions of ^{13}C and ^{14}C changes due to anthropogenic disturbances. Stuiver and Quay (1981) modelled the Suess effect, i.e. the decrease of ^{14}C due to the emission of fossil CO_2, using a similar model. A more detailed analysis was later presented by Peng *et al.* (1983), in which case the diffusion model for the ocean was modified as described in Section 3.5.6 and a more detailed treatment of the terrestrial ecosystems as outlined in Section 3.6.4 was included.

Emanuel *et al.* (1984) also pursued the same problem. Although their treatment of the terrestrial ecosystems is somewhat more sophisticated, their model suffers from the fact that the oceans are described by merely two reservoirs. By assuming a rather deep surface mixed layer they, however, arrive at general conclusions similar to those deduced by Peng *et al.* (1983). The analysis by Goudriaan and Ketner (1984) differs in important respects with regard to the treatment of the terrestrial ecosystems. We shall return to their results after having considered the analyses by the others in some detail. It is important to emphasize that for the time being the following processes are not being considered adequately or not at all.

— Charcoal formation during biomass burning, which might constitute a significant sink in the cycle (included by Goudriaan and Ketner, 1984).

— Enhancement of carbon storage in terrestrial ecosystems due to increased atmospheric CO_2 concentrations (included by Goudriaan and Ketner, 1984).

— Fertilization of terrestrial ecosystems due to deposition of nutrients (nitrogen and phosphorus compounds).

— Increase of carbon flux in rivers.

— Increase of the rate of sedimentation of organic carbon in lakes and oceanic coastal waters.

— Changes of the rate of photosynthesis in the surface mixed layer and of the exchange of water within the sea.

— Dissolution of carbonate sediments at or below the lysocline in the sea.

3.7.2 Simulation of Past Changes

Peng *et al.* (1983) in their study begin by attempting to reproduce the ^{14}C changes in the atmosphere that have been recorded in tree rings (cf Figure 3.5) before bomb-produced ^{14}C was emitted into the atmosphere (cf also Oeschger *et al.*, 1975). If we assume that the observed decrease of $\Delta^{14}C$ is due to the emission of ^{14}C free CO_2 from fossil fuel combustion, we can deduce these with due regard to the exchange with both terrestrial biota, soils and the oceans. The release of CO_2 to the atmosphere by deforestation and expanding agriculture (Section 3.6.3) may be ignored in this context, since $\Delta^{14}C$ of living biota and soils is only slightly below the atmospheric value and since therefore the ^{14}C content of the atmosphere is not significantly diluted . The solid curve in Figure 3.5 has been obtained by such model calculations. There is reasonable agreement with observations. We note, however, that significant variations of $\Delta^{14}C$ during the 19th century are not reproduced (and could not be, because model computations necessarily yield a monotonically decreasing ^{14}C concentration). The observed decrease in recent decades also seems more rapid than what the model yields. Some of the deviations may be due to variations in the rate of ^{14}C production by cosmic rays (cf Stuiver and Quay, 1981).

Bomb-produced ^{14}C emitted to the atmosphere after 1954 raised $\Delta^{14}C$ for atmospheric CO_2 rapidly (Figure 3.6) and a considerable transfer into the terrestrial biota and the oceans occurred. We may compare the integrated flux to the ocean 1954–1973 as deduced by the model with the increase as observed by the GEOSECS measurements 1972–1974. The observed increase of $\Delta^{14}C$ in surface water agrees within about 15% with computed values (cf Broecker *et al.*, 1980). We may conclude that the ^{14}C data are in general agreement with the global carbon cycle model developed by Peng *et al.* (1983), although the way the global ocean data have been averaged for validation has not been justified.

We may next ask the question: How much carbon must have been emitted into the atmosphere by fossil fuel combustion, deforestation and changing land use with known ^{13}C concentrations (cf Section 3.3.2) to explain the observed ^{13}C changes since early last century? Before considering this problem, it should be emphasized that in this way we can only determine the net exchange between the atmosphere and the terrestrial biosphere. The question was preliminarily analysed already by Siegenthaler *et al.* (1978) and Wagener (1978). Later Peng *et al.* (1983), Emanuel *et al.* (1984) and Stuiver *et al.* (1984) have used their somewhat different models to answer this question. Peng *et al.* (1983) use the ^{13}C data given by Freyer (1979) and Freyer and Belacy (1983), which probably show too large changes (cf 3.3.2), and so do Emanuel *et al.* (1984). Stuiver *et al.* (1984) on the other hand employ their own data, which by and large agree with the direct measurements reported by Friedli *et al.* (1984). Figure 3.18 shows the total emissions and also those from forests and soils alone according to Peng *et al.* (1983), which have been obtained by subtraction of the fossil emissions. Peng *et al.* (1983) estimate the emissions from forests and soils during the last 120 years to $260 \cdot 10^{15}$ g C, while Emanuel *et al.* (1984) and Stuiver *et al.* (1984) in their calculations obtain $230 \cdot 10^{15}$ g C and $150 \cdot 10^{15}$ g C respectively, the last result being the most plausible one in the light of the discussion in Section 3.6.3. The Peng *et al.* (1983) model has the most efficient exchange with the oceans, which means more rapid equalization between the different reservoirs. Larger emissions are therefore needed to explain the observed ^{13}C change in the atmosphere. Figure 3.18 further shows that release from deforestation and soils should have reached a maximum at the end of the

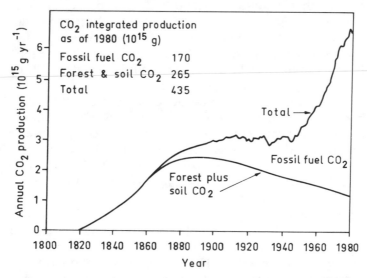

Figure 3.18 Emissions of carbon dioxide into the atmosphere due to deforestation and changing land use, derived from tree ring based atmospheric δ^{13}C concentrations (see Figure 3.4). The result is obtained by deducting the fossil fuel CO_2 input history (also shown) according to Rotty (1981) from the total emissions computed using the carbon cycle model developed by Peng *et al.* (1983)

last century, $2 \cdot 10^{15}$ g C yr^{-1}, and then decreased to about $1 \cdot 10^{15}$ g C yr^{-1}. Emanuel *et al.* (1984) do not find such a time course, and it is hardly supported by the direct estimates that were summarized in Section 3.6.3.

Having in this way deduced approximately the total emissions as a function of time, we may finally analyse the past changes of atmospheric CO_2 concentrations that the models predict if subject to such external disturbances. It is useful to impose the constraint on the solution that concentrations in 1957 agree with the Mauna Loa observations, i.e. 315.5 ppmv (cf Section 3.3.1). Peng *et al.* (1983) deduce CO_2 concentrations early last century to have been 243 ppmv and Emanuel *et al.* (1984) derive an even lower value (235 ppmv), both well below the most plausible value of 275 ± 10 ppmv (see Section 3.3.1). Also the increase between 1958 and 1979 is more rapid in the models (27 ppmv for Peng *et al.* (1983)) than the observed value (21 ppmv). Stuiver *et al.* (1984) do not present a computation of this kind, but it is questionable if agreement with the data would be achieved unless the emissions from forests and soils during recent decades as summarized in Section 3.6.3 were reduced significantly.

The carbon cycle model developed by Goudriaan and Ketner (1984) yields quite different results because of the uptake due to CO_2 fertilization of the terrestrial ecosystems and charcoal formation. In spite of the fact that the emissions due to deforestation and expanding agriculture during the period 1860–1980 are assumed to have been considerably larger than given in Section 3.6.3, the net global decrease during this period is small (totally only 15 $\cdot 10^{15}$ g C while the atmospheric CO_2 concentration only increases from 290 to 339 ppmv). The model has not been tested against ^{13}C and ^{14}C changes in the way Peng *et al.* (1983) and Emanuel *et al.* (1984) have done, nor are the assumptions about the response characteristics of the terrestrial ecosystems supported by independent data. Although it is of interest to explore the implications of different carbon cycle models in this manner, the model presented by Goudriaan and Ketner (1984) cannot be used with confidence for projections of future changes of atmospheric CO_2 concentrations.

It is difficult to explain the discrepancies between the different models without rather extensive sensitivity analyses which are not available, although they might be explained, if the observed changes of $\delta^{13}C$ in the atmosphere during the last 120 years are overestimates of the real changes. It should be emphasized, however, that no detailed verification of the carbon cycle models has been made. They have rather been tested with regard to changes of integral properties as revealed by the temporal changes in the atmosphere and the well-mixed surface layer of the oceans. The verification of the ocean submodel using basic oceanographic data (cf Section 3.5.6) might clarify the role of the oceans, which obviously would be of great help in trying to resolve the remaining inconsistencies of the present global carbon models, often referred to as the problem of 'the missing sink'. Some maintain that the emissions due to deforestation and changing land use have been considerably overestimated, while others are of the opinion that the processes listed in Section 3.7.1, which mostly have not been included, must not be ignored.

3.7.3 The Concept of Airborne Fraction

Before a discussion of how to make use of the carbon-cycle models to assess likely future atmospheric CO_2 concentrations as a result of given emission scenarios, it is useful to express past changes in terms of the so-called airborne fraction. A comparison with the projections of future atmospheric concentration presented by Nordhaus and Yohe (1983) then also becomes possible (cf also Section 3.5.6).

First it is important to distinguish clearly between total airborne fraction (α), which is the ratio between increase of CO_2 in the atmosphere $\Delta (CO_2)_a$ and total emissions to the atmosphere ($\Delta (CO_2)_{tot}$:

$$\alpha = \frac{\Delta(CO_2)_a}{\Delta(CO_2)_{tot}}$$

and marginal airborne fraction (α_m), which has traditionally been defined as the ratio between increase of CO_2 in the atmosphere and emissions of CO_2 due to fossil fuel combustion ($\Delta CO_2)_{ff}$:

$$\alpha_m = \frac{\Delta(CO_2)_a}{\Delta(CO_2)_{ff}}$$

The latter concept is useful when comparing model computations of the ocean response to emissions (usually a scenario of fossil fuel emissions, cf Section 3.5.6) without the complicating influence of changes of the terrestrial biosphere. The total airborne fraction, however, determines the rate of CO_2 increase in the atmosphere and is useful in predicting likely future CO_2 concentrations in the atmosphere.

Many projections of future atmospheric CO_2 concentrations have been based on simple extrapolations of past changes (cf WMO, 1981; Nordhaus and Yohe, 1983) usually assuming the most likely future value of α to be about 50%. We therefore first estimate the value of α in the past on the basis of available measurements and choose to do so for the two periods 1860–1980 and 1958–1982 (cf Bolin *et al.*, 1985).

Fossil fuel emissions are well known (Chapter 2) and so are recent changes in atmospheric CO_2 concentrations. We do not know accurately pre-industrial atmospheric concentrations nor the emissions due to deforestation and changing land use. It is important to assess the uncertainty of past values for α due to these uncertainties. Figure 3.19 shows the range of values obtained depending on the atmospheric CO$_2$ concentrations in 1860 and the net emissions from the terrestrial biosphere since then due to deforestation and changing land use. Accepting the ranges to have been 275 $\pm$ 10 ppmv and $(150 \pm 50) \cdot 10^{15}$ g C we find $\alpha = 43^{+17}_{-11}$. Also for the period 1958–1982 the net emissions from forests and soils due to man's impact are uncertain. Figure 3.20 shows the dependence of α on the magnitude of these emissions. Accepting a value $1.6 \pm 0.8 \cdot 10^{15}$ g G yr^{-1} (cf Section 3.6.3) we obtain $\alpha = 39^{+7}_{-5}$. The most plausible estimates are below the commonly accepted value of 50%, although not excluded in the former case.

We may also compute the airborne fraction that the global carbon cycle models yield, in response to the external disturbances as described above. It should be recalled that the models by Peng *et al.* (1983) and Emanuel *et al.* (1984) consider the oceans to be the only net sink for the anthropogenic emissions into the atmosphere, which on the other hand is not the case in

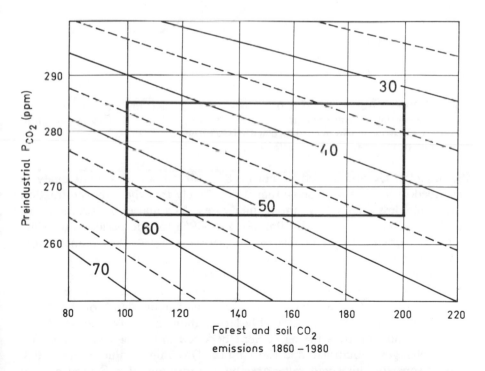

Figure 3.19 The average airborne fraction of CO$_2$ emissions into the atmosphere 1860–1980 as dependent on the emissions due to deforestation and changing land use and pre-industrial atmospheric CO$_2$ concentrations. Fossil fuel emissions are assumed to have been as reported by Rotty (1981)

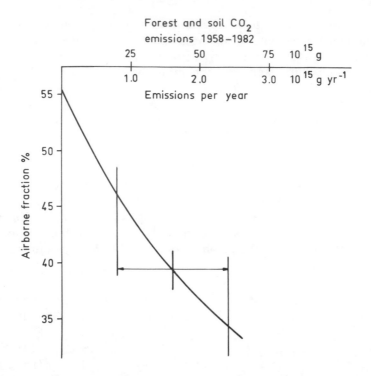

Figure 3.20 The average airborne fraction 1958–82 as dependent on the emissions due to deforestation and changing land use during this period. Change of atmospheric CO_2 assumed in accordance with Bacastow and Keeling (1981) and fossil fuel emissions as given by Rotty 1981)

the model of Goudriaan and Ketner (1984). Already the sensitivity analyses described in Section 3.5.6 show that it is unlikely that the airborne fraction α would be less than 50% in the former cases. As a matter of fact the result obtained by Peng et al. (1983) implies $\alpha = 49\%$ for the period 1860–1980 while that obtained by Emanuel et al. (1984) is 55%. The comparatively small value obtained in the former study means that the modifications of the ocean model by changing the vertical eddy diffusivity and considering deep water formation separately have increased the ocean uptake capacity. Emanuel et al. (1984) have chosen a value for the depth of the mixed layer (≈ 300 m) which is considerably larger than in reality, and which decreases α and improves the agreement between model behaviour and observed changes. Goudriaan and Ketner (1984) obtain a value of about 0.3. The reasons for this small value are the assumptions of enhanced growth due to the increasing atmospheric CO_2 concentrations and charcoal formation. As previously pointed out (Section 3.6.3), it is questionable if these processes are dealt with quantitatively in an adequate way.

3.8 PROJECTIONS OF FUTURE ATMOSPHERIC CO_2 CONCENTRATIONS

3.8.1 Using the Concept of Airborne Fraction for Extrapolation

It has been common in the past to estimate future atmospheric CO_2 concentrations by simply assuming that the airborne fraction of projected emissions remains constant. Usually a value of 50% has been assumed. On the basis of an assumption that emissions will increase to $13.6 \cdot 10^{15}$ g C yr^{-1} in 2025 the first joint ICSU/UNEP/WMO CO_2 assessment (WMO, 1981) estimated in this way that atmospheric CO_2 concentrations would reach a value between 410 and 490 ppmv in 2025. Net emissions of $50 \cdot 10^{15}$ g C from the terrestrial ecosystems were assumed also to occur during this period.

A similar approach has been used by Nordhaus and Yohe (1983). It should be noted, however, that their definition of the airborne fraction is different from the one used here because of the inclusion of a seepage factor to account separately for the slow absorption of atmospheric CO_2 into the deep oceans. Furthermore, the values quoted in their paper, 38, 47, and 59% for the low, medium, and high cases respectively, are incorrect and were in reality about 10% higher (personal communication). On the basis of their published graphs showing the total emissions and associated increases of atmospheric CO_2 concentrations for the period 1975–2100 (their Figures 2.3 and 2.4) we compute that the airborne fraction (as defined here), which they have used implicitly during this period, rather is between 55 and 75%, much above what now seems reasonable. Nordhaus and Yohe (1983) have not considered future emissions due to deforestation and changing land use. Further, the airborne fraction depends on the rate of CO_2 emissions into the atmosphere in another way than implied by their formulation of the oceanic uptake. Although their approach is of considerable methodological interest, the results of their computations cannot be used directly for projections of future atmospheric CO_2 concentrations.

3.8.2 The Use of Carbon Cycle Models

It was concluded in Section 3.7.3 that present models for the global carbon cycle are unable to account fully for past changes. Although we thus are not yet able to use such models explicitly for more precise projections of future changes of atmospheric CO_2 concentrations as a result of given emission scenarios, they are useful as a guide in the following discussion. It is important to consider possible reasons for model discrepancies (cf Section 3.7.1). We shall analyse in turn

— Change of the buffer factor, ξ, for sea water due to carbonate sediment dissolution.

— Charcoal formation during biomass burning.

— Increase of organic carbon flux to the sea by rivers.

— Sedimentation of organic carbon.

— Fertilization of terrestrial ecosystems due to increase of atmospheric CO_2 and to nitrogen and phosphorus deposition.

It is possible that the increase of atmospheric CO_2 in the past has been limited by these (and possibly other) processes. We assess qualitatively their role and whether it is likely that they will increase or decrease in the future.

(1) The buffer factor will slowly increase with time (cf Section 3.4.2) and may reach values of 13–17 (i.e. increase by 30–50%) if atmospheric CO_2 concentrations reach 600–1200 ppmv provided the alkalinity of sea water does not change. Such changes are included in present models. Since, however, the acidity of sea water simultaneously increases, some dissolution of calcium carbonate in the sediments may take place, whereby the alkalinity is increased and the buffer factor increase is slowed down. It does not seem likely that these processes can explain the present rather small α-value that seems to prevail ($\approx 40\%$). The dissolution of calcium carbonate would probably not prevent ξ and accordingly not α either from increasing slowly in the future. The effect due to carbonate dissolution will only become of importance very gradually since transfer of acidified water down to below the lysocline and return flow of more alkaline water back to the surface is required before the rate of oceanic uptake of CO_2 is influenced.

(2) Charcoal formation in the burning process might be contributing to the rather low α-values observed in the past (cf Seiler and Crutzen, 1980; Goudriaan and Ketner, 1984). To the extent this is so, α would increase if destruction of terrestrial biomass by burning would decline in the future and vice versa.

(3) There may have been an increase of dissolved and particulate organic matter in rivers in recent decades. Since most such compounds are rapidly oxidized, it does not seem likely that such a transfer of organic matter from land to sea would be a significant sink for excess atmospheric CO_2. Even if these organic compounds were decomposed slowly, as is the case for some dissolved organic matter in the sea, it is probable that their decomposition in the soil would also have been slow. Therefore this pathway does not seem important unless sedimentation has become increasingly important (see further below).

(4) Increasing eutrophication of lakes and marine coastal waters has been observed and some net withdrawal of carbon from the atmosphere and the surface layers of the sea may occur because of increased sedimentation. Undoubtedly this process would represent a sink for excess atmospheric CO_2.

Reductions of the release of nutrients to the environment would decrease its importance.

(5) It seems likely that photosynthesis has been enhanced because of increasing atmospheric CO_2. This is most probably so in agricultural areas where there is adequate nutrient supply, but also in natural terrestrial ecosystems the improved water use efficiency might have increased the formation of organic matter (cf Chapters 9 and 10). It does not necessarily follow, however, that storage is increased (cf Section 3.6.4). There are presently no data available that could tell if this has happened or not. We may roughly assess possible changes in the soil as follows. Most detritus annually incorporated into the soil reservoir decomposes rapidly and does not contribute to an accumulation. The average turnover time of the slowly decomposing organic constituents in the soil is at least a few hundred years, which means an annual global turnover of 4–$8 \cdot 10^{15}$g C yr^{-1}. The input due to enhanced photosynthesis and detritus formation has probably not exceeded about 10% ($\beta = 0.3$–0.4, cf Section 3.6.4 and Chapter 10). Therefore it does not seem likely that such a sink at present would exceed $0.5 \cdot 10^{15}$g C yr^{-1} and it would probably decrease in the future due to the reduction of virgin forests where terrestrial biomass could be enhanced significantly.

It is difficult to assess what these processes and their possible changes will mean with regard to future atmospheric CO_2 concentrations. On the basis of the results obtained from carbon cycle modelling and the discussion above we conclude: If emissions increase rapidly, e.g. by about 2% per year (or more), it seems likely that the airborne fraction will increase above its present value of approximately 40% (cf Section 3.5.6). For a more modest rate of emission increase, e.g. 1% or less, it seems more plausible that the airborne fraction remains unchanged, while it may decrease significantly if emissions remain constant or decrease. Although these conclusions are approximate and uncertain, we will make use of them in our final projections.

3.8.3 Estimates of the Range of Future Atmospheric CO₂ Concentrations

A number of projections of likely future CO_2 concentrations have been made by using the carbon cycle models described above (Budyko, 1972; Keeling and Bacastow, 1977; Siegenthaler and Oeschger, 1978; Laurmann and Spreiter, 1983; Oeschger and Heimann, 1983). Although they represent good first approximations, most of them overestimate changes of atmospheric CO_2 concentrations as a result of likely past emission scenarios. We shall here give a plausible range of the increase of future atmospheric CO_2 concentrations by consideration of these model computations and their deficiencies and also on the basis of simple estimates using the concept of airborne fraction.

Future CO_2 emissions due to fossil fuel combustion were discussed in some detail in Chapter 2, where a 'lower bound' (LB) and an 'upper bound' (UB) scenario were defined (cf Figure 2.10). They were determined on the basis of the large number of scenarios of this kind that have been presented in recent years and in addition subjective judgements of what seems possible in the light of other constraints (political, social, logistic, etc.) that usually have not explicitly been considered in the models used for such projections. We shall adopt these two bounding scenarios for the assessment of future atmospheric CO_2 concentrations. Although it is questionable to extend such projections beyond 2050, we shall do so for later comparison with corresponding analyses by Nordhaus and Yohe (1983). We simply assume that for the UB scenario emissions will continue to increase linearly with unchanged rate, while constant emissions of $2 \cdot 10^{15}$ g C yr^{-1} will characterize the LB scenario during the latter part of the 21st century.

We also need some plausible projections of likely future emissions from terrestrial ecosystems. In a long-term perspective it seems clear that they will be of less importance than those due to fossil fuel burning, for the simple reason that these carbon pools are limited. Some further reduction of the tropical forests seems likely, but the total storage of carbon as living matter in these ecosystems is merely $170 \pm 80 \cdot 10^{15}$ g C (cf Table 3.1). The biomass of ecosystems at higher latitudes on the other hand is presently not changing much. The further exploitation of forests for timber and pulp may lead to reduction of organic carbon in the soils, but probably not to the same extent as is the case when forest soil is being transformed into farm land.

In the light of (1) past changes of the terrestrial ecosystems due to man's exploitation of land for his needs, (2) (the total amounts of carbon in these systems and (3) the rates of present changes, it seems plausible to assume that an additional amount of the same magnitude may be transferred to the atmosphere in the period 1980–2100 as was emitted before 1980, i.e. approximately 1–$2 \cdot 10^{15}$ g C yr^{-1} or totally 120–$240 \cdot 10^{15}$ g C for this period. As will be seen this assumption is not crucial for our final conclusions. Accordingly we add $2.0 \cdot 10^{15}$ g C yr^{-1} to the UB and $1.0 \cdot 10^{15}$ g C yr^{-1} to the LB scenario.

We first produce scenarios for future atmospheric CO_2 concentrations by simply applying fixed airborne fractions in the future assuming a value 50% for the UB scenario, while we present two alternatives for the LB scenario, namely 40% and the lower projections obtained by using the Siegenthaler model (cf Section 3.5.6 and Svenningsson, 1985). The computed atmospheric CO_2 concentrations until 2100 are shown in Figure 3.21 (heavy solid lines). It should be emphasized that the buffer factor gradually will increase, when atmospheric CO_2 concentrations increase and so will the airborne fraction. The assumed value of 50% will therefore be too small when CO_2 concentrations reach well above 600 ppmv, i.e. in the UB scenario towards the end

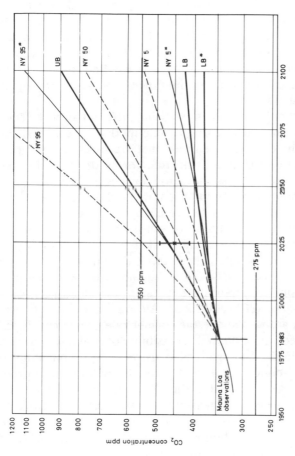

Figure 3.21 Scenarios for future atmospheric CO_2 concentrations. UB is based on emissions from fossil fuel combustion according to the upper bound given in Chapter 2 from terrestrial sources being constant at a level of $2 \cdot 10^{15}$ g C yr^{-1} and the airborne fraction of the total emissions being 50%; correspondingly LB is based on the lower bound fossil fuel emissions, terrestrial emissions being 10^{15} g C yr^{-1} and the airborne fraction 40%. In the case of LB* the assessment of atmospheric CO_2 concentration was based on the carbon cycle model developed by Siegenthaler (1983) with enhanced uptake by the intermediate ocean waters (Svenningsson, 1985). NY 95, 50 and 5 are 95th, 50th and 5th percentiles of the atmospheric CO_2 concentration as given by Nordhaus and Yohe (1983), NY 95* and NY 5* are the corresponding scenarios if using 50% and 40% airborne fractions respectively rather than the computations presented by the authors. For the year 2025 the previous UNEP/WMO/ICSU assessment is shown (WMO, 1981). Pre-industrial concentration (275 ppmv) and twice that value (550 ppmv) are also shown

of the next century. For comparison we also reproduce in Figure 3.21 the future changes as computed by Nordhaus and Yohe (1983) and show the 95th, 50th and 5th percentiles of the set of scenarios they deduce (dashed lines). See further the legend to Figure 3.21.

We notice that the modified CO_2 concentrations using Nordhaus and Yohe (1983) emission scenarios are considerably lower than those that appeared in their original publication. The range of future CO_2 concentrations deduced on the basis of the UB and LB scenarios are still lower and also below the previous ICSU/UNEP/WMO assessment for the year 2025. Further the LB emissions beyond 2050 (totally $3.0 \cdot 10^{15}$ g C yr^{-1}) only increase the atmospheric concentrations very slowly to about 370 ppmv in 2100, if a carbon cycle model is used rather than assuming a fixed airborne fraction (Svenningsson, 1985). The UB scenario on the other hand yields 580 ppmv in 2050 and about 900 ppmv in 2100.

A comparison of these simple extrapolations using different airborne fractions and projections with the aid of carbon cycle models reveals that the results depend more on the emission scenarios that are used than the particular response of the natural carbon cycle that forms the basis for the projections. When attempting to assess the consequences of different emission strategies, model computations are indispensable (cf Section 3.8.4). We also need improved models to interpret future changes of the carbon cycle and to improve projections and the design of emission strategies. For the time being we conclude that it seems quite unlikely that doubling of the atmospheric CO_2 concentration would occur before the middle of the next century and it may not even take place before the end of the next century.

Before closing this section it is important to recall briefly the basis for the UB and LB emission scenarios (cf Section 2.5). Although a number of scenarios have appeared that are well above the former one, it was concluded that these seem unlikely because of limited accessibility of coal reserves, lack of readily available water in the mining areas, constraints of terminal development and resource distribution issues and general environmental problems in addition to those directly related to the CO_2 issue. The LB scenario, on the other hand, although technically feasible, requires sustained efforts to reduce energy demands in most sections of society and a gradual expansion of other energy sources (solar, wind, biomass, nuclear). Still lower emission scenarios have been proposed, but seem unrealistic.

3.8.4 Emission Strategies

It is of interest to attempt to outline the emission strategies that are necessary in order not to exceed certain prescribed atmospheric CO_2 concentrations in the future. The problem may become of particular importance if it becomes possible to ascribe a climatic change to the emissions of CO_2

or other greenhouse gases (cf Chapters 4 and 6). To analyse this question it is necessary to employ a carbon cycle model, since we cannot prescribe the airborne fraction of the emissions. Although available carbon cycle models still are deficient, an approximate idea of the necessary constraints on emissions can be obtained by using the model developed by Siegenthaler and Oeschger (1978) and Siegenthaler (1983), cf Svenningsson (1985).

As referred to in the preceding section atmospheric concentrations will not exceed about 400 ppmv in a foreseeable future if the emissions decrease to about $3 \cdot 10^{15}$ g C yr^{-1} during the next 65 years and thereafter are kept constant at that level. On the other hand, emissions according to the UB scenario would yield such high atmospheric CO$_2$ concentrations by 2020–2030 that unrealistically rapid reductions of the emissions thereafter would be necessary in order not to exceed 550 ppmv later during the century, e.g. a doubling as compared to pre-industrial conditions.

We next design two intermediate emission scenarios for the next 40-50 years: (a) continuation of the present (1973–1983) modest increase by about 1.5% per year and (b) a break of the present trend by reducing the increase to somewhat less than 0.5% per year. This latter scenario is similar to the one developed by Goldemberg (see Chapter 2.4). Different rate development assumptions were tested during the remainder of the next century to identify the effects of various policies on atmospheric CO$_2$ concentrations. Gas and oil reserves, as known today, were used up to depletion after 2030 modelled by a decreasing part of a logistic function. Coal rise was given to produce a smooth curve of total fossil fuel rise. Two ceilings towards the end of the next century of 430 and 540 ppmv were chosen, i.e. about 60% and 100% above pre-industrial conditions.

Figure 3.22a shows the necessary reductions of fossil energy use for the scenario (a) after 2030 in order not to exceed 540 ppmv, while the 430 ppmv ceiling necessarily will be exceeded. The distribution of the use of oil, gas and coal as a function of time is also given. With the low scenario case (b) there is considerable freedom as to what path to choose after 2020 even if the ceiling is placed at 430 ppmv (Figure 3.22b). There is even room for a modest expansion of coal utilization after 2020 lasting until late next century to replace the depleted oil and gas reserves. We show also in Figure 3.22b the results of a similar computation by Siegenthaler and Oeschger (1978), in which case a simpler ocean model (Oeschger et al., 1975) was used. The different concentration scenarios are in principle about the same.

3.9 CONCLUSIONS

— During the period 1860–1984 altogether $183 \pm 15 \cdot 10^{15}$ g C has been emitted into the atmosphere by fossil fuel combustion and the present (1984) rate is $5.2 \cdot 10^{15}$ g C yr^{-1}.

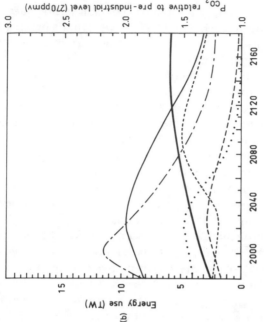

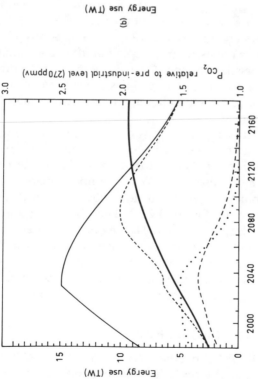

Figure 3.22 Development of fossil energy use (thin solid line, left scale) and atmospheric CO_2 concentrations, based on the Siegenthaler (1983) model of the global carbon cycle (heavy solid line, right scale). The short-dashed line shows the energy contribution by coal combustion, dotted line is that due to oil combustion and long-dashed line gas combustion. (a) Comparatively high CO_2 emission scenario until 2030 and thereafter the decline necessary not to exceed atmospheric CO_2 concentrations of 540 ppmv. (b) Low CO_2 emission scenario until 2020 and thereafter one possible emission scenario which yields atmospheric CO_2 concentrations not exceeding 430 ppmv (Svenningsson, 1985). The dash–dotted curve is a similar scenario produced by Siegenthaler and Oeschger (1978) in which case CO_2 concentrations were not permitted to rise by more than 50% above the pre-industrial level

— During the same time $150 \pm 50 \cdot 10^{15}$ g C has been released due to deforestation and changing land use, the present rate being $1.6 \pm 0.8 \cdot 10^{15}$ g C yr^{-1}.

— Atmospheric CO_2 concentrations have increased from 275 ± 10 ppmv in the middle of the last century to 343 ± 1 ppmv in 1984.

— We understand the basic features of the global carbon cycle quite well. It has been possible to construct quantitative models which can be used as a general guide for the projection of future CO_2 concentrations in the atmosphere as a result of given emission scenarios.

— The uncertainties of projections of likely future CO_2 changes on the basis of a given emission scenario are considerably less than those of the emission scenarios themselves.

— On the basis of a range of emission scenarios developed in Chapter 2, we conclude that it is unlikely that atmospheric CO_2 concentrations will double (i.e. exceed about 550 ppmv) before the middle of the next century and such concentrations may not be reached until after the end of the next century.

— The major uncertainties of these projections are due to inadequate knowledge about the importance of the following processes:

  • rate of water exchange between surface, intermediate and deep layers of the sea.

  • the sensitivity of marine primary production to changes of nutrient availability in surface waters.

  • burying of organic matter in the sediments in coastal regions (and lakes).

  • change of alkalinity, and thus the buffer factor of sea water due to increasing amounts of dissolved inorganic carbon (DIC) (in a long-term perspective also due to carbonate dissolution on the continental shelves).

  • fertilization and increase of biomass and organic matter in soils in terrestrial ecosystems due to increasing atmospheric CO_2 concentrations and possibly deposition of nutrients emitted from anthropogenic sources.

  • enhanced rate of decay of organic matter in soils, particularly in the process of forest exploitation.

  • charcoal formation during biomass burning.

— A slow-down of the increase of fossil fuel use during the next decades would greatly widen the range of options in the future use of fossil fuels without reaching atmospheric CO_2 concentrations twice pre-industrial levels.

ACKNOWLEDGEMENT

The preparation of this chapter has benefited greatly by the viewpoints from the participants in the expert group meeting on the carbon cycle and the projection of future concentration of atmospheric CO_2 held in Stockholm, 28 November–2 December, 1983: W.S. Broecker, D. Lal, B. Moore, V. Siegenthaler, and M. Whitfield.

Most valuable and detailed comments on a first draft of this paper have been received from J. Goudriaan, H. Oeschger, U. Siegenthaler, and J.D. Woods.

They have led to some significant modifications and I am grateful for having been able to take advantage of the views and information forwarded.

I am also grateful for numerous suggestions for modifications by: M.I. Budyko, W. Böhme, E. Degens, W. Ellsaesser, H.L. Ferguson, R.M. Gifford, V.A. Gorbachev, V.G. Gorshkov, I.G. Gringof, M.R. Kiangi, D. Lal, H. Landsberg, J.A. Laurmann, A.S. Monin, B. Moore, G. Preining, A. Rebello, G. deQ. Robin, R. Rotty, E. Salati, G. White, M. Whitfield, M.M. Yoshino.

3.10 REFERENCES

Ajtay, G. L., Ketner, P., and Duvigneaud, P. (1979) Terrestrial primary production and phytomass, in Bolin, B., Degens, E., Kempe, S., Ketner, P. (eds) *The Global Carbon Cycle*, SCOPE 13, Chichester, Wiley, 129–182.

Bacastow, R., and Björkström, A. (1981) Comparison of ocean models for the carbon cycle, in Bolin, B. (ed.) *Carbon Cycle Modelling*, SCOPE 16, Chichester, Wiley, 29–79.

Bacastow, R., and Keeling, C. D. (1981) Atmospheric carbon dioxide concentration and the observed airborne fraction, in Bolin, B. (ed.) *Carbon Cycle Modelling*, SCOPE 16, Chichester, Wiley, 103–112.

Baes, F.C. (1982) Effects of ocean chemistry and biology on atmospheric carbon dioxide, in Clark, W. C. (ed.) *Carbon Dioxide Review (1982)*, 187–204.

Baes, C. F., Goeller, H. E., Olson, J. S., and Rotty, R. M. (1976) *The Global Carbon Dioxide Problem*, ORNL-5194, Oak Ridge National Laboratory, Oak Ridge, Tenn, USA, 1–72.

Bathen, K. H. (1972) On the seasonal changes in the depth of the mixed layer in the North Pacific Ocean, *J. Geophys. Res.*, **77**, 7138–7150.

Bazilevich, N. I., Rodin, L. Ye., and Roznov, N. N. (1970) *Geographical Aspects of Biological Productivity*, Pap. V Congr. USSR Geogr. Soc. USSR, Leningrad.

Bazilevich, N. I. (1974) Energy flow and biogeochemical regularities of the main world ecosystems, in *Proceedings First International Congress of Ecology*, Purdoc, Wageningen, Netherlands.

Bischof, W., Fabian, P., and Borchess, R. (1980) Decrease in CO_2 mixing ratio observed in the stratosphere, *Nature*, **288**, 347–348.

Björkström, A. (1979) A model of CO_2 interaction between atmosphere, oceans, and land biota, in Bolin, B., Degens, E., Kempe, S., Ketner, P. (eds) *The Global Carbon Cycle*, SCOPE 13, Chichester, Wiley, 403–457.

Bolin, B. (1981) Steady state and response characteristics of a simple model of the carbon cycle, in Bolin, B. (ed.) *Carbon Cycle Modelling*. SCOPE 16, Chichester, Wiley, 315–331.

Bolin, B. (1983) C, N, P and S cycles. Major reservoirs and fluxes 2.2 The carbon cycle, in Bolin, B., and Cook, R. (eds) *The Major Biogeochemical Cycles and their Interactions*, SCOPE 21, Chichester, Wiley, 41–45.

Bolin, B., and Eriksson, E. (1959) Changes of the carbon dioxide content of the atmosphere and sea due to fossil fuel combustion, in Bolin, B. (ed.) *Atmosphere and Sea in Motion*, 130–142. The Rockefeller Institute Press.

Bolin, B., Degens, E. T., Duvigneaud, P., and Kempe, S. (1979) The global biogeochemical carbon cycle, in Bolin, B., Degens, E. T., Kempe, S., Ketner, P. (eds) *The Global Carbon Cycle*, SCOPE 13, Chichester, Wiley, 1–56.

Bolin, B., Keeling, C. D., Bacastow, R., Björkström, A., and Siegenthaler, U. (1981) Carbon cycle modelling, in Bolin, B. (ed.) *Carbon Cycle Modelling*, SCOPE 16, Chichester, Wiley, 1–28.

Bolin, B., Björkström, A., Holmen, K., and Moore, B. (1983) The simultaneous use of tracers for ocean circulation studies, *Tellus*, 35B, 206–236.

Bolin, B., Houghton, R. A., and Moore, B. (1985) *Changing Forests and the CO_2 Concentration in the Atmosphere*, United Nations University (in press).

Brewer, P. G. (1978) Direct observations of the oceanic CO_2 increase, *Geophys. Res. Lett.*, **5**, 997–1000.

Broecker, W. S. (1973) Factors controlling CO_2 content of the oceans and atmosphere, in Woodwell, G. M., and Pecan, E. V. (eds) *Carbon in the Biosphere*, US Atomic Energy Commission, 32–50.

Broecker, W. S. (1981) Glacial to interglacial changes in ocean and atmospheric chemistry, in Berger A. (ed.) *Climatic Variations and Variability: Facts and Theories*, Dordrecht, D. Reidel, 109–120.

Broecker, W. S., Gerard, R., Ewing, M., and Heezen, B. C. (1960) Natural radiocarbon in the Atlantic Ocean, *J. Geophys. Res.*, **65**, 2903–2931.

Broecker, W. S., and Takahashi, T. (1977) Neutralization of fossil fuel CO_2 by marine calcium carbonate, in Andersson, N., and Malahoff, A. (eds) *The Fate of Fossil Fuel CO_2 in the Oceans*, New York, Plenum Press, 213–241.

Broecker, W. S., Peng, T.-H., and Engh, R. (1980) Modelling the carbon system, *Radiocarbon*, **22**, 565–598.

Broecker, W. S., and Peng, T.-H. (1982) *Tracers in the Sea*. Lamont-Doherty Geological Observatory, Columbia University, Palisades, N.Y. 10964.

Brown, S., and Lugo, A. E. (1982) The storage and production of organic matter in tropical forests and their role in the global carbon cycle, *Biotropica*, **14**, 161–187.

Brown, S., and Lugo, A. E. (1984) Biomass of tropical forests: A new estimate based on volumes, *Science*, **223**, 1290–1293.

Budyko, M. I. (1972) *Man's Impact on Climate*, Leningrad Hydrometeoizdat.

Buettner, E. K., and Zacharova, O. K. (1983) Modelling the carbon cycle by using the fraction derivative model, *Meteorology and Hydrology*, **4**, 14–20.

Bullister, J. L., and Weiss, R. F. (1982) Anthropogenic chlorofluoromethanes in the Greenland and Norwegian Seas, *Science*, **221**, 265–268.

Buringh, B. (1984) Decline of organic carbon in soils of the world, in Woodwell, G. M. (ed.) *The Role of Terrestrial Vegetation in the Global Carbon Cycle*, SCOPE 23, Chichester, Wiley.

Chen, C. T. (1982) Oceanic penetration of excess CO_2 in a cross section between Alaska and Hawaii, *Geophys. Res. Lett.*, **9**, 117–119.

Craig, H. (1957a) The natural distribution of radiocarbon and the exchange time of carbon dioxide between atmosphere and sea, *Tellus*, **9**, 1–17.

Craig, H. (1957b) Isotope standards for carbon and oxygen and correlation correction factors for mass spectrometric analysis of CO_2, *Geochem. Cosmochim. Acta*, **12**, 133–149.

Delmas, R., Ascencio, J. M., and Legrand, M. (1980) Polar ice evidence that atmospheric CO_2 20,000 years B.P. was 50% of present, *Nature*, **284**, 155–157.

Detwiler, R. P., Bogdonoff, P., and Hall, C. A. S. (1984) Landuse change and carbon exchange in the tropics II Preliminary simulations for the forests as a whole. *Environmental Management* (in press).

De Vooys, C. G. N. (1979) Primary production in aquatic environments, in Bolin, B., Degens, E.T. Kempe, S., Ketner, P. (eds) *The Global Carbon Cycle*, SCOPE 13, Chichester, Wiley, 259–292.

Druffel, E. M., and Suess, H. E. (1983) On the radiocarbon record in banded corals: Exchange parameters and net transport of $^{142}CO_2$ between atmosphere and surface ocean, *J. Geophys. Res.*, **88C**, 1271–1280.

Emanuel, R. E., Killough, G. G., Post, W. M., and Shugart, H. H. (1984) Modeling terrestrial ecosystems in the global carbon cycle with shifts in carbon storage capacity by land-use change, *Ecology*, **65**, 970– 983.

FAO (1981) Tropical forest resources assessment project, *Forest Resources of Tropical America, Africa and Asia*, Rome.

Farquhar, G. D., O'Leary, M. H., and Berry, J. A. (1982) On the relationship between carbon isotope discrimination and the intercellular carbon dioxide concentration in leaves, *Austr. J. Plant Physiol.*, **9**, 121–137.

Francey, R. J., and Farquhar G. D. (1982) An explanation of the $^{132}C/^{122}C$ variations in tree rings, *Nature*, **297**, 28–31.

Freyer, H. D. (1979) On the 132C record in tree rings, Part I. ^{132}C variations in northern hemisphere trees during the last 150 years, *Tellus*, **31**, 124–137.

Freyer, H. D., and Belacy, N. (1983) 13C/12C records in northern hemisphere trees during the past 500 years: anthropogenic impact and climate superposition, *J. Geophys. Res.*, **88**, 6844–6852.

Friedli, H., Moor, E., Oeschger, H., Siegenthaler, U., and Stauffer, B (1984) $^{13}C/^{12}C$ ratios in CO_2 extracted from Antarctic ice, *Geophys. Res. Lett.*, **11**, 1145–1148.

Gorshkov, V. G. (1982) The possible global budget of carbon dioxide, *Il Nuovo Cimento*, **5**, 209–222.

Goudriaan, J., and Ketner, P. (1984) A simulation study for the global carbon cycle including man's impact on the biosphere, *Clim. Change*, **6**, 167–192.

Hasselmann, K. (1982) An ocean model for climate variability studies, *Progr. in Oceanogr.*, **11**, 69–92.

Hoffert, M. I., Callegari, A. J., and Hsieh, C.-T. (1981) A box-diffusion carbon cycle model with upwelling polar bottom water formation and a marine biosphere, in Bolin, B. (ed.) *Carbon Cycle Modelling*, SCOPE 16, Chichester, Wiley, 287–306.

Holdridge, L. R. (1967) *Life Zone Ecology*, San José, Tropical Science Center.

Holmén, K. (1985) The global carbon cycle, in Butcher, S. (ed.) (to be published).

Honjo, S (1980) Material fluxes and modes of sedimentation in the mesopelagic and bathypelagic zones, *J. Marin Res.*, **38**, 53–97.

Houghton, R. A. (1984) Estimating changes in the carbon content of terrestrial ecosystems from historical data, in *Global Carbon Cycle*, Sixth ORNL Life Science Symposium.

Houghton, R. A., Hobbie, J. E., Melillo, J. M., Moore, B., Peterson, B. J., Shaver, G. R., and Woodwell, G. M. (1983) Changes in the carbon content of terrestrial biota and soils between (1860) and (1980). A net release of CO_2 to the atmosphere, *Ecolog. Monogr.*, **53**, 235–262.

Keeling, C. D. (1973) The carbon dioxide cycle: Reservoir models to depict the exchange of atmospheric carbon dioxide with the ocean and land plants, in Rasool, S. I. (ed.) *Chemistry of the Lower Atmosphere*, New York, Plenum Press, 251–329.

Keeling, C. D., and Bacastow, R. B. (1977) Impact of industrial gases on climate, in *Energy and Climate*, Washington, D.C., National Academy of Sciences, 72–95.

Keeling, C. D., Mook, W. G., and Tans, P. P. (1979) Recent trends in the 13C/12C ratio of atmospheric carbon dioxide, *Nature*, **277**, 121–123.

Keeling, C. D., Carter, A. F., and Mook, W. G (1984) Seasonal, latitudinal and secular variations in the abundance and isotope ratios of atmospheric CO_2, *J. Geophys. Res.*, **89**, D3, 4615–4628.

Keeling, C. D., and Heimann. M. (1985) Meridional eddy diffusion model of the transport of atmospheric carbondioxide: 2. The mean annual carbon cycle. *J. Geophys. Res.* (in press).

Killough, G., and Emanuel, W. (1981) A comparison of several models of carbon turnover in the ocean with respect to their distributions of transit time and age, and responses to atmospheric CO_2 and ^{14}C, *Tellus*, **33**, 274–290.

Knox, F., and McElroy, M. B. (1984) Changes in atmospheric CO_2: Influence of the marine biota at high latitudes, *J. Geophys. Res.*, **89**, 4629–4637.

Laurmann, J. A., and Spreiter, J. R. (1983) The effects of carbon cycle model error in calculating future atmospheric carbon dioxide levels, *Clim. Change*, **5**, 145–181.

Machta, L. (1972) The role of the oceans and the biosphere in the carbon dioxide cycle, in Dryssen, D., and Jagner, D. (eds) *The Changing Chemistry of the Oceans*, Stockholm, Almqvist & Wiksell, 121–146.

Maier-Reimer, E. (1984) Towards a global ocean carbon model, *Biometeorology*, **3**, 295–310.

Moore, B., Boone, R. D., Hobbie, J. E., Houghton. R. A., Melillo, J. M., Peterson, G. R., Shaver, G. R., Vörösmarty, C. J., and Woodwell, G. M. (1981) A simple model for analysis of the role of terrestrial ecosystems in the global carbon budget, in Bolin B. (ed.) *Carbon Cycle Modelling*, SCOPE 16. Chichester, Wiley, 365–385.

Neftel A., Oeschger, H. Schwander, J. Stauffer, B., and Zumbrunn, R. (1982) Ice core sample measurements give atmospheric CO_2 content during the past 40,000 years, *Nature*, **295**, 220–223.

Neftel, A, Moor, E., Oeschger, H., and Stauffer, B. (1985) The increase of atmospheric CO_2 in the last two centuries. Evidence from polar ice cores, *Nature* (in press).

Nordhaus, W. D., and Yohe, G. W. (1983) Future carbon dioxide emissions from fossil fuel, in *Changing Climate*, Washington D.C., National Academy Press, 87–153.

Nydal, R., and Lövseth, K. (1983) Tracing bomb ^{14}C in the atmosphere (1962–1980), *J. Geophys. Res.*, **88**, 3621–3642.

Oeschger, H., Siegenthaler, U., Schotterer, U., and Gugelmann, A. (1975) A box diffusion model to study the carbon dioxide exchange in nature, *Tellus*, **27**, 168–192.

Oeschger, H., and Heimann, M. (1983) Uncertainties of predictions of future atmospheric CO_2 concentrations, *J. Geophys. Res.*, **88**, C2, 1258–1262.

Olson, J. S., Watts, J. A., and Allison, L. J. (1983) *Carbon in Live Vegetation of Major World Ecosystems*, United States Department of Energy, TROO4.

Östlund, H. G., Doresy, H. G., and Brescher, R. (1976) *GEOSECS Atlantic Radiocarbon and Tritium Results*, Tritium Lab. Data Report No 8, Rosenstiel School of Marine and Atmospheric Science, University of Miami, 14 pages.

Östlund, H. G., and Stuiver, M. (1980) *GEOSECS Pacific Radiocarbon*, **22**, 25–53.

Pearman, G. I., and Hyson, P. (1980) Activities of the global biosphere as reflected in atmospheric CO_2 Records, *J. Geophys. Res.*, **85**, 4468–4474.

Peng, T.-H., Broecker, W. S., Mathieu, G., and Li, Y.-H. (1979) Radon evasion rates in the Atlantic and Pacific oceans as determined during the GEOSECS program, *J. Geophys. Res.*, **84**, 2471–2486.

Peng, T.-H. Broecker, W. S., Freyer, D. D., and Trumbore, S. (1983) A deconvolution of the tree ring based C^{13} record, *J. Geophys. Res.*, **88**, 3609–3620.

Peterson, B. J., and Melillo, J. M. (1985) The storage of carbon caused by man's eutrophication of the biosphere, *Tellus* (to be published).

Pierrou, U. (1976) The global phosphorus cycle, in Rosswall and Söderlund (eds) *N, P, S Global Cycles*, SCOPE 7, Ecol. Bull., 22, Editorial Service, NFR, Stockholm.

Post, W. M., Emanuel, W. R., Zinke, P. J., and Stangenberger, A. G. (1983) Soil carbon pools and world life zones, *Nature*, **298**, 156–159.

Raynaud, D., and Barnola, J. M. (1985) An Antarctic ice core reveals atmospheric CO_2 variations over the past few centuries, *Nature*, **315**, 309–311.

Redfield, A. C. (1958) The biological control of chemical factors in the environment, *Am. Sci.*, **46**, 206–226.

Revelle, R., and Suess, H. (1957) Carbon dioxide exchange between atmosphere and ocean and the question of an increase of atmospheric CO_2 during the past decades, *Tellus*, **9**, 18–27.

Revelle, R., and Munk, W. (1977) The carbon dioxide cycle and the biosphere, in *Energy and Climate*, Stud. Geoph., Washington D.C., National Academy of Sciences, 140–158.

Richards, J. P., Olson, J. S., and Rotty, R. M. (1981) *Development of a Data Base for Carbon Dioxide Releases Resulting from Conversion of Land to Agricultural Use*, ORAU/IEA-81-10 (M), ORNL/TM-8801, Oak Ridge National Laboratory, Oak Ridge, Tenn., USA.

Rotty, R. M. (1981) Data for global CO_2 production from fossil fuels and cement, in Bolin, B. (ed.) *Carbon Cycle Modelling*, SCOPE 16, Chichester, Wiley, 121–125.

Sarmiento, J. L., and Toggweiler, J. R. (1984) A new model for the role of the oceans in determining atmospheric P_{CO2}, *Nature*, **308**, 621–624.

Schlesinger, W. H. (1977) Carbon balance in terrestrial detritus, *Ann. Ecol. Syst.*, **8**, 51–81.

Schlesinger, W. H. (1984) Soil organic matter: A Source of atmospheric CO_2, in Woodwell, G. M. (ed.) *The Role of Terrestrial Vegetation in the Global Carbon Cycle. Methods of Appraising Changes*, SCOPE 23, Chichester, Wiley, 111–127.

Schlesinger, W. H., and Melack, J. M. (1981) Transport of organic carbon in the world's rivers, *Tellus*, **33**, 172–187.

Seiler, W., and Crutzen, P. (1980) Estimates of gross and net fluxes of carbon between the biosphere and the atmosphere from biomass burning, *Clim. Change*, **2**, 207–248.

Siegenthaler, U. (1983) Uptake of excess CO_2 by an outcrop-diffusion model of the ocean, *J. Geophys. Res.*, **88**, C6, 3599–3608.

Siegenthaler, U. (1984) 19th century measurements of atmospheric CO_2—a comment, *Clim. Change*, **6**, 409–411.

Siegenthaler, U., Heimann, M., and Oeschger, H. (1978) Model responses of the atmospheric CO_2 levels and $^{13}C/^{12}C$ ratio to biogenic CO_2 input, in. Williams J. (ed.) *Carbon Dioxide, Climate and Society*, Oxford, Pergamon Press.

Siegenthaler, U., and Oeschger, H. (1978) Predicting future atmospheric CO_2 levels, *Science*, **199**, 388–395.

Siegenthaler, U., and Wenk, Th. (1984) Rapid atmospheric CO_2 variations and ocean circulation, *Nature*, **308**, 624–626.

Stuiver, M. (1980) ^{14}C distribution in the Atlantic Ocean, *J.Geophys. Res.*, **85**, 2711–2718.

Stuiver, M., and Polach. H. (1977) Reporting of ^{14}C data, *Radiocarbon*, **19**, 355–363.

Stuiver, M., and Östlund. H. G. (1980) GEOSECS Atlantic radiocarbon, *Radiocarbon*, **22**, 1–24.

Stuiver, M., and Quay, P. D. (1981) Atmospheric ^{14}C changes resulting from fossil fuel CO_2 release and cosmic ray flux variability, *Earth and Planetary Science Letters*, **53**, 349–362.

Stuiver, M., Burk, R. L., and Quay. P. D. (1984) $^{13}C/^{12}C$ ratios in tree-rings and the transfer of biospheric carbon to the atmosphere, *J. Geophys. Res.*, **89**, D7, 11,731–11,748.

Svenningsson, P. (1985) *Global Atmospheric CO_2 Concentration and Fossil Energy Use*, Environmental Studies Programme, Univ. of Lund, Lund, Sweden.

Takahashi, T., Broecker, W. S., Werner, S. R., and Bainbridge, A. E. (1980) Carbonate chemistry of the surface waters of the world oceans, in Goldberg, E. D. et. al. (eds) *Isotope Marine Chemistry*, Tokyo, Uchida Rokakuho Publ. Co.

Takahashi, T., Broecker, W. S., and Bainbridge, A. E. (1981) The alkalinity and total carbon dioxide concentration in the world oceans, in Bolin, B. (ed.) *Carbon Cycle Modelling*, SCOPE 16, Chichester, Wiley, 271–286.

Tans, P. (1981) A compilation of bomb 142C data for use in global carbon model calculations, in Bolin, B. (ed.) *Carbon Cycle Modelling*, Scope 16, Chichester, Wiley, 131–157.

Viecelli, J A. (1984) The atmospheric carbon dioxide response to oceanic primary productivity fluctuations, *Clim. Change*, **6**, 153–166.

Vitousek, R. M. (1983) The effects of deforestation on air soil and water, in Bolin, B., and Cook, R. (eds) *The Major Biogeochemical Cycles and their Interactions*, Chichester, Wiley, 223–245.

Wagener, K. (1978) Total anthropogenic CO_2 production during the period 1860–1975 from carbon 13 measurements in tree rings, *Rad. and Environ. Biophys.*, **15**, 101–111.

Walsh, J. J., Rowe, G. T., Iverson, R. L., and McRoy, C. P. (1981) Biological export of shelf carbon is a sink of the global CO_2 cycle, *Nature*, **291**, 196–201.

Weiss, W., and Roether, W. (1980) The rates of tritium input to the world oceans, *Earth and Planetary Sci. Lett.*, **49**, 435–446.

Whittaker. R. H., and Likens, G. E. (1975) The biosphere and man, in Lieth, H., and Whittaker, R. H. (eds) *Primary Productivity of the Biosphere*, Ecol. Studies 14, Berlin, Heidelberg, New York. Springer-Verlag, 305–328.

Wigley, T. M. L. (1983) The preindustrial carbon dioxide level, *Clim. Change*, **5**, 315–320.

Wigley, T. M. L., and Mueller, A. B. (1981) Fractionation corrections in radiocarbon dating, *Radiocarbon*, **23**, 173–190.

Williams, P.J. LeB. (1975) Biological and chemical aspects of dissolved organic material in sea-water, in Riley, J. P., and Skirrow, G. (eds) *Chemical Oceanography*, 2nd edn, II, London, Academic Press, 301–363.

WMO (1981) *On the Assessment of the Role of CO_2 on Climate Variations and their Impact*, Geneva, WMO.

WMO (1983) *Report on the WMO (CAS) Meeting of Experts on the CO_2 Concentrations from Preindustrial Times to IGY*, ICSU/WMO World Climate Programme, Geneva.

CHAPTER 4

Other Greenhouse Gases and Aerosols
Assessing Their Role for Atmospheric Radiative Transfer

H.-J. BOLLE, W. SEILER, AND B. BOLIN

4.1 INTRODUCTION

It has become increasingly evident during the last decade that in addition to CO_2 other atmospheric gases that interact with the radiative fluxes in the atmosphere are increasing in abundance due to anthropogenic sources. Their radiative characteristics and possible influence on climate need careful study. They will in the following be called 'other greenhouse gases'. The possible role of the atmospheric aerosol is also considered.

In principle, all constituents which interact with the radiation field of the atmosphere must be regarded as part of the climate system, but only those which show strong enough absorption features and distinct trends in their abundances are of direct importance in the present context. The most significant gases in this regard are chlorofluorocarbons ($CFCl_3$–F11 and CF_2Cl_2–F12), methane (CH_4), nitrous oxide (N_2O) and ozone (O_3). The concentrations of the first three of these are increasing quite rapidly, tropospheric ozone equally so, and they are, therefore, of particular importance. A possible future decrease of stratospheric ozone is also assessed.

In addition, there are some minor atmospheric constituents which may influence climate indirectly. The chemical balance may be affected due to changes of their concentration, leading to changes of the concentrations of the greenhouse gases (Hameed et al., 1980). Cooling of the stratosphere due to an increase of some greenhouse gases (e.g. CO_2) will change the chemical reaction rates and, accordingly, the concentrations of other gases.

Some reject the idea that aerosols might have a significant effect on climate because their normal residence times in the atmosphere are short and because even major volcanic eruptions are transient events which last for

157

some years at most. Also, no long-term global trend in the optical depth of the atmosphere has yet been detected. Since the current total aerosol contribution to the greenhouse effect is only about 1 °K, it can also be argued that expected concentration changes may only result in climatic effects which are less significant than those due to other greenhouse gases. However, the aerosol distribution may change very significantly on the regional scale and in this way have an impact on climate. Further, the role of aerosols must be considered in designing strategies for the early detection of man-induced climatic changes.

Only the direct radiative effects of aerosols are discussed in this chapter, although indirect effects due to the change of the optical properties of clouds by embedded aerosol (Twomey, 1977b) or deposition of highly absorbing particulate material in polar regions, have also been discussed in the literature (e.g. Rosen *et al.*, 1981; Cess, 1983).

4.2 TRACE GASES IN THE ATMOSPHERE

4.2.1 The Role of Trace Gases in Climatic Studies

The direct radiative effects of trace gases in the atmosphere are primarily due to their absorption in the infrared part of the spectrum. Most effective in this regard are gases with absorption in the parts of the spectrum where water vapour and CO_2 are almost transparent (8–13 μm), the atmospheric window. If absorption bands for different gases overlap, the effects of a changing concentration on the radiative transfer are reduced. Gases with no change in their abundance do not directly contribute to a climatic change, but may affect the magnitude of the role of other constituents due to such overlapping of absorption bands. They must therefore be included in radiative transfer computations with their fixed concentrations.

Measurements of concentration trends and sensitivity studies by means of radiation models have resolved that the strongest climatic change signals are to be expected from $CFCl_3$ (F 11), CF_2Cl_2 (F 12), CH_4, N_2O and O_3 (see further Section 4.5.1). A number of other halogenated compounds and hydrocarbons altogether probably are less important than any one of the above mentioned constituents and will not be considered specifically in the present context (cf Ramanathan *et al.*, 1985).

In the following sub-sections we shall consider in some detail our knowledge about present atmospheric concentrations of these gases and the processes that regulate them. It should be stressed that the natural as well as anthropogenic sources and sinks of methane and nitrous oxide are still quite uncertain. In the account given below an attempt is made to be as internally consistent as possible, but considerable modifications may turn out to be necessary as more data become available.

4.2.2 Methane

4.2.2.1 Past and Present Concentrations

The presence of CH_4 in the atmosphere has been known since the 1940s when Adel (1939) and Migeotte (1948) observed strong absorption bands in the infrared region of the solar spectrum which they attributed to the presence of atmospheric CH_4. The first *in-situ* measurements of tropospheric CH_4 mixing ratios were made in the late 1960s when sensitive analysis techniques had become available. Measurements indicated average tropospheric CH_4 mixing ratios of about 1.41 ppmv for the Northern Hemisphere and about 1.30 ppmv for the Southern Hemisphere (Ehhalt and Schmidt, 1978). Significantly higher values (1.6–1.8 ppmv) were observed in 1976 in the free atmosphere above Europe by Seiler *et al.* (1978), who attributed the difference to a possible ongoing increase of CH_4. The existence of an upward trend was clearly established by Rasmussen and Khalil (1981a) and has lately been confirmed by numerous measurements at different baseline stations (Fraser *et al.*, 1981; Fraser *et al.*, 1984) and by measurements of the meridional CH_4 distribution between the two hemispheres (Blake *et al.*, 1982; Mayer *et al.*, 1982). Figure 4.1 shows the long-term trend of atmospheric CH_4 in clean air at midlatitudes in the Northern Hemisphere. The individual data points

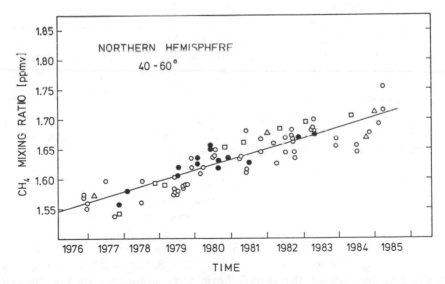

Figure 4.1 Trend of tropospheric CH_4 mixing ratios observed at midlatitudes of the Northern Hemisphere. Data are obtained from measurements carried out on aircrafts (circles), on ships (squares), and at different land based stations during clean air conditions (triangles). The figure includes data (dots) measured by Rowland and colleagues (see Blake, 1984) at similar latitudes in air from the Pacific Ocean

have been obtained from measurements during aircraft and ship missions between 40–60° N over Europe and the Northern Atlantic (W. Seiler, unpublished). Data obtained by Blake *et al.* (1982) and Blake (1984) at similar latitudes on the west coast of the United States in clean air from the Pacific Ocean have also been included. Despite the different sampling locations and sampling altitudes, these observations agree very well suggesting that they are representative for middle latitudes in the Northern Hemisphere.

The data summarized in Figure 4.1 clearly demonstrate the existence of an average temporal increase of atmospheric CH_4 during the last ten years of about 18 ppbv per year or 1.1% per year. The deviation of the individual values from the average increase is probably due to seasonal variations (Rasmussen and Khalil, 1981b) and to meridional transport. Unpublished CH_4 data over the Atlantic from ship cruises between 50° N and 40° S (W. Seiler) show an average increase of 23 ppbv per year between 1977 and 1984 north of the Inter-Tropical Convergence Zone (ITCZ) and 18 ppbv per year for the region south of the ITCZ, i.e. an increase of 1.4% and 1.2% per year, respectively. Similar trends have been found by Rasmussen and Khalil (1981a), Fraser *et al.* (1984), Blake (1984) showing that the long-term trend of atmospheric CH_4 is a global phenomenon.

Information on CH_4 trends cannot be obtained from direct measurements prior to 1975, because of a change of calibration (Heidt and Ehhalt, 1980). Using spectroscopic data, Ehhalt *et al.* (1983) concluded that the average CH_4 increase in the troposphere between 1948 and 1975 cannot have been larger than 0.5% per year. Considering the large uncertainties involved in the calculation of absolute CH_4 mixing ratios from absorption in the solar spectrum (see Fink *et al.*, 1964), this conclusion should be considered with caution.

Better information on long-term trends of CH_4 can be obtained from analysis of air bubbles trapped in the ice sheets of Greenland and Antarctica. Figure 4.2 summarizes results reported by Robbins *et al.* (1973), Craig and Chou (1982) and Rasmussen and Khalil (1984). The age of the trapped air has been corrected for the time needed to close the firn by densification, i.e. about 90 years. Despite the uncertainties involved in the analysis of very small air volumes, the individual data sets reported by the different groups agree reasonably well and show relatively constant CH_4 mixing ratios of about 0.7 ppmv before 1700 AD. Samples from polar ice cores taken in Greenland (e.g. at Camp Century, Crete, Dye 33) show an approximately exponential increase of the tropospheric CH_4 abundance during the last 300 years (Figure 4.3). This increase correlates remarkably well with the increase of human world population which is shown in Figure 4.3 by filled dots, which indicates that the increase of the tropospheric CH_4 abundance is most likely related to anthropogenic activities, primarily agriculture.

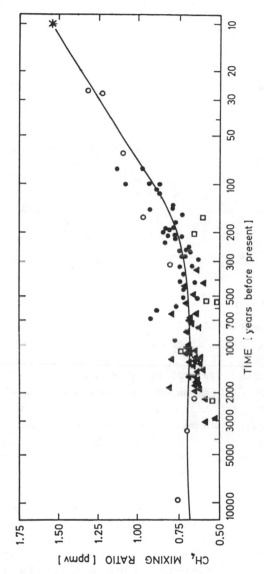

Figure 4.2 CH$_4$ mixing ratios measured in air trapped in ice cores as function of time. Dots and filled triangles are data taken from Rasmussen and Khalil (1984) and represent values obtained from ice cores in Greenland and Antarctica, respectively. Open circles are data published by Craig and Chou (1982) and squares are data published by Robbins *et al.* (1973)

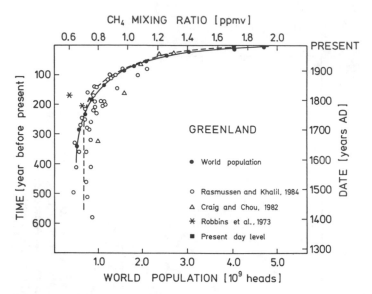

Figure 4.3 Growth of human population and increase of atmospheric CH_4 mixing ratios during the last 600 years

4.2.2.2 Sources and Sinks

Methane is produced by microbial activities during the mineralization of organic carbon under strictly anaerobic conditions, e.g. in waterlogged soils and within the intestines of herbivorous animals. CH_4 is also released by anthropogenic activities such as exploitation of natural gas, biomass burning, coal mining. The CH_4 emission rates have been estimated by different authors, e.g. Koyama (1963), Ehhalt (1974), Ehhalt and Schmidt (1978), Sheppard *et al.* (1982), Khalil and Rasmussen (1983a), Seiler (1984), Blake (1984), and Crutzen (1983, 1985). The emission rates of the individual sources reported in the literature differ by more than one order of magnitude (see Table 4.1) which reflects the large uncertainties in estimating production rates. This uncertainty is primarily due to the limited data base from the individual ecosystems but also to the insufficient information on the size of individual ecosystems that contribute to the CH_4 emissions. Estimates are also difficult because of the complexity of methanogenic food chains which implies large spatial and temporal variations of the CH_4 emission rates within individual ecosystems. Thus, the CH_4 budget estimated for 1980 and shown in the last column of Table 4.1 should still be considered as tentative.

Most of the biogenic methane is released by ruminants, from rice paddies, and freshwater swamps and marshes. Only the releases due to ruminants

Table 4.1 CH₄ emission rates from individual ecosystems (in Tg per year = 10^{12} g per year). The last column presents new estimates made in this volume (W. Seiler)

Sources	Sheppard et al. (1982)	Khalil and Rasmussen (1983a)	Seiler (1984)	Blake (1984)	Crutzen (1985)	Seiler (this volume)
Ruminants	90	120	72–99	71–160[a]	60	70–100
Paddy fields	39	95	30–75	142–190	120–200	70–170
Swamps/Marshes	39	150	13–57	121–190	70–90	25–70
Ocean/Lakes	65	23	71–7	18–34	–	}15–35
Other biogenic	817[b]	100[c]	6–15	60–397[d]	–	
Biomass burning	60	25	53–97	25–110	20–70	55–100
Natural gas	50	–	18–29	–	33	30–40
Coal mining	–	40	30	62–100	34	35
Other nonbiogenic	50	–	1–2	–	–	1–2
Total	1210	553	225–395	500–1160	400	300–550

(a) including herbivorous insects

(b) including CH₄ production from organic solid waste and natural ecosystems

(c) including CH₄ production by termites

(d) including CH₄ production from seasonal and tropical rain forests

appear to be reasonably well known. Early estimates by Hutchinson (1949) yielded a value for emissions by large herbivores in the 1940s of 45 Tg per year. Seiler (1984) has recently estimated that the production was 72–99 Tg per year in 1975. This figure is based on the world population of domestic and non-domestic ruminants and accounts for different emission rates for different species.

The estimates of the CH_4 emission from rice paddies show large variations. The maximum value, i.e. 220 Tg per year (Ehhalt and Schmidt, 1978) was based on emission rates obtained from laboratory experiments (Koyama, 1963), whereas the minimum values, 30–75 Tg per year (Seiler, 1984) were based on *in-situ* measurements of the CH_4 release rates in Spanish and Californian rice paddies (Seiler *et al.*, 1984a; Cicerone *et al.*, 1983). Higher emission rates (70–170 Tg per year) have recently been obtained by Holzapfel-Pschorn and Seiler (1985) by evaluating semi-continuous, *in-situ* measurements of CH_4 fluxes from Italian rice paddies during a whole vegetation period. The lower CH_4 emission rates observed in the Spanish rice paddies are explained by the inflow of Mediterranean water containing sulphate which may have caused a reduction of the methanogenesis. Since more than 95% of the total harvested rice paddy area of 1.5×10^{12} m^2 (FAO, 1983) is located in the Far East from where data on CH_4 emission rates are not available, the figure of 70–170 Tg per year remains uncertain.

Estimates of the CH_4 emission from swamps and marshes are even more uncertain and in some cases speculative. Most of the reported figures are based on CH_4 emission rates measured during short time periods from small eutrophic ponds or swamps at midlatitudes which may not be representative of the large areas of swamps and marshes in tropical regions. The few measurements which have been carried out in natural, undisturbed freshwater wetlands show CH_4 emission rates that differ by more than two orders of magnitude (see e.g. Harriss *et al.*, 1982) which makes a reliable estimate of the global CH_4 emission difficult. A more reliable figure may be obtained by making use of the annual CH_4 flux rates measured in rice paddies which on the average are 45–110 g per m. This seems reasonable since most rice paddies and natural freshwater wetlands are located in similar climatic regions and show comparable net primary production rates.

The total area of swamps and marshes in 1980 was about 1.6×10^{12} m^2 (Clark, 1982), which is comparable with the total harvested rice paddy area. About 75% of this area or 1.2×10^{12} m^2 is located in the tropical and subtropical regions. These areas are only temporarily flooded and, therefore, only active in methanogenesis during part of the year. During the dry season, some fraction of these areas may even act as sinks for atmospheric CH_4 (see Harriss *et al.*, 1982). The duration of the flooding period and the extension of the flooded area depend on the annual rainfall pattern and differ considerably between climatic regions. Large parts of the marshlands in the Amazon and Congo Basins are only flooded during half of the year. Even shorter flooding periods may occur in some swamp areas (e.g. Pentanal, Brazil; Okawengo, Botswana; Junk, private communication). As a first approximation, it is assumed that the total freshwater swamp and marsh areas are flooded on the average during 5–6 months, equivalent to an area of $0.5–0.6 \times 10^{12}$ m^2 being flooded throughout the year. Applying the average CH$_4$ emission rate of 45–110 g m^{-2} per year, we derive a total CH$_4$ production from swamps and marshes of 25–70 Tg per year. This figure does not take into account the fact that significant portions of these areas consist of unvegetated open water (Walter, 1973), which show considerably lower CH$_4$ emission rates than vegetated areas (Cicerone *et al.*, 1983; Seiler *et al.*, 1984a).

Blake (1984) recently reported an average CH$_4$ emission from swamps and marshes of 121 Tg per year. This value was calculated by applying the temperature dependence of CH$_4$ emission from a swamp at northern mid-latitudes to a total swamp area of 2.6×10^6 km^2. Using rather the value of 1.6×10^6 km^2 reported by Clark (1982) and assuming that tropical swamps and marshes are only flooded during 5–6 months per year, the annual emission from these ecosystems is reduced to 31–37 Tg per year which is in good agreement with our estimate of 25–70 Tg per year. Similarly, figures ranging between 150 and 300 Tg per year (Koyama, 1963; Baker-Blocker *et al.*, 1977; Ehhalt and Schmidt, 1978) seem too large.

Methane is also formed by methanogenesis in oceans and lakes, in water-logged tundra soils, and in human intestines. Together with the production of CH$_4$ by herbivorous insects these sources emit 7–22 Tg per year (Seiler, 1984, 1985). Another source is the production and release of CH$_4$ from decomposition of solid waste which might contribute as much as 10 Tg per year. This figure is based on an estimated total disposal of 250×10^6 ton per year of organic municipal wastes and a CH$_4$ release rate of about 40 g CH$_4$ per kg (EPA, 1973; Cheremisinoff *et al.*, 1980). Methane may also be released from digestors which are commonly used in Asian countries for production of biogas. Production rates and emission factors, however, are not known, so that their contribution to the global CH$_4$ budget cannot be quantified. Sheppard *et al.* (1982) proposed an additional CH$_4$ production from natural aerobic soils and estimated the total CH$_4$ emission from this source to be about 700 Tg per year. This figure is much biased by the experimental procedure used, which does not allow extrapolation to natural conditions. There is rather clear evidence that natural aerobic soils do not produce but decompose atmospheric methane (Keller *et al.*, 1983; Seiler *et al.*, 1984a).

Methane is also produced by abiogenic processes such as biomass burning (55–100 Tg per year), leakage and venting of natural gas (30–40 Tg per year), coal mining (about 35 Tg per year), fossil fuel combustion (1–2 Tg per year), i.e. 120–180 Tg per year (Seiler, 1984). This figure agrees reasonably well with other estimates as summarized in Table 4.1.

The present total CH_4 release rate into the atmosphere in 1980 was thus 300–550 Tg, the average being about 425 Tg per year. The global average tropospheric CH_4 mixing ratio of 1.55 ppmv in 1980 corresponds to a total mass of CH_4 of about 4,400 Tg, which implies an average residence time for CH_4 of about 10 years. This production must be equal to the total sink strength plus the amount of CH_4 needed to raise the CH_4 abundance at the observed rate. The following approximate balance between sources and sinks seems possible. The ongoing rise corresponds to about 50 Tg per year so that the required sink strength to balance the CH_4 cycle is about 375 Tg per year. The most important sink mechanism is the photochemical oxidation of CH_4 by tropospheric OH. Adopting an average OH concentration of $5 \cdot 10^5$ molecules per cm^3 (Crutzen and Gidel, 1983) yields a sink strength in the troposphere of 260 Tg per year. About 60 Tg CH_4 is transported annually into the stratosphere where it is oxidized. Furthermore, CH_4 is decomposed in the soil by methanotrophic organisms (Keller *et al.*, 1983; Seiler *et al.*, 1984a). The decomposition rate has been estimated to be about 20 Tg per year (Seiler *et al.*, 1984a). These CH_4 sinks altogether account for 340 Tg per year, which is about the destruction required to balance the CH_4 budget.

4.2.2.3 Temporal Changes of Sources and Sinks

The observed increase of atmospheric concentrations of CH_4 due to an imbalance between sources and sinks has apparently increased over time. The sinks may to a first approximation be considered as first order reactions, i.e. they are approximately proportional to the atmospheric CH_4 concentrations. In the case of methane, the most dominant sink mechanism is tropospheric oxidation by the OH-radical. It is possible that the OH-abundance has decreased with time, particularly due to increasing concentrations of carbon monoxide, which has a significant influence on the distribution of tropospheric OH (Crutzen, 1983). Measurements carried out during more than one decade in the Southern Hemisphere (Seiler *et al.*, 1984b) and in the free Northern Hemisphere troposphere over Germany and at Mauna Loa Observatory at $20°N$ (Seiler, 1985, unpublished data) indicate an upward monoxide trend of 0.6–1% per year which may have caused an average reduction of OH by about 0.2% per year. This is much less than the observed increase in atmospheric CH_4 indicating that increasing CH_4 production probably is the dominant cause for the observed increase. This also seems plausible in view of the observed ^{13}C concentrations in atmospheric CH_4 (Craig and Chou, 1982).

There is strong direct evidence that the total CH_4 source has increased during the last decades due to increasing anthropogenic activities (see e.g. Seiler,

Table 4.2   Estimated trend of CH_4 emission rates from individual ecosystems (Tg per year)

Sources	1940	1950	1960	1970	1975	1980
Ruminants	53.5	58	68.5	78.5	84	86
Rice paddies	64.5	74.5	89	105	115	117
Swamps/Marshes	79	73	63	54	51	47
Other biogenic	15.5	18	20.5	23	24	25
Biomass burning	49	57	65	71	75	79
Natural gas	2	5	11	24.5	30	34.5
Coal mining	19	19	26	29	31	35
Total	283	305	343	385	410	423

1984). Table 4.2 summarizes the production from different CH_4 sources calculated for selected years. The accuracy of the individual values in the table is of course less than indicated by the digits retained, which is, however, done to illustrate the trends as well as possible.

The trend of CH_4 emission by ruminants and rice paddies has been based on data for the increase of the ruminant population and the annually harvested rice paddy area, respectively, as published in the yearbooks of the Food and Agriculture Organization (FAO). In the case of swamps and marshes it has been assumed that the total surface area of these wetlands has decreased with time due to draining and dredging. We assess that the total area has decreased by about 1.2% per year from about 2.5×10^6 km^2 in the middle of the 1940s (Twenhofel, 1951) to 1.6×10^6 km^2 in 1980 (Clark, 1982). This may have caused a reduction of the CH_4 emission from swamps and marshes from about 79 Tg per year in 1940 to the present value of 47 Tg per year.

The emissions from 'other biogenic sources' (cf Section 4.2.2) were 7–22 Tg per year in 1980 and probably changed only slightly during previous decades. The CH_4 emission from waste disposal on the other hand may have tripled between 1940 and 1980.

The temporal change of the CH_4 emission by biomass burning is based on statistics published by the Food and Agriculture Organization (FAO, 1983), and data reported by Seiler and Crutzen (1980). We conclude that the burning of industrial and fuel wood has increased by about 1.1% per year. The burning of agricultural wastes probably increases at the same rate as food production (1.5% per year) and the burning due to shifting agriculture and deforestation may change at the rate of increase of rural population in tropical areas, i.e. 1.2% per year. In contrast, burning of savannahs and wild fires in temperate and boreal forests have been assumed not to change with time. On this basis, we deduce that the CH_4 emission from biomass burning has changed from about 49 Tg per year in 1940 to about 79 Tg per year in 1980.

The estimate of CH_4 emission because of transmission losses during distribution and use of natural gas has been based on the rate of natural gas production (Clark, 1982) assuming loss rates of natural gas to be 3–4%. Similarly the CH_4 emission by coal mining is assumed to be related to the total coal production. Increasing usage of natural gas and coal has then resulted in an increasing CH_4 emission from these two abiogenic sources from 21 Tg in 1940 to about 70 Tg in 1980.

The trend of the total CH_4 emission is shown in Figure 4.4. Although the absolute values of the individual annual CH_4 emission rates are uncertain, there is no doubt that the CH_4 emissions have increased during the last 40 years by about 140 Tg which corresponds to an average rate of about 3.5 Tg per year or 1% per year. The rate of CH_4 emission increase was highest between 1960 and 1975, when the cattle production, the harvested rice paddy area as well as the consumption of natural gas increased exponentially but has declined in recent years. If we assume that the atmospheric OH number density has not changed significantly during this period we may compute tropospheric CH_4 concentrations by assuming an approximate steady state at any time. The fact that the methane turnover time, about 10 years, is small compared with the doubling time of atmospheric CH_4 concentration implies that these values should be a reasonable first approximation of reality. We obtain values of about 1.04 ppmv for 1940, 1.12 ppmv for 1950, 1.26 ppmv for 1960 and 1.42 ppmv for 1970. The last figure is in fair agreement with the CH_4 amount given by Wilkniss *et al.* (1973), indicating that the calculated

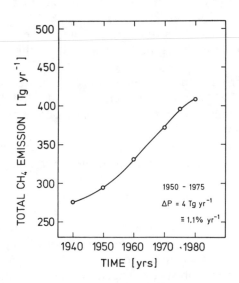

Figure 4.4 Trend of estimated total CH_4 emission rates between 1940 and 1980 (for detail see text)

CH_4 emission rates as shown in Figure 4.4 are reasonable. It should be pointed out, however, that the turnover time for CH_4 may be changing with time due to changes of the abundance of the OH-radical. Extrapolation of the CH_4 emission rates to times which were anthropogenically undisturbed yields a value of about 120 Tg per year which corresponds to a CH_4 mixing ratio of 0.5 ppmv compared to 0.7 ppmv as obtained from the analysis of air trapped in ice cores (Figure 4.3). Although approximate the above scenario of past changes is an internally consistent one.

4.2.2.4 Long-term Trends

A continued increase of tropospheric CH_4 may have an important influence on tropospheric and stratospheric chemistry by changing the distributions and concentrations of other gases, e.g. ozone which is another greenhouse gas. On the other hand other gases emitted into the atmosphere by man may affect the abundance of the OH-radical and thereby the methane concentration by decreasing the rate of methane destruction (cf Crutzen, 1985).

The reaction pathway for oxidation of CH_4 is strongly affected by the mixing ratio of NO. At NO mixing ratios higher than about 10 pptv in the lower troposphere, the oxidation pathway of CH_4 yields an average net gain of 2.7 ozone molecules and 0.5 OH-radicals per methane molecule oxidized (Crutzen, 1985). With an atmosphere containing less than about 10 pptv NO the main oxidation pathway is rather one which may destroy about 3 OH-radicals and 1.5 O_3 molecules per methane molecule oxidized to CO. The influence of NO may, therefore, lead to the interesting situation of increasing OH and O_3 concentrations in air with high NO mixing ratios, e.g. in the polluted atmosphere of the Northern Hemisphere and decreasing OH and O_3 concentrations in the cleaner Southern Hemispheric air. In fact, measurements indicate a significant increase of the tropospheric O_3 in the Northern Hemisphere in recent decades, whereas the O_3 abundance in the Southern Hemisphere seems to be constant (cf Section 4.2.5.2).

Increasing CH_4 mixing ratios might also have an impact on stratospheric chemistry, particularly due to the reaction with Cl forming HCl which is a stable compound in the lower stratosphere. HCl diffuses into the troposphere where it is removed by precipitation. Increasing CH_4 mixing ratios would lead to a reduction of the Cl abundance in the stratosphere and thus reduce the effect of active chlorine species which destroy ozone.

The oxidation of CH_4 in the stratosphere also provides a source of water vapour in the stratosphere. Increasing CH_4 mixing ratios in the troposphere will therefore cause higher stratospheric water vapour concentrations. Increasing water vapour mixing ratios might in turn cause a temperature increase in the stratosphere due to the absorption of infrared radiation.

4.2.3 Nitrous Oxide

4.2.3.1 Past and Present Concentrations

A secular increase of tropospheric N_2O abundance has been observed although the rate is considerably less than that of CH_4. The first reliable data showing such an increase were reported by Weiss (1981) who measured the N_2O mixing ratios in a large number of air samples collected between 1976 and 1980 at different locations in both the Northern and Southern Hemispheres. These data show an increase from 298 to 301 ppbv in the Southern Hemisphere and 299 to 302 ppbv in the Northern Hemisphere during this five-year period, corresponding to an average annual increase of 0.2% per year. Rasmussen and Khalil (1981b) reported an increase of 0.3% per year and most recently, Khalil and Rasmussen (1983b) also determined an increase of 0.3% per year based on continuous measurements at Cape Meares (45° N) and Tasmania (42° S).

These results agree very well with measurements by Seiler (unpublished data; 0.33% per year), obtained from continuous measurements at Cape Point (35° S). Additional information is provided by measurements at 5 GMCC (Global Monitoring for Climatic Change) stations in both hemispheres which show an increase of 0.5–1.1 ppbv per year or 0.2–0.3% per year during a 6-year period between 1977 and 1982 (Harriss and Nickerson, 1984). 300 ppbv of N_2O is equivalent to 1.500 Tg N.

4.2.3.2 Sources and Sinks

Nitrous oxide is destroyed in the stratosphere almost exclusively by photolysis and reaction with $O(^1D)$. NO is formed, which in turn plays a major role in regulating the stratospheric ozone concentration and distribution. No tropospheric sink mechanisms of significance have been detected so far. The stratospheric destruction has been determined from the observed vertical N_2O profile in the stratosphere to be 8.4–9.4 Tg N per year (Johnston *et al.*, 1979; Crutzen, 1983), corresponding to an average residence time for N_2O of about 170 years. A similar value (159 years) has been derived by Ko and Sze (1982) on the basis of two-dimensional model calculations. The annual increase of N_2O by 0.2–0.3% implies that the total source strength should be 12–14 Tg N per year. Estimates of N_2O emission rates published during the last decade are summarized in Table 4.3.

The flux of N_2O into the atmosphere is primarily due to microbial processes in soil and water and is a part of the nitrogen cycle. Until recently, denitrification (i.e. reduction of NO_3^- to N_2) has generally been considered to be the major mechanism for N_2O production. Recent experiments, however, have shown that nitrification (i.e. oxidation of NH_4^+ to NO_3^-) also

Table 4.3   Summary of N_2O emission rates from individual ecosystems (in Tg per year). The last column presents new estimates for this volume (W. Seiler)

Source	Hahn and Junge (1977)	Khalil and Rasmussen (1983b)	Crutzen (1983)	Seiler (this volume)
Ocean/Freshwater	16–185	9.0	1–2	2
Natural soils	6–65	13.4	(?)	6
Fertilizer	6–20	—	<3	0.6–2.3
Gain of cultivated land	—	6.6	1–3	0.2–0.6
Fossil fuel burning	1–4	—	1.8	—
Biomass burning	—	—	1–2	1–2
Lightning	10–55	—	—	<0.1
Total	80–300[a]	29	8–12	12–15

(a) includes N_2O production from unknown sources

plays an important role or may even be the dominant source. Furthermore, N_2O is not only produced but also destroyed in the soil and water, whereby an N_2O equilibrium value is established which is determined by parameters such as temperature, redox potential, pH value, etc. (see e.g. Seiler and Conrad, 1985). Therefore, reliable flux rates between ocean/soil and atmosphere can only be obtained by *in-situ* measurements under natural conditions.

There is presently little information on the N_2O flux from natural, undisturbed soils. Measurements on unfertilized agricultural soils show high diurnal variations (Conrad *et al.*, 1983; Slemr *et al.*,1984) and high spatial and seasonal variability (see e.g. Mathias *et al.*, 1980; Bremner *et al.*, 1980). N_2O emissions obtained by different groups vary between 2 and 36 g N $m^{-2} h^{-1}$ in temperate climates (Keller *et al.*, 1983; Seiler and Conrad, 1981; Goodroad and Keeney, 1984) and 4–35 g N $m^{-2} h^{-1}$ in subtropical climates (Slemr *et al.*, 1984). Based on these data, Slemr *et al.* (1984) estimated the global N_2O flux from unfertilized soils to be about 4.5 Tg N per year. This figure increases to about 6 Tg N per year if we account for new information on the N_2O production in tropical rain forests which indicate flux rates of about 43 g N $m^{-2} h^{-1}$ (Keller *et al.*, 1983).

The application of mineral nitrogen fertilizers leads to enhanced N_2O flux rates indicating that part of the applied fixed nitrogen is converted into N_2O and released into the atmosphere. Until recently it was generally assumed that about 10–15% might be lost as N_2O. New data show, however, that this figure is much too high and that the loss as N_2O is strongly dependent on the mode of application and type of fertilizer. The highest loss rates were observed for anhydrous ammonia and ammonium fertilizers which supports the view that nitrification is the dominant N_2O production process. Results published by Breitenbeck *et al.* (1980), Conrad *et al.* (1983), Slemr *et al.* (1984) show average N_2O loss rates of 0.04% for nitrate, 0.15–0.19% for ammonium and urea, and 5% for anhydrous ammonia. These values seem to be independent of climate (Slemr *et al.*, 1984) and may thus be representative for global conditions. Based on the total production rates of the different types of fertilizers, the global loss of mineral fertilizers in the form of N_2O is estimated to 0.5–2.0%. The same amount may be emitted from denitrification and/or nitrification of mineral fertilizers leaching from the fields into groundwater or surface freshwater ecosystems (Conrad *et al.*, 1983; Kaplan *et al.*, 1978). The total loss rate of fertilizer nitrogen then becomes 1–4%. With a total mineral nitrogen fertilizer consumption of about 60 Tg in 1980, the total N_2O emission due to application of nitrogen fertilizers amounted to 0.6–2.3 Tg N per year in 1980 with an average of 1.5 Tg N per year.

N_2O may also be released from soils due to the increase of land area used for agriculture. During the last decade the cultivated land area has increased by about 0.3% per year (FAO, 1983) or 5×10^{10} m^2 per year which compares well with the area of seasonal and tropical forests (5.3×10^{10} m^2) cleared annually (Bolin *et al.*, 1983). Revelle and Munk (1977) estimated the total land area converted into cultivated land during the last 100 years to be on the order of 8.5×10^{12} m^2 which probably has caused an annual loss of soil carbon into the atmosphere of $0.1–0.4 \times 10^{15}$ g C. With a nitrogen to carbon ratio of about 1:20 for soil organic carbon, this carbon loss is equivalent to a nitrogen loss of 5–20 Tg per year of which about 3% (Terry *et al.*, 1980) or 0.2–0.6 Tg N may be lost as N_2O.

N_2O is also formed by nitrification in ocean water and emitted into the atmosphere. First estimates by Hahn and Junge (1977) indicated the total oceanic N_2O source to be 16–160 Tg N per year. Much lower values (4–10 Tg N per year) have been obtained by Cohen and Gordon (1979) and more recently Hahn (1981) arrived at a value of 14 Tg N per year by reconsideration of his earlier data. Extensive oceanic observations in the Atlantic Ocean between 50° N and 40° S (W. Seiler, unpublished data) indicate an average N_2O flux from the ocean into the atmosphere of about 2 Tg N per year which supports an earlier suggestion by Weiss (1981) that the oceanic N_2O source is small compared to the stratospheric sink.

N_2O is also emitted into the atmosphere by anthropogenic activities such as fossil fuel and biomass burning. The first estimates of the N_2O emission from fossil fuel burning were reported by Weiss and Craig (1976) and Pierotti and Rasmussen (1976) who deduced emissions of 1.6 and 2.2 Tg N per year, respectively, at the beginning of the 1970s. Production rates for 1980 can be calculated by applying the N_2O/CO_2 ratios reported by Weiss and Craig (1976) to be 2.05×10^{-4} for coal burning and 2.25×10^{-4} for oil burning. The ratio for burning of natural gas is about 1×10^{-4} and is calculated from data published by Pierotti and Rasmussen (1976). With an average figure of about 2×10^{-4}, the total N_2O emission from fossil fuel burning in 1980 becomes about 2.9 Tg N.

The N_2O emission from biomass burning was originally estimated by Crutzen *et al.* (1979) to be 8 Tg N per year, but was reduced to 1–2 Tg N per year when new N_2O emission factors from biomass burning became available (Crutzen, 1983).

Summarizing the N_2O production from all potential sources yields a likely total N_2O emission of 12–15 Tg N per year. Considering the uncertainties of the estimates of the individual sources, the agreement of these production rates with the total source strength deduced from the rate of increase of atmospheric N_2O may be fortuitous. The contributions of the different sources may turn out to be markedly different as more data become available.

4.2.3.3 Temporal Changes of Sources and Sinks

Using the N_2O/CO_2 ratio in emissions from fossil fuel burning (Weiss and Craig, 1976) we estimate that the N_2O emission from fossil fuel burning may have increased from about 0.6 Tg N per year in 1940 to the present value of about 1.9 Tg N per year which corresponds to an annual increase of about 0.03 Tg N or 3% per year as an average during these 40 years.

Based on the production rates of mineral fertilizers (FAO, 1983), the annual N_2O emission rate has probably increased from about 0.01 Tg in 1950 to about 1.4 Tg in 1980. Because of the upward trend of biomass burning, the N_2O emission from this source may have almost doubled during the last 40 years from 0.6–1.1 Tg N per year in 1940 to about 1–2 Tg N per year in 1980.

Increasing amounts of N_2O should also have been released due to the increasing demand of food and correspondingly increasing area of cultivated land. Because of the growth of world population, the N_2O emission of this source may have doubled during the last 50 years from 0.3 Tg N in 1930 to about 0.6 Tg N in 1980.

Summarizing all N_2O emission rates, the total N_2O flux into the atmosphere may have changed from 8–9 Tg N per year during anthropogenically undisturbed conditions to the present value of about 14 Tg N per year (see Figure 4.5). The upward trend of N_2O emission rate increased from about 0.1% per year at the beginning of the century to about 1.3% per year during the last 10 years, primarily due to the rapidly increasing emissions from fertilizer application and combustion of fuels. If we accept a residence time for N_2O in the atmosphere of about 170 years and first order sink mechanisms, we can deduce that the emission scenario shown in Figure 4.5 should have caused an increase of the N_2O mixing ratio by about 8% since late last century and a present annual increase by somewhat less than 0.3%. The latter figure is in good agreement with observations. The undisturbed N_2O mixing ratio therefore probably was about 280 ppbv as compared to 301 ppbv in 1980.

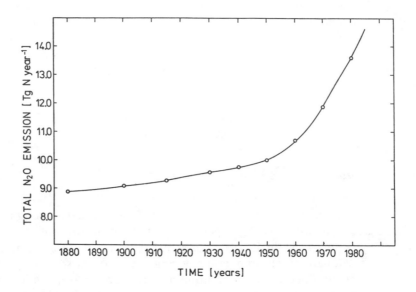

Figure 4.5 Trend of estimated total N_2O emission rates between 1880 and 1980 (for detail see text)

4.2.3.4 Interaction with Other Atmospheric Trace Gases

In contrast to CH_4, the influence of increasing N_2O on the atmospheric chemistry seems to be restricted to the stratosphere where N_2O is destroyed by reaction with atomic oxygen. This leads to formation of nitric oxide which in turn reacts with stratospheric O_3 by a catalytic reaction sequence, leading to a reduction of the overall O_3 abundance in the stratosphere by 3–5% for a doubling of the N_2O mixing ratio, assuming otherwise present atmospheric conditions.

4.2.4 Chlorofluorocarbons (CFCs)

Past and Present Concentrations

The presence of chlorofluorocarbons in the atmosphere was detected in the early 1970s and caught considerable attention in 1974 because of their being a possible source for chlorine in the stratosphere and accordingly a possible threat to the ozone layer (Molina and Rowland, 1974). Soon thereafter, Ramanathan (1975) and Wang *et al.* (1976) pointed out the possibility that due to their infrared absorption bands the CFCs enhance the atmospheric opacity and contribute to the greenhouse effect. Ramanathan *et al.* (1985) have recently considered a series of chlorinated and/or fluorinated hydrocarbons in the atmosphere, almost all of them exclusively anthropogenic, with regard to their role in the radiative balance of the atmosphere. It is clear that $CFCl_3$ (F11) and CF_2Cl_2 (F12) are the most important ones and we are justified in primarily considering these two gases in the following analysis of possible present and future changes of climate due to their emissions.

The CFCs are produced for a variety of uses such as solvents, refrigerator fluids and spray can propellants. Although atmospheric measurements are available for only about a decade, past concentrations can be deduced with reasonable accuracy on the basis of production and emission figures as provided by the Chemical Manufacturing Association (CMA, 1982). Figure 4.6 shows past emissions of F11 and F12. A rapid increase until about 1970 changed to a decline during the latter part of the 1970s, which was caused by restrictions on the use of CFCs introduced by some countries because of their possible threat to the ozone layer. It should be noted, however, that during this period the non-propellant use has continued to increase by about

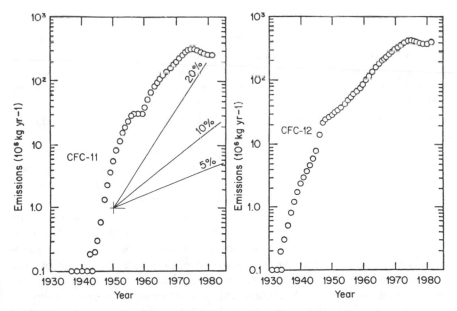

Figure 4.6 Historical emissions of CFC-11 ($CFCl_3$) and CFC-12 (CF_2Cl_2), (CMA, 1982). In the left figure the slopes corresponding to 5%, 10% and 20% annual increase are indicated (from CMA, 1982)

4% per year, while the propellant use has decreased from 56% to 34% of the total CFC production. A marked increase of the total CFC use has been reported for 1983.

The CFCs are photochemically decomposed almost exclusively in the stratosphere. On the basis of the temporal changes observed in recent years in relation to emissions as well as overall budget considerations the atmospheric lifetimes of F11 and F12 have been estimated to be about 80 and 170 years respectively (Ko and Sze, 1982; Cunnold *et al.*, 1983a, 1983b).

At the beginning of 1980 the average mixing ratio of F11 in the lower troposphere is estimated to have been 168 pptv and to have been increasing at an annual rate of 5.7%. Mid-latitude values for the Northern and Southern Hemispheres seem to have been about 10 pptv higher and lower respectively than the global average which is due to the fact that most emissions occur in the northern middle latitudes (Cunnold *et al.*, 1983a). The F12 concentration is estimated to have been 285 pptv with an annual increase of 6% and a difference between the Northern and Southern Hemispheres by about 30 pptv (Cunnold *et al.*, 1983b).

4.2.5 Ozone

4.2.5.1 Formation and Destruction

Atmospheric ozone varies considerably both in space and time as a result of interactions between atmospheric motions and chemical reactions. We know these reasonably well as a result of the development of complex models of relevant chemical reactions and long series of observations from the Earth's surface and since 1978 also from satellites (WMO, 1981; Mateer *et al.*, 1980; Zerefos and Ghazi, 1985; IAMAP, 1985; NASA, 1984; McPeters *et al.*, 1984).

It is now accepted that the concentration of tropospheric ozone is increasing due to photochemical processes. As was already discussed in Section 4.2.2.4, methane, carbon monoxide and nitrogen oxides (NO_x) play important roles in this context. Their increasing concentrations and reactions with the hydroxyl-radical are of prime importance for the ozone chemistry in the troposphere (Crutzen, 1985). Evidence for production of background O_3 in the troposphere has been reported by Fishman and Seiler (1983). Quantitative information is, however, difficult to obtain because of our present inadequate knowledge about the relative importance of the processes involved. From analysis of photochemical sink processes for ozone Fishman *et al.* (1979) conclude that production of ozone probably is more important for the ozone budget of the troposphere than the influx of ozone from the stratosphere.

4.2.5.2 Tropospheric Ozone

Hartmannsgruber, Attmannspacher and Claude (1985) have analysed the ozone sonde data from Hohenpeissenberg (Federal Republic of Germany) during the period 1967 to 1980 and found an increase of about 60%, i.e. an average annual increase of 4%. Feister and Warmbt (1985) have observed a 61% increase of tropospheric ozone between 1956 and 1983, i.e. 2.3% per year. Dütsch (1985) has found an annual increase of 0.7% per year from an analysis of ozone sonde ascents at levels between 1.5 and 7 km over Payerne (Switzerland) during the period 1968–1984. Reiter *et al.* (1983) have observed an increase of the ozone volume fraction by 32% from 1973 to 1982 for Garmisch-Partenkirchen (Federal Republic of Germany), most of which occurred between 1980 and 1982.

Bojkov and Reinsel (1985) and Logan (1985) have recently analysed the numerous observations (those quoted above and by others) from many parts of the world made by ozone sondes and Umkehr measurements. The uncertainties of these observations are still considerable (Hilsenrath *et al.*, 1985; WMO, 1982). Instrumental corrections, in exceptional cases more than 30%, need to be applied (De Muer, 1985). Many series of observations are from regions with considerable industrial activity and the global representativeness may be questioned.

Since the lifetime of the ozone molecule in the troposphere is comparatively short (a few weeks), considerable spatial variations occur and it is difficult to assess the possible global change of ozone. An increase has, however, clearly taken place at middle and high latitudes of the Northern Hemisphere during the last 2–3 decades, particularly during the summer months. Changes by 20–50% have been recorded and the present rate seems to be 1–2% per year. Few long records are available from the tropics, but in areas of extensive biomass burning (South America, Africa, India) an increase seems also to occur. Although probably not representative for the hemisphere as a whole (and certainly not for the globe), the average annual change of up to 100 mb ($\approx$16 km) at Resolute (Canada) and Hohenpeissenberg shows the typical upward trend of ozone as a function of elevation at middle and high latitudes during the last few decades (Figure 4.7).

4.2.5.3 Stratospheric Ozone

Dütsch (1985) has extracted the following trends from ozone soundings over Payerne for the period 1968–1984; −0.7% per year for the 12.5–16 km level, −0.5% per year at 19–21 km, +0.2% per year at 24 km, zero change at 27 km, and +0.1% per year at 30 km.

Measurements from 15 Umkehr-stations (see Figure 4.8) published by Reinsel *et al.* (1984) indicate a reduction of the ozone amount above 35 km

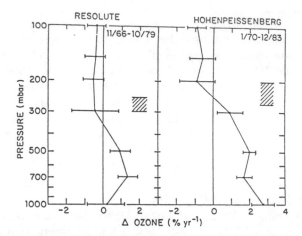

Figure 4.7 Vertical distribution of the trend of ozone (in % per year) at Resolute (75°N, Canada, 1966–1979) and Hohenpeissenberg (47°N, Germany, 1970–1983). The horizontal bars give the 90% confidence interval. The shaded area shows the range of the tropopause heights (from Logan, 1985)

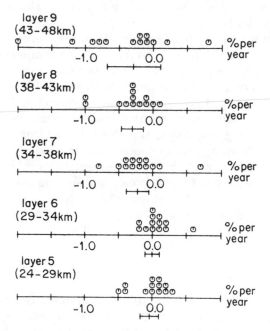

Figure 4.8 Estimated trends (in % per year) over the period 1970–1980 at 13 Umkehr-stations for each Umkehr layer, 5–9 using model which includes smoothed atmospheric transmission data. (Overall 95% confidence interval estimates are also indicated on diagrams.) From Reinsel *et al* (1984)

altitude and maybe a small increase below this level. A statistically significant trend for the period 1970–1980 of -0.2 to -0.3% per year is indicated in the layer 34–48 km. Changes at lower stratospheric levels do not seem to be statistically significant. In the data from Hohenpeissenberg a slight decrease is found between 14 and 24 km and a substantial decrease by 3–4% from 1977 to 1982 above 28 km corresponding to -0.7% per year. Satellite measurements between 1979 and 1982 also show a distinct change of roughly -5% at 1 mb (45 km) (Fleig *et al.*, 1985; Planet *et al.*, 1985). It should be emphasized, however, that rather rapid changes can occur at these upper levels, which should not necessarily be interpreted in terms of a trend.

4.2.5.4 Total Ozone

The world's longest Dobson observation record at Arosa indicates an increase of the total ozone by 2–3% between 1925 and 1942, a maximum around 1942, and a decrease by 1–2% between 1960 and 1981 before the El Chichon eruption suddenly decreased the total ozone by about 5% (Dutsch, 1985). From the Payerne soundings a change between 1940 and 1980 by about -3%, i.e. -0.75% per decade can be deduced. Komhyr *et al.* (1985) report a change of about -3% per decade since 1970 at Mauna Loa. Since about 90% of the ozone is found in the stratosphere, the trends of total ozone primarily reflect changes at these higher levels.

4.3 LIKELY FUTURE CONCENTRATIONS OF ATMOSPHERIC GREENHOUSE GASES

4.3.1 General Guidelines

Projections of likely future atmospheric greenhouse gas concentrations are very uncertain, partly because their generation, circulation and destruction in the atmosphere are not well understood, partly because the way man will disturb these natural cycles in the future cannot be foreseen very well. Although we shall attempt to define as 'realistic' scenarios as possible, these should be considered to be serving as a basis for assessments of the sensitivity of the climate system to anthropogenic disturbances, rather than describing most likely future changes.

As has been pointed out above there are interactions between the key greenhouse gases which we are considering, not only directly by chemical interaction, but also indirectly in that many natural processes are temperature dependent and thus sensitive to an induced climatic change. This fact needs to be carefully considered.

A few general guidelines for the estimation of likely future concentration should be noted

— The longer the mean atmospheric lifetime of an emitted gas is, the more probable it is that it will be accumulating in the atmosphere and that the increase will remain about proportional to the emissions.

— Gases with short residence times in the atmosphere will reach an equilibrium concentration comparatively soon , which is approximately proportional to the rate of emission.

— Although positive or negative feedback mechanisms often cannot be determined quantitatively, they may be considered qualitatively in assessing likely future changes.

In a recent paper Ramanathan *et al.* (1985) have considered altogether 47 atmospheric trace gases with regard to their possible contribution to the atmospheric greenhouse effect and its future change. Many of these only play a very minor role and will not be considered in any detail. We

Table 4.4 Pre-industrial and present volume fraction, residence time and estimated future volume fraction of atmospheric greenhouse gases (based on Ramanathan *et al.*, 1985)

Gas	Tropospheric volume fraction		Tropospheric residence time (yrs)	Projected volume fraction 2030 (ppbv)	
	Pre-industrial (ppbv)	1980 (ppbv)			
CO_2	$275 . 10^3$	$339 . 10^3$	($\approx$7)	$450 . 10^3$	
CH_4	700	1550	10	2340	1850–3300
N_2O	280	301	170	375	350–450
$CFCl_3$ (F11)	—	0.17	80	1.1	0.5–2.0
CF_2Cl_2 (F12)	—	0.28	170	1.8	0.9–3.5
$CHClF_2$ (F22)	—	0.06	20	0.9	0.4–1.9
CH_3CCl_2	—	0.14	8	1.5	0.7–3.7
CF_3Cl (F13)	—	0.007	400	0.06	0.04–0.1
CF_4 (F14)	—	0.07	>500	0.24	0.2–0.31

	Projected change in per cent from pre-industrial period
O_3 Tropospheric	+12.5
O_3 Stratospheric: 10km	+3.8
22km	+4.5
26km	+2.0
30km	−6.1
34km	−22.6
40km	−37.9
50km	−5.5

list in Table 4.4 the most important ones and also the estimates of future concentrations as given by Ramanathan *et al.* (1985). These will be discussed further below. Scenarios for CH_4, N_2O and CFCs have also been presented by Hoffman *et al.* (1983). Wuebbles *et al.* (1984) have published a Proposed Reference Set of Scenarios for Radiatively Active Atmospheric Constituents in which time-dependent as well as steady state scenarios are considered. For model studies it is important that some standardized scenarios are developed in order to intercompare the relative accuracy of different radiation models. The reader is referred to the report of Wuebbles *et al.* (1984) for detailed description of the proposed scenarios.

We make the following specific comments on the estimates of probable future concentrations of the greenhouse gases as given in Table 4.4.

4.3.2 Methane, CH_4

There is no doubt that the tropospheric CH_4 mixing ratio will continue to increase mainly due to anthropogenic activities in agriculture to meet the increasing demand for food. The rate is, however, difficult to predict as CH_4 is emitted by various sources whose individual future trends may differ considerably (cf Section 4.2.2). In addition, the future change of atmospheric CH_4 is also dependent on the anthropogenic emission of other trace constituents such as CO, NO_x, non-methane hydrocarbons, etc., which will affect the abundance and distribution of OH.

We can deduce an approximate value for future mixing ratios by extrapolating the linear regression between tropospheric CH_4 abundance and total human population as observed in the past (Figure 4.3) using estimates for the future world population. In this way we find that the CH_4 mixing ratios may reach values of 2.0 and 2.5 ppmv, for 2000 and 2050 respectively. At steady state these mixing ratios require CH_4 emission rates of 550 to 700 Tg per year. Whether or not these figures will be reached obviously depends on the population growth rate and on the possible introduction of new varieties of rice plants with higher yields, the influence of application of mineral fertilizer to rice paddies, increase of fossil fuel burning and future CH_4 destruction rates in turn depending on emissions of air pollutants.

The value given by Ramanathan *et al.* (1985) in Table 4.4 is in close agreement with the projection above. A more rapid increase has been proposed by Harriss and Nickerson (1984) which leads to a doubling (i.e. about 3.3 ppbv) in 2050. Also Hoffman *et al.* (1983) project a more rapid increase. An argument for a faster increase is the possible release of continental-slope sediment methane clathrates due to oceanic warming (Revelle, 1983). Since the arguments for increasing CH_4 production are not based upon very accurate knowledge of all processes involved, the projection by Ramanathan *et al.* (1985) seems reasonable, although possibly somewhat low.

4.3.3 Nitrous Oxide, N_2O

As was shown in Section 4.2.3, N_2O emissions seem to have increased rather rapidly during the last few decades from merely about 0.1% per year at the beginning of this century to about 1.3% per year during the last 10 years. This has been predominantly caused by the increase of N_2O from fertilizer application and combustion of fuel. If we accept a continued exponential growth of emissions we derive a value of 375 ppbv for 2030 and 500 ppbv for 2050 (Weiss, 1981). There is, however, strong evidence for believing that the upward trend of anthropogenic N_2O emission rates will slow down because of less rapid increase of the use of fossil fuels and limitations of the total acreage of land used for cultivation. We must keep in mind, however, that the atmospheric residence time of N_2O is long (about 170 years) so that the N_2O mixing ratios will approach constant steady state mixing ratios only after a period of about 200 years even if the N_2O release rates would stay constant. Such a steady state mixing ratio would be about 500 ppbv if the present production remains constant at 14 Tg N per year. If we assume a doubling of the fossil fuel consumption within the next 100 years (cf Chapter 2) and a likely maximum fertilizer consumption of 250 Tg N per year, the total N_2O production rate may become as high as 20 Tg N per year. The atmospheric concentration would still only reach about 360 ppbv in 2030 assuming a residence time of 170 years (cf Section 4.2.3). We accept the projections by Ramanathan *et al.* (1985) as reasonable, but consider the upper limit (450 ppbv) as unrealistically high.

4.3.4 Chlorofluorocarbons, $CFCl_3$, CF_2Cl_2

Because of the comparatively long residence times of these CFCs only about 10% of the amounts emitted into the atmosphere has so far been decomposed. During several decades to come a large fraction of the accumulated emissions will remain in the atmosphere. Because of the continued increase of the non-propellant use of CFCs by about 4% per year (cf Section 4.2.4) the slowdown of emission increase during the last decade most likely is only a temporary one, if no new restrictions on future uses of CFCs will be agreed upon. The projections made by Ramanathan *et al.* (1985) correspond to an annual emission increase by about 3% from prevailing conditions in 1980 and seem in this sense reasonable, although this is much less an increase than during the 1960s and early 1970s (cf Figure 4.6). Wuebbles *et al.* (1984) have used a model of Wuebbles (1983) to estimate the steady state CFC concentrations which would gradually be reached with constant emissions at 1983 rates and find the values 0.6 ppbv and 1.3 ppbv for F11 and F12 respectively. An exponential increase starting with the observed rates of emission increase in 1980 (about 5% per year) yields on the other hand the extreme values of 2.1 and 3.0 ppbv in 2030 and 5.6 and 7.8 ppbv in 2050, respectively.

A rapid increase of CFCs might gradually have a profound influence on stratospheric ozone. Concentrations above 1–2 ppbv may reduce ozone concentrations by 10% or possibly more. In view of the recent progress in agreement on an international ozone convention, it seems plausible that some restrictions on the use of CFCs will be agreed upon and that the emission scenarios given in Table 4.4 exaggerate likely future increases. It is still of considerable interest to carry out sensitivity analyses with regard to the role of F11 and F12 in changing the global radiation climate using the scenarios given in Table 4.4.

A few other halocarbons that may have a slight effect on radiative transfer through the atmosphere have also been listed in Table 4.4. Their likely effects are comparatively small and we need not consider their role further in the present context. The interested reader is referred to Ramanathan *et al.* (1985).

4.3.5 Ozone, O_3

The assessment of likely future changes of ozone is difficult because of the complex photochemical interactions that occur. All the gases discussed above play a role, and in addition other chlorinated species, since chlorine formed in their photolytic decomposition interacts with the ozone molecule. There is also a feedback through the temperature changes that the different greenhouse gases bring about. The results obtained by DeRudder and Brasseur (1985), Owens *et al* (1985) and Ramanathan *et al.* (1985) differ in some important respects. The former two arrive at considerably larger enhancement of tropospheric ozone than do Ramanathan *et al.* (1985). The differences depend both on the treatment of the chemistry and transport of the nitrogen catalysts and other possible chemical interactions, as well as different assumptions about likely future emissions of the whole series of the other trace gases that are of importance. We shall adopt their scenario (Table 4.4) for the following analysis. Tropospheric ozone is accordingly considered to increase on the average by 0.25% per year, which is considerably less than has been observed during the last two decades (cf Section 4.2.5). The upward trend is assumed to decrease above 9 km and becomes zero at 27 km. At higher elevation a substantial decrease is foreseen, primarily induced by CFCs and other chlorinated trace gases.

4.4 RADIATIVE EFFECTS OF GREENHOUSE GASES

4.4.1 Some Principal Considerations

The greenhouse gases we are concerned with affect the radiation field principally in the same way as does CO_2, i.e. increased amounts reduce the direct emission of infrared radiation from the Earth's surface to space and increase the emission from stratospheric levels. There are, however, some differences. Due to photochemical reactions the mixing ratio of most of these greenhouse gases decreases more rapidly above the tropopause than does CO_2. This leads to a more rapid decrease of the cooling rate between 15 and 35 km.

Ozone has a vertical distribution that is very different from those of the other greenhouse gases, which must be accounted for. Ozone also has an absorption band in the solar part of the spectrum. This absorption has the opposite effect of that due to radiation transfer in the infrared part.

In assessing the temperature changes that may result from changing concentrations of these greenhouse gases their contributions must be dealt with simultaneously because their absorption bands partly overlap. A radiation model that can be used for such computations necessarily becomes complex. It has not yet been possible to incorporate at the same time possible feedback mechanisms that are present in the real climate system, and which have been shown to be of importance by experiments with general circulation models (GCMs). So far only one-dimensional models have been employed. By comparison of experiments to assess the effects of increasing concentrations of CO_2 using GCMs and one-dimensional, radiative-convective models the latter can be corrected for ice-albedo and cloud feedback mechanisms not explicitly accounted for (cf Section 4.4.3 below).

4.4.2 Model Computations and the Importance of Spectral Overlapping

A most detailed analysis of the changes to be expected from the simultaneous increase of a number of greenhouse gases in the atmosphere has been presented by Ramanathan *et al.* (1985). They employ a one-dimensional, radiative-convective model which includes latent heat flux between the Earth's surface and the atmosphere, they improve the methods for transmission computations, and analyse in some detail the errors introduced by using different resolutions in describing the spectra for the different gases. In their analysis of the role of overlapping absorption bands, first suggested by Tannhuser (1968), they make use of the Malkmus narrow band model. They have, however, omitted the temperature dependence of hot band strength for gases other than CO_2, CH_4, and N_2O.

The effect of band overlapping on the computed temperature increase at the Earth's surface is shown in Table 4.5 for a series of gases considered in this analysis. Two different temperature increases are presented: the ('skin') temperature of the Earth's surface (ΔT_g) and the air temperature at the surface (ΔT_s). The table also shows ΔT_g computed for each gas separately, i.e. without consideration of overlapping. ΔT_g is generally about 10% larger than ΔT_s, which presumably is due to more direct heating of the atmosphere by radiative processes than by sensible and latent heat flux from the Earth's surface to the atmosphere. We further notice the marked differences between different gases with regard to overlapping absorption bands. The roles of CH_4 and N_2O are reduced by about 50% particularly due to overlap with water vapour. The CFCs on the other hand, have primarily absorption bands in the 'window region', where absorption by H_2O and CO_2 is weak. There is only 10–20% reduction of their role due to overlapping. For further details reference is made to the original publication by Ramanathan *et al.* (1985).

Table 4.5 The effect of overlapping of absorption bands on the computed surface warming. ΔT_s and ΔT_g are the changes of atmospheric temperature at the Earth's surface and of ground temperature respectively

Constituent	Change of greenhouse gas	With Overlap[a]		Without overlap
		ΔT_s (°C)	ΔT_g (°C)	ΔT_g (°C)
CH_4	1.25x	.09	.08	.19
N_2O	1.25x	.12	.10	.20
CF_2Cl_2 (F12)	0–1 ppbv	.16	.15	.19
$CFCl_3$ (F11)	0–1 ppbv	.14	.13	.16
CF_4 (F14)	0–1 ppbv	.06	.05	.12
CF_3Cl (F13)	0–1 ppbv	.22	.20	.25
CH_2Cl_2	0–1 ppbv	.03	.02	.04
$CHCl_3$	0–1 ppbv	.06	.06	.09
CCl_4	0–1 ppbv	.08	.07	.11
C_2H_2	0–1 ppbv	.02	.02	.07
C_2F_6 (F16)	0–1 ppbv	.13	.12	.20
CH_3CCl_3	0–1 ppbv	.02	.01	.03
PAN	0–1 ppbv	.04	.03	.06

a) the numbers have been rounded off. When the ΔT is less than .02 K the result may be uncertain by as much as ± 30%.

T_s is surface-air temperature change and ΔT_g is surface (or ground) temperature change.

4.4.3 Climatic Effects of Projected Increases of Other Greenhouse Gas Concentrations

The analysis of likely climatic effects of increasing greenhouse gas concentrations is a complex problem and requires the use of general circulation models (GCMs) as discussed in Chapter 5. No such experiments with enhanced concentrations of CH_4, N_2O or CFCs have, however, been made. We shall therefore express the climatic effects due to increasing concentrations of these gases in comparison with those deduced for increasing CO_2 concentrations by using the results from experimentation with a one-dimensional model of the atmosphere–ocean system (Ramanathan, 1980). Cloud and ice–snow–albedo feedbacks as well as other more complex interactions can, however, only be analysed by using three-dimensional general circulation experiments. As will be described in Chapter 5, such more elaborate experiments have been conducted to determine the likely importance of such processes in assessing the global temperature change due to doubling of CO_2. It is found that the most likely change becomes 3.5 ± 2.0 °C, which we compare with a warming by merely 2.1 °C, if employing a one-dimensional model (see Chapter 5). Based on this comparison and the assumption that similar modifications of the climate system would be found, if changing concentrations of the other greenhouse gases were introduced into GCMs, we shall adopt a factor 1.7 ± 0.9 to translate the results from one-dimensional models to changes expected in more complex GCMs, to take into account in an approximate way the feedback processes mentioned above.

The model results referred to above concern the likely change into a new equilibrium following an instantaneous change of greenhouse gas concentrations in the atmosphere. As is also pointed out in Chapter 5, a considerable delay of such a change can be expected because of the inertia of the climate system primarily due to the oceans. In the present situation, with gradually increasing concentrations of the greenhouse gases, merely about half of the equilibrium change may have been realized. For a more extensive discussion of this problem reference is made to Chapters 5 and 6.

With the aid of a one-dimensional model Ramanathan *et al.* (1985) have shown that CH_4, N_2O, $CFCl_3$, CF_2Cl_2 and ozone account for more than 90% of temperature changes due to other greenhouse gases than CO_2. The most important other gases have been listed in Table 4.4 with rough estimates of plausible future concentrations.

It is first of interest to assess the temperature changes that may have been induced by past increases of the concentrations of these other greenhouse gases. On the basis of the changes given in Table 4.4. (except that the preindustrial CH_4 concentration has been put equal to 1.15 ppmv) Ramanathan *et al.* (1985) deduce with their one-dimensional model an increase of 0.27 °C as compared with a CO_2-induced increase of 0.52 °C, i.e. a total increase of 0.79 °C. Referring to our discussion above, this would correspond to a total change of 0.6–2.1 °C if employing a GCM, of which 0.3–1.1 °C should have been realized if accounting also for the inertia of the climate system (cf Chapters 5 and 6). About one third of this change is due to the increase of the concentrations of other greenhouse gases.

Ramanathan *et al.* (1985) next compute the likely changes of the equilibrium temperature between 1980 and 2030, using the projected greenhouse gas concentrations (including uncertainties) as given in Table 4.4. The results Ramanathan .*et al.* obtained with their one-dimensional model are shown in Figure 4.9. The increase of CO_2 from 340 ppmv to 450 ppmv would bring about a temperature change of 0.7 °C and that due to the increase of other greenhouse gases becomes 0.3–1.8 °C with a most plausible value of 0.8 °C. If we add past changes as computed above and also consider the uncertainty of future CO_2 changes (cf Chapter 3, Figure 3.2) we obtain:

Computed equilibrium temperature change due to increasing greenhouse gases 1890–2030 (one-dimensional model). The indicated uncertainty is due to uncertainty of future projections of gas concentrations.

Carbon dioxide	−1980	0.5 °C	
	1980–2030	0.7 °C	(0.3–1.1 °C)
Other greenhouse gases	−1980	0.3 °C	
	1980–2030	0.8 °C	(0.3–1.8 °C)
Total		2.3 °C	(1.4–3.7 °C)

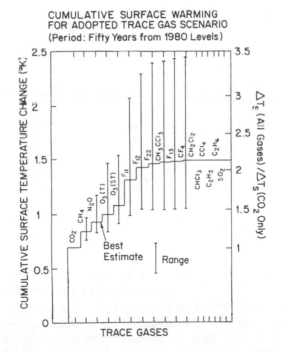

Figure 4.9 Cumulative equilibrium surface temperature warming due to increase in CO_2 and other trace gases, from 1980 to 2030. After Ramanathan *et al.*, 1985

We note that the value 2.3 °C is about the same as the expected temperature change for doubling of CO_2 (as deduced with the aid of the one-dimensional model). The uncertainty due to the inadequate knowledge of future greenhouse gas concentrations is considerable. Accounting also for other feedback mechanisms in the climate system that present GCMs are able to describe and as discussed above we deduce an increase of the global equilibrium temperature of at least 1.1 °C (slow increase of CO_2 as well as other greenhouse gases), more probably about 3.5–4.0 °C and possibly even more (if the high scenario projections of greenhouse gas concentrations are realized). The uncertainty range and delay due to the inertia of the climate system is discussed further in Chapters 5 and 6.

A few additional comments with regard to the role of the other greenhouse gases are of interest. CFCs may have a warming effect at the tropical tropopause which in turn might lead to an enhanced flux of water vapour into the stratosphere. The vertical distribution of the temperature change is also different from that due to CO_2 alone (Figure 4.10), and the implications should be further analysed.

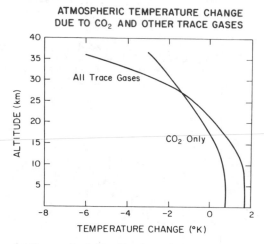

Figure 4.10 Change in the vertical distribution of temperature due to an increase of CO_2 alone and CO_2 with all other trace gases listed in Table 4.5. (After Ramanathan *et al.*, 1985)

The most important contributions to the heating come from $CFCl_3$, CF_2Cl_2, N_2O, and CH_4. We note also that both the decrease of ozone in the stratosphere and increase in the troposphere as projected contribute to warming at the Earth's surface. It should be emphasized, however, that the results with regard to the likely temperature change by a modified ozone distribution in the atmosphere are quite uncertain and should be interpreted with caution.

It is finally of some interest to translate the impact of the other greenhouse gases in the atmosphere to an equivalent additional CO_2 concentration, which would have about the same effect with regard to warming at the Earth's surface. This approach has been proposed by Flohn (1978) and Ramanathan (1980). In these terms the concentration of the other greenhouse gases in 1980 is equivalent to an additional atmospheric CO_2 concentration of about 40 ppmv and concentrations projected for 2030 equivalent to about 140 ppmv, i.e. the equivalent CO_2 concentration in 1980 was then about 380 ppmv and a plausible value for 2030 would be about 590 ppmv.

4.5 AEROSOLS

4.5.1 Some Basic Considerations

The principal difficulty encountered when trying to assess the possible climatic effects of increasing aerosol concentrations in the atmosphere due to anthropogenic activities is associated with the fact that their residence time in the troposphere is short (days to about a month) and that the global distribution therefore is very variable. Also the radiative characteristics of atmospheric aerosols vary significantly in space and time due to the great many different sources that are of importance. Although we know that local and regional changes have occurred, it is not yet possible to tell whether or not man has induced some globally averaged changes, nor do we know what future changes of this kind may occur. It has been estimated that natural aerosols probably reduce the global equilibrium temperature at the Earth's surface by about 1 °C.

It should be emphasized that temporary changes of climate due to increasing aerosol concentrations in the stratosphere from volcanic eruptions must be well understood in order to distinguish between these and possible changes due to anthropogenic emissions of both gases and aerosols.

Obviously systematic observations and further theoretical analyses are needed to improve our present understanding of the importance of atmospheric aerosols for the global climate. In the following we describe briefly present approaches to classify observations systematically and to develop appropriate models for analysis of the impact of changing aerosol distributions on climate.

4.5.2 Aerosol Types, Distributions and Variability

The relative magnitude of production rates from major sources of aerosols is listed in Table 4.6. As seen values are quite uncertain. The residence time of aerosols in the atmosphere depends on particle size, the height at which injection occurs and condensation processes by which aerosols are washed

Table 4.6   Estimated relative rates for aerosol production processes (%), after (a) Peterson and Junge (1971); (b) SMIC (1971); (c) Twomey (1977a); and (d) Dittberner (1978)

Processes	a	b	c	d
Sulphates (from H_2S and SO_2, gas-to-particle conversion including volcanic and man-made)	35	27–15	29–18	41–37
Ammonium salts	–	8–10	9–12	41–37
Sea-salt spray	33	31–11	33–14	23–27
Soil and rock debris	16	10–19	11–23	14–16
Organic volatiles from plants, forest fires, agricultural burnings, and hydrocarbons from natural decay	6	9–17	9–12	
Volcanic dust	2	3–6	3–7	7–1
Nitrates (from NO)	6	9–18	3–2	3
Man-made particulates (heating, industry, engine exhausts)	2	3–6	3–8	3–4

out. It is of the order of 10 days for tropospheric aerosols and several years for stratospheric aerosols consisting of submicron particles.

The chemical composition of the aerosols varies considerably depending on their source. Under humid conditions individual aerosol particles are covered by water and converted into droplets if the humidity is sufficiently high. Condensation generally starts at about 70% relative humidity (cf Deepak and Gerber, 1984; Harris, 1983; Kondratyev *et al.*, 1983).

Although we can distinguish between many aerosol types, we identify five major ones:

(a) coarse mechanically produced mineral dust (quartz, siliceous clays), >3 μm

(b) coarse oceanic sea-salt particles (water and sea spray particles), >3 μm

(c) fine directly-produced soot (elemental carbon), < 1 μm

(d) fine and medium sized products of gas-to-particle conversion (in the stratosphere: 75% solution of sulphuric acid; in continental air masses: hydrous ammonium sulphate; in maritime tropical air: sulphuric acid-ammonium bi-sulphate), $\leq$ 1 μm

(e) volcanic ash of varying composition, $\leq$ 1 μm.

The aerosols found in the atmosphere usually are mixtures of these major types. Five typical mixtures, aerosol models, can be defined and should suffice for the time being for an approximate global assessment, cf Table 4.7. Extinction coefficients, albedos and other optical properties can be derived for these models and used for radiative transfer studies.

Table 4.7 Basic aerosol models. The choice of aerosol components for the volcanic model depends on the height and time after an eruption. '75% H_2SO_4' refers to a 75% solution of sulphuric acid in water

Aerosol model	Aerosol component	Percentage by aerosol particle volume
Continental	Dust-like	70
	Water-soluble	29
	Soot	1
Urban/Industrial	Water-soluble	61
	Soot	22
	Dust-like	17
Maritime	Oceanic	95
	Water-soluble	5
Stratospheric	75% H_2SO_4	100
Volcanic	Volcanic Ash	up to 100
	75% H_2SO_4	remainder

The vertical distribution of optical properties must also be considered and six idealized vertical profiles have been defined on the basis of available observations (Table 4.8). For further details reference is made to the Standard Radiation Atmosphere, IAMAP (1985).

With the aid of these aerosol types and vertical distribution profiles a first attempt can be made to establish an aerosol climatology. Observations are not yet adequate to do this well. Particularly over the oceans, in polar regions and over desert areas present observations are not sufficient. Long (decadal) records are needed to detect global trends.

We note some observed features that may be of interest in the present context. Long distance transport of tropospheric aerosols has been observed. Thus the outbreak of Saharan dust over the Atlantic has been traced in satellite pictures as far as to the Caribbean and Europe (Carlson and Benjamin, 1980; Deepak and Gerber, 1984; Joseph and Wolfson, 1975) and, similarly, dust from Texan deserts could be traced to Europe. Aerosol particles have been collected in wintertime in the North Polar region which undoubtedly

Table 4.8 Atmospheric aerosol profile types

Name	Troposphere		Stratosphere	Mesosphere
	Boundary layer	Free troposphere		
I URB	0–2 urban	2–12 continental	undisturbed 'background' (75% H_2SO_4)	
II CONT-I	0–2 continental	2–12 continental		
III MAR-I	0–2 maritime	2–12 continental	or	
IV CONT II	0–6 continental (dense with exponential decrease)	6–12 continental		75% H_2SO_4 (upwards decreasing optical depth)
V MAR II	0–2 marit.	2–6 cont. (dense desert aerosol) 6–12cont.	volcanic ash + 75% H_2SO_4	
IV CON-VECTIVE	0–4.4 cont (exp. decrease)	4.4–12 cont.		

stem from automobile and diesel exhaust at middle latitudes (Rosen *et al.*, 1981). Elemental carbon originating from combustion at lower latitudes has been detected in the Arctic aerosol.

Our knowledge about stratospheric aerosols is more comprehensive. The variability is smaller and there are merely a few mechanisms that inject aerosols into the stratosphere. The background aerosol is a 75% H_2SO_4 solution. It is disturbed by volcanic ash from eruptions strong enough to inject volcanic debris to high elevations. The temporal and spatial distributions of these aerosols have been monitored since October 1978 by means of satellites and by lidar (McCormick and Brandl, 1983; Yue *et al.*, 1984; Russell *et al.*, 1981a and 1981b). Several major volcanoes erupted during this time and it has been possible to follow the aerosol transport. The average residence time has been determined to be a few years.

4.5.3 Radiative Effects

The range of radiative characteristics (extinction coefficient, scattering coefficient and volume scattering phase function) is reasonably well known both from measurements and theoretical analysis (Deepak and Gerber, 1984; Kondratyev and Prokofiev, 1984).

Non-absorbing aerosols increase the albedo of the atmosphere and reduce the amount of solar radiation that reaches the surface. If the aerosol absorbs in the shortwave range of the spectrum, energy is directly transferred to the atmosphere. The effect is heating of the atmosphere and cooling of the underlying surface. If the aerosol absorbs and consequently also emits radiation in the infrared part of the spectrum, energy is withdrawn from the upper troposphere due to emission to space but the greenhouse effect near the surface is increased. The net effect depends on the ratio of the absorption coefficients in the visible and infrared but also on the albedo of the surface and the altitude of the aerosol layer. The radiative fluxes due to the aerosol thus change the atmospheric temperature profile and surface temperature. The following examples illustrate these principles and may be of interest in considering problems of climatic change.

The bulk of the tropospheric aerosols is confined to the planetary boundary layer (PBL). Here their infrared radiative properties have only a modest impact on outgoing radiation fluxes because the temperature in this layer is close to that of the Earth's surface. The vertical temperature gradients are reduced by the presence of the aerosol and atmospheric cooling is strongest at the top of the boundary layer. Due to their shortwave radiative properties the PBL aerosols scatter back incoming solar radiation but also trap incoming radiation. More of the radiation is then finally absorbed at the ground. The heating due to this additional absorption partly compensates for the infrared cooling.

In the presence of tropospheric aerosols the outgoing shortwave flux depends on the surface albedo and the optical properties of the aerosol. If the albedo is less than about 0.3, as is the case for the oceans, the upward radiative flux is enhanced due to the presence of a continental aerosol, while the opposite is true if the surface is brighter, e.g. over deserts.

Desert aerosols often behave very differently from an aerosol in the planetary boundary layer. A transport upward into layers in the middle troposphere can occur, since heating takes place due to the absorption of shortwave solar radiation and thermal radiation from the Earth's surface. The amount of shortwave radiative energy reaching the surface is reduced resulting in cooling near the surface and heating aloft.

The stratospheric background aerosol does not absorb in the shortwave range but scatters part of the incoming solar radiation back to space. It absorbs and emits strongly in the infrared and since the radiation flux absorbed originates from the warm surface of the Earth and the emitted flux is determined by the stratospheric temperatures, this process results in warming of the stratosphere. During volcanic eruptions ash is injected into the stratosphere. This type of aerosol absorbs both in the visible and the infrared. Its absorption in the visible is not strong enough to compensate for the infrared cooling. It therefore enhances the cooling due to the background aerosol.

Also indirect effects of aerosols on the radiation field may be of interest. Aerosol particles may act as condensation nuclei and intensify the formation of clouds. The particles may also be embedded between cloud droplets without being involved in the condensation process. In this way the reflectivity of thin clouds can be enhanced while thick clouds with embedded aerosols absorb more and appear darker. As mentioned earlier, it is now evident that aerosols are transported from polluted areas to polar regions and that these and also volcanic aerosols finally sediment there. The deposition of aerosols in these regions may reduce the albedo of the snow.

4.5.4 Possible Future Changes of Climate Due to Anthropogenic Aerosols

In view of the many ways in which the global aerosol distribution may change and our incomplete knowledge about the geographical distribution of possible past changes, projections of future impacts on climate are not possible except in very general terms. Computations with one-dimensional models have demonstrated the likely effects of different scenarios for changing vertical profiles of aerosol concentrations (cf e.g. Reck, 1976; Charlock and Sellers, 1980). They show significant temperature changes and may be useful when a clearer picture of ongoing or possible future changes of the global aerosol distribution becomes available. Newiger (1985) has attempted to assess the present role of anthropogenic aerosols for the albedo of the Earth. While no significant change is deduced for the Southern Hemisphere, an increase of up to 0.3 to 0.5% in the Northern Hemisphere seems possible, which might significantly counteract the warming due to increasing greenhouse gas concentrations. The uncertainty of this result is, however, considerable since the spatial dispersion of aerosols is hardly adequately treated by zonal averaging and merely considering meridional and vertical transfer as done by Newiger (1985).

A few GCM experiments with specified aerosol distributions have been carried out (Joseph, 1977; Randall *et al.*, 1984; Tenre *et al.*, 1984). They were not designed to explore particularly long-term climate impacts, but yield some interesting results.

— The integral effect of an ensemble of aerosols in the atmosphere differs considerably from the sum of the effects of the individual aerosols.

— The stratospheric aerosol probably plays the most important role for the radiative climate of the atmosphere.

— The stratospheric aerosol seems to damp the Hadley circulation in the troposphere and to slow down the easterlies in the tropics and the westerlies in the subtropics.

— The Saharan aerosol is the only one which can produce noticeable upward motion in the middle of the troposphere with convergence below and divergence above.

— Radiative forcing in the planetary boundary layer due to the presence of an aerosol seems to be easily compensated for by more dominant processes.

The effect of the El Chichon volcanic eruption has also been studied by Tenre *et al.* (1984) with results that are in general agreement with the observations as analysed by Labitzke *et al.* (1983). These are of importance because of the necessity to know the natural variations and their causes in order to detect and determine the magnitude of possible cooling in the stratosphere due to the increasing concentrations of greenhouse gases.

Global changes of climate due to anthropogenic changes of the atmospheric aerosol distribution have not been detected. Although such changes may occur in the future we have at present no means to project any plausible scenarios. Nevertheless, to establish a global aerosol climatology and to monitor the variations that undoubtedly occur in order to detect possible future long-term trends remains an important research task.

4.6 CONCLUSIONS

— Global atmospheric concentrations of methane have probably increased from about 0.7 ppmv in the past when human activities played no significant role to about 1.55 ppmv in 1980. The present rate of increase is 1.1–1.3% per year and the concentration may well rise to 2.0–2.5 ppmv during the next fifty years.

— Similarly, nitrous oxide has increased from about 0.28 ppmv to 0.30 ppmv in 1980. The present rate of increase is about 0.3% per year and 0.35–0.40 ppmv is projected for the middle of the next century.

— Chlorofluorocarbons in the atmosphere are exclusively of anthropogenic origin, and their concentrations are all increasing. Present abundances of the two most important ones with regard to possible climate impacts, CCl_3F and CCl_2F_2, were 0.17 ppbv and 0.28 ppbv respectively in 1980. The present rate of increasing atmospheric concentrations is 5–6% per year, which is expected to decline. Future concentrations may still reach levels well above 1.0 ppbv, possibly 2–5 ppbv, towards the middle of the next century if no further restrictions on their use are imposed.

— Tropospheric ozone is increasing primarily in the Northern Hemisphere, while stratospheric ozone (above about 30 km) seems to be decreasing, although conclusive evidence is not available. With continued increasing emissions of methane, carbon monoxide, nitrogen oxides, chlorofluorocarbons and possibly other air pollutants, these present trends will be enhanced, particularly in the stratosphere, although quantitative predictions are uncertain.

— It seems plausible that the present enhanced concentrations of these greenhouse gases have an effect on the present global temperature that is about half of the effect caused by the past increase of CO_2, although no change has yet been unequivocally detected because of the natural variability of climate and the inertia of the climate system (cf Chapters 5 and 6).

— The scenarios of future increasing concentrations of these other greenhouse gases may imply changes of global surface temperatures that are as large as those envisaged for CO_2 during the next 50–70 years (cf further Chapter 5).

— Although global changes of climate due to increasing aerosol concentrations in the atmosphere probably have not been significant and future changes cannot be projected with any certainty, the possibility that they may become of some importance in the future cannot be excluded.

— Regional changes of the atmospheric aerosol distribution and changes of the associated regional radiation climate (e.g. in polar regions, over extensive industrial regions) have taken place and may well become of increasing significance.

— The development of methods to establish an aerosol climatology and to monitor future regional and possibly global changes is important.

NOTE ON AUTHORSHIP AND ACKNOWLEDGEMENTS

H.-J. Bolle is primarily responsible for Sections 4.2.5, 4.3.5 and 4.5; W. Seiler is the author of sections 4.2.2 and 4.2.3 and has contributed to 4.3.2 and 4.3.3. The remaining sections and the overall composition of the chapter is the joint responsibility of B. Bolin and H.-J. Bolle.

The chapter has been reviewed by P. Crutzen, K.Ya. Kondratyev, R. Cicerone, R.E. Dickinson and V. Ramanathan, and been modified accordingly. Their implicit contributions to the chapter are gratefully acknowledged.

4.7 REFERENCES

Adel, A. (1939) Note on the atmospheric oxides of nitrogen, *Astrophys. J.*, **90**, 627.

Baker-Blocker, A., Bonahue, T. M., and Mancy, K. H. (1977) Methane flux from wetland areas, *Tellus*, **29**, 245–250.

Blake, D. R. (1984) *Increasing Concentrations of Atmospheric Methane, 1979–1980*, Ph.D thesis, University of California, Irvine, USA.

Blake, D. R., Mayer, E. W., Tyler, S. C., Makide, Y., Montague, D. C., and Rowland, F. S. (1982) Global increase in atmospheric methane concentrations between 1978 and 1980, *Geophys. Res. Lett.*, **9**, 477–480.

Bojkov, R. D., and Reinsel, G. C. (1985) Trends in tropospheric ozone concentration, in Zerefos, C. S., and Ghazi, A. (eds) *Atmospheric Ozone*. Proceedings of the Quadrennial Ozone Symposium in Halkidiki, Greece, 3–7 September 1984, Dordrecht–Boston–Lancaster, D. Reidel, 775–781.

Bolin, B., Crutzen, P, J., Vitousek, P. M., Woodmansee, R. G., Goldberg, E. D., and Cook, R. B. (1983) Interactions of biogeochemical cycles, in Bolin, B., and Cook, R. B. (eds) *The Major Biogeochemical Cycles and Their Interactions*, SCOPE 21, Chichester, Wiley, 1–40.

Breitenbeck, G. A., Blackmer, A. M., and Bremner, J. M. (1980) Effects of different nitrogen fertilizers on emission of nitrous oxide from soil, *Geophys. Res. Lett.*, **7**, 85–88.

Bremner, J. N., Robbins, S. G., and Blackmer, A. M. (1980) Seasonal variability in emission of nitrous oxide from soil, *Geophys. Res. Lett.*, **7**, 641–644.

Carlson, T. N., and Benjamin, S. G. (1980) Radiative heating rates for Saharan dust, *J. Atmos. Sci.*, **37**, 193–213.

Cess, R. D. (1983) Arctic aerosols: Model estimates of interactive influences upon the surface-atmosphere clear sky radiation budget, *Atmos. Environ.*, **17**, 2555–2564.

Charlock, T. P., and Sellers, W. D. (1980) Aerosol effects on climate calculations with time-dependent and steady-state radiative–convective models, *J. Atmos. Sci.*, **37**, 1327–1341.

Cheremisinoff, N. P., Cheremisinoff, P. N., and Ellerbusch, F. (1980) *Biomass-application, Technology, and Production*, New York, Marcel Dekker, Inc.

Cicerone, R. J., Shetter, J. D., and Delwiche, C. C. (1983) Seasonal variation of methane flux from a Californian rice paddy, *J. Geophys. Res.*, **88**, 11022–11024.

Clark, W. C. (1982) *Carbon Dioxide Review*, Oxford, Clarendon Press, 1–469.

CMA, Chemical Manufacturing Association, 1982. *World Production and Release of Chlorofluorocarbons 11 and 12 through 1981*. Report of the Fluorocarbon Program Panel, 1982.

Cohen, Y., and Gordon, L. I. (1979) Nitrous oxide production in the ocean, *J. Geophys. Res.*, **84**, 347–353.

Conrad, R., Seiler, W., and Bunse, G. (1983) Factors influencing the loss of fertilizer nitrogen into the atmosphere as N_2O, *J. Geophys. Res.*, **88**, 6709–6718.

Craig, H., and Chou, C. C. (1982) Methane: The record in polar ice cores, *Geophys. Res. Lett.*, **9**, 1221–1224.

Crutzen, P. J. (1983) Atmospheric interactions in homogeneous gas reactions of C, N, and S containing compounds, in Bolin, B., and Cook, R. B. (eds) *The Major Biogeochemical Cycles and Their Interactions*, SCOPE 21, Chichester, Wiley, 67–112.

Crutzen, P. J. (1985) The role of the tropics in atmospheric chemistry, to be published in Dickinson, R. (ed.) *Geophysiology of Amazonia*, New York, Wiley.

Crutzen, P. J., Heidt, L. E., Krasnec, J. P., Pollock, W. H., and Seiler, W. (1979) Biomass burning as a source of atmospheric CO, H_2, N_2O, NO, CH_3Cl and COS, *Nature*, **282**, 253–256.

Crutzen, P. J., and Gidel, L. T. (1983) A two-dimensional photochemical model of the atmosphere, 2: The tropospheric budget of the anthropogenic chlorocarbons, CO, CH_4, CH_32Cl and the effect of various NO_x sources on tropospheric ozone, *J. Geophys. Res.*, **88**, 6641–6661.

Cunnold, D. M., Prinn, R. G., Rasmussen, R. A., Simmonds, P. G., Alyea, F. N., Cardelin, C. A., Crawford, A. J., Fraser, P. J., and Rosen, R. D. (1983a) The atmospheric lifetime experiment 3. Lifetime methodology and application to three years of $CFCl_3$ data, *J. Geophys. Res.*, **88**, 8379–8400.

Cunnold, D. M., Prinn, R. G., Rasmussen, R. A., Simmonds, P. G., Alyea, F. N., Cardelino, C. A., and Crawford, A. J. (1983b) The atmospheric lifetime experiment 4. The results for CF_2Cl_2 based on three years of data, *J. Geophys. Res.*, **88**, 8401–8414.

Deepak, A., and Gerber, H. E. (1984) *Aerosols and Their Climate Effects*, Hampton, Virginia, A. Deepak Publ.

DeMuer, D. (1985) Vertical ozone distribution over Uccle (Belgium) after correction for systematic distortion of the ozone profiles, in Zerefos, C. S., and Ghazi, A. (eds) *Atmospheric Ozone*. Proceedings of the Quadrennial Ozone Symposium in Halkidiki, Greece, 3–7 Sept. 1984, Dordrecht–Boston–Lancaster, D. Reidel, 330–334.

DeRudder, A., and Brasseur, G. (1985) Ozone in the 21st century: increase or decrease? in Zerefos, C. S. and Ghazi, A. (eds). *Atmospheric Ozone*, 92–97. Proceedings of the Quadrennial Ozone Symposium in Halkidiki, Greece, 3–7 Sept. 1984, Dordrecht–Boston–Lancaster, D. Reidel.

Dittberner, G. J. (1978) Climatic change: volcanoes, man-made pollution and carbon dioxide. *IEEE Trans. on Geosci. Electronics*, **GE-16**, 50–61.

Dütsch, H. U. (1985) Total ozone in the light of ozone soundings, the impact of El Chichon, in Zerefos, C. S., and Ghazi, A. (eds) *Atmospheric Ozone*, 263–268. Proceedings of the Quadrennial Ozone Symposium in Halkidiki, Greece, 3–7 Sept. 1984, Dordrecht–Boston–Lancaster, D. Reidel.

Ehhalt, D. H. (1974) The atmospheric cycle of methane, *Tellus*, **26**, 58–70.

Ehhalt, D. H., and Schmidt, U. (1978) Sources and sinks of atmospheric methane, *PAGEOPH*, **116**, 452–464.

Ehhalt, D. H., Zander, R. J., and Lamontagne, R. A. (1983) On the temporal increase of tropospheric methane, *J. Geophys. Res.*, **88**, 8442–8446.

EPA (1973) First report to Congress: *Resources, Recovery and Source Reduction*, EPA, Office of waste management programs, Feb 22, 1973.

FAO (1983) *Fertilizer Yearbook*, Vol. 32, Rome, FAO.

Feister, U., and Warmbt, W. (1985) Long-term surface ozone increase at Arkona (54,68° N. 13.43° E), in Zerefos, C. S., and Ghazi, A. (eds) *Atmospheric Ozone*, 782–787. Proceedings of the Quadrennial Ozone Symposium in Halkidiki, Greece, 3–7 Sept. 1984, Dordrecht–Boston–Lancaster, D. Reidel.

Fink, U., Rank, D. H., and Wiggins, T. A. (1964) Abundance of methane in the earth's atmosphere, *J. Opt. Soc. Am.*, **54**, 472–475.

Fishman, J., Solomon, S., and Crutzen, P.J. (1979) Observational and theoretical evidence in support of a significant *in-situ* photochemical source of tropospheric ozone, *Tellus*, **31**, 432–446.

Fishman, J., and Seiler, W. (1983) The correlative nature of ozone and carbon monoxide in the troposphere, *J. Geophys. Res.*, **88**, 3662–3670.

Fleig, A. J., Gille, J. C. McCormick, M. P., Rusch, D. W., Russell III, J. M., and Lindsay, J. M. (1985) Intercomparison of satellite ozone profile measurements, in Zerefos, C. S., and Ghazi, A. (eds) *Atmospheric Ozone*, 258–262. Proceedings of the Quadrennial Ozone Symposium in Halkidiki, Greece, 3–7 Sept. 1984, Dordrecht–Boston–Lancaster, D. Reidel.

Flohn, H. (1978) *Carbon Dioxide, Climate and Society*. Williams, J. (ed.), Pergamon Press, 227–238.

Fraser, P. J., Khalil, M. A. K., Rasmussen, R. A., and Crawford, A. J. (1981) Trends of atmospheric methane in the southern hemisphere, *Geophys. Res. Lett.*, **8**, 1063–1066.

Fraser, P. K., Khalil, M. A. K., Rasmussen, R. A., and Steele, L. P. (1984) Tropospheric methane in the mid-latitudes of the southern hemisphere, *J. Atmos. Chem.*, **1**, 125–135.

Goodroad, L. L., and Keeney, D. R. (1984) Nitrous oxide emission from forest, marsh and prairie ecosystems, *J. Environ. Qual.*, **13**, 448–452.

Hahn, J. (1981) Nitrous oxide in the oceans, in Delwiche (ed.) *Denitrification, Nitrification and Atmospheric Nitrous Oxide*, New York, Wiley, 191–227.

Hahn, J., and Junge, C. (1977) Atmospheric nitrous oxide: a critical review, *Z. Naturforsch.*, **32**, 190–214.

Hameed, S., Cess, R. D., and Hogan, J. (1980) Response of the global climate to changes in atmospheric chemical composition due to fossil fuel burning, *J. Geophys. Res.*, **85**, 7537–7545.

Harris, F. S. Jr (Scientific editor: H.E. Gerber), (1983) *Tropospheric Aerosols: Review of Current Data on Physical and Optical Properties*. WCP-43, Geneva.

Harriss, J. M., and Nickerson, E. C. (eds) (1984) *Geophys. Monitoring for Climatic Change* No 12, 184, NOAA Boulder, Colorado.

Harriss, R. C.. Sebacher, D. I., and Day Jr, F.P., (1982) Methane flux in the great dismal swamp, *Nature*, **297**, 673–674.

Hartmannsgruber, R., Attmannspacher, W., and Claude, H. (1985) Opposite behaviours of the ozone amount in the troposphere and lower stratosphere during the last years, based on the ozone measurements at the Hohenpeissenberg observatory from 1967–1983, in Zerefos, C. S., and Ghazi, A. (eds) *Atmospheric Ozone*, 770–774. Proceedings of the Quadrennial Ozone Symposium in Halkidiki, Greece, 3–7 Sept. 1984, Dordrecht–Boston–Lancaster, D. Reidel.

Heidt, L. E., and Ehhalt, D. H. (1980) Corrections of CH_4 concentrations measured prior to 1974, *Geophys. Res. Lett.*, **7**, 1980, 1023.

Hilsenrath, E., *et al.* (1985) Results from the balloon ozone intercomparison campaign (BIOC), in Zerefos, C. S. and Ghazi, A. (eds) *Atmospheric Ozone*, 454–459. Proceedings of the Quadrennial Ozone Symposium in Halkidiki, Greece, 3–7 Sept. 1984, Dordrecht–Boston–Lancaster, D. Reidel.

Hoffman, J. S., Keyes, D., and Titus, J.G. (1983) *Projecting Future Sea Level Rise.* U.S. Environmental Protection Agency, Washington, D.C.

Holzapfel-Pschorn, A., and Seiler, W. (1985) Methane emission during a vegetation period from an Italian rice paddy, submitted to *J. Geophys. Res.*

Hutchinson, G. E. (1949) A note on two aspects of the geochemistry of carbon, *Am. J. Sci.*, **247**, 27–32.

IAMAP (1985) A preliminary cloudless standard atmosphere for radiation computation. To be published in *WCP*, Geneva.

Johnston, H. S., Serang, O., and Podolske, J. (1979) Instantaneous global nitrous oxide photochemical rates, *J. Geophys. Res.*, **84**, 5077–5082.

Joseph, J. H. (1977) Effect of a desert aerosol on a model of the general circulation, in *Proc. Fut. Symposium on Radiation in the Atmosphere*, Princeton Science Press, **1**, 487–492.

Joseph, J. H., and Wolfson, N. (1975) The ratio of absorption to backscatter of solar radiation of aerosols during Khamin conditions and effects on the radiation balance, *J. Appl. Meteorol.*, **14**, 1389–1396.

Kaplan, W. A., Elkins, J. W., Kolb, C. E., McElroy, M. B., Wofsy, S. C., and Duran, A. P. (1978) Nitrous oxide in fresh water systems: An estimate for the yield of atmospheric N_2O associated with disposal of human waste, *PAGEOPH*, **116**, 423–438.

Keller, M., Wofsy, S. C., Goreau, T. J., Kaplan, W. A., and McElroy, M. B. (1983) Production of nitrous oxide and consumption of methane by forest soils, *Geophys. Res. Lett.*, **10**, 1156–1159.

Khalil, M. A. K., and Rasmussen, R. A. (1983a) Sources, sinks and seasonal cycles of atmospheric methane, *J. Geophys. Res.*, **88**, 5131–5144.

Khalil, M. A. K., and Rasmussen, R. A. (1983b) Increase and seasonal cycles of nitrous oxides in the earth's atmosphere, *Tellus*, **35B**, 161–169.

Ko, M. K. W., and Sze, N. D. (1982) A 2-D model calculation of atmospheric lifetimes for N_2O, CFC-11 and CFC-12, *Nature*, **287**, 317–319.

Komhyr, W. D., Oltmans, S. J., Chopra, A. N., Leonard, R. K., Garcia, T. E., and McFee, C. (1985) Results of Umkehr, ozonesonde, total ozone, and sulfur dioxide observations in Hawaii following the eruption of El Chichon volcano in 1982, in Zerefos, C. S., and Ghazi, A. (eds) *Atmospheric Ozone*, 305–310. Proceedings of the Quadrennial Ozone Symposium in Halkidiki, Greece, 3–7 Sept. 1984. Dordrecht–Boston–Lancaster, D. Reidel.

Kondratyev, K. Ya., Moskalenko, N. I., and Pozdnyakov, D. V. (1983) *Atmospheric Aerosols*, Leningrad, Gidrometioizdat.

Kondratyev, K. Ya., and Prokofiev, M. A. (1984) Typifying atmospheric erosol for assessments of its climatic impact, *Phys. Atmos. Ocean*, **5**, 339–349.

Koyama, T. (1963) Gaseous metabolism in lake sediments and paddy soils and the production of atmospheric methane and hydrogen, *J. Geophys. Res.*, **68**, 3971–3973.

Labitzke, K., Naujokat, B., and McCormick, M. P. (1983) Temperature effects in the stratosphere of the April 4, 1982 eruption of El Chichon, Mexico, *Geophys. Res. Lett.*, **10**, 24–26.

Logan, J. A. (1985) Tropospheric ozone: seasonal behaviour, trends and anthropogenic influence, *J. Geophys. Res.*, **89D**, 10463–10482.

Mateer, C. L., DeLuisi, J. J., and Porco, C. C. (1980) The Short Umkehr Method, Part I: Standard Ozone profiles for use in the estimation of ozone profiles by the inversion of short Umkehr observations. *NOAA Tech. Memo ERL ARL-86*, 20 pages.

Mathias, A. D., Blackmer, A. M.. and Bremner, J. M. (1980) A simple chamber technique for field measurements of emissions of nitrous oxide from soils, *J. Environ. Qual.*, **9**, 251–256.

Mayer, E. W., Blake, D. R., Tyler, S. C., Makide, Y., Montague, D. C., and Rowland, F. S. (1982) Methane: interhemispheric concentration gradient and atmospheric residence time. *Proc. Natl. Acad. Sci. USA*, **79**, 1366–1370.

McCormick, M. P., and Brandl, D. (1983) *SAM II Measurements of the Polar Stratospheric Aerosol*, NASA Ref. Publ. 1107, Washington, D.C.

McPeters, R. D., Heath, D. F., and Bhartia, P. K. (1984) Average ozone profiles for 1979 from the NIMBUS 7 SBUV instrument, *J. Geophys. Res.*, **89D**, 5199–5214.

Migeotte, M. J. (1948) Spectroscopic evidence of methane in the earth's atmosphere, *Phys. Rev.*, **73**, 519–520.

Molina, M. J., and Rowland, F. S. (1974) Stratospheric sink for chlorofluoromethanes: chlorine catalyzed destruction of ozone, *Nature*, **249**, 810–812.

NASA (1984) *Ozone Climatology Series Atlas of Total Ozone: April 1970–December 1976*, GSFC, Greenbelt, USA.

Newiger, M. (1985) Einfluss anthropogener Aerosolteilchen auf der Strahlungshaushalt der Atmosphäre. *Hamburger Geophysikalische Einzelschriften*, A 73, Hamburg, Max Planck Institut für Meteorologie.

Owens, A. J., Hales, C. H., Filkin, D. L., Miller, C., and McFarland, M. (1985) Multiple scenario ozone change calculations: the subtractive perturbation approach, in Zerefos, C. S., and Ghazi, A. (eds) *Atmospheric Ozone*, 82–86. Proceedings of the Quadrennial Ozone Symposium in Halkidiki, Greece, 3–7 Sept. 1984, Dordrecht–Boston–Lancaster, D. Reidel.

Peterson, J. T., and Junge, C. E. (1971) Sources of particulate matter in the atmosphere, in Matthews, W. H., Kellogg, W. W., and Robinson, G. D. (eds) *Man's Impact on Climate*, Cambridge, Massachusetts, MIT Press, 310–320.

Pierotti, D., and Rasmussen, R. A. (1976) Combustion as a source of nitrous oxide, *Geophys. Res. Lett.*, **3**, 265–267.

Planet, W. G., Lienesch, J. H., and Hill, M. L. (1985) Global total ozone from TIROS measurements: 1979–1983, in Zerefos, C. S., and Ghazi, A. (eds) *Atmospheric Ozone*, 234–238. Proceedings of the Quadrennial Ozone Symposium in Halkidiki, Greece, 3–7 Sept 1984, Dordrecht–Boston–Lancaster, D. Reidel.

Ramanathan, V. (1975) Greenhouse effect due to chlorofluorocarbons: climatic implications, *Science*, **190**, 50–52.

Ramanathan, V. (1980) Climatic effects of anthropogenic trace gases, in Williams J. (ed.) *Interactions of Energy and Climate*, D. Reidel, 269–280.

Ramanathan V., Cicerone, R. J., Singh, H. B., and Kiehl, J. T. (1985) Trace gas trends and their potential role in climate change, NCAR/0304/84-9, *J. Geophys. Res.*, **90**, D3, 5547–5566.

Randall, D., Carlson, T., and Mintz, Y. (1984) The sensitivity of a general circulation model to Saharan dust and heating, in Deepak, A., and Gerber, H. E. (eds) *Aerosols and Their Climate Effects*, Hampton, Virginia, A. Deepak Publ.

Rasmussen, R. A., and Khalil, M. A. K. (1981a) Increase in the concentration of atmospheric methane, *Atmos. Environ.*, **15**, 883–886.

Rasmussen, R. A., and Khalil, M. A. K. (1981b) Atmospheric methane: trends and seasonal cycles, *J. Geophys. Res.*, **86**, 9826–9832.

Rasmussen, R. A., and Khalil, M. A. K. (1984) Atmospheric methane in the recent and ancient atmospheres: concentrations, trends and interhemispheric gradient, *J. Geophys. Res.*, **89**, 11599–11605.

Reck, R. A. (1976) Thermal and radiative effects of atmospheric aerosols in the Northern Hemisphere calculated using a radiative-convective model, *Atmos. Environ.*, **10**, 611–617.

Reinsel, G. C., Tiao, G. C., DeLuisi, J. J., Mateer, C. L., Miller, A. J., and Frederick, J. E. (1984) Analysis of upper stratospheric Umkehr ozone profile data for trends and effects of stratospheric aerosol, *J. Geophys. Res.*, **89**, 4833–4840.

Reiter, R., Munzert, K., and Kantor, H.-J., (1983) Parameterization of the variation of CO_2 and O_3 in the lower troposphere based on 5-years recordings at 0.7, 1.8, and 3.0 km ASL with consideration of the most important magnitude of meteorology, biomass, and anthropogenic effects. *WMO Technical Conference on Observation and Measurement of Atmospheric Contaminants* (TECOMAC), Vienna, 17–21 October 1983.

Revelle, R. R. (1983) Methane hydrates in continental slope sediments and increasing atmospheric carbon dioxide. Section 3.5 (252–261). In *Changing Climate*, Washington, DC, National Academy Press, 496 pages.

Revelle, R., and Munk, W. (1977) The carbon dioxide cycle and the biosphere, in *Energy and Climate, Study Geophys.*, Washington, D.C., National Academy of Sciences, 140–158.

Robbins, R. C., Cavanagh, L. A., Salas, L. J., and Robinson, E. (1973) Analysis of ancient atmospheres, *J. Geophys. Res.*, **78**, 5341–5344.

Rosen, H., Novakov, T., and Bodhaine, B. A. (1981) Soot in the Arctic, *Atmos. Environ.*, **15**, 1371–1374.

Russell, P. B., Swissler, T. J., McCormick, M. P., Chu, W.P., Livingston, J. M., and Pepin, T. J. (1981a) Satellite and correlative measurements of the stratospheric aerosol, I, An optical model for data conversions, *J. Atm. Sci.*, **38**, 1279–1294.

Russell, P. B., McCormick, M. P., Swissler, T. J., Chu, W. P., Livingston, J. M., Fuller, W. H., Rosen, J. M., Hofmann, D. J., McMaster, L. R., Woods, D. C., and Pepin, T. J. (1981b) Satellite and correlative measurements of the stratospheric aerosol, II, Comparison of measurements made by SAM II, distsondes and an airborne lidar, *J. Atmos. Sci.*, **38**, 1295–1312.

Seiler, W. (1984) Contribution of biological processes to the global budget of CH_4 in the atmosphere, in Klug, M. J., and Reddy, C. A. (eds) *Current Perspectives in Microbial Ecology*, American Society of Meteorology, 468–477.

Seiler, W. (1985) *Increase of Atmospheric Methane: Causes and Impact on the Environments*, WMO Special Environmental Report No. 16.

Seiler, W., Muller. F, and Oeser, H. (1978) Vertical distribution of chlorofluoromethanes in the upper troposphere and lower stratosphere, *PAGEOPH*, **116**, 554–566.

Seiler, W., and Crutzen, P. (1980) Estimates of gross and net fluxes of carbon between the biosphere and the atmosphere from biomass burning, *Clim. Change*, **2**, 207–248.

Seiler, W., and Conrad, R. (1981) Field measurements of natural and fertilizer induced N_2O release rates from soils, *J. Air Poll. Contr. Assoc.*, **31**, 767–772.

Seiler, W., Conrad, R., and Scharffe, D. (1984a) Field studies of methane emissions from termite nests into the atmosphere and measurements of methane uptake by tropical soils, *J. Atmos. Chem.*, **1**, 171–186.

Seiler, W., Giehl, H., Brunke, E.G., and Halliday, E. (1984b) The seasonality of CO abundance in the Southern Hemisphere, *Tellus*, **36B**, 219–231

Seiler, W., and Conrad, R. (1985) Exchange of atmospheric trace gases with anoxic and oxic tropical ecosystems, to be published in Dickinson R. (ed.) *Geophysiology of Amazonia*, New York, Wiley.

Sheppard, J. C., Westberg, H., Hopper, J. F., Ganesea, K., and Zimmerman, P. (1982) Inventory of global methane sources and their production rates, *J. Geophys. Res.*, **87**, 1305–1312.

Slemr, F., Conrad, R., and Seiler, W. (1984) Nitrous oxide emissions from fertilized and unfertilized soils in a subtropical region (Andalusia, Spain), *J. Atmos. Chem.*, **1**, 159–169.

SMIC (1971) *Inadvertent Climate Modifications*, Cambridge, Massachusetts, MIT Press.

Tannhäuser, I. O. (1968) Uberlappungseffekten an infraroten Spektren, *Beitr. Phys. Atmos.*, **41**, 1–24.

Tenré, D., Geleyn, J. F., and Slingo, J. (1984) First results of the introduction of advanced aerosol-radiation interaction in the ECMWF low resolution global model, in Deepak, A., and Gerber, H. E. (eds) *Aerosols and Their Climate Effects*, Hampton, Virginia, A. Deepak Publ.

Terry, R. E., Tate III, R.L., and Duxbury, A. (1980) *Nitrous Oxide Emissions from Drained, Cultivated Organic Soils of South Florida*, presented at 23rd Annual Meeting of Air Pollution Control Assoc., Montreal, Quebec, June 22–27, 1980.

Twenhofel, W. H. (1951) *Principles of Sedimentation*, New York, McGraw-Hill, 78.

Twomey, S. (1977a) *Atmospheric Aerosols*, Amsterdam, Elsevier Sci. Publ.

Twomey, S. (1977b) The influence of pollution on the shortwave albedo of clouds, *J. Atmos. Sci.*, **34**, 1149–1152.

Walter, H. (1973) *Die Vegetation der Erde*, Band I, Stuttgart, Gustav Fischer Verlag, 355–369.

Wang, W. C., Yung, Y. L., Lacis, A. A., Mo, T., and Hansen, J. E. (1976) Greenhouse effect due to manmade perturbations of trace gases, *Science*, **194**, 685–690.

Weiss, R. F., and Craig, H. (1976) Production of atmospheric nitrous oxide by combustion, *Geophys. Res. Lett.*, **3**, 751–753.

Weiss, R. F. (1981) The temporal and spatial distribution of tropospheric nitrous oxide, *J. Geophys. Res.*, **86**, 7185–7195.

Wilkniss, P. E., Lamontagne, R. A., Larson, R. E., Swinnerton, J. W., Dickson, C. R., and Thompson, T. (1973) Atmospheric trace gases in the Southern Hemisphere, *Nature*, **245**, 45–47.

WMO (1981) The stratosphere 1981; theory and measurements, WMO global ozone research and monitoring project, *Report No. 11*, Geneva, WMO.

WMO (1982) Report of the meeting of experts on sources of errors in detecting of ozone trends. WMO global ozone research and monitoring project, *Report No. 12*, Geneva, WMO.

Wuebbles, D. J. (1983) Chlorocarbon emission scenarios. Potential impact on stratospheric ozone, *J. Geophys. Res.*, **88**, 1433–1443.

Wuebbles, D. J., MacCracken, M. C., and Luther, F. M. (1984) *A Proposed Reference Set of Scenarios for Radiatively Active Atmospheric Constituents*. U.S. Dept. of Energy Carbon Dioxide Research Division Technical Report (in press).

Yue, G.K., McCormick, M. P., and Chu, W. P. (1984) A comparative study of aerosol extinction measurements made by the SAM II and SAGE satellite experiments, *J. Geophys. Res.*, **89**, D4, 5321–5327.

Zerefos, C. S., and Ghazi, A. (eds) (1985) *Atmospheric Ozone*. Proceedings of the Quadrennial Ozone Symposium in Halkidiki, Greece, 3–7 Sept. 1984, Dordrecht–Boston–Lancaster, D. Reidel, 842 pages.

PART B

A Warmer Climate

CHAPTER 5

How Will Climate Change?
The Climate System and Modelling of Future Climate

R. E. DICKINSON

5.1 INTRODUCTION

Climate is popularly thought of as some sort of average weather and its fluctuations. More precisely, climate statistics are obtained by averaging weather over a period long compared to the deterministic limit of predictability for atmospheric motions, which is about two weeks. The climate system is now recognized to also include the oceans, ice sheets, and land-surface properties (Houghton, 1984) because of the close interactions between these and the atmosphere. Climate can vary from year to year, fluctuate on time scales of several years, or change on longer time scales. Detecting the effects of warming by CO_2 and other trace gases requires establishing the occurrence of long-term change that is statistically significant compared to past climate normals. Climate statistics can be obtained by averaging data over a large number of years, e.g., 30 years of data are usually used to define 'normals'. However, since climate is always changing, there are no fixed normals so that the statistics will always depend on the averaging period.

A more rigorous definition of climate is obtained by using, rather than a time average, an *ensemble* average, that is, the average over a hypothetical infinite set of Earths with the same external influences (i.e., the same solar input, the same atmospheric composition, etc.) but different detailed weather patterns.

This ensemble average idea introduces two new concepts, those of *weather noise* and of *external influences*. Weather noise is the uncertainty in the ensemble average arising from the sampling of unpredictable weather fluctuations, namely that part of climatic variability which arises from day-to-day weather variations. This noise is always present in a climate statistic. The term external influences implies that the climate can be affected either by external or by internal factors. What constitutes an external influence depends, in practice, on the time scale of interest. For example, the input of energy from the Sun will always be an external factor, but the extent and pattern of ice and snow cover (which help to determine how much solar radiation is reflected away from the Earth's surface) could be considered an external factor on short time scales, but an internal factor on long time scales.

5.2 CAUSES OF CLIMATIC CHANGE

External changes in global climate are forced by various processes that change the flows of radiative energy within the system. Either the absorption of solar radiation or the trapping of longwave radiation by atmospheric constituents may change.

Possible reasons for change include:

(1) A change in solar output (irradiance) or a change in the geometry of the Earth's orbit around the Sun.

(2) A change in the fraction of incoming (shortwave) energy at the top of the atmosphere which is absorbed by the surface or atmosphere.

(3) A change in the amount of net outgoing (longwave) energy at the top of the troposphere.

(4) A change in the amount of heat sequestered by the deep ocean.

The primary changes under headings (2) and (3) may result from:

(a) Changes in the fluxes of radiation caused by changing atmospheric composition.

(b) Changes in atmospheric transmissivity resulting from either variations in the amount of volcanic or anthropogenic aerosol in the atmosphere, or variations in cloudiness.

(c) Changes in the amount of radiation reflected by the Earth's surface (albedo changes).

(d) Changes in the amount of longwave radiation emitted by the surface and/or absorbed by H_2O in the atmosphere.

Solar Output

Solar output is known to vary on very long time scales (see, for example, Newkirk, 1983) and to vary markedly in the ultraviolet and higher frequency parts of the spectrum on short time scales (days to years) (Foukal, 1980). Variations in solar irradiance of $\pm 0.1\%$ occur with the 27-day equatorial solar rotation period (Smith *et al.*, 1983). There is, however, only indirect evidence for variations of climatological significance on time scales from 1 to

2,000 years. The influence of sunspots on climate is uncertain, but very few convincing statistical relationships have been demonstrated (Pittock, 1978, 1983), and any effects are likely to be small. Correlations between solar irradiance and sunspots have been demonstrated (Willson *et al.*, 1981; Eddy *et al.*, 1982; Smith *et al.*, 1983) and irradiance changes of 0.05% over an 11-year sunspot cycle are implied by analyses of short periods of satellite data. However, these solar-cycle related fluctuations could only have minor effects on global mean temperature, almost certainly below the limit of detectability.

Satellite data on solar variability also show a secular trend (amounting to 0.1% when extrapolated to a 40-year period) that may be more important. Since solar output should vary as the diameter of the Sun varies, and since there is astronomical evidence that the solar diameter varies on the 10- to 100-year time scale, we might expect solar irradiance to vary on these time scales. For example, there appears to be an approximately 80-year cycle in the diameter of the solar disc (Parkinson *et al.*, 1980; Gilliland, 1980, 1981) implying a similar cycle in irradiance. However, the theoretical diameter–output relationship has considerable uncertainty, up to 2 orders of magnitude (Gilliland, 1982). If the secular irradiance trend observed in satellite data is related to diameter changes, as Smith *et al.* (1983) have suggested, then the amplitude of the suggested 80-year cycle in solar irradiance would be about 0.1%, enough to cause a global mean temperature cycle of around 0.1 °C amplitude.

The satellite record of solar irradiance spans only a few years and it is not yet known how representative these data are of longer time scale variations. Some ground-based observations of solar features do, however, show marked variations on the decadal to century time scale. The analysis of accurately dated tree rings shows that the atmospheric concentration of the radioactive isotope carbon-14 has varied significantly in the past on these time scales. Since C-14 concentration is determined by its production rate in the stratosphere, which in turn is influenced by solar flare activity and the strength of the solar wind (Stuiver and Quay, 1980), we know that these solar parameters vary on the 10- to 100-year time scale. A statistically significant 200-year periodicity has been shown to exist in the atmospheric C-14 record over the past 8,500 years (Sonett, 1984). However, there is as yet no firm evidence that these C-14 fluctuations correlate with fluctuations in climate (Stuiver, 1980; Williams *et al.*, 1980).

Orbital Variations

Changes in the Earth's orbit around the Sun affect climate on time scales of 1,000 years or more by changing the latitudinal and seasonal distribution of incoming solar radiation at the top of the atmosphere. Locally, these changes can be 10% or more (Berger, 1979). North *et al.* (1983) have reviewed past modelling work and have suggested that with the present configuration of the land masses in the Northern Hemisphere a nonlinear ice–albedo feedback may enhance the development of large ice masses when the orbital elements are such that they favour cool Northern Hemisphere summers.

Orbital variations also affect the latitudinal and seasonal distribution of incoming solar radiation slightly on much shorter time scales (Borisenkov *et al.*, 1983).

Volcanoes

Volcanic eruptions inject into the stratosphere both dust and sulphur compounds which are converted to secondary aerosols. The latter is currently thought to be of more radiative importance (Rampino and Self, 1984). Individual volcanic eruptions may cause cooling up to around 0.3 °C in global mean surface air temperature by their screening effect on solar radiation (e.g., Hansen *et al.*, 1978; Newell and Deepak, 1982) and warming by several degrees in the absorbing layers in the stratosphere. Observational data show surface cooling on the monthly to annual time scale (e.g., Lamb, 1970; Oliver, 1976; Taylor *et al.*, 1980; Self *et al.*, 1981; Mitchell, 1983; Kelly and Sear, 1984), but in the tropical troposphere volcanic effects cannot be convincingly demonstrated, possibly because they are obscured by variability associated with the El Niño/Southern Oscillation phenomenon (Parker, 1985). Stratospheric data, however, offer convincing support of model predictions (e.g., Parker and Brownscombe, 1983; Labitzke *et al.*, 1983; Quiroz, 1983).

The effects of volcanic eruptions on past climate are difficult to assess because the observational record can be interpreted in alternative ways, and, equally importantly, because the volcanic forcing record is not well documented. At least four distinct types of volcanic records exist: the historical 'Dust Veil Index' of Lamb (1970) and others (e.g., Hirschboeck, 1980); the geological record of Simkin *et al.* (1981) (see also Newhall and Self, 1982); time series of atmospheric transmissivity (e.g., Pivovarova, 1977; Bryson and Goodman, 1980); and the Greenland ice core acidity record (Hammer, 1977; Hammer *et al.*, 1980). These different records show broad similarities, but also significant differences. The uncertainty in the volcanic forcing record has been highlighted by recent direct measurements of stratospheric sulphate aerosol concentration covering the period 1971–1981. These revealed a number of volcanic injection events that had not previously been reported (Sedlacek *et al.*, 1983).

Atmospheric Composition

Many atmospheric trace gases are radiatively active; they absorb and reradiate energy at both long and short wavelengths. The most important such gases are H_2O, O_3, CO_2, N_2O, CH_4 and chlorofluoromethanes (CFMs).

Water vapour is an internal factor (whose concentration varies widely in space and time) while CFMs, being of strictly anthropogenic origin, are purely external. Other trace gases, such as O_3, CO_2, N_2O, and CH_4, have concentrations which are influenced by the activities of Man, but may also vary naturally as parts of the internal feedbacks of the climate system as a whole.

Changes in the concentrations of any of these gases affect the way incoming and outgoing energy is distributed in the vertical, and increases may cause significant warming of the troposphere and cooling of the stratosphere.

The concentrations of several of these radiatively active gases are increasing within the atmosphere, and their concentrations can be projected into the future empirically or through modelling their sources (see Chapter 4). Of these gases, the increase of carbon dioxide (CO_2) is the most important from the viewpoint of its projected climate effect. It is also the best understood as a result of a relatively long data record for its concentration (since 1958) and relatively good accounting for its past sources, as reviewed in detail in Chapters 2 and 3 of this report.

Other trace gaseous atmospheric constituents are also of concern because their increases add to the CO_2 climate warming (World Meteorological Organization, 1982). Of particular importance are increases of the chlorofluorocarbons F-11 and F-12, nitrous oxide, methane, and ozone, listed in order of increasing ignorance as to the sources causing their increase and decreasing length of the data record. Ramanathan *et al.* (1985) have recently reviewed the questions of atmospheric trace gas concentrations and climate effects. See also Chapter 4.

The radiative effects of changes in atmospheric aerosol content due to human activities have been speculated on for many years but cannot yet be quantified (cf World Climate Research Programme, 1983, for a recent review of this subject; and Cess, 1983, for the question of Arctic aerosols). Possible climate effects of future land-use changes are another difficult topic that will not be treated in this review (cf Dickinson, 1981a; Henderson-Sellers and Gornitz, 1984, for treatments of the global effects of land-use change). Land-use changes are primarily of interest to global average climate because of their connections to the carbon cycle.

Internal causes of climatic change, e.g., as a result of changes in ocean temperatures and circulation, cloudiness, or sea-ice cover should be modelled as part of the internal climate system as discussed in later sections. However, we cannot exclude possible long-term natural trends in these parameters that would most usefully be regarded as external changes.

5.3 INTRODUCTION TO MODELLING THE CLIMATE SYSTEM

Models to better understand future climate resulting from changing atmospheric composition have been developed over the last two decades. The first component of any such model is the determination of the changes in atmospheric radiative fluxes which would result from the changed atmospheric composition. This determination requires good models for the transfer of atmospheric radiation through the relevant trace gases. The subject of atmospheric longwave radiation is complicated by the need to consider the tens of thousands of spectral absorption lines of the trace gases and their overlap with major absorbers, especially water vapour. However, such calculations can be done reasonably accurately for CO_2 and adequately for the other trace gases. Errors in radiative calculations introduce a relatively small (although not negligible) uncertainty into estimates of future climatic change and need not be reviewed further here (cf Dickinson, 1982; World Meteorological Organization, 1982; Kiehl and Ramanathan, 1983; Ramanathan *et al.*, 1985, for more detailed discussion of the uncertainties in the calculation of trace gas radiative effects).

Recent reviews of the question of modelling the climate effects of increasing CO_2 have been given by Dickinson (1982), CO_2/Climate Review Board (1982), Schlesinger (1983a, b), and Gilchrist (1983). More general recent treatments of the CO_2 problem include Carbon Dioxide Assessment Committee (1983), Jäger (1983), and Seidel and Keyes (1983).

5.4 TYPES OF MODELS—THEIR FEEDBACKS

5.4.1 Reasons for Different Models

The climate system involves transfers of energy between a three-dimensional turbulent and radiatively active atmosphere and spatially heterogeneous land, ocean, and cryosphere surfaces. This system is very complex and it has not yet been possible to produce models that use 'state-of-the-art' descriptions of either the atmospheric or surface processes. Furthermore, our 'state-of-the-art' understanding of some processes is still not satisfactory, as later discussed. Thus, various approximations and simplifications have been made to develop climate models, some based on empiricism. Because of lack of consensus as to which approximations do least damage to modelling results, many different modelling approaches have been developed. Some of these approaches have been superseded by more elaborate models that treat all physical processes essentially as well or better than they are treated in the simple models. The simple models still remain of interest because their relative computational economy allows consideration of a much wider range of parameter values. Furthermore, the simpler models can usefully provide insight and description of the dominant processes in the more elaborate models. That is, they are diagnostic and educational.

5.4.2 Zero-dimensional Model

One procedure used for simplification is spatial averaging. Averaging in all dimensions gives the simplest climate model of interest which can be written

$$C\frac{\partial \Delta T}{\partial t} + \lambda \Delta T = \Delta Q \tag{5.1}$$

where ΔT is the departure of the global average surface temperature from some climatological value, t is time, ΔQ is a perturbation in the net vertical flux of radiation at the tropopause that would take place due to some external change, such as increased CO_2, and in the absence of climatic change. Examples of ΔQ for CO_2 and other trace gases are given in Table 5.1. The term $\lambda \Delta T$ approximates the change of outward radiative energy flux, evaluated at the tropopause, resulting from global temperature change. The factor λ has units of Watts meter^{-2} $^\circ$C^{-1} and is referred to as the feedback parameter, whereas C is the system heat capacity.

Table 5.1   Estimates of ΔQ, the net heating of the troposphere system, for various hypothetical external changes of atmospheric constituents (based on World Meterorological Organization, 1982)

Gas	Change in Concentration Per Unit Volume		ΔQ
CO_2	300 ppm	$\rightarrow$ 600 ppm	4.0
CH_4	1.65 ppm	$\rightarrow$ 3.0 ppm	0.6
N_2O	0.3 ppm	$\rightarrow$ 0.5 ppm	0.5
$CFCl_3$	0 ppb	$\rightarrow$ 2 ppb	0.5
CF_2Cl_2	0 ppb	$\rightarrow$ 4 ppb	1.2

Equation (5.1) has no independent predictive value since λ and C are best obtained from more detailed models. For example, C depends on the depth to which thermal disturbances penetrate into the ocean which, in turn, depends on the time scale of the heating. Equation (5.1) is also of limited practical use because it only describes global average temperature change. More detailed models provide changes in surface temperature and hydrological processes which differ between locations and thus would be more useful to decision-makers.

However, Eq. (5.1) warrants attention for several reasons. First, it is straightforward enough that any reader with a modest calculus background can understand its physical description. In particular, the steady-state warming from increased CO_2 is simply given by $\Delta Q/\lambda$, and $t_c = C/\lambda$ defines a time scale required to approach steady state. Second, Eq. (5.1) is very convenient for interpreting the result of more detailed and complex models (as done by Climate Research Board, 1979), a task we shall return to later. Third, through use of Eq. (5.1) to summarize the results of more elaborate models, we can develop some idea as to the likely range of possible future climatic change implied by scenarios for future atmospheric concentrations of carbon dioxide.

5.4.3 One-dimensional Models

Two further popular general classes of simple climate models are the one-dimensional energy balance models (e.g., North *et al.*, 1981) and the radiative convective models (e.g. Ramanathan and Coakley, 1978). These models can be viewed as elaborations of Eq. (5.1) where, for the energy-balance models, ΔT in Eq. (5.1) becomes a function of latitude and in the radiative convective models a function of altitude. With either spatial dimension considered, it is necessary to add to Eq. (5.1) terms for the spatial coupling of temperature by energy transports.

5.4.4 Energy Balance Climate Models

Energy balance climate models are solved for temperature as a function of latitude or cosine of latitude, denoted y. The temperature feedback term, λ in Eq. (5.1), represents the variation with temperature of outgoing longwave radiation and reflected solar radiation. The longwave component λ_{LW} is most simply approximated by a constant term $\lambda_{LW} = B$, inferred from radiation models (Budyko, 1969) or from the empirical correlation of variations of longwave radiation at the top of the atmosphere observed by satellite with surface temperature (e.g., Cess, 1976). More elaborate treatments also consider the possible dependence of λ on other parameters, such as temperature, variations in cloudiness, variations in atmospheric water vapour, surface elevation, and changes in atmospheric lapse rate. Estimates of λ_{LW} have varied from between about 1.4 W m^{-2} °C^{-1} and 2.4 W m^{-2} °C^{-1}. The ice–albedo contribution is discussed separately below after consideration of transport.

The atmosphere does not locally reach a balance between radiation fluxes and vertical convection. Rather, it is heated in low latitudes and cooled in high latitudes by net radiative fluxes, i.e., the difference between absorbed solar and outgoing longwave radiation. This gradient in energy deposition provides the heat engine system that drives atmospheric winds and oceanic currents. The atmosphere transports thermal energy in sensible and latent form from warm to cold regions. The oceans may transport about as much thermal energy as the atmosphere. These transport processes are approximated in the energy balance models in various ways. The two simplest approaches are (1) to assume that energy divergence by transport is proportional to $(\overline{T}(y) - T)$, where $\overline{T}$ is global average temperature (Budyko, 1969); (2) to make a diffusion approximation, i.e., transport is proportional to the negative gradient of $T, -K \partial T / \partial y$, where K is a diffusion coefficient. The convergence of energy by transport is then given by $\partial / \partial y (K \partial T / \partial y)$. With solution by a low-order Legendre polynomial expansion (North, 1975), these two forms of transport are equivalent. More detailed treatments attempt to separate atmospheric and oceanic components of the climate system and make latent heat transport depend on the saturation vapour pressure of water.

The most realistic treatment of atmospheric transport is to solve explicitly dynamic equations of the time-varying atmospheric winds and use these to transport sensible and latent heat; that is the General Circulation Model (GCM) approach, to which we return later. Some intermediate models (e.g., Chou *et al.*, 1982) consider a separate atmospheric temperature with vertical transports determined as in the radiative convective models to be discussed later.

5.4.5 Ice–Albedo Feedback

Much of the solar radiation incident on the top of the atmosphere is absorbed at the Earth's surface (about half on a global average). The albedo at the Earth's surface thus is of major importance in determining the actual amount of absorbed solar radiation. Albedo can vary from as low as 0.02 for a clear calm water surface and overhead Sun to greater than 0.95 for fresh snow and visible wavelengths. In general, snow and ice surfaces, if exposed to solar radiation, have much higher albedos than do most land surfaces and liquid ocean under all conditions. Snow and ice can only be present for any length of time at temperatures which are freezing or below. In regions at the margins of ice or snow cover, decreases in surface temperature allow a larger area to be covered by ice and snow. Thus, the amount of absorbed solar radiation decreases with decrease in temperature. This temperature dependence of reflected solar radiation is referred to as ice–albedo feedback. In simple energy balance models, it is included by making albedo a function of temperature, e.g., by assuming a discontinuous increase in albedo where temperatures fall below some critical value (Budyko, 1969; Held and Suarez, 1974) or by making albedo negatively proportional to local temperature with some lower and upper limit representing zero and complete cover by ice and snow (Sellers, 1969).

The basic difference between one-dimensional energy balance models and Eq. (5.1) is the possibility of changes in the latitudinal variation of temperature in the former. If latitudinal temperature gradients are fixed or vary in a manner depending only on global average temperature, the one-dimensional model can be reduced to Eq. (5.1) (e.g., Wang and Stone, 1980). More generally, as already discussed, Eq. (5.1) can be used to interpret more detailed models. In particular, for any model we can always evaluate the ratio of change of global energy reflected by ice and snow to change of global average surface temperature and refer to this as ice–albedo feedback λ_A. This quantity is generally negative implying a destabilizing or positive feedback. For seasonally varying models, it is only meaningful to define ice–albedo feedback on an annual mean basis. The seasonal variation of albedo can become uncorrelated with seasonal variation of temperature if the albedo variation lags temperature by a season (Mokhov, 1981). Furthermore, while maximum changes in reflected radiation occur in the spring and summer seasons, maximum surface air temperature changes over sea ice are likely to occur in the autumn and winter as a consequence of changes in heat storage and release in the ocean (Manabe and Stouffer, 1980). This thermal inertial effect can swamp the explicit ice–albedo feedback over the seasonal cycle and may considerably amplify the annual average effective ice–albedo feedback (Robock, 1983a).

On longer time scales, additional feedbacks related to changes in continental ice sheets or atmospheric composition may become important. Feedbacks appropriate to past ice ages have been discussed by Hartmann and Short (1979), Held (1982), Bowman (1982), and Oerlemanns (1982).

5.4.6 Radiative–Convective Models

Radiative–convective models emphasize the effects of variation of temperature with altitude z and allow analysis of stratospheric radiative feedbacks on tropospheric temperatures. The simplest one-dimensional ones consider globally averaged atmospheric temperature $T(z)$. In the stratosphere, $T(z)$ is essentially in radiative equilibrium (e.g., as demonstrated by Fels *et al.*, 1980) and hence determined by a local balance between solar heating and net longwave cooling. The solar heating depends mostly on ozone concentrations. Above the lowest layers in the stratosphere, the longwave cooling is largely controlled by carbon dioxide and is primarily dependent on local temperature. Radiative balance just above the tropopause is complicated by the importance of absorption of longwave radiation originating from warmer tropospheric layers and from the Earth's surface. Not only carbon dioxide but also ozone and water vapour make important contributions to the opacity of the lower stratosphere. Radiative fluxes within and leaving the troposphere depend on the temperature profile, distribution of gaseous absorbers, especially water vapour and CO_2, and on the assumed distribution of cloud properties.

In the troposphere, the vertical variation of T is determined primarily not by local radiative balance but rather by the vertical energy redistribution by moist convective processes. Thus, one element of a radiative–convective model is a parameterization for tropospheric convection. The classical study by Manabe and Wetherald (1967) introduced the assumption that, if the lapse rate exceeds the 'critical rate' of 6.5 °C/km, a convective adjustment would ensue and maintain the 6.5 °C/km lapse rate. Many more recent radiative convective models have used this critical lapse rate. Other alternatives are to use the observed global average lapse rate or to use the lapse rate defined by the moist ascent of a saturated parcel (the moist–adiabatic lapse rate) (Rowntree and Walker, 1978; Ramanathan and Coakley, 1978; Lindzen *et al.*, 1982).

The moist–adiabatic assumption illustrates the possibility of lapse rate feedbacks (Schneider and Dickinson, 1974; Ramanathan and Coakley, 1978). If a temperature change at the surface is extended upward along a moist adiabat, the magnitude of the change increases with increasing altitude. Outgoing longwave fluxes largely originate within a tropospheric column with only about 0.3 of the total flux coming from surface or near-surface emission in the 8–12 μm window region. Thus, an increase of temperature change with increased altitude amplifies the consequent change of outgoing longwave flux. Thus, a smaller surface temperature change is required to balance a given external change in atmospheric composition or heat input. This process is a negative lapse rate feedback. Hence, models with moist adiabatic adjustment but fixed cloud properties have larger λ (cf Eq. (5.1)). In GCMs, a process of moist adiabatic adjustment controls to a large extent their tropical temperature profiles.

Alternatively, if a model is heated by large amounts at high elevations (e.g., as argued by Turco *et al.*, 1983, to be the consequence of nuclear war aerosol), then the lapse rate becomes less than the critical value, and moist adjustment no longer couples surface temperature to the tropospheric column. Under these conditions, large temperature changes can occur at the surface in response to changed surface heating with only small additional changes occurring at higher levels in the troposphere. In this case, a much larger surface temperature change would be required to balance some prescribed external change that warms the surface. The lapse rate feedback is positive. Positive lapse rate feedback applies especially to high latitudes for CO_2 warming (e.g., Manabe and Wetherald, 1975; Ramanathan, 1977).

Another important question which needs treatment in radiative–convective models is the amount of water vapour contained by the atmosphere. Manabe and Wetherald examined two assumptions. First, they considered fixed concentrations of water vapour. The temperature change they then inferred corresponds to $\lambda = 3.7$ W m^{-2} °C^{-1}, which would be the feedback for blackbody radiation from a planet with no atmosphere and albedo of 0.30, the presently observed value. Alternatively, they argued that a much more realistic assumption in the presence of climatic change is fixed relative humidity for which they found $\lambda = 2.2$ W m^{-2} °C^{-1}. The difference between these two values of 1.5 W m^{-2} K^{-1} is the positive water vapour feedback of their fixed relative humidity model. Increased water vapour increases the atmospheric trapping of longwave radiation and, to a much lesser extent, absorbed solar radiation, as does increased CO_2. For the actual warming, it is unlikely that relative humidity will stay strictly constant or change in the same way at all altitudes. For this reason and because of the three-dimensional variations of actual water vapour concentrations, changes in relative humidity would also be of considerable importance for determining cloud changes, the subject we turn to next.

5.4.7 Cloud-radiation Feedbacks

It is evident that the presence of clouds increases the longwave opacity of the atmosphere and hence intensifies the lapse rate feedbacks just discussed for a given temperature profile variation. One of the most important contributions of radiative–convective models has been to clarify other possible feedbacks between cloud cover and atmospheric radiation. Clouds can change in thickness, fractional cover, or in the altitude of their tops, as climate changes. Other more subtle changes such as in the size of cloud droplets or in the cloud horizontal scale may also significantly modify cloud radiative properties.

Manabe and Wetherald (1967) fixed the above cloud properties and obtained, as previously mentioned, a feedback $\lambda = 2.2$ W m^{-2} °C^{-1}. Alternatively, some models have assumed fixed cloud-top temperatures. This assumption provides a positive feedback since the longwave flux from the cloud tops cannot then change significantly and the fluxes from the surface, hence surface temperature, must change more. Typically, the assumption of fixed cloud-top temperature reduces λ by about 0.5 W m^{-2} °C^{-1} from its value for fixed cloud-top altitude. More generally, if cloud altitudes increase with a warmer temperature, the cloud feedback on surface temperature is positive, or vice versa if cloud altitudes decrease. Cloud feedbacks in energy balance models have been discussed by Golitsyn and Mokhov (1978).

The average fraction of the sky covered by clouds may also change with climatic change. Cloud fractional coverage is probably only weakly linked to global or perhaps even local surface temperatures and depends primarily on local atmospheric dynamical processes (e.g., Schneider *et al.*, 1978). For 'average' clouds, changing clouds by the same relative amount at all levels has about twice as large an effect on the amount of reflected solar radiation as it has on the trapping of longwave fluxes. Hence, decreased cloud fraction usually implies increased heating of the climate system. The exception is thin high cirrus clouds, which modulate longwave radiation to a greater extent than solar fluxes.

An increase in the thickness of clouds as described by cloud liquid water content would generally increase the albedo of the clouds but have little effect on the longwave emission of all but nonblack cirrus clouds. Greatest albedo change for given increase in liquid water is possible for relatively thin clouds since thick clouds are essentially saturated in brightness.

Various suggestions have been offered as to how clouds might change with a warmer climate. A warmer atmosphere would hold more water vapour and possibly also more liquid water, hence cloud thickness could well increase. A warmer climate would also probably provide a larger fraction of convective rainfall. Convective clouds tend to concentrate liquid water into small regions with a large amount of clear sky between clouds, as opposed to layer clouds which can uniformly cover a large area. Thus, more convective clouds would probably mean less low cloudiness. On the other hand, convective clouds can have relatively high tops and further increase high-level cloudiness through the cirrus shields formed by their outflow. More intense synoptic-scale vertical motions from the greater rainfall of a warmer climate could also reduce cloudiness, but could increase cloud heights (e.g., Schneider *et al.*, 1978; Wetherald and Manabe, 1980). One popular idea is that fractional changes would be compensated by height changes so that little net effect would be realized (e.g., Cess, 1976; Wetherald and Manabe, 1980; Cess *et al.*, 1982).

The net effect of the various possible changes in cloud properties can only be examined by physical models for clouds coupled to atmospheric dynamics in a three-dimensional context. We now turn to discussion of three-dimensional climate general circulation models.

5.4.8 General Circulation Models

General Circulation Models (GCMs) have evolved as a generalization and by-product of the development of numerical weather prediction models. They consider the atmosphere in three spatial dimensions and time. This domain is represented either in terms of a mesh of points or in terms of polynomial (e.g., spherical harmonic) basis functions. The GCMs solve jointly the equations of motion for atmospheric winds and equations for conservation of thermal energy and water vapour, including transport by the calculated atmospheric motions. The equation for thermal energy includes: detailed treatments of the vertical radiative transfer of solar and longwave radiation; parameterizations for moist and dry convective or turbulent redistribution of thermal energy and moisture. The equation for water vapour has an evapotranspiration source at the Earth's surface and a sink through formation of rain or snowfall.

The atmospheric model depends on boundary conditions or on other models at the Earth's surface. For example, an ocean model is required to provide ocean surface temperature since the atmosphere and climate, in general, are strongly coupled to ocean surface temperatures. Also necessary are models for sea ice and for various land-surface processes including snow cover, soil moisture, and evapotranspiration. Figure 5.1 shows schematically the various linkages treated by GCM climate models.

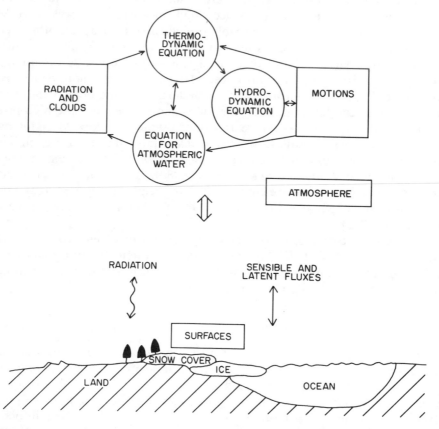

Figure 5.1 Schematic description of the structure and processes of GCM climate models

Only the GCMs among climate models can plausibly model such atmospheric surface variables as soil moisture, snow cover, and ground temperature. On the other hand, some of the current deficiencies and uncertainties in GCM climate projections stem from use of much cruder models for surface climate processes than are used for the atmosphere. Now we turn to this question of model deficiencies.

5.5 MODEL DEFICIENCIES

5.5.1 Relationships Between Different Kinds of Models

The radiative–convective models have traditionally been used to examine the most accurate and computationally expensive schemes for atmospheric radiative transfer. They have also been used to explore cloud-radiation feedbacks as discussed earlier. For clear-sky conditions, the radiative transfer aspects of GCMs are believed to be now more accurate than most other features in the models and are being further refined and validated. Furthermore, cloud formation and vertical convective energy transfers are essentially of a short time scale and three-dimensional in nature, and so in principle require (as a minimum) a GCM for their direct calculation. Likewise, the transport and ice–albedo parameterizations of energy-balance climate models are superseded by their potentially much more realistic treatments in GCMs. Furthermore, only GCMs are capable of providing information useful for evaluating potential impacts of future climatic change, that is, details regarding regional climatic change and changes in surface hydrological processes.

For these reasons, the simpler climate models, in principle, have no additional capabilities for projecting future climatic change besides that of the GCMs. In practice, some aspects of some GCMs may be less satisfactory than corresponding treatments in simpler models. The most obvious example of this principle has been past GCM studies of climatic change with fixed ocean temperatures. These studies found much smaller values for global warming than suggested by the one-dimensional climate models which more realistically allowed effective ocean temperatures to vary with climatic change.

Even today, most GCMs include 'swamp' or 'mixed-layer' ocean models that assume zero horizontal transport of heat by oceans. An energy balance model with a more realistic ocean heat transport characterization might in some ways be regarded as more correct than such GCMs for future climate projections. Furthermore, the developers of simple climate models often pay much more attention to observations constraining their models than do developers of GCMs, in part because there is such a vast range of possible data that could be used to validate a GCM. Thus, observational or physical constraints (e.g., Eq. (5.1)) have often been formulated first in simple models before application by GCM modellers.

In spite of the above considerations, it is most useful here to examine primarily the deficiencies of GCMs. One of the general deficiencies of GCMs is that their horizontal mesh scales are several hundred kilometres or larger because of computer expense. Thus, much regional climate detail cannot be resolved. Better climate model resolution might improve the simulation of larger scale features even more than features at the limit of resolution. Mesoscale climate models could be embedded within GCMs to attempt description of scales down to order of 100 km, but this procedure has not yet been applied to questions of future climatic change. It will not be discussed further here. The other presently most obvious deficiencies in GCM simulations of future climate are examined in the following sections.

5.5.2 Clouds

As discussed earlier, future changes in cloud properties could make either large increases or possibly large decreases in the net radiation budget of the climate system and hence could either amplify or diminish (but not cancel entirely) the effect of increased CO_2 and other trace gases. Regional changes in climate would result from changes in cloud radiative effects coupled to shifts in quasi-stationary planetary waves (e.g., Trenberth, 1983; Hartmann, 1984). General Circulation Models have attempted to generate cloudiness but with little confidence in the result. In particular, several models have simulated cloud formation for present and future climates by assuming that clouds form whenever the model precipitates, or whenever moist convection occurs in the model or according to the relative humidity (e.g., Wetherald and Manabe, 1980; Washington and Meehl, 1983, 1984; Hansen *et al.*, 1984). Wetherald and Manabe have suggested mechanisms for reduction of clouds in low latitudes and increased stratus in high latitudes in response to global warming.

There are several serious difficulties in including clouds in models. First, cloud properties are largely subgrid-scale. Even the most regular layered cloud systems have wide variations on a 100-km scale, and boundary-layer cumulus clouds can be of a scale less than 1 km. The thickness of stratus clouds may be much less than the vertical grid interval.

Second, there is a wide variety of cloud types, each of whose radiative properties varies widely from cloud system to cloud system. Thus, any binary yes–no description of cloud cover is too oversimplified to be useful. Different physical models are required for different cloud types. These models must satisfactorily give all the radiatively important cloud properties, in particular, fractional cover, height, liquid water content, drop sizes, and spatial scale. It is necessary to model these properties with enough attention to physical processes that their change with climatic change can be adequately projected. Modellers, in particular, recognize the need for separate models of high clouds, such as cirrus, boundary-layer clouds, trade-wind region stratocumulus, frontal-layer clouds, and clouds associated with deep cumulus convection. Cloudiness on the margins of the Arctic and Antarctic ice packs in spring and summer is especially important for determining the magnitude of ice–albedo feedback.

The third, and perhaps most serious difficulty, is that processes of cloud formation and cloud radiative properties have not been satisfactorily related observationally to other meteorological processes. Diurnal variations of cloudiness are poorly understood but may be important for determining changes in the radiation balance. The existing climatologies of cloud cover are of somewhat marginal value for the purpose.

As a consequence of the limitations discussed above, the current approach of modellers is to make up some 'cloud model' based on simple physical arguments, include this parameterization in a GCM simulation, and compare the resulting model cloud climatology with one or several observational climatologies. Such procedures are unsatisfactory because (1) they are usually only validated against fractional cloud cover, which is but one of the radiatively important cloud properties; (2) there can be large differences in cloud cover climatologies so that it is not difficult to find at least some data that the model clouds appear to agree with; and (3) since cloud parameterizations usually have one or more adjustable parameters, it is possible for models to agree with present data sets without the cloud model responding properly to climatic change.

5.5.3 Ocean Coupling

The climate warming by increased CO_2 depends in large part on increases in ocean temperatures. The oceans are also important for their uptake and storage of CO_2 as discussed further in Chapter 3. The oceans are as complex a dynamical system as the atmosphere and observationally more poorly characterized. Ocean surface temperatures are determined by the balance between spatially varying net surface heating and a variety of dynamical processes for the redistribution of thermal energy, including small-scale vertical mixing and large-scale horizontal transport by ocean currents (Bretherton, 1982; Woods, 1984). There are not yet available ocean GCMs with enough spatial resolution to resolve the energy-containing eddies as done in GCMs for the atmosphere. Rather, at present the most sophisticated ocean models for climate studies (e.g., Bryan *et al.*, 1982) are of coarse resolution (compared to ocean eddies) and heavily influenced by semi-empirical eddy diffusion parameterizations. Even the coarse-resolution ocean models are not yet being employed in most GCM studies of future climate.

Rather, GCM studies to date have largely assumed 'swamp' or 'simple mixed-layer' ocean models. Both kinds of models neglect entirely horizontal energy transport by the oceans. The swamp model characterizes the ocean surface as a wet surface of zero heat capacity, whereas the mixed-layer models assume an ocean reservoir for heat, uniformly mixed in the vertical. The 'swamp model' ocean evaporates water and conserves energy but is otherwise highly unrealistic. The heat capacity of the 'mixed layer' model can be adjusted to give a reasonable range of temperatures for the annual cycle. However, the lack of horizontal heat transport by the oceans is quite unrealistic and gives serious difficulty in modelling other important processes such as formation of polar sea ice. Since up to half the total heat transfer is in reality accomplished by the oceans, either the model heat transport is reduced and the pole–equator temperature difference in the model becomes larger than in reality, which seems not to be the case, or atmospheric transport is increased which must influence atmospheric flow patterns significantly.

A conceptual improvement upon the swamp and simple mixed-layer models that is computationally not much more complicated is to constrain the ocean horizontal heat transport or temperature gradients to correspond to observation or some other prescribed relationship. The study by Hansen *et al.* (1984) uses a fixed horizontal heat transport mixed layer ocean model. Hoffert *et al.* (1983) argue that ocean transport may act to minimize tropical surface temperature changes. Realistic patterns of ocean temperature change may be important for determining shifts in rainfall patterns, especially in the tropics.

5.5.4 Sea Ice

As discussed earlier, simple climate models show that the albedo decreases from reduction in the cover of snow and ice can significantly amplify the climate warming of increased CO_2, i.e., reduce λ in Eq. (5.1). This feedback mechanism has also been included in GCM studies but not necessarily with careful attention to physical realism. It appears (e.g., Robock, 1983a; Hansen *et al.*, 1984; Washington and Meehl, 1984) that changes in sea-ice cover are the major contributor in a seasonal cycle model to the albedo-feedback mechanism and depend on the seasonal variation of the effective thermal inertia at the sea-ice margins. Sea-ice cover persists into spring and summer or later, whereas most land snow cover melts sooner, and where shaded by tall vegetation, provides much less albedo contrast.

There have been at least four basic deficiencies in GCM treatments of sea-ice—(1) lack of realistic ocean heat transport, as already discussed, (2) unrealistic treatments of the albedo of sea ice, as discussed in the next section, (3) unsatisfactory models for cloud properties at the sea-ice margins as earlier discussed, and (4) oversimplified models of the thermodynamics and dynamics of the sea-ice itself (improvements are suggested by Semtner, 1984; Hibler, 1984; Hibler and Bryan, 1984). Treatments of the effects of leads and turbulent transport at the water–ice interface are inadequate. Other questions of possible deficiencies, such as whether GCMs give a realistic description of sensible fluxes between sea ice and atmosphere, also need examination.

5.5.5 Surface Albedos

Both the numerical and conceptual bases for specifying surface albedos in GCMs have been deficient. That is, numbers have been used for model albedos inconsistent with the available observational studies, and possible changes in surface albedo with climatic change have not been included beyond a crude treatment of albedo change from change in snow and sea-ice cover. Furthermore, the observational basis for parameterizing surface albedos in GCMs is inadequate.

Except for the effects of snow and ice, surface albedo changes may be small but may still need consideration. For example, ocean albedos depend on bubbles and turbidity of the surface water and on the angle between solar beam and water interface. Thus, ocean albedos vary with solar zenith angle, scattering of the solar beam by cloud particles, and wave spectrum, as well as varying with the amounts and nature of particles in the water, such as phytoplankton. At least some of these dependencies may imply significantly different ocean albedos with future climatic change.

Land surface albedos are much more complex and dependent on atmospheric conditions and surface microclimates. There are large variations of these albedos with wavelength of the solar beam as well as with the solar zenith angle. Theoretical characterization of the albedo of vegetation canopies includes a complicated problem in radiative transfer. Soil albedos depend on soil moisture and on soil chemical and physical structure. As an example of the level of detail regarding land albedos actually used in current GCM studies of future climate, Washington and Meehl (1984) assume a value of 0.13 for all non-snow-covered land surfaces, except deserts for which they use 0.25.

Perhaps improved model descriptions of most aspects of surface albedo are currently not warranted considering the large uncertainties regarding changes in cloud optical properties. The same argument could be made to justify current model neglect of atmospheric aerosol radiative effects. However, more refined treatments of snow and ice albedo properties would certainly increase confidence in the model sensitivities to ice–snow feedbacks. For example, Manabe and Stouffer (1980) do not even distinguish between the albedos of snow-covered and snow-free ice. Except for Hansen *et al.* (1984), GCM studies of future climate have not even included the reduction of snow surface albedo by tall vegetation.

5.5.6 Land Surface Hydrology

Water in liquid and solid form is deposited by actual and GCM atmospheres on the land surface. The subsequent disposition of this water is a major factor in determining regional climates. Snow surfaces melt and change their albedos in complicated ways that have been studied by snow hydrologists. Yeh *et al.* (1983) have shown with a GCM simulation that removal of snow cover in early spring would give greater continental dryness that persists into late summer.

Some of the water incident at the surface is intercepted and re-evaporated by the foliage of vegetation (e.g., Shuttleworth and Calder, 1979), some runs off on the surface, some infiltrates into the soil where it provides water for plant transpiration, some re-evaporates at the soil surface, and some percolates to below the root zone where it supplies the subsurface water reservoirs and hence the 'base flows' of drainage basins. The effects of vegetation changes on evapotranspiration have recently been reviewed by McNaughton and Jarvis (1983). A model for a vegetated surface coupled to a planetary boundary layer model has been discussed by DeBruin (1983). Shukla and Mintz (1982) have demonstrated that changing GCM land surfaces from completely wet to completely dry gives a large decrease in rainfall over land. Many other recent studies demonstrating the sensitivity of rainfall to evaporation parameterizations have been reviewed by Mintz (1984).

The maximum rate of infiltration is limited by the soil hydraulic conductivity and so can vary widely over small distances. The amounts of water infiltrated depend strongly on the distribution of precipitation in time and space. That is, surface runoff may be very different for an intense localized convective storm than it is for a uniform drizzle, although these two situations have the same rainfall over a GCM grid square and averaged over a day. From the viewpoint of climate impacts, this difference could determine whether a farmer has a well-watered crop or, conversely, whether much of his soil ends up in the river and his land is subject to drought. From the viewpoint of atmospheric climate, the subgrid-scale structure of rainfall has a significant influence on how much water the land surface can return to the atmosphere through evapotranspiration.

All the above questions are neglected in current GCM treatments of surface hydrology. The conventional GCM treatment of soil water is to assume that the soil acts as a bucket which after being filled adds its surplus to runoff. The full bucket evaporates as would a moist surface until it has lost some fraction of its capacity. At lower levels of water in the bucket, evaporation is assumed to be linearly proportional to its remaining water content. Variations in surface roughness because of differing vegetation cover are usually neglected. The model of Hansen *et al.* (1984) elaborates on this scheme somewhat by use of two soil reservoirs whose capacity depends on vegetation type. However, the details of their formulation are somewhat arbitrary.

It may be that the present GCM treatments of land-surface processes are adequate for making first estimates of future regional climatic change. On the other hand, it could turn out that more refined treatments (e.g., Dickinson, 1984) may be necessary. It must be kept in mind that land-surface processes can only be properly modelled in GCMs which give realistic simulations of precipitation and which include other realistic details, in particular, a good model of the planetary boundary layer and a diurnal cycle of radiation. Both these latter factors are lacking in most of the GCMs currently used to study future climate. Furthermore, the dependence of surface hydrology on the subgrid-scale details of rainfall intensities is of especial concern since these details could change significantly with a warmer climate. Better descriptions of river discharge may be important, especially for modelling changes in the Arctic Ocean. More attention must be given to surface radiation depending on cloudiness as well as temperature, and to relative humidity in the models since these not only are important for evapotranspiration but for plant growth in general.

5.5.7 Transient Response

Most GCM studies of the climate response to a CO_2 warming have considered the steady-state response for a given increase in CO_2, in particular, a doubling or quadrupling of CO_2. It is fortunate that this has been the tack, for standard benchmark calculations are required to make comparisons between different models and to help focus attention on various model deficiencies. However, the results of a steady-state calculation may be misleading in application to estimating the actual climatic change caused by human activities at present or in the future. Emissions of CO_2 into the atmosphere are not fixed but growing, since 1973 at a rate of somewhat less than 2% a year (Carbon Dioxide Assessment Committee, 1983), and atmospheric concentrations have been growing by about 0.4% per year.

The oceanic surface mixed layers with depths of 50–100 m equilibrate with atmospheric temperatures on a time scale of several years. However, high-latitude oceanic convection mixes surface waters in contact with the atmosphere to great depths in the ocean. Consequently, an ocean reservoir of order of 1 km in depth, depending on the time scale of the warming, must be warmed before ocean surface waters can equilibrate with the atmosphere. Roughly 50 years are required for the atmosphere to nearly equilibrate with this reservoir; equivalently, the actual temperature response to CO_2 warming may lag the steady-state response by up to several decades, or at any one time have a magnitude not much larger than half the steady-state response. Obtaining the details of the transient response to CO_2 warming would require a satisfactory ocean general circulation model, a tool not yet available as discussed earlier. However, estimates of the time scale involved can be made using studies of ocean tracers (e.g., Broecker *et al.*, 1979) if buoyancy effects of thermal anomalies can be neglected. Energy balance model studies of the transient response have been discussed by Robock (1978), Schneider and Thompson (1981), Cess and Goldenberg (1981), Dickinson (1981b), Michael *et al.* (1981), Schlesinger (1983a), Hansen *et al.* (1984), Wigley and Schlesinger (1985), and Harvey (1985).

The transient response to CO_2 warming differs from the steady-state response not only in overall magnitude but also in the latitudinal and regional details. This assertion follows from the different, probably slower, response times for high latitudes compared to low latitudes. Both the deep oceanic convection and the ice–albedo feedbacks act to retard the high-altitude response more than elsewhere. The question of how far the transient response departs from the steady-state one as a function of latitude is controversial and different conclusions have been drawn (e.g., Schneider and Thompson, 1981; Bryan *et al.*, 1982; Thompson and Schneider, 1982; North *et al.*, 1984). Better estimates of ocean heat uptake will likely require further progress in the development of detailed models of ocean circulation.

5.6 MODEL LIMITATIONS

5.6.1 Validation and Performance of Control Runs

The GCMs are the most complete simulation tools for the climate system. However, the models that have so far been used to study CO_2 warming are crude and incomplete descriptions of many important climate processes. These difficulties arise in part because of the limits of the models' spatial resolution. There will probably always remain some questionable aspects of model formulations. Yet many deficiencies may make little difference in projections of future climate. Furthermore, physically more complete descriptions are usually more complicated and more difficult to understand than are simpler parameterizations, hence also more susceptible to serious logical or programming errors in their formulation.

Thus, it is necessary to find means to obtain more precise estimates of the reliability of the results of the limited models that are available now or would be at any time, and so, hopefully, to increase our confidence in them. One approach for doing this is to compare model simulations with observed climate. In particular, the models should be able to reproduce the observed spatial and seasonal variations of climate.

The GCMs are able to reproduce the large seasonal variation surface air temperature changes of the Northern Hemisphere continents, e.g., Manabe and Stouffer (1980). Their simulation of global average surface temperatures and pole-to-equator temperature differences has also been in surprisingly good agreement considering the lack of ocean heat transport. The control model of Hansen *et al.* (1984) apparently gives a good simulation of the observed seasonal variation of sea ice in both hemispheres (Barry *et al.*, 1984), presumably because of the adjustment of prescribed ocean heat transport to provide such agreement.

Many parameters in the GCMs have somewhat arbitrary magnitude and are 'tuned' to satisfy observational constraints. Thus, for example, agreement between modelled and observed cloud cover or radiation fluxes at the top of the atmosphere is usually in part the result of adjustment of cloud parameterization parameters to achieve such agreement. While such tuning is a necessity, the consequent agreement with data does not help validate the model's performance as would agreement with data that had not been used to develop the model. Furthermore, model results may deteriorate with physical improvements in other parameterizations because of removal of compensating errors. Another good check of a model's surface physics and planetary boundary-layer parameterizations in response to large radiation changes would be an agreement for different land surfaces between model and observed diurnal cycle of temperature. Most current GCMs, however, do not have a diurnal cycle of radiation. Agreement between modelled and observed diurnal cycle would likely be poor for GCMs without adequately realistic treatments of the planetary boundary layer and surface energy transfer processes.

Another approach to model validation is to see the extent to which past climates can be modelled. Unfortunately, the data regarding past climates are not sufficiently complete to allow any unambiguous model testing. Times of large difference from present conditions may be most appropriate for such tests. Several studies have been made of GCM simulations of the last ice age at 18K BP (i.e., 18,000 years ago). However, these studies failed to provide much model validation since they used mostly available data for boundary conditions. Hansen *et al.* (1984) considered the question of global energy balance in their 18K BP simulation, with prescribed ocean surface temperatures and ice sheet. They found the radiative fluxes at the top of the atmosphere to be somewhat negative relative to current conditions. They concluded that either the positive cloud feedback in their model was somewhat exaggerated or that some of the boundary conditions they used, perhaps the assumption of 200 ppmv CO_2 concentration or tropical ocean temperatures, were somewhat in error. Broecker and Takahashi (1984) have reviewed possible mechanisms for changing atmospheric CO_2 on a time scale of ice ages.

The most documented periods of past climates which were much warmer than present are the early Holocene post-glacial epoch 6000 to 9000 years ago, with observations indicating local summer temperatures at times at least 1 to 2 °C warmer than present, and the Cretaceous period about 100 million years ago with temperatures 10–20 °C warmer. GCM simulations of 9K BP with changed orbital parameters have been discussed by Kutzbach and Otto-Bliesner (1982).

Barron and Washington (1984) have discussed simulation of Cretaceous climate. The latter study found that changes in model geography were inadequate to explain the observed Cretaceous warmth and hence invoked some additional heating, possibly that due to more atmospheric CO_2 then than now. Berner *et al.* (1983) have shown that atmospheric CO_2 could vary by at least an order of magnitude on a time scale of tens of millions of years as a result of shifting balances between geophysical sources and sinks. Studies of past climates do not provide much direct model validation because of the presence of additional very long time scale feedbacks, such as natural changes of CO_2 concentrations which may be invoked for removing differences between model and observations. However, simulation of the climate of the last ice age may be useful in validating overall model feedback since estimates of CO_2, ocean temperatures, and ice sheet geography are available. More generally, any studies which require comparisons between model output and observations are likely to improve our understanding of model capabilities.

5.6.2 Signal-to-noise Problem

GCMs simulate climate as an average of day-to-day weather fluctuations and other aspects of natural variability. Thus, just as actual climatic change is obscured by natural variability (e.g., Leith, 1973; Madden and Shea, 1978), so are the climatic changes obtained by GCMs for increased CO_2. The problem is one of obtaining the climatic change signal in the presence of model noise. In principle, models can generate much more extensive climatic statistics than are available for the past actual climate. However, since GCMs require large computing resources, in practice, the length of time period for which they can simulate climate is limited to at most a few decades. Over such a time period, only relatively large climatic change (e.g., that associated with doubling or quadrupling atmospheric CO_2) can be established over the model noise background. The climatic change up to 1985 due to the previous increase of CO_2 is at least a factor of five smaller than that due to doubling. To estimate this relatively small change with reasonable statistical significance would likely require at least ten model simulations of climate starting perhaps 50 years ago.

Perhaps when the speed of supercomputers has increased 100-fold over current values, such integrations might be practical. Such increases in computing power may be available in the 1990s. However, it is also desirable to develop GCM climate models with higher spatial resolution, which is also now limited by lack of computing resources. Thus, the problem of distinguishing model signals from noise will probably not diminish greatly with future computers.

The most widely applied statistical approaches to the GCM noise problem have been univariate. That is, changes in the value of each variable at each gridpoint are separately tested for statistical significance (e.g., Chervin and Schneider, 1976; Chervin, 1980, 1981; Hayashi, 1982; Katz, 1982, 1983). There are various recognized difficulties with this approach (e.g., as discussed by Hasselmann, 1979), the most obvious of which is that, if testing for significance at a 95% confidence level, 5% of the gridpoints will on the average be falsely identified as significant. Hayashi (1982) suggests an emphasis on confidence limits rather than on significance testing. Livezey and Chen (1983) have discussed possible approaches for testing the statistical significance of changes in meteorological fields accounting for their multiplicity and spatial correlations.

5.6.3 Regional Continental Scale Details

It has long been recognized that the most useful information on future climate would necessarily involve changes on a regional–continental scale of typically 1000 km or so over the continental areas where the bulk of population and economic activities is found. If GCM climatic change integrations are carried out for a sufficient time to overcome the signal-to-noise problem, regional-scale climatic change features can be found. However, at present, we can have little or no confidence in the reality of such features as a description of expected future climate, for several reasons. First, careful validation of GCM simulations of present regional-scale features is lacking. Second, shifts in continental-scale features are dependent on dynamical and physical details poorly treated in GCMs. These might include, for example, shifts in the stationary planetary wave structure and associated shifts in radiative sources and sinks (e.g., as discussed by Hartmann and Short, 1979; Hartmann, 1984).

Considerable effort will be needed to establish the physical, as contrasted to statistical, validity of GCM simulations of regional climatic change. One possible approach is to compare similarities and differences between different model simulations. There are considerable duplications between models in questionable parameterizations, but there are also many differences. Disagreements between different models can be used to help flag uncertainties of model projections of future climate.

5.7 REVIEW OF GCM RESULTS FOR INCREASED CO$_2$

5.7.1 The More Realistic GCMs

Here we compare results from recent GCM simulations of the climatic change from increased CO$_2$. It is useful to distinguish between exploratory GCMs and the more realistic GCMs. The most realistic GCMs have annual cycles and realistic continents and orography within the constraints of model resolution. They should also have adequate horizontal and vertical resolution. What is adequate resolution needs to be established by comparison of model simulations not just with observations but also with higher resolution simulations with the same model. The most satisfactory models, according to these considerations, are the models of Manabe and Stouffer (1980) and Washington and Meehl (1984). Also included is the model of Hansen *et al.* (1984) which, although coarse in its horizontal resolution, appears to have adequate baroclinic wave variability (Hansen *et al.*, 1983) and possibly does a superior job on some aspects of climate simulation. All the models suffer from probably inadequate planetary boundary-layer parameterizations and resolution. Their vertical resolution around the tropopause is also questionable. They may extend insufficiently far into the stratosphere for realistic planetary wave structures. Other model features are summarized in Table 5.2.

All the models referenced above are formulated to maintain global average radiative equilibrium as required to be able to estimate global average temperature change. Another informative approach is to use observed ocean temperatures for a control simulation and to take ocean surface temperature changes as prescribed rather than calculated, but changing with increased CO_2 as inferred from previous integrations and energy balance criteria and consistent with GCM simulations that do calculate ocean temperature change. This has been done by Mitchell (1983), who assumed a constant ocean temperature change of 2 °C along with doubled CO_2, and by Mitchell and Lupton (1984) who assumed quadrupling of CO_2 and prescribed ocean temperature changes for each latitude belt that would seem to imply approximately no change in oceanic heat transport. Continental-scale variations in the climate response in these studies may be less confused by errors in the control simulation than in studies which assume mixed-layer ocean models for both control and perturbation simulations.

Large quantities of diagnostic information are generally reported by authors of GCM studies. We consider here only some features that are of practical importance in interpreting the surface climate simulations. Note that Manabe and Stouffer assumed a quadrupling of CO_2 whereas the other studies a doubling. Figures 5.2 to 5.6 compare the calculated temperature change for increased CO_2. Figures 5.2 and 5.3 show the zonal average temperature change as a function of altitude for the Washington and Meehl and the Manabe and Stouffer models, respectively. The results differ most in the stratosphere where the Manabe and Stouffer model shows about twice as large a temperature decrease. This is not surprising considering that it assumed twice as large CO_2 increase. What is more surprising is the similarity of the troposphere change. The Manabe and Stouffer model surface temperature is evidently about half as sensitive to external heating (λ twice as

Table 5.2 Comparison of some features of the models of Manabe and Stouffer (1980), Hansen *et al.* (1984), Washington and Meehle (1984)

	Manabe and Stouffer	Hansen *et al.*	Washington and Meehle
Atmospheric model	Nine-layer, 15 wave rhomboidal spectral.	Nine-layer. 8° latitude 10° longitude, second-order, Arakawa B-grid.	Nine-layer, 15 wave rhomboidal, spectral.
Ocean model	Simple 68 m mixed layer.	65 m or less mixed layer with constant horizontal transport tuned to present ocean temperatures.	50 m simple mixed layer.
Cloud prescription	Fixed zonal.	Layer clouds depend on relative humidity; cumulus depend on vertical mass flux.	Clouds with precipitation.
Sea-ice model	One slab, thermodynamic.	Two slab, thermodynamic	One slab, thermodynamic
Snow albedo	For deep snow 0.6 equatorward of 55°, 0.8 poleward of 66° S, linearly interpolated in between; albedo reduced for snow of less than 1 cm liquid water.	For deep snow 0.5 to 0.85 depending on snow age. Reduced for less than 1 cm liquid water. Shading of vegetation accounted for.	0.8 independent of depth or latitude.
Sea ice albedo	0.5 equatorward of 55°, 0.7 poleward of 66.5°, linearly interpolated in between.	Snow albedo if snow covered. 0.45 for bare ice.	Snow albedo if snow covered, 0.70 for bare ice.

large) as the Washington and Meehl model, as shown by the fact that global average surface temperature increase for both models was about 4 °C. Both models show largest warming in the Northern Hemisphere in winter in high latitudes. In the Southern Hemisphere, however, the Washington and Meehl model shows largest warming at the edge of Antarctica, whereas the Manabe and Stouffer model shows a less intense maximum warming centred over the South Pole. Figure 5.4 bottom frame, shows annual average tempera-

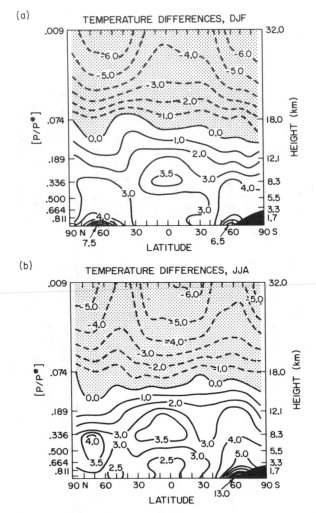

Figure 5.2 Zonal average temperature change in °C versus altitude and latitude for a steady-state doubling of CO_2 according to the model of Washington and Meehl (1984). Top frame is for December to February; bottom frame is for June to August. Cooling in the stratosphere is shaded

ture change contours for the Hansen *et al.* model. This model gives greater warming in the tropical troposphere, nearly double that of the other two models but shows relatively small annual mean polar amplification. Their global average warming is also about 4 °C. The model details responsible for significant differences between outputs are not known. However, 'causes' for differences in global temperature change can be diagnosed in terms of differences in feedbacks as discussed further in Section 5.8.

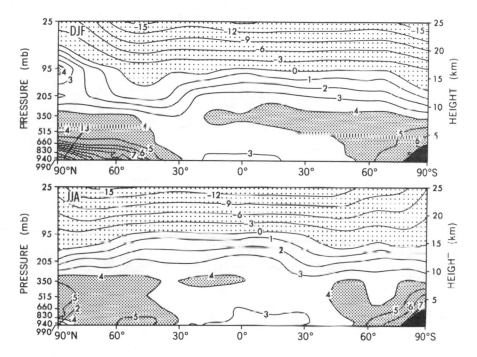

Figure 5.3 Same as Figure 5.2, except for quadrupling of CO_2, according to Manabe and Stouffer (1980). Cooling in the stratosphere is indicated by a dot pattern. Regions of warming greater than 4 °C are shaded

The upper frame of Figure 5.4 shows the geographical distribution of the annual mean Hansen *et al.* surface warming. No obvious pattern is seen beyond the fact that the high-latitude warming appears to be somewhat larger than that seen in the bottom frame, which only extends to the lowest model layer and does not include surface air temperatures. The 2 ° change contained in the bottom frame over the South Pole which especially gives this impression is not actually found but is the result of a minor glitch in

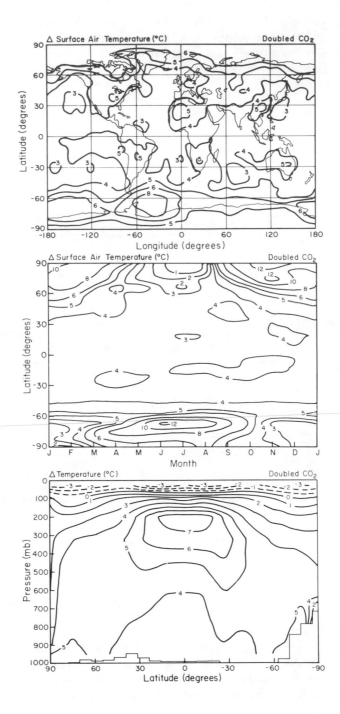

the plotting program (Hansen, personal communication). The middle frame shows the seasonal variation of the zonal average warming. Largest warming is seen in high latitudes in winter of both hemispheres, in qualitative agreement with the other models being discussed, and very little warming at the same latitudes in summer. For comparison with the upper two frames of Figure 5.4, Figures 5.5 and 5.6 show the geographical distribution of surface warming averaged over December–February and June–August for the Washington and Meehl and the Manabe and Stouffer models, respectively. Figure 5.5 shows very large temperature changes in the winter season in the North Atlantic and North Pacific and the Southern Ocean around Antarctica. By contrast, the Manabe and Stouffer model has a Northern Hemisphere maximum winter temperature increase occurring in the Arctic Ocean and high-latitude continental areas, especially Northern Siberia. Their Southern Ocean temperature increase is less than that over the Antarctic continent which, in turn, is less than the Washington and Meehl Southern Ocean warming. Elsewhere, Figure 5.5 shows the largest temperature increase of up to 8 ° over summertime Australia, whereas Figure 5.6 shows summertime maximum increases greater than 6 ° in Central Asia and Eastern North America.

One climate feature of high latitudes of considerable importance for determining model response is the extent of sea ice. Sea ice determines in large part the ice–albedo feedback in a GCM since sea ice persists into the spring and summer seasons of greatest solar irradiance. Figure 5.7 shows the sea ice obtained by the Manabe and Stouffer model for their control study in comparison with observed coverage. In the Northern Hemisphere, their modelled sea ice is close to that observed in summer and is in considerable excess of that observed in winter. By contrast, in the Southern Hemisphere their sea ice is less than half of that observed in winter and practically disappears in summer. Figure 5.8 shows the sea-ice extent obtained by Washington and Meehl for their control and doubled CO_2 calculation. Their control sea-ice coverage is higher than observed in both hemispheres and both seasons and appears to have less seasonal variation than that observed, as shown in Figure 5.7. In particular, the model Norwegian Sea and all the Arctic Ocean are completely filled with ice in summer, whereas in reality the summer sea ice recedes at least 15° poleward into the Barents Sea. It should be noted that this comparison is somewhat unfair since the model results show sea-

Figure 5.4 Air temperature change in °C for doubling of CO_2, according to Hansen *et al.* (1984). The upper graph shows the geographical distribution of annual mean surface air warming; the middle graph shows the seasonal variation of the surface air warming averaged over longitude, and the lower graph shows the altitude distribution of the temperature change in model layers averaged over season and longitude

(a)

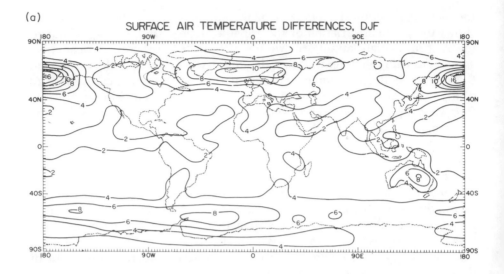

(b)

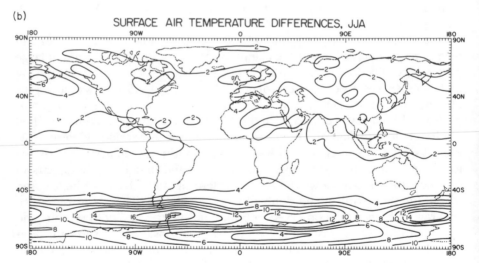

Figure 5.5 Same as Figure 5.4 (upper frame) for steady-state doubling of CO_2 (Washington and Meehl, 1984), except showing winter and summer of surface air temperature change in °C

ice extent averaged over the three summer months rather than for August when sea ice is near its minimum. Nevertheless, the model summer Arctic ice appears to be at least as extensive as the observed winter maximum extent. In the Southern Hemisphere, the model sea ice in both seasons extends to about 55° S, about 5° farther than observed in winter and 10° to 15° farther than observed in summer. Hansen *et al.* use two control simu-

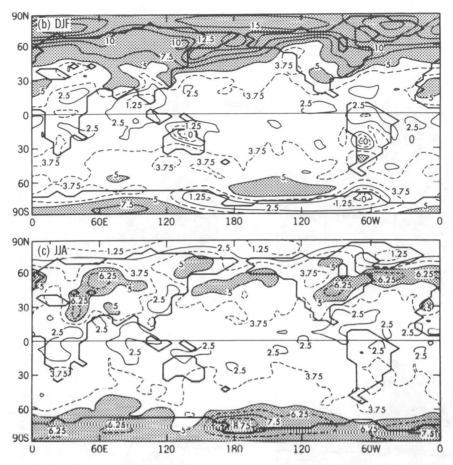

Figure 5.6 Same as Figure 5.5 but for quadrupling CO_2 (Manabe and Stouffer, 1980), showing a geographical distribution of surface air temperature change. Regions of warming greater than 5° are shaded

lations, whose annual mean sea ice is compared with observations in Figure 5.9. Their standard control was reported to have 15% less sea ice than seen in observations with an especially noticeable deficit at longitudes around 100° W and 50° E in the Southern Hemisphere. They produced an alternate control with a different assumption concerning sea-ice melting that gives 23% greater sea-ice cover than observed.

The role of sea ice in ice–albedo feedback for model CO_2 warming is indicated by Figures 5.10 and 5.11. Figure 5.10 shows the change in planetary albedo for the Washington and Meehl model. Extensive areas of albedo decrease of greater than 0.05 are seen in both hemispheres and for both seasons along the model sea-ice margins but nowhere else. During the summer

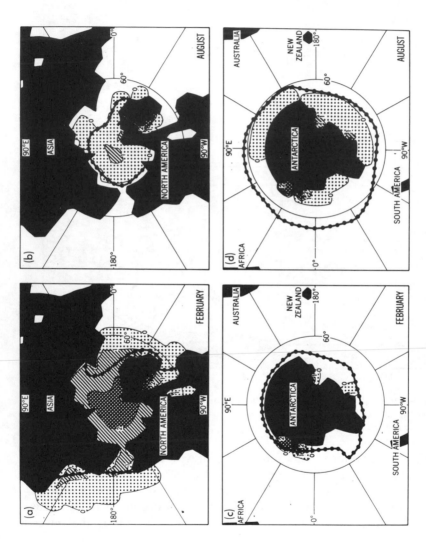

Figure 5.7 Distribution of sea-ice cover for February and August for the $1 \times CO_2$ control simulation from Manabe and Stouffer (1980) shown by dotted and hatched areas, with the observed distribution quoted by them indicated by the heavy dashed–dotted lines

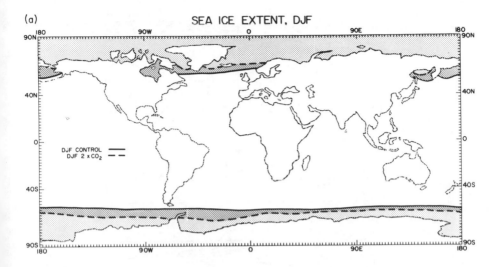

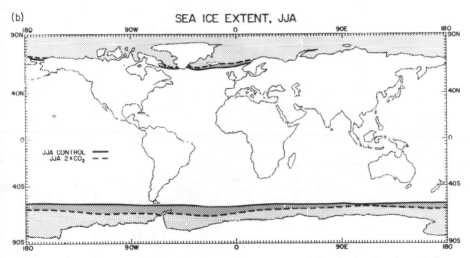

Figure 5.8 Distribution of sea-ice extent for $1 \times CO_2$ (solid lines) and $2 \times CO_2$ (dashed lines) for December–February (top frame) and June–August (bottom frame), according to Washington and Meehl (1984)

period, a belt of albedo change greater than 0.2 at the top of the atmosphere surrounds the Antarctic continent at a latitude of about $60°$ S, apparently equatorward of any real sea ice at that time. Figure 5.11 shows separately for continent and ocean the seasonal variation of increase in absorbed solar radiation for the Manabe and Stouffer model. Over the oceans, they find the most extensive large change in the Northern Hemisphere centred

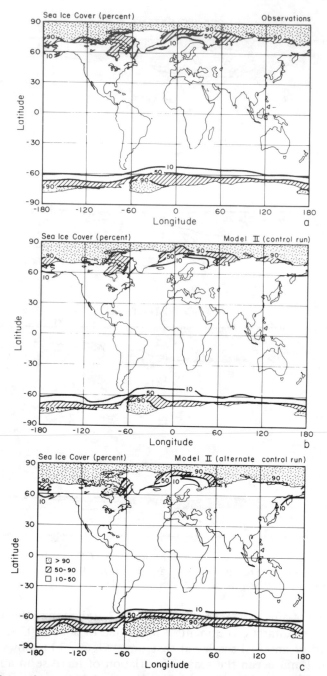

Figure 5.9 Annual mean sea-ice cover according to Hansen *et al.* (1984). Frame (a) is observations, (b) is the standard 1 × CO₂ control, and (c) is the alternate control

at 70° N. In the Southern Hemisphere, their largest change is at the edge of the Antarctic continent in the spring season. By summer, their change in absorbed solar radiation has dropped to small values consistent with the absence of control sea ice at that time.

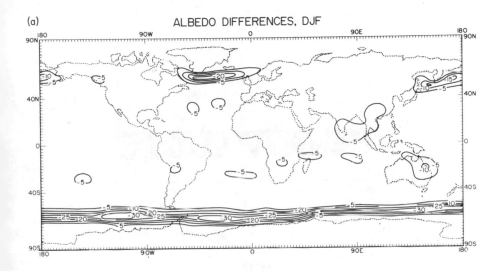

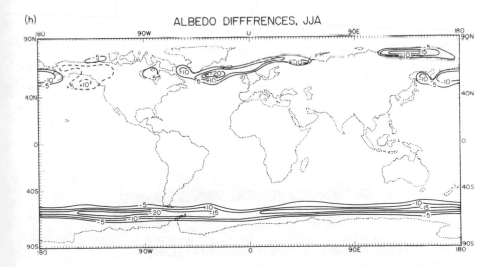

Figure 5.10 Top of the atmosphere albedo change in per cent for $2 \times CO_2 - 1 \times CO_2$, December–February and June–August, according to Washington and Meehl (1984)

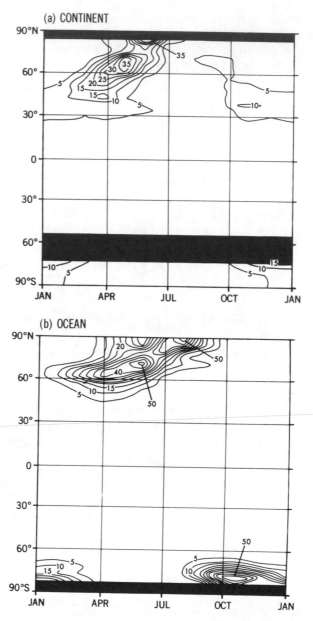

Figure 5.11 Latitude–season plots of change in absorbed solar radiation for $4 \times CO_2 - 1 \times CO_2$, according to Manabe and Stouffer (1980)

Planetary and surface albedo annual average change calculated by the Hansen *et al.* model is shown in Figure 5.12. Their surface albedo plot indi-

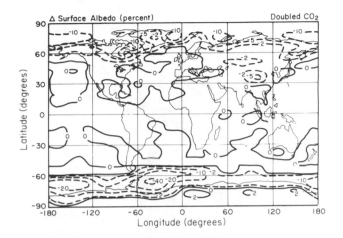

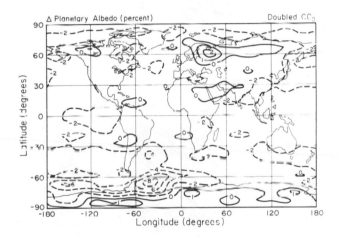

Figure 5.12 Annual average albedo changes in per cent for $2 \times CO_2 - 1 \times CO_2$, according to Hansen *et al.* (1984). Top frame is surface albedo; bottom frame is top of the atmosphere albedo

cates a large sea-ice-associated surface albedo change in both hemispheres. However, their planetary albedo plot shows hardly any sea-ice-associated concentration of albedo change in the Northern Hemisphere. Along Antarctica in the Western Hemisphere, there is a stretch of albedo decreases greater than 0.05 at the top of the atmosphere, a much smaller change than that seen in the Washington and Meehl plot, Figure 5.10.

Another climate feature of considerable importance for inferring changes in surface heating is the calculated change in cloudiness. The Manabe and

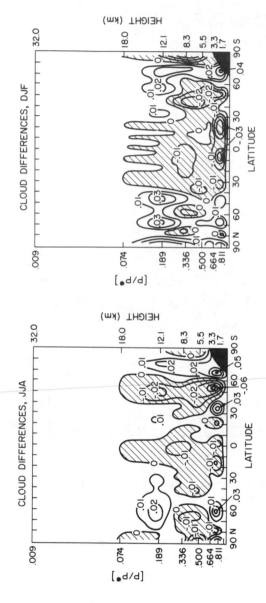

Figure 5.13 Same as Figure 5.2 for doubling CO_2 (Washington and Meehl, 1984).
except the plot shows fractional changes in cloud cover

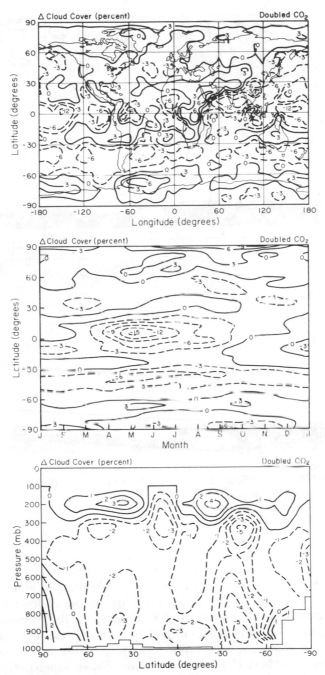

Figure 5.14 Same as Figure 5.4 for doubling CO_2 (Hansen *et al.*, 1984), except the plot shows per cent change of cloud cover

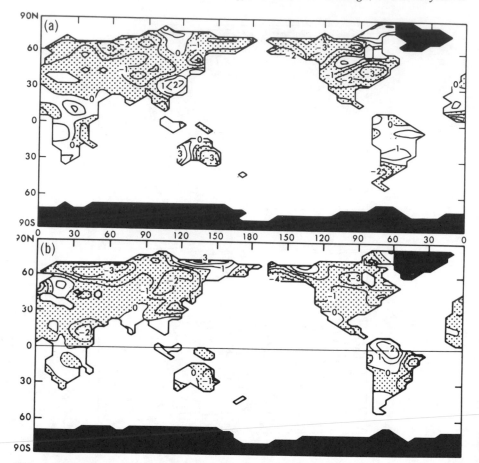

Figure 5.15 Geographical distribution of soil moisture change in cm during June–August in the Manabe and Stouffer model for $4 \times CO_2 - 1 \times CO_2$. The upper figure shows the result of a simulation with a comparatively low resolution (15 waves for both longitude and latitude) and the lower figure a simulation with a higher resolution (21 waves). (Manabe *et al.*, 1981)

Stouffer study avoided this issue by prescribing clouds to be independent of longitude with latitudinal variation based on observation. Figure 5.13 shows the change in cloudiness calculated by Washington and Meehl for the December-to-February and June-to-August periods. They find a slight decrease of cloudiness in low latitudes at all levels in the troposphere and a predominance of small increases in high latitudes. These high-latitude increases become relatively large around the tropopause at the top model level with clouds. These changes in cloudiness have some positive effect on the model global radiation balance and hence global surface warming, as discussed later. Figure 5.14 shows the cloudiness change found by Hansen *et*

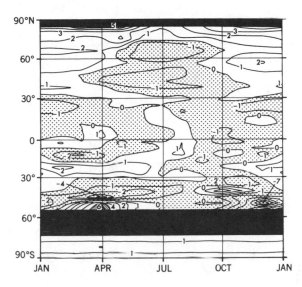

Figure 5.16 From Manabe and Stouffer (1980). Latitude–time distribution of zonal mean difference in soil moisture in cm over continents for the $4 \times CO_2 - 1 \times CO_2$ experiments. Note that the maximum moisture storage in this study is assumed to be 15 cm everywhere

al. The top frame shows the annual mean geographical distribution of total cloudiness change, the bottom frame the annual mean vertical distribution of cloudiness change, and the middle frame the seasonal variation of total cloud-cover change. They find decreases in cloudiness of several per cent or more at most levels, latitudes, and seasons. Increases are seen for low-level clouds in high latitudes and high clouds in middle latitudes. This latter change resembles somewhat the high-level cloud change found by Washington and Meehl. The cloud changes found by Hansen *et al.* are evidently large enough to be responsible for a major amplification of their surface warming as discussed later.

Another important climate feature calculated by the GCMs is change in soil moisture. Although obtained very crudely, it does represent the balances between model precipitation versus evaporation and runoff. Manabe *et al.* (1981) have analysed the soil moisture change of the Manabe and Stouffer study and that of a somewhat higher resolution version of the same model. Their results for the Manabe and Stouffer model are shown in Figure 5.15 for the Northern Hemisphere summer period. Although there is some variation between models and considerable sampling error, there is a predominance of soil moisture decrease in the Northern Hemisphere during the summer season and in middle and high latitudes. They find some high-latitude regions of increased soil moisture. This summertime drying of land

(a)
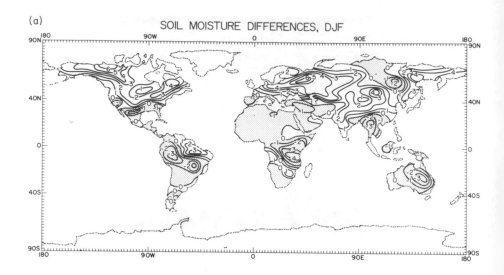

(b)
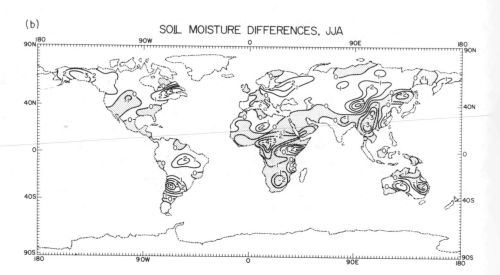

Figure 5.17 Geographical distribution of soil moisture differences in cm for December–February and June–August obtained by Washington and Meehl (1984) for $2 \times CO_2 - 1 \times CO_2$

north of $30°$ N is also seen in Figure 5.16 which shows their seasonal variation of longitudinally averaged soil moisture change. Qualitatively similar results were obtained using an idealized geography sector model discussed further below (Manabe *et al.*, 1981; Wetherald and Manabe, 1981).

The results of Washington and Meehl (Figure 5.17), on the other hand, show a predominance of increased soil moisture over most extratropical continental areas in the Northern Hemisphere for all seasons. They do find a small tongue of drying reaching from Southern California to the Central Great Plains of the United States and reduced soil moisture over Southeast Asia and much of Africa south of the Sahara. The seasonal variation of longitudinally averaged soil moisture according to Washington and Meehl is shown in Figure 5.18. This plot shows increases in continental soil moisture between 30° and 60° N with greatest increases in the winter and spring seasons. Decreased soil moisture is found for most latitudes and seasons in tropical latitudes and the Southern Hemisphere. The model results of Hansen *et al.* do not indicate a substantial drying in spring and summer in middle-to- upper midlatitudes (Rind and Lebedeff, 1984, p. 54). The calculation of Mitchell and Lupton (1984), shown in Figure 5.19, does give the same general soil moisture patterns as obtained by Manabe and Stouffer. Thus, there is some but not general model agreement as to summer drying in middle and high latitudes. Furthermore, all the models suggest significant high-latitude increases in precipitation, and hence runoff.

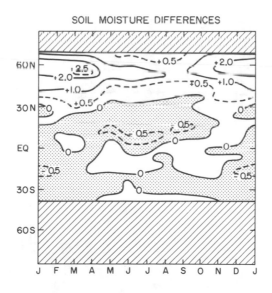

Figure 5.18 Same as Figure 5.17, except longitudinal average values of soil moisture in cm plotted versus month

CHANGE IN SOIL MOISTURE DUE TO QUADRUPLING CO_2
AND ENHANCING SSTs (mm)
JUNE   JULY   AUGUST

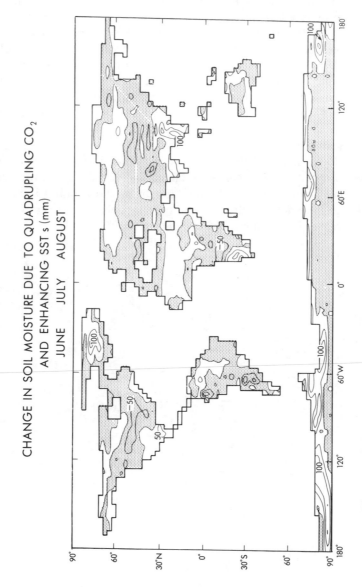

Figure 5.19 Changes in soil moisture content in mm for the June to August period
for $4 \times CO_2$ according to Mitchell and Lupton (1984)

5.7.2 Exploratory GCMs

It is difficult to judge the validity of the climatic change given by the more detailed GCMs unless it is possible to interpret their results in terms of well understood physical processes. Such interpretation is not readily developed using only the more realistic models both because of limitations in computer time and because of the inherent complexity of these models. Thus, it is necessary to study the climatic change question with a hierarchy of simpler models (Schneider and Dickinson, 1974) including simple GCMs.

Manabe and co-workers have developed a model consisting of 120° of longitude, half flat land and half ocean, and have used this sector model for a number of interesting studies; e.g., Wetherald and Manabe (1980) studied the role of interactive cloudiness in climatic change, and Manabe and Wetherald (1980) studied the differences between annual mean and seasonal cycle simulations. It is not possible to further summarize here the results from these or several other interesting studies from other groups.

Simulations involving coupling of a sector GCM to an ocean model have been discussed by Bryan *et al.* (1982) and Spelman and Manabe (1984). It is usually easier to analyse and interpret the results from a sector model than those from a global model because of the simplicity of the geography of the sector model. With an impulsive change in CO_2 global heating, Bryan *et al.* found (Figure 5.20) that after 25 years their model surface temperatures had nearly the same latitudinal distribution as a steady-state calculation and about 0.7 the amplitude of the steady-state response. Thompson and Schneider (1982) have argued that Bryan *et al.* could have come to a different conclusion if their high-latitude ocean had a larger heat capacity as appropriate to the Southern Hemisphere. Figure 5.21 shows the vertical distribution for the Bryan *et al.* model of the ratio of transient to steady-state temperature increase. Evidently, temperature increase in the ocean is largely confined to the upper 300 metres. The increased stable stratification of the upper layers of the ocean and thermal insulation by sea ice help to reduce the large effective thermal inertia of high latitudes.

Spelman and Manabe (1984) found a much smaller temperature increase for quadrupling of CO_2 using their ocean GCM which simulates poleward transport of heat by ocean currents than they found with an otherwise similar mixed-layer model. Through further integrations, they discovered that this difference could be associated with differences in global average control temperature as shown in Figure 5.22. This dependence on global mean temperature is presumably a result of differences in ice–albedo feedbacks, e.g., as discussed in Schneider and Dickinson (1974).

Washington and Meehl (1983) studied the climatic change from CO_2 doubling using an annual mean swamp model with realistic geography. Their calculation had essentially no ice–albedo feedback and no indication of high-

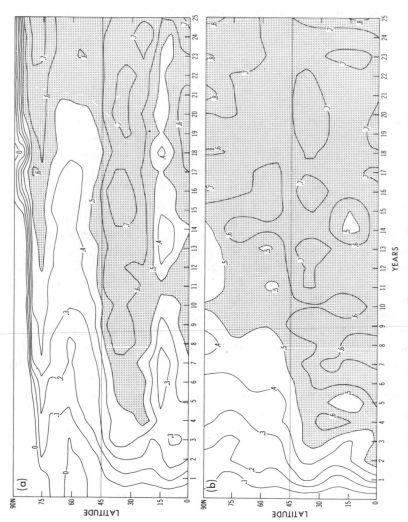

Figure 5.20 Time–latitude variation of the zonally averaged normalized response in °C of (a) sea surface temperature and (b) surface air temperature averaged over continent and ocean for an instantaneous 4 × increase of CO_2 (Bryan et al., 1982)

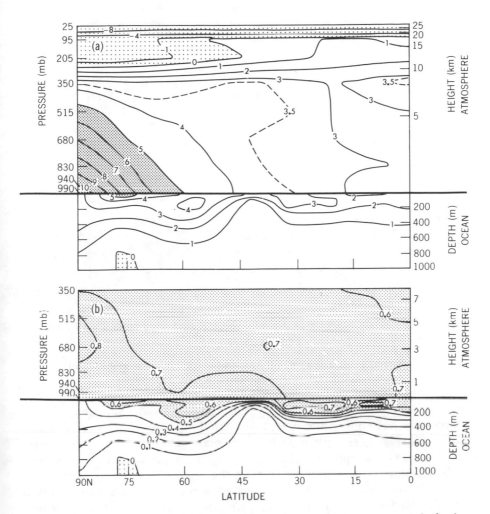

Figure 5.21 Latitude–height distributions from the transient response study for instantaneous $4 \times CO_2$ showing (a) the zonally averaged temperature at 25 years minus intitial temperature (°C) and (b) the fractional response of zonally averaged temperature relative to steady state (from Spelman and Manabe, 1984)

altitude intensification of the temperature increase. Their global average warming of 1.3° is the smallest warming obtained by any GCM for doubling CO_2 simulations and only one-third the global warming obtained using a seasonal version of essentially the same model, i.e., Washington and Meehl (1984). They suggest that this small response is due to weak ice–albedo and water vapour feedback. Their model indeed appears to have essentially no albedo feedback, which they interpret to result in part from their control simulation having a mean surface air temperature 1.7 °C warmer than that

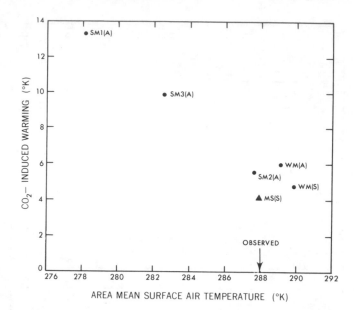

Figure 5.22 The change of global mean surface air temperature (°C) in various GFDL models for $4 \times CO_2 - 1 \times CO_2$ versus global mean control temperatures. Solid dots are plotted for the models having limited computational domain with idealized geography; the triangle indicates the model that has global domain with realistic geography. From Spelman and Manabe (1984)

of their seasonal cycle model and in part from weak water vapour feedback. There does not appear, however, to be anything anomalous in their water vapour feedback, as both models have 'maximum increases of moisture' in the lower troposphere of 0.57 g kg^{-1} per °C of warming. Thus, why their warming was significantly less than the usual 2 °C warming of radiative convective models is not known. Their weak warming could have involved negative feedbacks from lapse rate, water vapour profile or cloud changes.

5.8 RELIABILITY OF MODEL RESULTS

5.8.1 Global Mean

The scenarios presented in Chapter 3 suggest that it is unlikely that the atmospheric CO_2 concentration will double before 2050, while a modest increase of CO_2 emissions would lead to about a doubling of the concentration towards the end of the next century. Since, furthermore, increased atmospheric heating grows logarithmically with CO_2 concentration, the climate around the year 2100 could be near-equilibrium with the CO_2 concentration present at that time. I shall now argue that the global average equilibrium warming for doubled CO_2 would lie between 1.5 and 5.5 °C and probably between 2.5 and 4.5 °C. The largest known sources of uncertainty from the modelling viewpoint are the feedbacks due to ice albedo and cloud radiative properties. Recent studies, including those reviewed here, have suggested that these terms may either provide large positive feedback to forced climatic change or alternatively give very little feedback. A significant negative feedback for clouds has also not been excluded. In the present assessment of uncertainty, the confidence limits have been broadened from those previously arrived at by the Climate Research Board (1979) of 3 ± 1.5 °C. Two GCM studies have been recently published with global temperature increase in the upper range of the Climate Research Board's stated confidence limits (i.e., Washington and Meehl, 1984; Hansen *et al.*, 1984), but with different large positive feedbacks primarily responsible for their large estimates. Evidently, we cannot exclude the possibility of both large ice–albedo and large cloud feedbacks occurring together. Furthermore, Washington and Meehl (1983) found a temperature increase for doubled CO_2 of 1.3 °C which is below the Climate Research Board's lower limit of 1.5 °C.

One formal procedure to evaluate the uncertainty in global warming is to estimate the uncertainty in the feedback factor λ in Eq. (5.1). The different contributions to λ are assumed to be independent stochastic variables with normal distribution about some expected value. We could attempt to approach the uncertainty analysis mechanically by viewing different GCM simulations as providing samples from the ensemble of possible climates. However, I believe that in this situation a more satisfactory approach is a Bayesian one, that is, using all the available information, including also judgmental estimation to obtain expected values and standard deviations.

A standard value for λ from one-dimensional radiative convective models, in the absence of ice and cloud feedback with fixed cloud top altitude and fixed relative humidity, is 2.0 W m^{-2} °C^{-1} (Ramanathan and Coakley, 1978). To this must be added an ice–albedo feedback term which is estimated to be -0.4 ± 0.3 W m^{-2} °C^{-1} based on previous GCM model simulations, empirical estimates, and the current status of sea-ice modelling capabilities. The confidence limits are intended to refer to two standard deviations. The following information has been considered in arriving at this judgmental estimate. Estimates of this term inferred from different studies are Washington and Meehl (1983), -0.1 W m^{-2} °C^{-1}; Lian and Cess (1977) using a seasonal energy balance model with a snow-ice parameterization based on seasonal variation of satellite-measured albedo, -0.3 W m^{-2} °C^{-1}; Hansen *et al.* (1984), -0.3 and -0.4 W m^{-2} °C^{-1}; Robock (1983a) from measured seasonal variation of snow and ice cover, -0.6 W m^{-2} °C^{-1}; Manabe and Stouffer (1980), -0.3 W m^{-2} °C^{-1}; Wetherald and Manabe (1981), -0.4 to -0.5 W m^{-2} °C^{-1}; Spelman and Manabe (1984), -0.45 W m^{-2} °C^{-1}; Washington and Meehl (1984), -0.6 W m^{-2} °C^{-1}. The ice–albedo feedback in the Manabe and Stouffer model is probably too low because of too little sea ice in the Antarctic and model snow–ice albedos that may be too low, whereas the ice–albedo feedback of Washington and Meehl (1984) is probably too high because of excessive sea ice and too high albedo for summer pack ice in their control in both hemispheres. However, it might also be argued that the latter study could conceivably have too weak an ice–albedo feedback because of the weakness of its seasonal cycle response to annual forcing. Without further analysis, neither of these two studies can be regarded as much beyond a standard deviation of a best value. Since about 80% of the Washington–Meehl ice–albedo feedback is in the Southern Hemisphere and about 80% of the Manabe Stouffer ice–albedo feedback is in the Northern Hemisphere, an ice–albedo term between -0.2 and -0.7 can be obtained by using one hemisphere from one model and another hemisphere from the other model. The Washington and Meehl (1983) result is taken to be two standard deviations on the low side because of lack of any plausible physical mechanism to provide so little ice–albedo feedback. One large source of uncertainty in modelling the sea-ice–albedo feedback is the amount and thickness of clouds covering the sea ice and their possible change with changed sea-ice cover. The ice–albedo feedback of Hansen *et al.* (1984) appears to have been strongly masked by high-latitude cloudiness which increased with CO_2 warming.

The second serious source of uncertainty in the question of global climatic change is that of cloud feedback. The Hansen *et al.* model estimates this parameter to be -0.8 W m^{-2} °C^{-1} (i.e., positive feedback), whereas Manabe and Wetherald (1980) find it for annual mean model to be only -0.05 W m^{-2} °C^{-1} as a result of cancellation of positive feedback in low latitudes and negative feedback from stratus clouds in high latitudes. The study by Washington and Meehl (1984) shows cloud changes qualitatively similar to those of Hansen *et al.* but weaker. Their cloud feedback is estimated to be about -0.3 W m^{-2} °C^{-1}. Cess *et al.* (1982) provide another estimate of 0.0. Formally combining these four estimates gives a mean of -0.3 and a standard deviation of 0.35, which doubled gives 0.7 as an 'error bar'. None of the above three-dimensional studies has considered the probable negative feedback from increasing cloud liquid water with a warmer climate. Hansen *et al.* examined the effect of cloud optical thickness change with a one-dimensional radiative convective model and found it could cancel all the positive feedback found in their three-dimensional model. Other one-dimensional studies have also demonstrated the possibility of large negative feedbacks from cloud optical thickness change (e.g., Somerville and Rener, 1984, and references therein). We take as a judgmental estimate -0.3 ± 0.7 W m^{-2} °C^{-1} for the cloud feedback contribution to (the same as obtained by formally combining model results above).

We estimate the contribution of all other feedbacks, e.g., from changes in lapse rate, vertical distribution of humidity, horizontal transport by atmosphere and oceans to be 0.0 ± 0.4 W m^{-2} °C^{-1}. The analysis of Hansen *et al.* found lapse rate and humidity vertical distribution feedbacks to be individually large but mostly cancelling.

The estimated means for various feedbacks as discussed above are linearly averaged and the error bars are root-mean-square-averaged to give $\lambda = 1.3 \pm 0.86$ W m^{-2} °C^{-1}. The confidence limits of 0.45 to 2.16 W m^{-2} °C^{-1} are divided into 4.2 W m^{-2} for CO_2 doubling to give a range of temperature increase of from 2.0 to 9 °C. Such a large change as the upper limit just obtained cannot be excluded, but I believe it to be extremely unlikely. Given the range of past climate variations, the present weak evidence for global warming (as discussed in Chapter 6), and the questionableness of the present analysis for such large changes, a more reasonable upper limit is believed to be 5.5 °C. The largest GCM warming yet reported for doubling CO_2 is 4.8 °C obtained by Hansen *et al.*, for their alternative control with more sea ice. I have also judgmentally dropped the lower limit to 1.5 °C in recognition of the 1.3 °C temperature increase obtained by Washington and Meehl (1983). A best estimate of temperature response is proportional to the expected value of $1/\lambda$ which cannot be obtained directly from the expected value of λ. Assuming that λ has a Gaussian distribution with mean of 1.3 and standard deviation of 0.43 but truncated to exclude values less than 0.3, I find that $\overline{1/\lambda} = 1.14/\overline{\lambda}$ (where the bars refer to expected values). Roughly speaking the possibility of small values of λ becomes weighted more heavily when we invert λ. Thus, the expected temperature change is $4.2 \times 1.14/1.3 = 3.7$ °C. In round numbers, the estimated ΔT for steady-state CO_2 doubling is 3.5 ± 2 °C.

It is evident that the above discussion, although couched in terms of a zero-dimensional climate model, obtains most of its quantitative information from GCM studies. Only by narrowing the uncertainties and errors in the GCM simulations can we hope to improve significantly our estimates of future climatic change. Values of warming for equilibrium CO_2 doubling greater than 4 °C are suggested by two recently published studies and by some GCM studies yet unpublished (and so not reviewed here) which were not available to previous assessments. These studies indicate that either large positive cloud feedbacks or, alternatively, large ice–albedo feedbacks and modest positive cloud feedbacks can give warmings of at least 4 °C. The possibility cannot be excluded that both large positive cloud feedback and large ice–albedo feedback might be realized. For example, the Hansen *et al.* (1984) simulations but with an ice–albedo feedback as strong as that of Washington and Meehl (1984) would give a 6 °C warming. Ice–albedo feedback necessarily weakens with a warmer climate and would become much weaker with the disappearance of polar sea ice during the summer. Likewise, cloud feedbacks would greatly weaken with the near disappearance of cloud cover. However, disappearance of summer sea ice and large reductions in cloudiness would increase absorbed solar radiation by at least several tens of W m^{-2}.

Weak cloud feedbacks and relatively small ice–albedo feedbacks are also not unlikely and together would imply CO_2 doubling temperature changes in the range 2 to 3 °C. While the assumption of a statistical model for the distribution of feedback factors is not justified by any real information we have for the feedback terms, it provides guidance for the combination of uncertainties and suggests that the possibility of a relatively small total feedback should be given somewhat more weight than the possibility of a relatively large total feedback. In summary, the CO_2 doubling equilibrium warming is likely (estimated probability of 68%) to be in the range 2.5 to 4.5 °C with only a small possibility, judged to have a 5% probability, for values to lie outside this range by more than 1 °C. A warming of as little as 1 °C or as large as 7 °C cannot be entirely excluded from the modelling evidence for CO_2 doubling but would seem extremely unlikely.

To estimate the temperature increase at a given date requires scaling the above estimates to the expected CO_2 concentrations at that date, adding an estimate of the net warming from increases of other anthropogenic gases, and combining the above uncertainties with the uncertainties in the concentrations of CO_2 and other trace gas increases expected by that date. In addition, for rapid growth of heating as at present, a further correction is needed for the effects of ocean heat storage.

The increase of global rainfall is likely to scale like the increase of saturation vapour pressure and thus follow global temperature change but weighted toward the temperature change in moist tropical regions where evapotranspiration is normally largest.

5.8.2 Consideration of Transient Lag Due to Ocean Heat Uptake

Most GCM studies have concentrated on the equilibrium response to doubling or quadrupling CO_2. The heating by other trace gases can be expressed in terms of equivalent concentrations of CO_2. The present-day equilibrium response is about one-third of that which would correspond to a doubling of CO_2. As evident from Eq. (5.1), we must in general consider also the effect of the time-dependent ocean heat storage, that is, the rate of heat taken up by the oceans. This term might become sufficiently small in a century or two to be negligible. However, as a result of the rapid growth in fossil fuel emissions over the recent past, the ocean heat uptake term is of comparable magnitude to the net feedback term that has been extensively discussed in this chapter. Under these conditions, it is evident from Eq. (5.1) that the warming now resulting from the past increases of CO_2 should vary more weakly with λ than as $1/\lambda$, as pointed out by Hansen *et al.* (1984). Indeed, in the limit of very rapid growth of the heat input, almost all the heating would go into the ocean and the temperature change would be nearly independent of λ as seen analytically, e.g., if an exponential growth of ΔQ is assumed with an exponent large compared to λ/C.

A family of more realistic zero-dimensional transient models with a surface mixed layer coupled to an underlying diffusive ocean has been analysed by Wigley and Schlesinger (1985). For a plausible range of parameters, they show that an increase of CO_2 from 270 ppmv to current values of 340 ppmv implies the present warming to lie between about 0.3 and at least 0.8 °C, assuming that the steady-state doubling response lies between 1.5 and 4.5 °C. The warming from other trace gases was estimated to be an additional 0.2 °C, but this would be offset if the initial CO_2 level were raised to 290 ppmv. Hansen *et al.* (1984) earlier found similar results. Wigley *et al.* (Chapter 6) suggest that the transient warming to date due to CO_2 and other trace gas changes should be in the range 0.3–1.1 °C, given all model uncertainties. Harvey (1985) has discussed parameterization changes that can modify the time scale of the calculated transient ocean temperature response to CO_2 warming. Particularly important seems to be whether energy imbalances over land are compensated locally or are coupled to ocean temperature changes. He also emphasizes the possible effect of including vertical advection which, for fixed eddy diffusion, shortens the response time. However, with inclusion of vertical advection, it is presumably necessary to increase eddy diffusion coefficients to maintain agreement with tracer data and such an adjustment would lessen the response time change.

The detailed mechanisms of ocean heat uptake are poorly known, especially in regions of deep and bottom water formation so that the ratio of actual to steady-state warming could be more uncertain than indicated by the analyses of Hansen *et al.*, Wigley and Schlesinger, and Harvey.

5.8.3 Regional Patterns

Only the global average temperature change can be constrained by physical arguments such as given above. Thus, any changes in regional patterns of climate are much more speculative. Several possible features of regional climatic change have been suggested. First, there is the indication that warming will be enhanced in high latitudes. All the models discussed here appear to agree qualitatively with the hypothesis of Manabe and Stouffer (1980) that large wintertime warming would be expected over polar sea ice. However, the extent of sea-ice recession is quite uncertain for the reasons discussed earlier. Another plausible result, reported in models of Manabe and co-workers and Mitchell and co-workers but not in the other models discussed, is a preponderance of summertime soil moisture decrease in middle and high latitudes of the Northern Hemisphere. High latitude precipitation increases in all the models.

5.8.4 Model Results as a Guide to Detecting CO_2 Climatic Change Over the Next Few Decades

As discussed above and in Chapter 6, allowance for ocean heat uptake indicates that the present CO_2 warming signal should lie approximately between 0.3 and 1.1 °C. None of the GCM model results for regional pattern is yet sufficiently certain as to suggest useful regional 'fingerprints' for the warming signal. Thus, as discussed further in the next chapter, global average temperature is the most useful quantity to monitor. This term should show an increase with time following, with some time lag, the logarithm of the atmospheric CO_2 concentration. Because of the considerable uncertainty in the expected CO_2 signal, a temperature increase of about 1 °C or larger maintained over at least a decade may be required to provide convincing evidence of the long-awaited CO_2 temperature increase. If means were found not only to identify but also to monitor some of the causes of natural variability such as volcanic aerosol, the task of obtaining evidence of the CO_2 signal would be made easier.

5.9 CONCLUSIONS

- Results are now available from several independent GCM modelling groups which all show a warming of at least several degrees for the climate response to doubling the atmospheric concentrations of CO_2.
- The global equilibrium temperature change expected from increases of CO_2 and other trace gases equivalent to a doubling of CO_2 is expected to be in the range 1.5 to 5.5°.
- The largest sources of uncertainty in modelling global average temperature change appear to be the level of feedback from clouds, ice–albedo, and possibly lapse-rate and water-vapour changes.
- Storage of heat in the oceans reduces the warming presently expected for an equilibrium response to carbon dioxide heating (and that of other trace gases) and may significantly modify the geographical distribution of climatic change.
- The global warming in 1985 suggested by models to result from previous increases of CO_2 and other trace gases is in the range of 0.3 to 1.1 °C.
- Continental-scale climatic change, especially that involving conditions required for agriculture, is extremely important but not yet modelled with any degree of confidence. There is qualitative agreement among all current three-dimensional models that largest equilibrium temperature increases would occur in regions of high-latitude winter. There is some but not compelling evidence for mid-latitude mid-continent summer drying. High-latitude precipitation increases in winter are likely.
- Recommendations:

(a) The modelling of cloud radiative interactions and ice–snow albedo in GCMs as used to model future climatic change needs to be significantly improved using better physical parameterizations and based on current and future data sets.

(b) Model simulations of climatic change from increases of CO_2 are most readily interpreted in terms of global feedback factors. Future model studies should attempt to explicitly obtain the feedback effects of cloud property changes, ice–albedo feedback, and lapse-rate feedbacks.

(c) More generally, modellers should attempt to present the most important model results in as uniform a graphical and numerical format as possible.

ACKNOWLEDGEMENTS

Part of Section 5.1 and most of Section 5.2 were originally drafted by Tom Wigley. Helpful suggestions to improve the manuscript were given on the initial draft by James Hansen, Brian Hanson, Gerald Meehl, Alan Robock, and Warren Washington. Further extensive suggestions were given by the International Meteorological Institute review committee: Georgi Golitsyn, Eero Holopainen, Syukuro Manabe, John Mitchell, and David Parker. On a later draft valuable comments were provided by Stephen Schneider.

The author also wishes to acknowledge comments made by T. Asai, A. Berger, W. Böhme, H. W. Ellsaesser, H. L. Ferguson. R. M. Gifford, J. Goudriaan, H. Landsberg, J. A. Laurmann, G. McBean, F. Mesinger, A. S. Monin, O. Preining, R. W. Stewart, J. Tukey, J. Woods, and Du-Zheng Ye.

5.10 REFERENCES

Barron, E. J., and Washington, W. M. (1984) The role of geographic variables in explaining paleoclimates: Results from Cretaceous climate model sensitivity studies, *J. Geophys. Res.*, **9**, 1267–1279.

Barry, R. G., Henderson-Sellers, A., and Shine, K. P. (1984) Climate sensitivity and the marginal cryosphere, in Hansen, J. E., and Takahashi, T. (eds) *Climate Processes and Climate Sensitivity*, Maurice Ewing Series 5, American Geophysical Union, Washington, D.C., 221–237.

Berger, A. (1979) Insolation signature of quaternary climate changes, *Il Nuovo Cimento*, **2C(1)**, 63–87.

Berner, R. A., Lasaga, A. C., and Garrels, R. M. (1983) The carbonate–silicate geochemical cycle and its effect on atmospheric carbon dioxide over the last 100 million years, *Am. J. Sci.*, **283**, 641–683.

Borisenkov, Ye. P., Tsvetkov, A. V., and Agaponov, S. V. (1983) On some characteristics of insolation changes in the past and the future, *Climatic Change*, **5**, 237–244.

Bowman, K. P. (1982) Sensitivity of an annual mean diffusive energy balance model with an ice sheet, *J. Geophys. Res.*, **87**, 9667–9674.

Bretherton, F. P. (1982) Ocean climate modeling. *Progress in Oceanography*, **11**, 93–129.

Broecker, W. S., Takahashi, T., Simpson, H. S., and Peng, T.-H. (1979) Fate of fossil fuel carbon dioxide and the global carbon budget. *Science*, **206**, 409–422.

Broecker, W. S., and Takahashi, T. (1984) Is there a tie between atmospheric CO_2 content and ocean circulation?, in Hansen, J. E., and Takahashi T. (eds) *Climate Processes and Climate Sensitivity*, Maurice Ewing Series 5, American Geophysical Union, Washington, D.C., 314–326.

Bryan K., Komro, F. G., Manabe, S., and Spelman, M. J. (1982) Transient climate response to increasing atmospheric carbon dioxide, *Science*, **215**, 56–58.

Bryson, R. A., and Goodman, B. M. (1980) Volcanic activity and climatic change, *Science*, **207**, 1041–1044.

Budyko, M. I. (1969) The effect of solar radiation variations on the climate of the earth, *Tellus*, **21**, 611–619.

Carbon Dioxide Assessment Committee (1983) *Changing Climate*, Washington, D.C., National Academy Press, 490 pages.

Cess, R. D. (1976) Climatic change: An appraisal of atmospheric feedback mechanisms employing zonal climatology, *J. Atmos. Sci.*, **33**, 1831–1843.

Cess, R. D. (1983) Arctic aerosols: Model estimates of interactive influences upon the surface-atmosphere clear-sky radiation budget, *Atmos. Environ.*, **17**, 2555–2564.

Cess, R. D., and Goldenberg, S. D. (1981) The effect of ocean heat capacity upon global warming due to increasing atmospheric carbon dioxide, *J. Geophys. Res.*, **86**, 498–502.

Cess, R. D., Briegleb, B. P., and Lian, M. S. (1982) Low-latitude cloudiness and climate feedback: Comparative estimates from satellite data, *J. Atmos. Sci.*, **39**, 53–59.

Chervin, R. M. (1980) On the simulation of climate and climate change with general circulation models, *J. Atmos. Sci.*, **37**, 1903–1913.

Chervin, R. M. (1981) On the comparison of observed and GCM simulated climate ensembles, *J. Atmos. Sci.*, **38**, 885–901.

Chervin, R. M., and Schneider, S. H. (1976) On determining the statistical significance of climate experiments with general circulation models, *J. Atmos. Sci.*, **33**, 405–412.

Chou, M.-D., Peng, L., and Arking, A. (1982) Climate studies with a multi-layer energy balance model, Part II: The role of feedback mechanisms in the CO_2 problem, *J. Atmos. Sci.*, **39**, 2657–2666.

Climate Research Board (1979) *Carbon Dioxide and Climate: Scientific Assessment*, Washington, D.C., National Academy of Sciences, 22 pages.

CO_2/Climate Review Board (1982) *Carbon Dioxide and Climate: A Second Assessment*, Washington, D.C., National Academy of Sciences, 72 pages.

DeBruin, H. A. R. (1983) A model for the Priestley–Taylor parameter α, *J. Clim. Appl. Meteorol.*, **22**, 572–578.

Dickinson, R. E. (1981a) Effects of tropical deforestation on climate, in *Studies in Third World Societies*, Publ. No. 14, Department of Anthropology, College of William and Mary, 515 pages.

Dickinson, R. E. (1981b) Convergence rate and stability of ocean–atmosphere coupling schemes with a zero-dimensional climate model, *J. Atmos. Sci.*, **38**, 2112–2120.

Dickinson, R. E. (1982) Modeling climate changes due to carbon dioxide increases, in Clark, W. C. (ed.) *Carbon Dioxide Review 1982*, New York, Oxford University Press, 101–133.

Dickinson, R. E. (1984) Modeling evapotranspiration for three-dimensional global climate models, in Hansen, J. E., and Takahashi, T. (eds) *Climate Processes and Climate Sensitivity*, Maurice Ewing Series 5, American Geophysical Union, Washington, D.C., 58–72.

Eddy, J. A., Gilliland, R. L., and Hoyt, D. V. (1982) Changes in the solar constant and climatic effects, *Nature*, **300**, 689–693.

Fels, S. B., Mahlman, J. D., Schwarzkopf, M. D., and Sinclair, R. W. (1980) Stratospheric sensitivity to perturbations in ozone and carbon dioxide: Radiative and dynamical response, *J. Atmos. Sci.*, **37**, 2265–2297.

Foukal, P. V. (1980) Solar luminosity variation on directly observable time scales: observational evidence and basic mechanisms, in *Sun and Climate*, CNES/CNRS/DGRST Conference Proceedings, 275–284.

Gilchrist, A. (1983) Increased carbon dioxide concentrations and climate: The equilibrium response, in Bach, W. *et al.* (eds) *Carbon Dioxide: Current Views and Developments in Energy/Climate Research*, D. Reidel.

Gilliland, R. L. (1980) Solar luminosity variations, *Nature*, **286**, 838.

Gilliland, R. L. (1981) Solar radius variations over the past 265 years, *Astrophys. J.*, **248**, 1144–1155.

Gilliland, R. L. (1982) Modeling solar variability, *Astrophys. J.*, **253**, 399–405.

Golitsyn, G. S., and Mokhov, I. I. (1978) Sensitivity estimates and the role of clouds in simple models of climate. *Izvestiya, Atmospheric and Oceanic Physics*, **14**, 569–576.

Hammer, C. U. (1977) Past volcanism revealed by Greenland Ice Sheet impurities, *Nature*, **270**, 482–486.

Hammer, C. U., Clausen, H. B., and Dansgaard, W, (1980) Greenland icesheet evidence of post-glacial volcanism and its climatic impact, *Nature*, **288**, 230–235.

Hansen, J. E., Wang, W. C., and Lacis, A. A. (1978) Mount Agung eruption provides test of a global climate perturbation, *Science*, **199**, 1065–1068.

Hansen, J., Russell, G., Rind, D., Stone, P., Lacis, A., Lebedeff, S., Ruedy, R., and Travis, I. (1983) Efficient three-dimensional global models for climate studies: models I and II, *Mon. Weather Rev.*, **111**, 609–662.

Hansen, J., Lacis, A., Rind, D., Russell, G., Stone, P., Fung, I., Ruedy, R., and Lerner, J. (1984) Climate sensitivity: Analysis of feedback mechanisms, in Hansen, J. E., and Takahashi, T. (eds) *Climate Processes and Climate Sensitivity*, Maurice Ewing Series 5, American Geophysical Union, Washington, D.C., 368 pages.

Hare, K. F. (1983) Climate and Desertification: A Revised Analysis, *World Climate Programme*, 44, 149 pages.

Hartmann, D. L. (1984) On the role of global-scale waves in ice–albedo and vegetation-albedo feedback, in Hansen, J. E., and Takahashi, T. (eds) *Climate Processes and Climate Sensitivity*, Maurice Ewing Series 5, American Geophysical Union, Washington, D.C., 18–28.

Hartmann, D. L., and Short, D. A. (1979) On the role of zonal asymmetries in climate change, *J. Atmos. Sci.*, **36**, 519–528.

Harvey, L. D. D. (1985) Effect of ocean mixing on the transient climate response to a CO_2 increase: Analysis of recent model results, *J. Geophys.*, **90**, (in press).

Hasselmann, K. (1979) On the signal-to-noise problem in atmospheric response studies, in *Meteorology over Tropical Oceans*, Royal Meteorological Society, 251–259.

Hayashi, Y. (1982) Confidence intervals of a climatic signal, *J. Atmos. Sci.*, **39**, 1895–1905.

Held, I. M. (1982) Climate models and the astronomical theory of the ice ages, *Icarus*, **50**, 449–461.

Held, I. M., and Suarez, M. J. (1974) Simple albedo feedback models of the icecaps, *Tellus*, **26**, 613–629.

Henderson-Sellers, A., and Gornitz, V. (1984) Possible climatic impacts of land cover transformations, with particular emphasis on tropical deforestation, *Climatic Change*, **6**, 231–256.

Hibler, W. D., III (1984) The role of sea ice dynamics in modeling CO_2 increases, in Hansen, J.E., and Takahashi, T. (eds) *Climate Processes and Climate Sensitivity*, Maurice Ewing Series 5, American Geophysical Union, Washington, D.C., 238–253.

Hibler, W. D., and Bryan, K. (1984) Ocean circulation: Its effects on seasonal sea-ice simulations, *Science*, **224**, 489–491.

Hirschboeck, K. K. (1980) A new worldwide chronology of volcanic eruptions, Palaeogeography, Palaeoclimatology, *Palaeoecology*, **29**, 223–241.

Hoffert, M. I., Flannery, B. P., Callegari, A. J., Hsieh, C. T., and Wiscombe, W. (1983) Evaporation-limited tropical temperatures as a constraint on climate sensitivity, *J. Atmos. Sci.*, **40**, 1659–1668.

Houghton, J. T. (ed.) (1984) *The Global Climate*, Cambridge, Cambridge University Press, 233 pages.

Jäger, J. (1983) *Climate and Energy Systems. A Review of their Interactions*, Wiley, 231 pages.

Katz, R. W. (1982) Statistical evaluation of climate experiments with general circulation models: A parametric time series modeling approach, *J. Atmos. Sci.*, **39**, 1446–1455.

Katz, R. W. (1983) Statistical procedures for making inferences about precipitation changes simulated by an atmospheric general circulation model, *J. Atmos. Sci.*, **40**, 2193–2201.

Kelly, P. M., and Sear, C. B. (1984) The climatic impact of explosive volcanic eruptions, *Nature*, **311**, 740–743.

Kiehl, J. T., and Ramanathan, V. (1983) CO_2 radiative parameterization used in climate models: Comparison with narrow band models and with laboratory data, *J. Geophys. Res.*, **88**, 5191–5202.

Kutzbach, J. E., and Otto-Bliesner, B. L. (1982) The sensitivity of the African–Asian monsoonal climate to orbital parameter changes for 9000 years B.P. in a low-resolution general circulation model, *J. Atmos. Sci.*, **39**, 1177–1188.

Labitzke, K., Naujokat, B., and McCormack, M. P. (1983) Temperature effects on the stratosphere of the April 4 1982 eruption of El Chichon, Mexico, *Geophys. Res. Lett.*, **10**, 24–27.

Lamb, H. H. (1970) Volcanic dust in the atmosphere; with a chronology and assessment of its meteorological significance, *Phil. Trans. R. Soc. (London)* Series A, 266, 425–533 (See also the updates given by Lamb in *Clim. Monit.*, **6**, 57–67, 1977 and *Clim. Monit.*, **12**, 76–90, 1983.).

Leith, C. E. (1973) The standard error of time-average estimates of climatic means, *J. Appl. Meteor.*, **12**, 1066–1069.

Lian, M. S., and Cess, R. D. (1977) Energy balance climate models: A reappraisal of ice-albedo feedback, *J. Atmos. Sci.*, **34**, 1058–1062.

Lindzen, R. S., Hou, A. Y., and Farrell, B. F. (1982) The role of convective model choice in calculating the climate impact of doubling CO_2, *J. Atmos. Sci.*, **39**, 1189–1205.

Livezey, R. E., and Chen, W. Y. (1983) Statistical field significance and its determination by Monte Carlo techniques, *Mon. Weather Rev.*, **111**, 46–59.

Madden, R. A., and Shea, D. J. (1978) Estimates of the natural variability of time-averaged temperatures over the United States. *Mon. Weather Rev.*, **106**, 1695–1703.

Manabe, S., and Wetherald, R. T. (1967) Thermal equilibrium of the atmosphere with a given distribution of relative humidity, *J. Atmos. Sci.*, **24**, 241–259.

Manabe, S., and Wetherald, R. T. (1975) The effect of doubling the CO_2 concentration on the climate of a general circulation model, *J. Atmos. Sci.*, **32**, 3–15.

Manabe, S., and Stouffer, R. J. (1980) Sensitivity of a global climate model to an increase of CO_2 concentration in the atmosphere, *J. Geophys. Res.*, **85**, 5529–5554.

Manabe, S., and Wetherald, R. T. (1980) On the distribution of climate change resulting from an increase in CO_2-content of the atmosphere, *J. Atmos. Sci.*, **37**, 99–118.

Manabe, S., Wetherald, R. T., and Stouffer, R. J. (1981) Summer dryness due to an increase of atmospheric CO_2 concentration, *Climatic Change*, **3**, 347–386.

McNaughton, K. G., and Jarvis, P. G. (1983) Predicting effects of vegetation changes on transpiration and evaporation, in *Water Deficits and Plant Growth*, Vol. VII, New York, Academic Press, 1–47.

Michael, P., Hoffert, M., Tobias, M., and Tichler, J. (1981) Transient climate response to changing carbon dioxide concentration, *Climatic Change*, **3**, 137–153.

Mintz, Y. (1984) The sensitivity of numerically simulated climates to land-surface boundary conditions, in Houghton, J. T. (ed.) *The Global Climate*, Cambridge, Cambridge University Press, 79–105.

Mitchell, J. F. B. (1983) The seasonal response of a general circulation model to changes in CO_2 and sea temperatures, *Q. J. R. Meteorol. Soc.*, **109**, 113–152.

Mitchell, J. F. B., and Lupton, G. (1984) A $4 \times CO_2$ integration with prescribed changes in sea surface temperatures, *Progress in Biometeorology*, **3**, 353–374.

Mitchell, J. M., Jr. (1983) Empirical modeling of effects of solar variability, volcanic events and carbon dioxide on global-scale average temperature since AD 1880, in McCormac, B. M. (ed.) *Weather and Climate Responses to Solar Variations*, Colorado Associated University Press, 265–273.

Mokhov, I. I. (1981) Effect of CO_2 on the thermal regime of the earth's climatic system, *Meteorologiya i Gidrologiya*, **4**, 24–34.

Newell, R. E., and Deepak, A. (eds) (1982) Mount St Helens Eruptions of 1980: atmospheric effects and potential climatic impact, *NASA SP-458*, NASA Scientific and Technical Information Branch, Washington, D.C., 119 pages.

Newhall, C. G., and Self, S. (1982) The Volcanic Explosivity Index (VEI): an estimate of explosive magnitude for historical volcanism, *J. Geophys. Res.*, **87**, 1231–1238.

Newkirk, G., Jr. (1983) Variations in solar luminosity, *Ann. Rev. Astron. Astrophys.*, **21**, 429–467.

North, G. R. (1975) Theory of energy balance climate models, *J. Atmos. Sci.*, **32**, 3–15.

North, G. R., Cahalan, R. F., and Coakley, J. A. (1981) Energy-balance climate models, *Rev. Geophys. Space Phys.*, **19**, 91–122.

North, G. R., Mengel, J. G, and Short, D. A. (1983) Simple energy balance model resolving the seasons and the continents: application to the astronomical theory of the ice ages, *J. Geophys. Res.*, **88**, 6576–6586.

North, G. R., Mengel, J. G., and Short, D. A. (1984) On the transient response patterns of climate to time-dependent concentrations of atmospheric CO_2, in

Hansen, J. E., and Takahashi, T. (eds) *Climate Processes and Climate Sensitivity*, Maurice Ewing Series 5, American Geophysical Union, Washington, D.C., 164–170,

Oerlemanns, J. (1982). Glacial cycles and ice-sheet modeling, *Climatic Change*, **4**, 353–374.

Oliver, R. C. (1976) On the response of hemispheric mean temperature to stratospheric dust: an empirical approach, *J. Appl. Meteorol.*, **15**, 933–950.

Parker, D. E. (1985) The influence of the Southern Oscillation and volcanic eruptions on temperature in the tropical troposphere, *J. Clim.*, **5**, 273–282.

Parker, D. E., and Brownscombe, J. L. (1983) Stratospheric warming following the El Chichon volcanic eruption, *Nature*, **301**, 406–408.

Parkinson, J. H., Morrison, L. V., and Stephenson, F. R. (1980) The constancy of the solar diameter over the past 250 years, *Nature*, **288**, 548–551.

Pittock, A. B. (1978) A critical look at long-term sun–weather relationships, *Rev. Geophys. Space Phys.*, **16**, 400–420.

Pittock A. B. (1983) Solar variability, weather and climate: an update, *Q. J. R. Meteorol. Soc.*, **109**, 23–57.

Pivovarova, Z. I. (1977) *Radiation Characteristics of Climate in the USSR* Leningrad, 355 pages, Gidrometeoizdat (in Russian).

Quiroz, R. S. (1983) The climate of the 'El Niño' winter of 1982–1983—a season of extraordinary climatic anomalies, *Mon. Weather Rev.*, **111**, 1685–1706.

Ramanathan, V. (1977) Interactions between ice-albedo, lapse-rate, and cloud-top feedbacks: An analysis of the nonlinear response of a GCM climate model, *J. Atmos. Sci.*, **34**, 1885–1897.

Ramanathan, V., and Coakley, J. A., Jr. (1978) Climate modeling through radiative-convective models, *Rev. Geophys. Space Phys.*, **16**, 465–489.

Ramanathan, V., Singh, H. B., Cicerone, R. J., and Kiehl, J. T. (1985) Trace gas trends and their potential role in climate change, *J. Geophys. Res.*, **90**, D3, 5547–5566.

Rampino, M. R., and Self, S. (1984) Sulphur rich volcanic eruptions and stratospheric aerosols, *Nature*, **310**, 677–679.

Rind, D., and Lebedeff, S. (1984) *Potential Impacts of Increasing Atmospheric CO_2 with Emphasis on Water Availability and Hydrology in the United States*, Report Prepared for the Environmental Protection Agency, NASA Goddard Space Flight Center Institute for Space Studies, New York, NY, 96 pages.

Robock, A. (1978) Internally and externally caused climate change, *J. Atmos. Sci.*, **35**, 1111–1122.

Robock, A. (1983a) Ice and snow feedbacks and the latitudinal and seasonal distribution of climate sensitivity, *J. Atmos. Sci.*, **40**, 986–997.

Robock, A. (1983b) The dust cloud of the century, *Nature*, **301**, 373–374.

Rowntree, P., and Walker, J. (1978) The effects of doubling the CO_2 concentration radiative-convective equilibrium, in Williams, J. (ed.), *Carbon Dioxide, Climate and Society*, New York, Pergamon Press, 181–192.

Schlesinger, M. E. (1983a) A review of climate models and their simulation of CO_2-induced warming, *Int. J. Environ. Stud.*, **20**, 103–114.

Schlesinger, M. E. (1983b) *A Review of Climate Model Simulations of CO_2-Induced Climatic Change*, Climate Research Institute Report No. 41, Oregon State University, 135 pages.

Schneider, S. H., and Dickinson, R. E. (1974) Climate modeling, *Rev. Geophys. Space Phys.*, **12**, 447–493.

Schneider, S. H., Washington, W. M., and Chervin, R. M. (1978) Cloudiness as a climatic feedback mechanism: Effects on cloud amounts of prescribed global and regional surface temperature changes in the NCAR GCM, *J. Atmos. Sci.*, **35**, 2207–2221.

Schneider, S. H., and Thompson, S. L. (1981) Atmospheric CO_2 and climate: Importance of the transient response, *J. Geophys. Res.*, **86**, 3135–3147.

Sedlacek, W. A., Mroz, E. J., Lazrus, A. L., and Gandrud, B. W. (1983) A decade of stratospheric sulphate measurements compared with observations of volcanic eruptions, *J. Geophys. Res.*, **88**, 3741–3776.

Seidel, S., and Keyes, D. (1983) Can we delay a greenhouse warming?, in *The Effectiveness and Feasibility of Options to Slow a Build-up of Carbon Dioxide in the Atmosphere*, Strategic Studies Staff, Office of Policy Analysis, Office of Policy and Resources Management, EPA, Washington, D.C. 20460.

Self, S., Rampino, M. R., and Barbera, J. J. (1981) The possible effects of large 19th century volcanic eruptions on zonal and hemispheric surface temperatures, *J. Volcanology and Geothermal Res.*, **11**, 41–60.

Sellers, W. D. (1969) A global climate model based on the energy balance of the earth–atmosphere system, *J. Appl. Meteor.*, **8**, 392–400.

Semtner, A. J., Jr. (1984) On modelling the seasonal thermodynamic cycle of sea ice in studies of climate change, *Climatic Change*, **6**, 27–38.

Shukla, J., and Mintz, Y. (1982) Influence of land-surface evapotranspiration on the earth's climate, *Science*, **215**, 1498–1501.

Shuttleworth, W. J., and Calder, I. R. (1979) Has the Priestley–Taylor equation any relevance to forest evaporation? *J. Appl. Meteorol.*, **18**, 639–646.

Simkin, T., Siebert, L. McClelland, L., Bridge, D., Newhall, C., and Latter, J. H. (1981) *Volcanoes of the World*, Stroudsburg, Pennsylvania, Hutchinson Ross Publishing Company, 233 pages.

Smith, E. A., von der Haar, T. H., Hicket, J. R., and Maschhoff, R. (1983) The nature of short-period fluctuations in solar irradiance received by the Earth, *Climatic Change*, **5**, 211–235.

Somerville, R. C. J., and Remer, L. A. (1984) Cloud optical thickness feedbacks in the CO_2 climate problem, *J. Geophys. Res.*, **89**, 9668–9672.

Sonett, C. P. (1984) Very long solar periods and the radiocarbon record, *Rev. Geophys. Space Phys.*, **22**, 239–254.

Spelman, M. J., and Manabe, S. (1984) Influence of oceanic heat transport upon the sensitivity of a model climate, *J. Geophys. Res.*, **89**, 571–586.

Stuiver, M. (1980) Solar variability and climatic change during the current millennium, *Nature*, **286**, 868–871.

Stuiver, M., and Quay, P. D. (1980) Changes in atmospheric carbon-14 attributed to a variable sun, *Science*, **207**, 11–19.

Taylor, B. L., Gal-Chen, T., and Schneider, S. H. (1980) Volcanic eruptions and long-term temperature records: an empirical search for cause and effect, *Q. J. R. Meteorol. Soc.*, **106**, 175–199.

Thompson, S. L., and Schneider, S. H. (1982) Carbon dioxide and climate: The importance of realistic geography in estimating the transient temperature response, *Science*, **217**, 1031–1033.

Trenberth, K. E. (1983) Interactions between orographically and thermally forced planetary waves, *J. Atmos. Sci.*, **40**, 1126–1153.

Turco, R. P., Toon, O. B., Ackerman, T., Pollack, J. B., and Sagan, C. (1983) Nuclear winter: Global consequences of multiple nuclear explosions, *Science*, **222**, 1283–1292.

Wang, W. C., and Stone, P. H. (1980) Effect of ice–albedo feedback on global sensitivity in a one-dimensional radiative–convective climate model, *J. Atmos. Sci.*, **37**, 545–552.

Washington, W. M., and Meehl, G. A. (1983) General circulation model experiments on the climatic effects due to a doubling and quadrupling of carbon dioxide concentration, *J. Geophys. Res.*, **88**, 6600–6610.

Washington, W. M., and Meehl, G. A. (1984) Seasonal cycle experiment on the climate sensitivity due to a doubling of CO_2 with an atmospheric general circulation model coupled to a simple mixed layer ocean model, *J. Geophys. Res.*, **89**, 9475–9503.

Wetherald, R. T., and Manabe, S. (1975) The effects of changing the solar constant on the climate of a general circulation model, *J. Atmos. Sci.*, **32**, 2044–2059.

Wetherald, R. T., and Manabe, S. (1980) Cloud cover and climate sensitivity, *J. Atmos. Sci.*, **37**, 1485–1510.

Wetherald, R. T., and Manabe, S. (1981) Influence of seasonal variation upon the sensitivity of a model climate, *J. Geophys. Res.*, **86**, 1194–1204.

Wigley, T. M. L., and Schlesinger, M. E. (1985) Analytical solution for the effect of increasing CO_2 on global mean temperature, *Nature*, **315**, 649–652.

Williams, L. D., Wigley, T. M. L., and Kelly, P. M. (1980) Climatic trends at high northern latitudes during the last 4000 years compared with 14C fluctuations, in *Sun and Climate*, CNES/CNRS/DGRST Conference Proceedings, 11–20.

Willson, R. C., Gulkis, S., Janssen, M., Hudson, H. S., and Chapman, G. A. (1981) Observations of solar irradiance variability, *Science*, **21**, 200–202.

Woods, J .D. (1984) The upper ocean and air–sea interaction, in Houghton, J. T. (ed.) *The Global Climate*, Cambridge, Cambridge University Press, 79–106.

World Climate Research Programme (1983) Report of the Experts Meeting on Aerosols and Their Climatic Effects, *World Climate Programme*, **55**, 101 pages.

World Meteorological Organization (1982) WMO Global Ozone Research and Monitoring Project, *Report No. 14 of the Meeting of Experts on Potential Climatic Effects of Ozone and Other Minor Trace Gases*, Boulder, Colorado, 13–17 September 1982, 35 pages, Geneva, WMO.

Yeh, T.-C., Wetherald, R. T., and Manabe, S. (1983) A model study of the short-term climatic and hydrologic effects of sudden snow-cover removal, *Mon. Weather Rev.*, **111**, 1013–1024.

CHAPTER 6

Empirical Climate Studies
Warm World Scenarios and the Detection of Climatic Change Induced by Radiatively Active Gases

T. M. L. WIGLEY, P. D. JONES and P. M. KELLY

6.1 INTRODUCTION

In order to understand, and eventually predict, how increasing atmospheric CO_2 concentration might alter the Earth's climate, both modelling studies and empirical analyses of observational data are required. The current state of the modelling art has been reviewed by Dickinson in Chapter 5. Here we describe a number of empirical approaches to the CO_2 problem. Before beginning, however, we note that empirical studies do not yet distinguish the effects of CO_2 from those of other radiatively active trace gases, such as methane (CH_4), nitrous oxide (N_2O), ozone (O_3) and chlorofluorocarbons (CFCs). These gases are therefore considered together under the broad heading 'greenhouse gases'. Since the different gases do have spatially distinct effects on the atmosphere's radiative balance, distinguishing their separate influences is theoretically possible, but this is a problem which has practical difficulties well beyond those applying to the characterization of their combined effects on climate.

Empirical studies of the greenhouse gas problem may be grouped under three broad headings: analysis of past climatic change, production of warm-world scenarios, and the detection of the effects of these radiatively active gases on climate. In Section 6.2 we review past climatic changes and fluctuations. Although the bulk of this section deals with the past 100 years (since this is the period for which we have the most abundant and highest quality data, and it is the period over which most anthropogenic changes have occurred), longer time scale climatic change is also considered in order to place the more recent changes in a proper perspective. Section 6.3 covers the scenario approach to the estimation of future climatic conditions. In Chapter 5, Dickinson has shown that present climate GCMs cannot yet predict the *regional* details of future climates, largely because of uncertainties in the way these models parameterize (i.e. simplify) many important climate processes, and the fact that they model the oceans so crudely. Because of these model limitations a number of authors have used the past climate record to construct 'scenarios' for the future, i.e. geographically detailed empirical projections of future possibilities. In Section 6.4 we consider the question: has the climatic effect of greenhouse gases been detected in the observational data? Detection of greenhouse gas effects is the ultimate test of model results and so is a research area of considerable importance, one that requires a careful and statistically rigorous approach. Finally, in Section 6.5, we summarize the detection issue and present a number of recommendations for future research.

6.2 PAST CLIMATIC CHANGE

6.2.1 Relevance to the Greenhouse Gases Issue

The record of past climate is relevant to the study of anthropogenic CO_2 and trace gas effects for four reasons. First, it provides information on the natural variability of the climate system which allows us to evaluate possible future changes in a broader context. A quantification of this natural variability, which constitutes the background 'noise' level above which anthropogenic effects must be identified, is central to the detection problem. Second, in trying to understand the causes of past climatic change, we obtain insights into the way the climate system responds to external forcing and into the system's complex internal interactions and feedback mechanisms. Third, climatic data provide an important source of information for testing climate models, both in the control mode (i.e. in simulating present climatic conditions) and in the perturbed mode (i.e. in simulating past conditions for some assumed change in external forcing). Fourth, some aspects of past climatic change are undoubtedly related to past changes in atmospheric CO_2 level, so the paleoclimatic record provides direct information on the effects of CO_2 forcing.

6.2.2 CO_2 Effects on the 10^6–10^9 Year Time Scale

On the very longest time scales, changes in CO_2 concentration are thought to be one of the primary controls on climatic variability. When the Earth first formed, the Sun was much fainter than today, only about 75% as bright, yet we know that the early Earth was not ice-covered. This 'Faint Sun Paradox' has been explained by the much higher CO_2 concentration that existed then, with an enhanced greenhouse effect counteracting the lower solar output (Owen *et al.*, 1979; Roxburgh, 1980). Life on Earth owes its existence to CO_2, both because of the greenhouse effect and, perhaps, because the chemistry of the oceans in a high-CO_2 world provided a more favourable environment for the chemical precursors of life (Wigley and Brimblecombe, 1981). Since those early days, billions of years ago, it appears that life has had a major effect on atmospheric CO_2 concentration, with photosynthesizing plants slowly reducing CO_2 levels as the Sun's irradiance increased, maintaining approximate homeostasis for the planet. Superimposed on this very long time scale decreasing CO_2 trend, there have been noticeable fluctuations (revealed by analysis of the geological sediment record; see Berner *et al.*, 1983, and Schneider and Londer, 1984, pp. 241–7) which almost certainly had major impacts on the climate. For example, the warmth of the Cretaceous period some 100 million years ago, given the vastly different positions of the continents in this era, can most easily be explained by an elevated CO_2 level (Barron and Washington, 1984; Lloyd, 1984). Budyko (1982) has considered the role of CO_2 as a climatogenic factor on these long time scales in considerable detail.

6.2.3 The 10^3–10^5 Year Time Scale

On the 100,000 year time scale, the Earth has experienced a fairly regular series of cold glacial periods which become most apparent in the climate record roughly 1.7 million years ago. Fluctuations are clearly evident prior to 1.7×10^6 BP, but an increase in the amplitude of these fluctuations seems to have begun at that time in association with the appearance of seasonally permanent Arctic ice (Shackleton *et al.*, 1984). These glacial/interglacial cycles occur approximately every 100,000 years. While the causes are not yet known in detail, the primary forcing factor appears to be changes in the Earth's radiation budget due to orbital parameter effects (Milankovitch, 1941; CLIMAP, 1976, 1984; Hays *et al.*, 1976). The most convincing evidence for this is that the main orbital parameter periodicities (at approximately 100,000, 40,000, 23,000 and 19,000 years, see Berger, 1977) are also observed in the proxy climate record from deep sea sediments (albeit with different relative strengths).

In spite of this strong evidence, independent climate models (i.e. those with few empirically tuned parameters) have, to date, been unable to reproduce the observed equilibrium climate using orbital forcing alone (see North *et al.*, 1983, for a recent review). Important feedback mechanisms may be missing in these models and changing CO_2 levels may be a significant factor (Hansen *et al.*, 1984; Pisias and Shackleton, 1984; Kerr, 1984). There is direct evidence to support this possibility, and large glacial/interglacial time scale changes in CO_2 concentration have been measured using fossil CO_2 in bubbles from ice cores (Delmas *et al.*, 1980; Neftel *et al.*, 1982) and inferred from carbon isotope data from deep-sea sediments (Shackleton *et al.*, 1983). Atmospheric CO_2 concentration was around 200 ppmv at the peak of the last glaciation (20–15,000 years ago). This low level was probably a response to orbitally forced climate changes in the oceans (Broecker and Takahashi, 1984; Knox and McElroy, 1984; Sarmiento and Toggweiler, 1984; Siegenthaler and Wenk, 1984) and changing CO_2 concentration may constitute an important positive feedback process which amplifies the orbital effects. Oceanically mediated, natural CO_2 changes may operate to amplify climatic change no matter what the forcing mechanism.

The last interglacial/glacial cycle can be considered to have begun around 120,000 BP. The period 120,000 to 80,000 BP (known in Europe as the Eemian) was a time of general warmth, similar to today. This was interrupted by two cold intervals which began and ended relatively abruptly (i.e. over a period of order 100 to 1,000 years) and which lasted for only a few thousand years (see, for example, Shackleton and Opdyke, 1973; Woillard, 1978, 1979). Earlier warm periods tended to last for less than the 40,000 years of the last interglacial, but their durations varied widely. The last interglacial is particularly interesting because, for at least some of this period, mean sea level may have been considerably (around 6 m) above present-day levels indicating a substantially lower ice volume than today. This higher sea level is thought by some to be due to partial melting of the West Antarctic ice sheet (Mercer, 1978). This is an important observation since such an event has also been suggested as a possible consequence of the warming induced by an increased concentration of greenhouse gases (see Robin, Chapter 7, for further details).

The period 70,000 to 15,000 BP was globally cool, with global mean temperature probably ranging between 2 and 5 °C below today's level (Gates, 1976a,b; Peterson *et al.*, 1979; Kutzbach and Guetter, 1984; Hansen *et al.*, 1984). Maximum coldness occurred around 20,000 to 15,000 BP, associated with maximum extents of continental ice masses in the Northern Hemisphere over North America and Eurasia, and maximum extent of sea ice around Antarctica. At this time, mean sea level was roughly 100 m below today's level due to the mass of water locked up in the continental ice sheets.

Although orbital changes are the most likely primary cause of these past glaciations, it is of interest to note that the orbital insolation signature 18,000 years ago was quite similar to that prevailing today. The cold conditions at 18,000 BP are presumed to have developed from earlier orbital changes, and were maintained partly by the large Northern Hemisphere ice masses, by prevailing sea surface temperatures, and by reduced CO_2 levels. The relative importance of these factors is still under study (see, for example, Hansen *et al.*, 1984, and Manabe and Broccoli, 1984). These studies suggest that

CO_2 played a relatively minor role in maintaining the glacial equilibrium state, but CO_2 changes may well have been more important in initiating glacial/interglacial climatic changes (Knox and McElroy, 1984; Pisias and Shackleton, 1984). The 70,000 to 15,000 BP interval, while generally much colder than today, was, nevertheless, punctuated by a number of rapid warming and cooling events. Of particular relevance to the CO_2 issue are a series of fluctuations between about 33,000 and 22,000 BP, when ice core data show that virtually simultaneous, rapid changes in both climate and CO_2 level occurred (Stauffer *et al.*, 1984).

The period 15,000 to 10,000 BP is known as the late-glacial period. The recovery from full-glacial conditions was geographically and temporally complex. Warming appears to have begun earlier in the Southern Hemisphere. In the Northern Hemisphere, particularly around the eastern side of the North Atlantic, very rapid warming occurred around 13,000 BP. Warming in western Europe was so rapid that conditions were almost as warm at 12,700 BP as today. This relatively warm period prevailed until around 11,000 BP when a sharp and severe cooling episode occurred. Warm conditions returned with a rapid warming around 10,000 BP. Over a period of only a few centuries, winter temperatures warmed by around 15 °C and summer temperatures by 8 °C, mirroring the ~13,000 BP warming event (Atkinson *et al.*, 1985). These rapid changes in European climate appear to have occurred in phase with similar changes in Greenland (based on the oxygen isotope record from the Dye 3 ice core; see Siegenthaler *et al.*, 1984, for details) and may relate to changes in the influx of cold fresh water into the North Atlantic from the melting continental ice sheets (Mercer, 1969; Ruddiman and McIntyre, 1981) and changes in the rate of formation of deep water in the North Atlantic (Broecker *et al.*, 1985). There is ice core evidence that changes in atmospheric CO_2 concentration may have occurred at about the same time as these rapid late-glacial climatic changes (Oeschger *et al.*, 1984).

Although these rapid changes in climate were probably not global in extent (Mercer, 1969; Broecker *et al.*, 1985), the possible contemporaneous CO_2 changes and the link with changes in the production rate of North Atlantic Deep Water (NADW), make them highly relevant to the anthropogenic CO_2 issue. Because NADW represents a sink of water at ~2 °C in a region where the balancing influx of water has a considerably higher temperature, the net effect of NADW formation is to provide a major heat source for the atmosphere. Lower NADW formation rate should, therefore, result in atmospheric cooling (for further details see Hoffert *et al.*, 1985, and Watts, 1985a,b). This same ocean circulation change should also, on the basis of carbon cycle considerations, result in a reduction in atmospheric CO_2 concentration. Broecker *et al.* (1985) hypothesize that NADW formation rate changes are the main mechanism for the observed rapid late-glacial climatic fluctuations. Similar NADW changes may well occur today (indeed, there is observational evidence for such changes). Although the magnitude of the changes possible today would probably be less than in the late-glacial, they could be of considerable importance in modifying any anthropogenic greenhouse gas-induced changes in climate (Broecker *et al.*, 1985). NADW effects are discussed further in Section 6.4.4.

By 10,000 BP, the globe as a whole had warmed to approximately its present-day mean temperature, but large ice masses still existed in the Northern Hemisphere. In Scandinavia, ice retreat, which had begun at around 13,000 BP, was particularly rapid from 10–9,000 BP and ice had virtually disappeared by 8,000 BP. On the North American continent, the ice retreat chronology was initially similar to this, but a substantial amount of ice was left at 8,000 BP. Between approximately 8,000 and 7,500 BP, the North American ice sheet became two separate ice sheets. At 6,000 BP, only three small ice vestiges were left in northern Canada and these soon disappeared. Ice retreat in Eurasia and North America, however, was nowhere smooth, and details of retreat timing are still sketchy (Andrews and Barry, 1978). It is likely that the decay was not simply by melting, but also by mechanical processes such as surging and rapid movement along ice streams (Denton and Hughes, 1981).

The period 10,000 BP to the present is known as the Holocene. The early Holocene (9,000–6,000 BP) was a time of possibly global and certainly regional warmth compared with today (Webb and Wigley, 1985). For example, at 6,500 BP, in spite of the existence of small residual ice masses, forests were present significantly further north in parts of North America compared with their limits today. In addition, many of the world's desert regions were substantially wetter than today, with fewer sand dunes, more vegetation and higher lake levels (Sarnthein, 1978; Street and Grove, 1979; Kutzbach and Street-Perrott, 1985).

This early Holocene warm period has been used as an analogue for a CO_2-induced warmer world (see Section 6.3.2). The cause of this warmth appears to be, at least in part, orbital variations (Kutzbach, 1981; Kutzbach and Otto-Bliesner, 1982; Kutzbach and Guetter, 1984; Kutzbach and Street-Perrott, 1985). At 9,000 and 6,000 BP, the spatial and seasonal distribution of solar radiation input was quite different from that prevailing today. In the Northern Hemisphere, July insolation at 9,000 BP was 8% above the present-day level (+6% at 6,000 BP), while Northern Hemisphere January insolation values were below those prevailing today by roughly the same amount. There is some (debatable) evidence of CO_2 levels higher than the nineteenth century 'pre-industrial' level (Neftel *et al.*, 1982).

Even though there are quantitative details yet to be understood, the importance of orbital effects in determining past climates on the 1,000-year time scale is undeniable. On the basis of present knowledge, predictions can be made about future long time scale climatic change. In the absence of anthropogenic effects, further glacial periods are inevitable, but the next is unlikely to occur until far into the future. Over the next 5,000 years or so, we can expect, on the basis of orbital variations alone, only a slight global-mean cooling (Imbrie and Imbrie, 1980; Kukla *et al.*, 1981), a trend which is very small on the century time scale and insignificant relative to the projected effects of greenhouse gases.

6.2.4 The interval 6000 BP to 1850 AD

While the whole of the Holocene was a time of warmth relative to the previous 60–70,000 years, global climate was far from static during this period. Many warming and cooling episodes occurred on the 100-year timescale. Because these events are only observable through indirect or proxy evidence, which is both local and often poorly dated, we do not know how global mean climate changed, or even *if* global mean temperature has changed since, say, 6,000 BP. Regionally, however, substantial changes *have* occurred, as evidenced by vegetation changes, glacial fluctuations, isotopic data from ice cores, and many other proxy climate indicators.

Over the last 2,000 years the most widespread changes observed in the climate record are those associated with the Medieval Warm Epoch (approximately 800 to 1200 AD) and the Little Ice Age (approximately 1400 to 1800 AD). The former may well be restricted to the North Atlantic Basin region, and, even over this large area, changes in different locations show large differences in timing on century and shorter time scales (Williams and Wigley, 1983). The Little Ice Age, however, seems to have been more nearly global, but, again, substantial differences can be seen when time series from different regions are compared. The cause of the Little Ice Age is not known. Changes in solar activity associated with the period of minimum sunspot activity known as the Maunder Minimum have been suggested (Eddy, 1976a,b), but the evidence is equivocal. Another possible causal factor is a change in the frequency of explosive volcanic activity. Some support for this hypothesis is available from ice cores (Hammer *et al.*, 1980; Mosley-Thompson and Thompson, 1982), but even this evidence is open to question. There is also ice core evidence that atmospheric CO_2 levels may have been depressed during the Little Ice Age (Raynaud and Barnola, 1985), and atmospheric methane concentration was apparently much less at this time than during the twentieth century, perhaps as low as 700 ppbv compared with a 1980 level of around 1650 ppbv (Craig and Chou, 1982).

6.2.5 Recent Climatic Change: 1850 AD to Present

Suggestions of CO_2 effects on climate can be found on almost all time scales in the past climatic record. The evidence, however, is insufficient to quantify the CO_2 influence. It is only for the past century or so, the era of instrumental meteorology, that we have sufficient and suitable data to be able to make meaningful quantitative analyses. The potential effects of CO_2 and other trace gases embrace the whole climate system; temperatures, precipitation, cloudiness, winds, and so on. Here we consider recent changes in the most well documented weather variables, temperature and precipitation, in order to provide a background to the discussion of the detection issue given in Section 6.4.

It is only since about 1850 AD that we can begin to make detailed quantitative statements about past climatic change. However, the confidence that can be placed in such statements depends critically on the particular variable considered, on the size of the area covered, and on location. Precipitation, temperature and surface (or mean sea level) pressure data extend back into the nineteenth century and earlier, but spatial coverage is poor in the early years of these records (Lamb, 1977; Bradley and Jones, 1985). Even today, spatial coverage is relatively poor over the oceans, particularly in the southeastern Pacific and in high southern latitudes. This deficiency may eventually be overcome using satellite data. Plentiful upper air data, essential in obtaining a three-dimensional picture of climate, exist only since the late 1940s.

In describing past temperature or precipitation changes it is of greatest value to consider large scale area averages. Local or small scale regional changes may be of relatively minor consequence in understanding the climate system as a whole. Moreover, in the present context, the uncertainty in the expected signal due to CO_2 and other trace gases increases as the spatial scale decreases. For these reasons, we concentrate here on large scale spatial averages. Greatest attention will also be given to temperature, since this is the best documented climate variable and since the signal for temperature is always likely to be better established than that for other variables.

A. Surface Temperature Changes

Although this had been attempted earlier by Köppen (1873, 1914), the first reasonably reliable estimates of large scale area average temperatures were those made by Callendar (1961), Mitchell (1961, 1963) and Budyko (1969). These and other works have been reviewed by a number of authors (e.g. Jones *et al.*, 1982; Wallén, 1984; Ellsaesser *et al.*, 1985; Wigley *et al.*, 1985). The best documented records are those of Jones *et al.* (1982, 1986a,b; see also Kelly *et al.*, 1982, and Raper *et al.*, 1984). At least four other groups have recently published important work in this area. For convenience, these five groups will be referred to below as Jones (Jones *et al.*, 1982, 1986a,b,c; Kelly *et al.*, 1982; Raper *et al.*, 1983, 1984), Vinnikov (Borzenkova *et al.*, 1976; Vinnikov *et al.*, 1980), Hansen (Hansen *et al.*, 1981), Yamamoto (Yamamoto, 1981; Yamamoto and Hoshiai, 1979, 1980) and Folland (Folland and Kates, 1984; Folland *et al.*, 1984a,b). The work of Folland and colleagues and of Jones *et al.* (1986c) differs from the others in using marine (i.e. ship-based) data. The other analyses are based on land station data with sparse coverage over ocean areas. Only Jones *et al.* (1986c) have produced averages which incorporate both land and marine data.

Large scale averages for the land masses are based on individual station observations. Such records may be influenced by various sources of inhomogeneity. Variations may be caused by non-climatic factors such as site changes, changes in the type of instrument and/or its exposure, changes in observation times and/or the method of producing daily and monthly averages, or local anthropogenic effects such as urban warming (see Mitchell, 1953, and Bradley and Jones, 1985). Provided the data are carefully screened for possible inhomogeneities, these problems do not appear to have a noticeable effect on large scale averages (Jones *et al.*, 1986a). The reliability of these averages will be discussed further below.

Three of the four land-based studies give extremely similar results, a necessary consequence of their reliance on the same basic data sources. Minor differences exist due to small differences in the area of coverage, the data sources and the different methods of producing area averages. In the case of Yamamoto, averages were based on the incorrect assumption that no changes had occurred in regions with no data, so their results show much smaller amplitude changes and are not directly comparable with the others. Of the analyses of land-based data, only Hansen claims an estimate of *global mean* temperature. Since data coverage in the Southern Hemisphere is extremely poor (approximately 80% is ocean), this claim is difficult to justify.

The Northern Hemisphere land-based records of Jones, Vinnikov and Hansen are shown in Figure 6.1. The three curves correlate highly on annual and longer time scales and show changes of similar amplitude. The general trends of warming to around 1940, cooling to the mid-1960s, and warming since around 1970, are common to all analyses. When individual seasons are considered, similar trends can be seen, but with inter-seasonal phase differences of a decade or more (see Jones *et al.*, 1982, and Wigley *et al.*, 1985). The curves in Figure 6.1 have often been used as indicators of global mean changes. While a true global average can only be obtained by including data from the ocean areas and all of the Southern Hemisphere, recent comparisons of land and ocean data and of Southern and Northern Hemisphere data show remarkable parallels (Wigley *et al.*, 1985; Jones *et al.*, 1986b; see also Figure 6.2).

A number of workers have attempted to study ocean, or ocean-plus-land

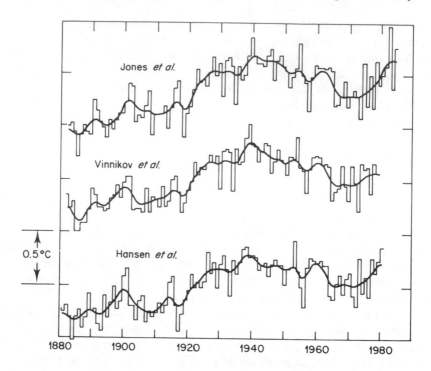

Figure 6.1 Comparison of three estimates of Northern Hemisphere temperature changes. The top curve is from Jones *et al.* (1982) and the middle curve is from Vinnikov *et al.* (1980). The lower curve is Hansen *et al.*'s (1981) data for the whole hemisphere (personal communication, J. E. Hansen). In Hansen *et al.* (1981), only the average for 23.6–90°N is given, and the data are smoothed with a 5-year running mean. The curves given here show both annual mean values and values smoothed with a 10-year Gaussian filter (padded at the ends to cover the whole period of record)

temperature changes (Folland and co-workers; Barnett, 1978, 1984; Chen, 1982; Paltridge and Woodruff, 1981; Fletcher, 1984; Jones *et al.*, 1986c). There are, however, great difficulties in obtaining a strictly homogeneous time series of ship-based data due to changes in instrumentation and problems related to instrument exposure (reviewed by Barnett, 1984, 1985a, and by Folland *et al.*, 1984a). Either sea surface temperatures (SSTs) or ship-based (marine) air temperatures (MATs) may be used; but both types of data require adjustments to produce a homogeneous record. SST and MAT data are highly correlated.

   In spite of these instrumentation problems, the marine data appear to be reliable at least back to the early twentieth century. The MAT data correlate highly with the land-based data (Jones *et al.*, 1986a,b,c). Such high correlations are expected on physical grounds (Wigley *et al.*, 1985). The fact that

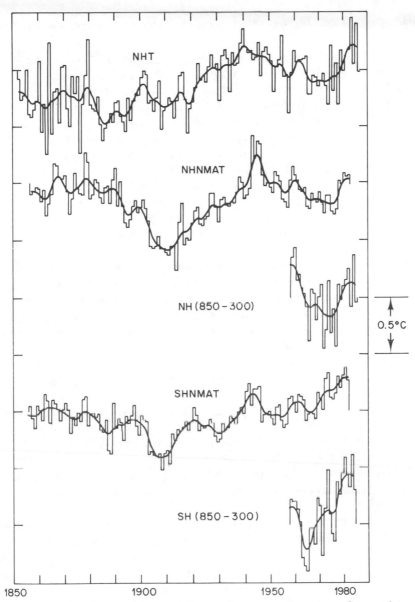

Figure 6.2 Northern and Southern Hemisphere annual mean surface and tropospheric air temperature changes. Top curve (NHT): land-based record from Jones *et al.* (1986a), a revision and extension of the data given by Jones *et al.* (1982). Second and fourth curves (NHNMAT and SHNMAT); Folland *et al.*'s night-time marine air temperature data (data from C. K. Folland and D. E. Parker, personal communication). Third and fifth curves (NH(850–300) and SH(850–300)); Angell and Korshover's 850–300 mb column mean temperatures (data from J. K. Angell, personal communication). Smooth curves are 10-year Gaussian filtered values. Note that the NMAT values in the early 1940s may be too high due to measurement uncertainties (see Folland *et al.*, 1984a)

they occur between independently derived data sets and are similar in both hemispheres is evidence supporting the reliability of the data.

In Figure 6.2 we compare Folland *et al.*'s night-time marine air temperature (NMAT) records for both hemispheres with the Jones *et al.* (1986a) land-based Northern Hemisphere record. Similarities between the hemispheres on decadal and longer time scales are clearly evident. Both hemispheres show general warming from the late nineteenth century to around 1940 and a decline to the mid-1960s. Since the mid-1960s, the globe as a whole appears to have warmed substantially. However, this warming trend seems to have been delayed in the Northern Hemisphere (see Figure 6.2 and also Jones *et al.*, 1986c). The reason for this difference between the hemispheric trends is not yet known (but see Section 6.4.4).

Prior to about 1900, marine data coverage was generally less extensive, although it was still good in some periods. Because of the greater spatial and temporal autocorrelation of marine temperatures, data requirements are less than over land. Nevertheless, the Northern Hemisphere land data correlate less well with Folland *et al.*'s marine data in the nineteenth century than in the twentieth century. A visual comparison of the land-based and marine-based (NMAT) Northern Hemisphere time series in Figure 6.2 shows that, as one moves back in time, the curves begin to diverge around the beginning of the twentieth century, although they show similar decadal time scale fluctuations back to around 1875. Similar discrepancies exist between the Southern Hemisphere land and marine data (see Jones *et al.*, 1986b). In view of the overall similarity for the twentieth century data, we must suspect that either the land record or marine record is unreliable prior to about 1900. Lowering the NMAT data by about 0.3 °C would largely remove the discrepancy. This issue is discussed in more detail by Jones *et al.* (1986a,b,c).

If one accepts the reliability of the land data in the nineteenth century, then, for Folland *et al.*'s long time scale changes to be correct, there would have to be quite large differences between the temperature trends over land and sea in the nineteenth century, but not in the twentieth century. Although this seems unlikely, the possibility of differences in land and marine temperature trends requires further comment, since it has also been suggested by the work of Paltridge and Woodruff (1981). These authors showed an apparent lag of twentieth century temperature changes over the oceans behind those over the land by approximately 20 years. The reality of this lag is highly suspect because Paltridge and Woodruff used uncorrected marine data. Use of the more carefully homogenized twentieth century data fails to reveal any noticeable lag between SST data and either MAT data or Northern Hemisphere land-based data.

Jones *et al.* (1986c) argue that the land-based data are intrinsically more reliable than the marine data and that the best way to correct the marine data is by reference to the land-based data. In making such corrections they have produced hemispheric-mean MAT and SST time series that differ noticeably from those of Folland *et al.* on time scales greater than decadal, especially prior to 1900. Their global mean temperature curve is shown in Figure 6.3. It should be noted that there is still some dispute over the pre-1900 temperature trends because of the differences between the marine time series of Jones *et al.* (1986c) and Folland *et al.* (1984a,b). These differences have yet to be fully resolved.

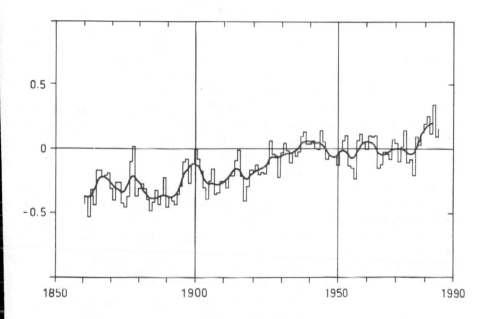

Figure 6.3 Global mean annual surface temperature changes from Jones *et al.* (1986c). Smooth curve shows 10-year Gaussian filtered values.

B. Upper Air Temperature Changes

By 'upper air', we mean the whole of the atmosphere, although useful observations are only available for the troposphere and low to middle stratosphere. World-wide, three-dimensional observations of the atmosphere began only in the late 1940s.

Upper air temperatures may be obtained either directly from temperature observations at the mandatory constant-pressure observing levels, or indirectly from thickness data (i.e. the height difference between constant pressure levels, which is proportional to the column mean temperature). Both types of data, especially the former, are affected by changes in instruments and changes in correction procedures, both of which may produce spurious non-climatic trends and variations (Bradley and Jones, 1985). Large scale area averages of these data may be based either on station records or on data interpolated on to a regular grid. Additional inhomogeneities in the latter data may arise through changes in the gridding analysis procedure.

Many analyses of upper air temperatures have used gridded thickness data. These analyses (reviewed in Wigley *et al.*, 1985) show little evidence of any long-term temperature trend. In some records, the main event is a rapid drop in tropospheric temperatures around 1963/64. This has been associated with the eruption of the volcano Mt. Agung by a number of workers. However, the evidence is equivocal (see Parker, 1985) and the drop in temperature may be partly due to changes in analysis procedures (see below).

In this regard, Parker (1980) has exposed a major problem with grid-point thickness data. Values determined by different meteorological agencies are in considerable disagreement, particularly over the oceans and in the subtropics. The differences, which represent a basic data uncertainty, are of similar magnitude to inferred temperature fluctuations based on the individual agency analyses. They can be attributed partly to instrumental problems and partly to changes in the methods of analysis. Grid-point analyses were originally produced manually and are currently produced by interpolation techniques that use model-generated error statistics. Recent data are, therefore, to some degree model dependent (see, for example, Leith, 1984). Because of these problems with grid-point data, we concentrate here on the results of station data analyses.

Extensive analyses of station data based on a 63-station network distributed fairly evenly over the globe have been published by Angell and Korshover (1977; 1978a,b; 1983a,b). Their analyses probably avoid most of the problems associated with grid-point data pointed out by Parker (1980), but there is some doubt about the representativeness of global (or large area) averages because of the limited station network (Oort, 1978). Furthermore, with a limited station network, spurious trends may arise if significant changes occur in the long wave pattern of the general circulation (see Parker, 1981).

Angell and Korshover's hemispheric mean values for the 850–300 mb layer are shown in Figure 6.2, where they can be compared with the surface data discussed above. The correlations between the surface and these middle tropospheric air temperature changes are striking ($r = 0.80$ with Northern Hemisphere NMAT and $r = 0.81$ with Southern Hemisphere NMAT), especially given the fairly limited upper air station network.

Angell and Korshover also document stratospheric temperature changes. Figure 6.4 shows the global mean change for the 100–50 mb layer, with

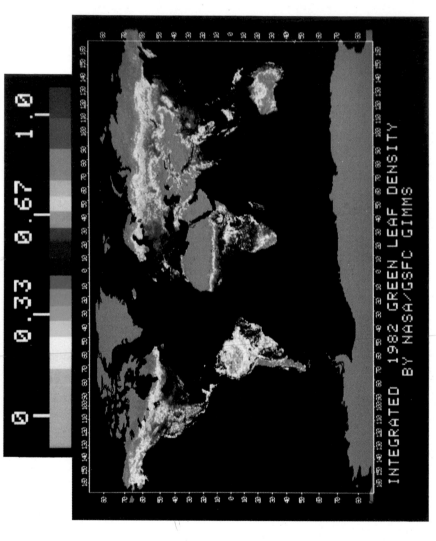

Plate 1 The global pattern of the integral of the photosynthetically active radiation absorbed by terrestrial vegetation for 1982 obtained from the NOAA-7 advanced very high resolution radiometer sensor system. The intergral image approximately represents net primary production for the year

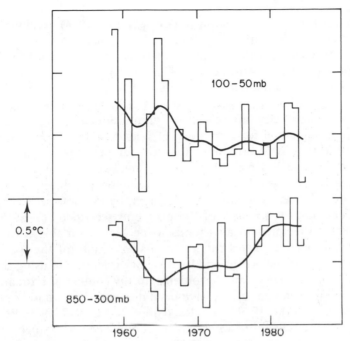

Figure 6.4 Global mean temperature changes for 850–300 mb (bottom) and 100–50 mb (top) from Angell and Korshover (data from J. K. Angell, personal communication)

their 850–300 mb global mean shown for comparison. None of their results up to around 20 km (50 mb) show any strong trends. Above 20 km there are strong cooling trends, but data coverage is restricted to only part of the Northern Hemisphere and even these data may have severe instrumentation problems (Wigley *et al.*, 1985). The lack of clear trends in lower stratospheric temperatures may be partly the result of the shortness of the record (less than 30 years) and the quite large decadal time scale variability (see also Figures 6.2 and 6.4). Even if a CO_2-induced trend did exist, it would be obscured by this medium time scale noise because of the shortness of the record.

C. Precipitation Changes

Model results suggest that increasing CO_2 may cause an increase in the intensity of the global hydrologic cycle and, therefore, a slight increase in global mean precipitation. Estimates of the global mean change in precipitation due to a doubling of CO_2 vary widely, from +3% (Manabe and Stouffer, 1980) to +11% (Rind and Lebedeff, 1984). Even the lower value may be an overestimate because of deficiencies in the way current climate models model the oceans, a point noted by Stone (1984). Regional changes are expected to differ noticeably from the global mean. Although the details of these future changes are unknown, there is strong evidence for increases in precipitation in high latitudes (see Chapter 5). There is also evidence for possible increases in mid-latitude dryness (Manabe et al., 1981), but these changes are not necessarily associated with reductions in precipitation.

Little work has been carried out on temporal changes in large area averages of precipitation for a number of reasons. The spatial variability of precipitation is considerably more than for temperature and it is generally thought that a much denser observational network is required in order to establish area averages. In addition, precipitation is subject to more measurement difficulties than temperature. Precipitation time series often show inhomogeneities; for a review of the problems see Rodda (1969) and Barnett (1985b). Precipitation also tends to be highly variable from year-to-year. Global mean estimates cannot be made because of the sparseness of data coverage over the oceans and because of difficulties in using ship-based precipitation data. These latter problems, in the context of estimating long-term mean precipitation values, have been discussed by a number of authors (Dorman and Bourke, 1979, 1981; Reed and Elliott, 1977, 1979; Reed, 1979; Elliott and Reed, 1984). No work has been published on changes in precipitation over the oceans. There have, however, been a number of important regional studies for continental areas (e.g. Diaz, 1981; Diaz and Fulbright, 1981; Fleer, 1981; Nicholson, 1980, 1981; Tabony, 1981).

Although precipitation data show high spatial variability, inter-annual precipitation variations do show large scale (order 1,000 km or more) coherent patterns which account for significant amounts of overall precipitation variance (Barnett, 1985b). The most spatially comprehensive studies of precipitation (i.e. those taking a global or, at least, hemispheric view) have been the works of Corona (1978, 1979), Gruza and Apasova (1981; see also Angell and Gruza, 1984) and Barnett (1985b). These studies have shown that large area average precipitation data exhibit considerable medium time scale variability (order 10 years) together with a few significant long time scale trends (order 100 years). For example, Barnett (1985b) notes long-term trends of increasing precipitation over Europe/western Asia and over India, and a long-term decreasing trend over Africa. Barnett's analysis is only preliminary, however, and is based on rather limited numbers of stations with little quality control.

Barnett concludes that there is no evidence of any overall (i.e. global mean) trend and that, if any global mean trend were to result from increasing CO_2, it would almost certainly fail to reach statistical significance due to the high inter-annual and inter-decadal variability of precipitation data. The lack of any overall trend is supported by the work of Gruza and Apasova (1981), although these authors note a slight increase in Northern

Hemisphere continental precipitation in January and a slight decrease in July over the past century. These studies of hemispheric-scale precipitation trends, however, have deficiencies, the major problem being that the data used have not been critically examined for homogeneity.

6.3 CLIMATE SCENARIOS

6.3.1 Introduction

In order to assess the effects of changes in climate on Man's activities, we require detailed regionally and seasonally specific simulations of future anthropogenic climatic change. Such information may be obtained either by modelling the future climate with appropriate general circulation climate models (see Chapter 5) or by using past climate data to provide analogues for the future. Both approaches have limitations. At their present stage of development, climate models cannot produce reliable predictions of climatic change at the regional and seasonal level. Similarly, because we do not yet fully understand the causes of past changes in climate, analogues for the future based on the past cannot be considered as reliable predictions. Nevertheless, both methods do produce internally consistent representations of climatic conditions that could reasonably be expected to occur in a high-greenhouse-gas world. Such simulations are referred to as scenarios. Climate model results are discussed in detail in Chapter 5. In this section, we consider scenarios based on past climate data, either data from the distant past (paleoclimatic analogues) or data from the twentieth century (instrumentally based scenarios).

The value of these analogue scenarios is based partly on the argument that past climate patterns can be used as future climate analogues, even if the causes leading to these past patterns are unknown. The underlying assumption is that, for similar atmospheric boundary conditions (as represented by the oceans, the land surface and the cryosphere), the general circulation of the lower atmosphere responds in a similar way to different forcing mechanisms (Wigley *et al.*, 1980). This assumption can be tested using both model results and empirical data (see Section 6.3.4).

6.3.2 Potential Analogues

Pittock and Salinger (1982) have distinguished three different approaches to analogue-based scenario development. The first is to use a suitably defined ensemble of warm years from the recent instrumental record and to compare this, either with the long-term mean, or with a similarly defined cold-year ensemble (Wigley *et al.*, 1980; Williams, 1980; Namias, 1980). The second is to use regional reconstructions of paleoclimate during past warm periods. Flohn (1977) has suggested a number of possible paleo climatic analogue periods: the Medieval Warm Epoch (c. 800–1200 AD); the time of maximum early Holocene warmth (c. 6,000–9,000 BP, referred to variously as the Altithermal, the Hypsithermal or the post-glacial climatic optimum); and the last (Eemian) interglacial around 100,000 BP. A third alternative is to use atmospheric dynamical arguments together with a knowledge of empirical climate relationships and correlations to develop an educated guess (Bryson, 1974; Flohn, 1979; Pittock and Salinger, 1982). Of these possibilities, scenarios based on instrumental data have been most extensively developed.

A number of papers have been written about the early Holocene analogue, notably by Kellogg (1977, 1978), Kellogg and Schware (1981) and Butzer (1980). The early Holocene is generally thought to have been a globally warm period, but this could be open to doubt. Conditions were up to 2 °C warmer than today in many regions of the globe, but the global mean temperature cannot be reliably estimated because of incomplete data coverage. Until recently (see Webb and Wigley, 1985), there has been an insufficient amount of well-dated, reliably interpreted paleoclimatic data from this period for much confidence to be placed in derived analogues. Scenarios that have been produced have generally used the available data uncritically, particularly with regard to dating uncertainties. Furthermore, the seasonal and spatial distribution of incoming solar radiation in the early Holocene differed radically from today (see Section 6.2.3) and the boundary conditions for the atmospheric circulation were also substantially different, particularly in North America where the decaying Laurentide ice sheet did not finally disappear until around 6,000 BP. Clearly, with such large spatial differences in solar forcing and with large residual ice masses, the early Holocene is hard to justify as a suitable analogue for a warm, high-greenhouse-gas world.

The other two possible paleoclimatic warm-world analogues mentioned above (the Medieval Warm Epoch and the Eemian) have not been extensively discussed in the literature, largely because of incomplete data coverage. Williams and Wigley (1983) have shown that the former period was not a time of spatially uniform warmth even in the Northern Hemisphere. Warm conditions (relative to today) were largely confined to the North Atlantic Basin region and, even in this region, warm intervals were interspersed by shorter time scale (<100 yr) changes in climate which show little large scale spatial coherence (see Section 6.2.4).

Flohn (1980, 1981) has considered the Eemian as a warm-world analogue, and the recently published CLIMAP (1984) data may allow more detailed interpretations of this period. The main interest in this period stems from the evidence of higher sea levels that probably prevailed at this time. The period may, therefore, provide insight into the possibility of partial melting of the Antarctic ice mass, an event which has been suggested as a possible extreme future consequence of global warming (Mercer, 1978).

More distant analogues for a future warmer world have been considered by Flohn (1980, 1981) in trying to assess the consequences of complete melting of the Arctic ice. The possibility that atmospheric CO_2 concentration increases might cause the Arctic ice to disappear, at least seasonally, was first suggested by Budyko (1962, 1969). Complete disappearance is now thought to be unlikely for any anticipated future CO_2 level, but the possibility cannot be ruled out entirely because all studies of the problem have been based on over-simplified models (see Chapter 5, and also Semtner, 1984).

6.3.3 Instrumental Scenario Construction

The basic method is to select a set of warm years and a set of cold years and to produce spatially detailed composites of the differences in pressure, temperature and precipitation between the two sets. (Some workers have used the long-term mean as a baseline rather than the mean of a set of cold years.) In this way, scenarios can be produced down to the monthly time scale (although published work has considered only seasonal or annual scenarios).

There are a number of different factors which should be considered when selecting years for scenario construction, imposed by the need to simulate, as closely as possible, conditions in a warmer world (see Palutikof *et al.*, 1984). Scenarios based on composites of isolated individual warm and/or cold years, the method used by Wigley *et al.* (1980), Williams (1980), Namias (1980) and Pittock and Salinger (1982), are not compatible with the slow evolution of anthropogenic climatic change. A better way to simulate the effects of the gradual atmospheric greenhouse gas increase is to use blocks of warm and cold years rather than individual extreme years. Studies by Jäger and Kellogg (1983) and Palutikof *et al.* (1984) show that scenarios obtained by compositing data from isolated individual years do differ from those based on groups of consecutive years.

Even when groups of consecutive years are considered, an appropriate data-set must be chosen from which to select the years. Scenarios should ideally be based on differences between past warm and cold periods, where warm and cold apply to the global average surface air temperature. In the absence of global mean data, the years used to construct most published scenarios have been based on Northern Hemisphere land data, either the hemispheric or high-latitude, annual or winter mean values. The various possibilities have been discussed and compared by Palutikof *et al.* (1984).

An alternative approach to instrumental scenario construction has been employed by Budyko *et al.* (1978), Groisman (1981) and Vinnikov and Kovyneva (1983), based partly on earlier work by Drozdov (1966, 1974) and on the empirical modelling results of Vinnikov and Groisman (1979)

and Groisman (1979). The method used is to derive linear relationships between local, seasonally specific climate data and the Northern Hemisphere surface air temperature (17.5–87.5° N) record of Borzenkova et al. (1976) and Vinnikov et al. (1980). This method may be expressed as

$$C_i = aT_i + b + e_i \qquad (1)$$

where C_i is the year-i value of a chosen local climate variable (e.g. temperature or precipitation), T_i is the Northern Hemisphere temperature averaged over the 12 preceding months, a and b are regression coefficients and e_i is an error term. Having determined a and b, the change in the chosen climate variable can be estimated for any given Northern Hemisphere temperature change, and regionally and seasonally specific scenarios can be developed.

The results of this approach depend on the period over which the regression equations are developed, and Vinnikov, Groisman and colleagues' use of as long a period as possible is clearly advisable. However, for the case where C_i is local temperature, Brinkmann (1979) and Jones and Kelly (1983) have shown that the correlation coefficient between C_i and T_i (and, hence, the regression coefficient, a, in equ. (1)) varies with time. This temporal instability casts doubt on the validity of equ. (1) as a forecasting tool.

The most detailed scenarios produced to date are those of Lough et al. (1983) and Palutikof et al. (1984). These workers place greatest confidence on scenarios which exploit the early twentieth century warming. Warm and cold year composites were calculated using the warmest and coldest twenty-year periods from the Northern Hemisphere surface air temperature record of Jones et al. (1982), viz. 1934–53 and 1901–20. (Note that there was also a warming trend in the Southern Hemisphere over this period, see Figure 6.2.) Lough et al. (1983) present regional scenarios for Europe and discuss the implications of these for energy demand and for agriculture. Their most surprising result is that winter temperatures over a substantial part of Europe were colder and showed greater inter-annual variability during the warm period, probably as a result of increased blocking (Figure 6.5). Rainfall patterns showed overall decreases in spring and summer, and increases in autumn and winter.

Further details of these European scenarios are given by Palutikof et al. (1984), along with similarly derived scenarios for North America. The North American scenarios for temperature and pressure exhibit much less inter-seasonal contrast than is the case for Europe. Temperatures are shown to be generally higher and less variable throughout the year in a warm world, although there is a band of cooler conditions which runs across the continent between about 50° and 60° N in all seasons. Most of the continent south of 50° N shows considerable warming, especially in summer (Figure 6.6). Precipitation patterns are complex with substantial areas of increase and decrease (Figure 6.6).

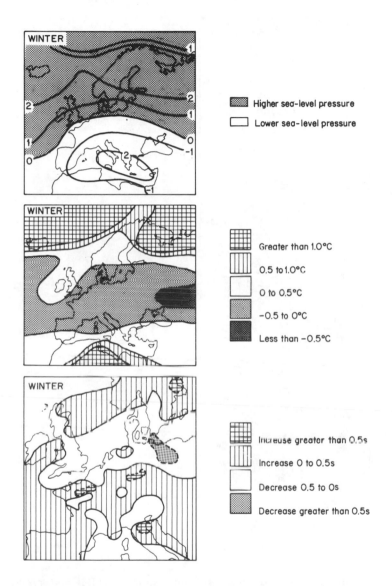

Figure 6.5 Winter pressure (upper), temperature (middle) and precipitation (lower) scenarios for Europe showing higher pressure in the north (pressure changes in mb) and a band of cooler conditions across central Europe. Precipitation changes show an irregular pattern (shown in multiples of the local standard deviation, *s*). The values shown here are differences between the 1934–53 and 1901–20 mean values and correspond to a warming of the Northern Hemisphere of about 0.5 °C. (From Palutikof *et al.*, 1984)

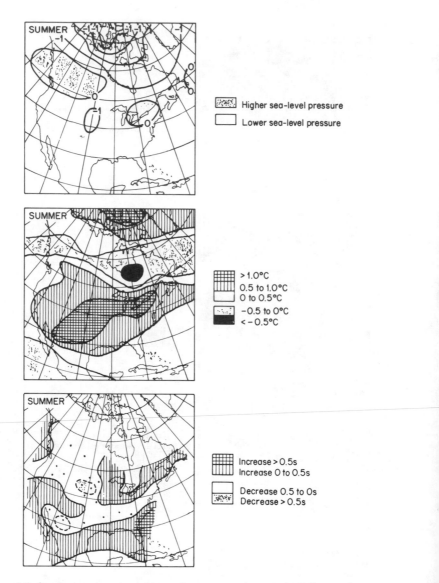

Figure 6.6 Summer pressure (upper), temperature (middle) and precipitation (lower) scenarios for North America. Pressure (changes in mb) is generally lower over the continent. The main part of the continent shows substantial warming, but a band of cooling exists between 50 and 60° N. The precipitation changes are shown in multiples of the local standard deviation (*s*). The dots on the lower diagram show the locations of precipitation stations used. The values shown here are differences between the 1934–53 and 1901–20 mean values and correspond to a warming of the Northern Hemisphere of about 0.5 °C. (From Palutikof *et al.*, 1984)

A disadvantage of the instrumental scenario method is that the scenarios are based on relatively small temperature changes compared with those expected to result from future increases in atmospheric CO_2 and other trace gas concentrations. Furthermore, a CO_2-doubling is likely to produce substantial changes in the oceanic and cryospheric boundary conditions, changes which may be well beyond the range experienced in the twentieth century. Thus, instrumental scenarios can only be taken as indicative of conditions during the early phase of a warming, changes which are expected to take place by the early decades of the twentyfirst century.

6.3.4 The Relevance of Scenarios

As noted above, instrumentally based scenarios have been justified on the grounds that, no matter what the cause of a change in climate, the spatial patterns of surface changes will be approximately the same. There is evidence both for and against this assumption. Some of this material also relates to the relevance of equilibrium GCM results as scenarios for a future changed climate.

The equilibrium modelling studies of Manabe and Wetherald (1980) and Hansen *et al.* (1984) provide the most direct supporting evidence. They applied two different forcing mechanisms to their general circulation models: an increased carbon dioxide concentration and an increased value of the solar constant. In both studies, the *latitudinal* character of the near-surface response of the model climate was similar for both types of forcing.

Similarities between equilibrium modelling results and the patterns of recent surface air temperature changes provide empirical support for instrumentally based climate scenarios. In the Northern Hemisphere, the latitudinal and seasonal patterns of the early twentieth century warming are remarkably similar to the patterns predicted by the equilibrium response general circulation model study of Manabe and Stouffer (1979, 1980) for a quadrupling of CO_2 concentration (see Wigley and Jones, 1981, their Figure 6.3). If this similarity were a result of early twentieth century CO_2 forcing then it may indicate that equilibrium models can give realistic results, at least when one considers seasonal data and zonal averages covering around $30°$ of latitude. If CO_2 change were not the dominant forcing mechanism then this similarity supports the idea that the latitudinal patterns of climatic change are largely independent of the type of forcing.

Evidence against the value of instrumentally based scenarios can be found in the observational record. Work by Jones and Kelly (1983) indicates that the regional details of a scenario may depend strongly on the time period(s) used for scenario construction. These authors compared the patterns of annual mean temperature change during the early twentieth century warming

294 *The Greenhouse Effect, Climatic Change, and Ecosystems*

with the patterns of change during two other periods, the cooling from 1940 to the mid-1960s and the warming that occurred subsequently. They found noticeable differences between these periods. These differences arise mainly from longitudinal changes in regions of warming and cooling which in turn reflect changes in meridional heat transport associated with changes in the stationary and transient eddies of the general circulation.

Contrary evidence also comes from modelling studies of the transient response of the climate system. Most model studies of CO_2 effects, including the work cited above, have considered only the steady-state response to a specified step-function increase in atmospheric CO_2 concentration. In reality, CO_2 (and trace gas) levels are changing continually and, because of spatial variations in the response time (Hansen *et al.*, 1984), the spatial patterns of the transient response to steadily increasing CO_2 may differ from those for the steady-state response to a step function CO_2 change. This was first pointed out by Schneider and Thompson (1981). Bryan *et al.* (1982, 1984) and Spelman and Manabe (1984) have considered this possibility in greater detail. They used coupled ocean-atmosphere general circulation models to show that the *latitudinal* signature of the transient response was similar to the steady-state response, provided that the rate of change of CO_2 was not too rapid. Thompson and Schneider (1982), however, in a response to the Bryan *et al.* (1982) paper, still attest that the general spatial character of the transient response (i.e. including longitudinal detail) might differ noticeably from the steady-state response.

These results are doubly important because they determine, not only the validity of instrumental climate scenarios, but also the relevance of equilibrium GCM results as warm-world scenarios. If, as transient response studies suggest, the patterns of climatic change for time-varying forcing differ from those for the equilibrium response, then the above conclusions regarding the effects of different forcing mechanisms based on equilibrium model studies may simply not be relevant to the case of a continual increase in greenhouse gas concentrations. Furthermore, equilibrium GCM experiments may, themselves, produce information which is largely irrelevant as far as the geographical detail of future anthropogenic climatic change is concerned. Of course, equilibrium model studies can provide valuable physical insights into the models themselves and hence, indirectly, into the possible future behaviour of the climate system at the regional level.

In summary then, the scenario approach (both analogue-based and equilibrium-model-based) is neither convincingly supported nor contradicted by available theoretical evidence or observational data. It is pertinent to recall, however, that scenarios are not meant to be predictions of future climate; rather they are meant to be internally consistent pictures of a plausible future climate, a basis for other workers to evaluate the possible impacts of climatic change on Man and society.

6.4 DETECTION OF CLIMATIC CHANGE

6.4.1 Background

A doubling of atmospheric CO_2 concentration is expected to raise the global annual mean surface air temperature by 1.5 to 5.5 °C (see Chapter 5). As indicated by this range of values, there is considerable uncertainty about the magnitude of the CO_2 warming effect. Detection of such a warming and other associated greenhouse-gas-induced changes of climate in the observational record has, therefore, become a high priority issue (MacCracken and Moses, 1982; Kellogg and Bojkov, 1982). Detection would not only test model predictions, but would, ideally, also allow one to place reliable, empirically determined bounds on the magnitude of the greenhouse effect.

As noted earlier, the concentrations of other radiatively active trace gases are also increasing. These increases will add to the CO_2 effect and recent estimates suggest that, over the past few decades and in the future, the trace gas contribution to global warming may be of similar magnitude to the CO_2 contribution (Lacis et al., 1981; WMO, 1982; Ramanathan et al., 1985; Wigley, 1985; see also Chapter 4). It is virtually impossible to distinguish the climatic effects of CO_2 from those of other trace gases on the basis of observational data alone. As a consequence, statements about the detection of CO_2 effects should be considered as applying to the detection of the combined effects of increases in CO_2 and trace gases.

The detection of these effects on climate requires, first, some idea of what we are seeking to detect. The simplest indicator of large scale climatic change is global mean surface air temperature. Most model experiments have estimated the *equilibrium* temperature change corresponding to a doubling of the atmospheric CO_2 concentration (viz. ΔT_{2r}). This is a good indicator of the sensitivity of the climate system to CO_2 changes, but it does not immediately tell us what changes to expect at any given time (viz. $\Delta T(t)$). To find $\Delta T(t)$, we need to study the transient response of the climate system to time-dependent forcing. To date, only very simple models have been applied to this problem. Such models can, however, give considerable insight into the 'signal' that we seek to detect, and, in particular, can give us some idea of the magnitude of the important damping effect of oceanic thermal inertia.

One of the simplest models that can be used is a box-diffusion energy balance model (Siegenthaler and Oeschger, 1984; Hansen et al., 1984; Wigley and Schlesinger, 1985). More sophisticated variants employ an advective term to account for deep water production (e.g. Hoffert et al., 1980; Harvey and Schneider, 1985a,b; Watts, 1985a,b; see also the review by Hoffert and Flannery, 1985). Here, to illustrate certain key points, we will use the box-diffusion energy balance model of Wigley and Schlesinger (1985). For this model, the main uncertainties in $\Delta T(t)$ arise through the diffusivity term, which parameterizes mixing processes below the oceanic mixed layer, and the climate sensitivity, determined by ΔT_{2x}.

In Figure 6.7, we show $\Delta T(t)$ for one diffusivity value (2 cm^2 sec^{-1}) and for $\Delta T_{2x} = 1.5$, 3.0 and 4.5 °C. We have used the ice-core-based CO$_2$ history from 1765 to 1958 recently published by Neftel *et al.* (1985, their Figure 1), the accurately measured Mauna Loa values for 1958 to 1984, and the values employed by Wigley (1985) for the period to 2050. For the trace gases (CH$_4$, N$_2$O and the various CFCs), we have used measured values for recent decades and the future projections employed by Wigley (1985); see Figure caption for further details. Ozone changes have not been considered, although these may well be important (Ramanathan *et al.*, 1985).

Figure 6.7, therefore, shows estimates of past and future changes in global mean surface air temperature. The biggest uncertainty arises through uncer-

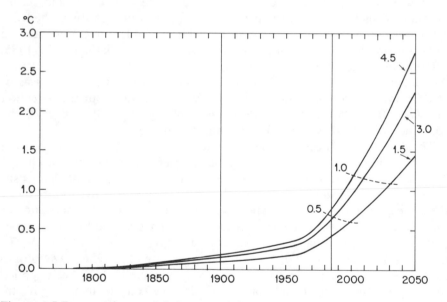

Figure 6.7 Past and future global mean surface temperature changes due to atmospheric CO$_2$ and trace gas concentration changes based on the transient response model of Wigley and Schlesinger (1985) for three different values of the equilibrium CO$_2$-doubling temperature change ($\Delta T_{2x} = 1.5$, 3.0 and 4.5 °C). CO$_2$ values from Neftel *et al.* (1985) to 1958 and Mauna Loa data to 1984. Projections to 2050 based on $C = 260.1 + 9.9$ exp(($Y - 1850$)/63), which gives concentrations of 338 ppmv for Y(year) = 1980, 367 ppmv for $Y = 2000$ and 497 ppmv for $Y = 2050$. Trace gas concentrations are measured values to 1980 and linear projections to 2050 based on estimates given by W.M.O. (1982), Hansen *et al.* (1984) and Ramanathan *et al.* (1985) (see Wigley, 1985, for details). The two transverse dotted lines show when the modelled global mean temperature change since 1900 reaches 0.5 °C and 1.0 °C

tainties in ΔT_{2x}, but a considerable degree of additional uncertainty exists because of uncertainties in oceanic mixing processes and because of the simplicity of the model. Nevertheless, Figure 6.7 is sufficiently realistic to illustrate a number of important points. First, there is a change in the modelled warming rate around 1960, largely due to the influence of trace gases which began to increase around this time. Between 1960 and 1985, CO_2 and trace gases contributed almost equally to the warming. Second, the warming at any given time is substantially less than the instantaneous equilibrium warming because of the damping effect of oceanic thermal inertia (around 50–75% depending on how the ocean is modelled). Third, the warming becomes less sensitive to ΔT_{2x} variations for higher ΔT_{2x} (note the larger separation between the 1.5 °C and 3.0 °C lines than between the 3.0 °C and 4.5 °C lines). Finally, the expected warming between 1900 (the earliest date for which a reasonable estimate of global mean surface air temperature can be made; see Section 6.2.5) and 1985 is between about 0.5 and 0.9 °C (roughly 0.2 °C of this is the trace gas contribution). A wider range of possible values is obtained if one accounts for the uncertainty in oceanic mixing effects. This particular element of uncertainty cannot be reliably quantified because current ocean models are inadequate, but some idea can be gained by varying the diffusivity in the box-diffusion model used here. Such calculations suggest that the range should be increased by $+0.2$ °C to 0.3–1.1 °C. The high end of this range corresponds to high climate sensitivity ($\Delta T_{2x} \sim 4.5$ °C) and slow ocean mixing, while the low end corresponds to low sensitivity ($\Delta T_{2x} \sim 1.5$ °C) and rapid ocean mixing.

We will discuss Figure 6.7 further in Section 6.4.4.

6.4.2 The Signal-to-noise Ratio Concept

It is useful to view the detection problem in terms of the signal-to-noise ratio concept. In broad terms, one can claim to have detected a change in climate once the signal has risen appreciably above the background noise level. The distinction between 'signal' and 'noise', however, depends on the particular application. For convenience, we will consider surface air temperature as the detection variable. The signal here is the slow warming shown schematically in Figure 6.7 that is expected to result from the gradual increases in CO_2 and other trace gases. All other aspects of temperature variability can be considered as noise. Contributions to this noise occur on all time scales, and components of the noise can be divided into climatic and non-climatic effects. We consider the latter first.

Non-climatic noise in individual site records may arise from changes in instrumentation, changes in times of measurement or methods of calculating averages, or from local anthropogenic effects such as urban warming. These noise elements, which we can refer to as *site inhomogeneity noise*, can be minimized by careful examination and inter-comparisons of site records and the subsequent production of homogeneous time series. Non-climatic noise may also arise in area averages if these averages are based on a small, or unrepresentative, or changing number of site records. This noise component can be referred to as the *spatial sampling noise*. The problem of spatial representativeness has been considered in some detail by Oort (1978) using general circulation model output data and has also been discussed by Wigley *et al.* (1985), but no systematic attempt has yet been made to estimate the magnitude of this noise component in the various data series considered in Section 6.2.5. Studies by Jones *et al.* (1985) of the representativeness of their land-based Northern Hemisphere data indicate that, for these data, this particular noise component is small, at least for the period after the late 1870s.

Climatic noise arises in a number of different ways. We will discuss noise according to time scale, but note that this is a purely artificial separation since even the shortest time scale processes may have noticeable longer time scale effects. One contribution to climatic noise comes from *weather noise*, the uncertainty in a time-averaged quantity resulting from unpredictable day-to-day weather variability. Note that, in some applications (for example, in studies of predictability on the monthly to seasonal time scale), the terms weather noise and climate noise have been used interchangeably. Here, climate noise has a more general meaning. Most studies of weather noise have been concerned with its fractional contribution to monthly or seasonal time scale variability at individual sites (e.g. Madden, 1976; Madden and Shea, 1978). Stefanick (1981) has shown theoretically that the weather noise component of a spatially and temporally averaged climate parameter reduces as the averaging period and/or the averaging area increases. Weather noise may contribute appreciably to the inter-annual variability of the climate both directly and through the modulating effect of the ocean, along the lines described by Hasselmann (1976).

Many climate data time series show considerable inter-annual variability over and above the weather noise (see, for example, the weather noise publications cited above; and Weare, 1979). This derives from poorly understood short time scale climate processes, including quasi-biennial oscillations (e.g. Trenberth and Shin, 1984) and fluctuations arising from, or linked with, sea surface temperature anomalies (the Southern Oscillation/El Niño phenomenon is an important example of the latter). Present understanding of the causes of such inter-annual climatic variability is limited, and its study is an important part of the World Climate Research Programme (see, for example, J.S.C., 1984).

The most important noise component, so far as detection of CO_2 and other trace gas effects is concerned, is that which is manifest on time scales longer than the inter-annual time scale. All temperature time series show significant variability on decadal and longer time scales (see Figures 6.1–6.3), i.e. the time scales on which the main effects of increasing CO_2 and trace gases should occur. Such fluctuations can be most troublesome in attempts to detect the CO_2 influence. Unless they can be factored out in a statistically and physically convincing way, variables used in detection studies must have records spanning many decades in order to minimize the effects of decadal time scale noise.

6.4.3 Statistical Strategies

Two statistical detection strategies may be distinguished: first, detecting a significant change in the mean value of a given climate variable or detecting a significant trend in a variable over a specified time interval; and second, the use of regression techniques to associate a statistically significant part of past variations of a variable with past variations in atmospheric CO_2 and trace gas concentrations.

The first method has been employed by Madden and Ramanathan (1980) and by Wigley and Jones (1981, 1982). From a purely statistical viewpoint, detecting a trend is closely related to detecting a change in the mean, although the appropriate t-test is slightly more efficient in the former case (in that less data are required to achieve a significant result). Since CO_2 and other trace gas increases should cause a continual increase in lower tropospheric temperature (with superimposed decadal and shorter time scale natural fluctuations), searching for a significant trend may be more appropriate than searching for a change in the mean. Work published to date, however, has used the latter approach.

It is clear from signal-to-noise ratio arguments that detection will be hastened if the noise level can be reduced. This can be achieved using some form of regression analysis with CO_2 variations and other climate forcing factors as predictors. The form of the regression equation may be guided by model results. By identifying a proportion of past climate fluctuations with factors other than increasing CO_2, and removing these effects, this method effectively reduces the noise level and increases the signal-to-noise ratio. A number of authors have used this approach in works which were not necessarily aimed at detection of CO_2 effects *per se*, but towards explaining recent changes in global climate (see Weller *et al.*, 1983, for a comprehensive summary of such work). The method, however, requires an accurate record of past variations in the various possible forcing factors (which we do not have), and is fraught with statistical difficulties which will be discussed below.

A. *Simple Signal-to-noise Ratio Studies*

All studies to date have considered only temperature data. The analyses of Madden and Ramanathan (1980) and Wigley and Jones (1981, 1982) effectively assumed that all observed variations are noise and that this noise is random with a simple first-order autoregressive structure. The observed noise level was taken to be the square root of the variance of the sampling distribution of the mean. For data with no serial correlation, this is given by $S(N) = \hat{s}/\sqrt{N}$ where $S(N)$ is the standard deviation of the mean of N observations and $\hat{s}$ is the standard deviation of the individual observations. However, since the temperature data are autocorrelated, due allowance must be made for this, with $S(N)$ appropriately inflated. Madden and Ramanathan achieved this using a frequency-domain approach, while Wigley and Jones used a time-domain approach.

The assumption that the noise has a simple autoregressive structure is an oversimplification. Although there may well be an autoregressive component (associated with the weather noise discussed earlier), it is difficult to separate this stochastic element from medium to long time scale trends that are probably deterministic in nature. Nevertheless, these analyses provide extremely useful insight into the detection problem.

Madden and Ramanathan were primarily concerned with identifying a significant signal in the observational temperature record around $60°$ N. They concluded that the signal cannot yet be detected. Wigley and Jones were concerned with choosing the right combination of season and latitude band to maximize the ratio of theoretical signal to observed noise. They found the highest ratio for summer, mid-latitude temperatures; but, given the uncertainties in both signal and noise levels, all seasons and latitude bands are similar and the best detection variable is probably mean annual temperature averaged over as large an area as possible. Madden and Ramanathan also concluded that summer data were better for detection. Wigley and Jones used data from Manabe and Stouffer's (1980) GCM experiment to define the signal. In this model, the summer signal is much less than in other seasons, yet summer still yields the highest signal-to-noise ratios because observed summer noise levels are so small. Like Madden and Ramanathan, Wigley and Jones also concluded that the CO_2 signal cannot yet be detected in the available data.

Bell (1982) has considered an interesting modification of this approach. Instead of using a simple area average, he shows that the signal-to-noise ratio can be increased by appropriate weighting, a result previously demonstrated (in a different context) by Hasselmann (1979). The weighting factors, however, depend on the spatial character of both the expected signal and the observed noise, neither of which is well defined (especially the former).

B. Noise Reduction Studies

The noise level can be reduced by relating part of the past variations to specific forcing factors, removing these effects, and considering only that which remains. This approach has been used indirectly by many authors in attempts to explain past temperature changes (Bryson and Dittberner, 1976; Miles and Gildersleeves, 1977, 1978; Hoyt, 1979a; Robock, 1979; Bryson and Goodman, 1980; Hansen *et al.*, 1981; Gilliland, 1982; Mitchell, 1983; Gilliland and Schneider, 1984). These studies range from the purely empirical through to strongly model-oriented work. Most of them conclude that the past record is consistent with theoretical estimates of the magnitude of CO_2-induced warming, but none provides statistically convincing and conclusive evidence that increasing CO_2 is the cause of the warming that has already occurred. Gilliland (1982) and Gilliland and Schneider (1984), in particular, stress the lack of statistical significance in their results. Vinnikov and Groisman (1981, 1982) have also used this approach. Their work is more directly concerned with detecting CO_2 effects and they state quite categorically that this exercise has been successful, not only in detecting CO_2-induced global warming, but even in identifying the spatial and seasonal patterns of warming. These conclusions are not statistically convincing.

To illustrate the statistical problems involved, let us consider a rather simple direct regression approach which encapsulates the main difficulties. Suppose surface air temperature in year i can be expressed in the form

$$T_i = aV_i + bS_i + cC_i + A_i + e_i \qquad (2)$$

where V, S and C are past variations in volcanic activity, solar output and atmospheric CO_2 level (none of the analyses cited above considered trace gases), A_i is an autoregressive term (i.e. $A_i = f(T_{i-1})$) and e_i is a residual error term. Fitting this equation to observed temperatures produces estimates of a, b and c with associated confidence limits. If the limits on the CO_2 coefficient, c, exclude zero, this might be taken as proof of a significant CO_2 effect. It is, however, an extremely difficult task to assign confidence limits to these regression coefficients, and standard formulae are inappropriate for two reasons.

First, the values of the forcing functions V, S and C are uncertain. (So too is the response variable, T; it must involve some spatial sampling noise.) Forcing function uncertainties are manifest in the different functions used by different authors. For example, the volcanic forcing functions used by Vinnikov and Groisman (1981, 1982), Hansen *et al.* (1981) and Gilliland (1982) are quite different. Vinnikov and Groisman used Pivovarova's (1977) actinometric data on atmospheric transmissivity (P), Hansen *et al.* used Lamb's (1970) Dust Veil Index (L), and Gilliland used the ice core acidity record of Hammer *et al.* (1980) (H). The correlations between these three volcanic forcing functions are small: $r_{PL} = -0.46$, $r_{PH} = -0.32$ and $r_{LH} = 0.41$. While these correlations are all statistically significant at the 5% level, the time series clearly have only a small amount of variance in common ($<22\%$).

The solar and CO_2 forcing functions used by these authors are also quite different. Gilliland used the 80-year cycle in solar diameter (see Parkinson *et al.*, 1980; Gilliland, 1980, 1981; and Section 5.2 in Dickinson), tuning the phase and amplitude of the implied solar irradiance cycle to best fit the observed temperature data. Hansen *et al.* (1981) used the umbral–penumbral ratio of Hoyt (1979a,b) as an indicator of solar irradiance changes (an indicator that has since been largely discredited by satellite data; see Eddy *et al.*, 1982). Vinnikov and Groisman have no solar term, although a solar effect may be implicit in their transmissivity data. Finally, the assumed CO_2 forcing functions differ noticeably from author to author and trace gases have not been considered.

It is obviously impossible for a single record of past climate to be correctly explained by three different sets of forcing function records; yet all studies claim to explain a large fraction of the temperature variance. The only way this can occur is for all results to be subject to considerable statistical uncertainty; i.e. to have wide confidence bands for the regression coefficients and for the total explained variance.

The second reason why confidence levels are difficult to assign is purely statistical. Because the response and predictor variables show medium to long (>10 yr) time scale trends, a good fit could be obtained if some of these trends happened to match by chance. For example, the umbral–penumbral ratio solar index used by Hansen *et al.* has rising and falling trends either side of (approximately) 1940 which are similar to those in the temperature record. This is sufficient to ensure a good correlation between T_i and S_i. However, since the strength of this correlation depends mainly on one factor, the similar turning point around 1940, the link is almost certainly not statistically significant. Building an arbitrary and variable time lag into any relationship, as in Gilliland's analysis, introduces the statistical problem of multiplicity (namely, that if enough experiments are performed, different phases and amplitudes in this case, a statistically significant result will eventually arise by chance). This further reduces the statistical significance of any results.

The central statistical problem in these analyses is that the number of degrees of freedom that should be used in testing significance cannot easily be determined. In addition to the reasons for this already alluded to, further reductions in the number of degrees of freedom may arise through autocorrelation in the data and from multicollinearity. The effect of autocorrelation in both the response and predictor data is difficult to account for, and Bartlett's method, for example, (see Quenouille, 1952) cannot be applied because the autocorrelation does not arise solely through an inherent (and stable) autoregressive process. Multicollinearity (i.e. inter-correlated predictor variables) is an issue because, on decadal and longer time scales, all three forcing variables may show similar low frequency fluctuations. Apart from making it impossible to assign a causal role to any particular variable in such circumstances, correlations between the predictors can produce unstable regression results, which can further undermine estimates of statistical significance.

In addition to their failure to properly address these problems, in none of the studies cited above is the transient response of the system (i.e. the effects of oceanic thermal inertia) adequately modelled. Thus, most of the noise reduction studies cited above are deficient because of their neglect of one or more of the factors mentioned here. Other less detailed evaluations and reviews of the detection problem (e.g. Kellogg and Bojkov, 1982) have been similarly deficient.

Given these uncertainties, the noise reduction method at present can only give supporting evidence in the detection problem. The possibility of removing part of the low-frequency variance by ascribing it confidently (and with *known* confidence) to other causes is an important consideration. However, because our understanding of other climate forcing mechanisms is probably at an equal or lower level than our understanding of the effects of CO_2 and other trace gases, a statistically rigorous application of this method is extremely difficult.

C. The Fingerprint Method

At a U.S. Department of Energy workshop on the first detection of CO_2 effects (Moses and MacCracken, 1982; MacCracken and Moses, 1982), it was suggested that detection would be facilitated by developing 'a unique CO_2-specific "fingerprint" for the CO_2 response involving a set of parameters, distinctive from responses that would be caused by all other known influences, and to search for this correlated pattern of changes, not just for a change in one isolated parameter' (MacCracken and Moses, 1982, p. 1172). The main purpose of the fingerprint method is to aid in the attribution aspect of detection; the association of a statistically identified change or set of changes specifically with CO_2 (or greenhouse gas) forcing. As an illustration, we might consider both tropospheric and stratospheric temperatures as a simple two-element fingerprint. Since a solar irradiance increase should warm both the troposphere and stratosphere, while a greenhouse gas increase should cause stratospheric cooling, this would be sufficient to distinguish between solar and greenhouse gas forcing, although it would not allow greenhouse gas effects to be readily isolated from, for example, the effects of changing stratospheric aerosol concentration. The statistical aspects of this simple type of fingerprint have been explored fully by Epstein (1982).

Such 'limited discrimination' markers can be expected to be useful before complete fingerprints can serve us directly. However, even limited discrimination fingerprints are beset by practical problems. Difficulties in detecting the effects of changing CO_2 and other trace gas concentrations using a single variable must also apply to both limited discrimination and complete fingerprint methods. The uncertainty inherent in both may be as great as in their most ill-defined part.

Numerous variables have been suggested as detection parameter candidates and, therefore, as candidates for part of either a discriminator or a fingerprint. Some of these are listed in Table 6.1 (see Section 6.5.2). Others that have been proposed include mean values of daily temperature range, temperatures near the seasonal snow-ice boundary, regional precipitation, and so on. Problems with using these and other suggested discriminator or fingerprint candidates arise because, in most cases, the expected signal is not well-defined and/or our knowledge of the natural variability is meagre. Very few variables are currently suitable as fingerprint components.

In order to be able to distinguish greenhouse gas effects from other possible causes of climatic change, we require a much better understanding of the natural variability of the whole climate system and its responses to forcing. In particular, since a useful discriminator or fingerprint probably requires a knowledge of the greenhouse gas signal at the regional spatial scale (because of the paucity of global-scale data), we need to know more about the spatial details of climatic change. Both discriminator and complex fingerprint methods, therefore, can be viewed at present only as goals that modellers and data analysts should strive towards. It is likely that relatively sophisticated multivariate statistical tests will be required to implement fingerprint detection strategies and, even though no useful results have yet been obtained, such techniques are currently being developed (e.g. Hasselmann, 1979; Bell, 1982; Epstein, 1982). A continuation of this effort is essential.

6.4.4 A Simple Analysis of the Recent Surface Air Temperature Record

If model results are correct, we should already have experienced a substantial global mean surface warming due to the increases in the concentrations of CO_2 and other trace gases, although this warming may well be partly or wholly obscured by natural climate fluctuations. In Section 6.2.5, we showed that the globe has warmed noticeably since the late nineteenth century, by about 0.5 °C, with quite large amplitude decadal and shorter time scale fluctuations superimposed on this overall warming trend. In Section 6.4.1, we used a simple model to show that, between 1900 and 1985, CO_2 and trace gas changes should have warmed the globe by 0.3–1.1 °C. Can we, therefore, claim to have detected the anthropogenic greenhouse effect in a statistically rigorous way? The short answer to this question is 'no'; but, before explaining why, let us first look more closely at the uncertainties in model predictions.

The predicted warming due to greenhouse gas increases since pre-industrial times depends on their initial concentrations, the size of the signal for a CO_2 doubling, and the damping effect of oceanic thermal inertia. The pre-industrial CO_2 concentration is not known with any certainty and the recently accepted range of 260–280p pmv (W.M.O., 1983; Wigley, 1983; Siegenthaler, 1984) has, yet more recently, been revised upward by some of the newest ice core data (Neftel *et al.*, 1985, see also Chapter 3). There is also considerable uncertainty in the equilibrium temperature change due to a CO_2 doubling. We consider 1.5 to 5.5 °C to be a more realistic estimate of the range of uncertainty than the Carbon Dioxide Assessment Committee's 1.5–4.5 °C range (N.R.C., 1983); see Chapter 5. Additional uncertainties accrue from uncertainties in the oceanic thermal inertia effect. It is the sum of all these uncertainties that leads to the range of $T(1900–1985)$ values, 0.3–1.1 °C, given in Section 6.4.1.

Two further comments are pertinent. This range was based on ΔT_{2x} values of 1.5–4.5 °C, so a slight upward revision of the upper limit would be required to cover the 5.5 °C upper limit for ΔT_{2x} recommended in Chapter 5. An upward revision of the whole range would be required if the Neftel *et al.* (1985) CO_2 values are too high. An overestimate by 15 ppmv would raise the $T(1900–1985)$ range by 0.1–0.2 °C. In this regard, it is worth recalling some details of Neftel *et al.*'s data. These show a smooth upward trend in CO_2 concentration from around 280 ppmv in the mid-eighteenth century, through 285 ppmv in 1850 and 297 ppmv in 1900. The 1850 value is 15 ppmv higher than the mid-point of the previously accepted 260–280 ppmv range.

If model results suggest a global warming of 0.3–1.1 °C since 1900, and if the observational data show a warming of 0.5 °C, are these results compatible? We note first that the 0.5 °C observed warming is also subject to some uncertainty due to the various problems discussed in Section 6.2.5. The post-1900 warming could be anywhere in the range 0.3–0.7 °C due to statistical uncertainties in estimating the trend, uncertainties in data quality, and gaps in coverage. It is clear that the observational and model predicted ranges overlap so they must be judged as compatible (or, at least, not inconsistent). Since observations are in the low end of the model range, this implies, either that the climate sensitivity (as quantified by ΔT_{2x}) is in the low end of the range 1.5–5.5 °C, or that ocean mixing processes are relatively rapid, or both. If the Neftel *et al.* CO_2 values are too high, then (as noted by Wigley and Schlesinger, 1985) this would reduce the implied climate sensitivity.

On the other hand, the observed warming since 1900 may well include substantial natural variations, and it may be entirely wrong to assume, as has been done above, that the 0.3–0.7 °C observed warming is largely a greenhouse gas effect. Since the mid-nineteenth century there have been marked decadal and longer time scale temperature fluctuations in both directions, warming and cooling. For example, between 1940 and 1965 the

Northern Hemisphere cooled by approximately 0.3 °C, while the Southern Hemisphere showed little overall change (see Figures 6.1 and 6.2). These changes are contrary to the predicted warming effect.

Explaining the 1940–1965 cooling is clearly important, but it is unlikely that we will ever be able to do so convincingly, simply because the appropriate data are not available. If model estimates of ΔT_{2x} are correct, then this cooling must be due either to some external factor other than the greenhouse gases, or to a major internal climatic oscillation (possibly associated with a change in ocean circulation). Changes in stratospheric volcanic aerosol loading and/or changes in solar irradiance have been suggested as causal factors by the noise reduction studies described in Section 6.4.3. However, proof of these contentions is virtually impossible because the aerosol and solar forcings are so poorly known.

The contrast between the temperature trends in the Southern and Northern Hemispheres (see Figure 6.2) is clearly an important factor that must be considered. While this does not necessarily eliminate the aerosol and solar forcing hypotheses, it does suggest that changes in some hemispherically specific factor (such as the rate of formation of North Atlantic Deep Water) could be involved. Evidence that NADW changes might have affected global climate in the past has been presented in Section 6.2.3. We also know that important changes have occurred in North Atlantic water masses in recent years (Brewer *et al.*, 1983; Swift, 1984; Roemmich and Wunsch, 1984; Bennett *et al.*, 1985), and that these changes are compatible with significant changes in NADW formation rate. Furthermore, simple modelling studies (Hoffert *et al.*, 1985; Watts, 1985a,b) and heat budget calculations (Broecker *et al.*, 1985) indicate that NADW formation rate changes are a physically realistic forcing mechanism. Further analyses are required to test this hypothesis.

To summarize, three points should be noted. First, if the post-1900 global warming is assumed to be due to changes in atmospheric CO_2 and other trace gas concentrations, then the observed data are compatible with model predictions, possibly favouring the lower half of the ΔT_{2x} range of 1.5–5.5 °C. However, uncertainties in modelling the transient response of the climate system are such that higher ΔT_{2x} values cannot be ruled out. Second, because the observational record shows large variations in global mean temperature that we cannot explain, we cannot yet claim to have unequivocally detected the signal due to increases of CO_2 and other trace gases.

A third and final point should be stressed. Because of oceanic thermal inertia effects, the present global mean temperature may well be quite far removed from equilibrium (Hansen *et al.*, 1984; Wigley and Schlesinger, 1985). Thus, even if greenhouse gas increases could be halted today, it is possible that the globe would warm substantially over coming decades as the system tended towards a new equilibrium with the prevailing greenhouse gas level. If ΔT_{2x} were in the upper half of the range 1.5–5.5 °C, this residual warming could currently exceed 0.5 °C. Its value is highly sensitive to uncertainties in ΔT_{2x} (Wigley, 1985; Wigley and Schlesinger, 1985) and to the way the oceans are modelled.

6.5 MONITORING REQUIREMENTS

For monitoring in general, there is a basic need for spatial averaging in defining variables suitable for detecting CO_2 and other trace gas effects. There are two reasons for this. First, much of the variability at a point is specific to that location and is not coherent with variations on larger spatial scales (thus, spatial averaging reduces the noise level). Second, the reliability of available model projections is generally less for smaller spatial scales (i.e. the signal is best defined for the largest spatial scales).

For a particular climate variable to be useful for detection, we require the available data to be accurate (i.e. minimum site inhomogeneity noise), representative (i.e. minimum spatial sampling noise) and of sufficient length to allow for appropriate statistical tests to be applied and to ensure that climatic noise levels are well-defined on time scales relevant to possible greenhouse gas effects. We also require the expected signal to be at least qualitatively well-defined. Potential detection variables can be selected on the basis of, and rated in terms of these criteria (see Table 6.1). For a more extensive discussion of possible detection variables see Weller *et al.* (1983).

In addition to monitoring climate variables which respond to changes of CO_2 and other trace gases, it is important to monitor the greenhouse gas changes themselves, and, where possible, other climate forcing factors. A knowledge of these is essential for future noise reduction studies. Table 6.1, therefore, also includes important climate forcing factors.

Definition of the noise level is primarily dependent on the amount of data available; i.e. on the length of the available observational record. At present, only surface temperature, precipitation and pressure have sufficiently long records to be able to establish their noise levels reliably, but continued monitoring of other variables will rapidly enhance their value in detection studies.

For confidence in the expected signal, variables such as global mean annual surface air temperature and annual mean lower stratosphere temperature have reasonably well-defined signals, but variables that depend on the regional or seasonal details of climate or on poorly understood physical links within the climate system have, at present, poorly defined signals. Further developments in climate models should improve this situation.

Table 6.1 also gives an estimate of the accuracy required for particular variables to be useful in detection studies. In some cases, this is within present capabilities, in others it is far beyond present capabilities. Accuracy levels have been assigned on the basis of the expected magnitude of the sig-

Table 6.1   Variables which could or should be monitored to detect the effects of atmospheric greenhouse gas concentration increases on climate

Variable	Reliable record length	Confidence in signal	Required monitoring accuracy	Priority
Climate variables				
Surface air temperature[1] (land)	~100 yr	high[2]	±0.1 °C	high
Surface air temperature[1] (marine)	~100 yr	high[2]	±0.1 °C	high
Sea surface temperature[1]	~100 yr	high[2]	±0.1 °C	high
Tropospheric temperature[1]	~30 yr	high[2]	±0.1 °C	high
Lower Stratospheric[1] temperature (to ~50 mb)	~30 yr	high[2]	±0.1 °C	high[3]
Mid/upper stratospheric[1] temperature	—[4]	high[2]	±0.3 °C	high[3]
Precipitation[1]	—[5]	moderate[2]	±0.2%	low
Sea-ice extent	~20–80 yr(N.H.) ~10–20 yr(S.H.)	low	±0.2°[6]	moderate[7]
Global mean sea level	~100 yr[8]	moderate	±1 cm	moderate
Ocean circulation[9,10]	—[11]	low	—[12]	high[13]
Global cloud cover	—	low	—[14]	moderate
Forcing factors				
Atmospheric CO_2 and trace gas concentrations	~100 yr[15]	—	—[16]	high
Solar irradiance	~10 yr[17]	—	±0.1%	high
Stratospheric aerosol concentration	—[18]	—	—[19]	moderate[20]
Tropospheric aerosol concentration	—[21]	—	—[19]	low

Footnotes for Table 6.1

[1] Large-scale area averages of seasonal and/or annual means.

[2] Less for smaller spatial scales and for seasonal compared with annual means.

[3] Particularly important in fingerprint studies in distinguishing greenhouse gas effects from those of other forcing factors.

[4] Present spatial coverage limited, and data have possible instrumental errors.

[5] Reliable hemispheric-scale averages not yet available, but suitable data exist.

[6] Shift in longitudinally-averaged latitude of ice boundary.

[7] Lower priority as a detection variable, but an important boundary condition for models (with snow cover and surface albedo).

[8] Subject to some uncertainty (see Barnett, 1983a,b).

[9] Both vertical and horizontal, and including circulation-change indicators such as temperature and salinity.

[10] Can be considered as a forcing factor in some instances.

[11] No spatially comprehensive data set yet available, but numerous short time series and regional studies exist.

[12] Strongly dependent on the particular ocean variable.

[13] Low priority as a detection variable, but high priority as a modulator of past and future climatic change.

[14] Difficult to estimate due to the complex ways that clouds affect climate, but, based solely on albedo change, accuracy required is $< \pm 1\%$.

[15] Data prior to 1958 less reliable.

[16] Present accuracy adequate.

[17] Homogeneous records available for only part of this period.

[18] Considerable differences exist between various data sets.

[19] Not estimated.

[20] Reduced priority due to difficulties in obtaining globally representative values.

[21] Only available for a few locations whose representativeness is uncertain.

nal between the years 1980 and 2000, on the assumption that the year 2000 is a detection threshold. In order to account for transient response effects we have based values for all climate variables on the global mean temperature change calculations illustrated in Figure 6.7. These show that $T(1980–2000)$ is between one fifth and one seventh of the equilibrium CO_2-doubling temperature change. We have used a factor of one-sixth for all variables. We have further assumed that the accuracy required to detect such a change is $\pm 20\%$ of this figure.

For the forcing factors listed in Table 6.1, accuracy levels were estimated on the basis of the change required to produce a global mean temperature change of ± 0.1 °C. Unfortunately, this is only possible for solar and greenhouse gas forcing because our understanding of the decadal and longer time scale climatic effects of volcanic or tropospheric aerosols is inadequate to make reasonable estimates. In both cases, however, monitoring requirements are almost certainly beyond our current measuring capabilities.

It should be noted that the figures given in Table 6.1 are only a guide, based on rather simple arguments. However, given the uncertainty in estimates of the various aspects of the signal, a more sophisticated method for defining accuracy figures (e.g. incorporating information about noise levels and statistical significance) does not seem warranted at this stage.

Table 6.1 also lists subjective evaluations of priorities for each detection variable. These are based on data length and signal confidence, and on required monitoring accuracy relative to what is currently possible. Priorities for the forcing parameters are based on their relative importance as forcing factors and on the current feasibility of adequate monitoring.

6.6 CONCLUSIONS

(1) Although the observed global-scale warming experienced over the past ~ 100 years is compatible with model estimates of the magnitude of the greenhouse effect, unequivocal, statistically convincing detection of the effects of changing CO_2 and trace gas levels on climate is not yet possible. An important problem in the positive identification of a greenhouse gas effect on climate is to explain the medium to long time scale ($\sim$decades or more) fluctuations in the past record. Attempts to model such changes have, to date, suffered from a number of deficiencies.

(2) Our understanding of the role of the oceans in explaining decadal and longer time scale climatic fluctuations is particularly poor. The World Ocean Climate Experiment (WOCE) will help to improve this understanding, and continued efforts are required both to improve the oceanic observational network and, through modelling studies, to improve physical insight into the way the oceans modulate the climate.

(3) Further examinations of existing surface and upper air data are required in order to improve the quality of these data, to try to extend climate records backwards in time with high reliability, and to establish more carefully the large scale spatial representativeness of limited-coverage data sets. Inconsistencies between marine-based and land-based surface air temperatures in the nineteenth century need to be resolved in order that we may be more confident of global mean temperature trends prior to 1900.

(4) In order to accurately determine large scale area-averages, more extensive data coverage is required. Existing meteorological networks are adequate for this purpose in many regions, but conspicuous gaps exist in high southern latitudes, and over the oceans in general. Any cut-back of the existing network should be strongly resisted.

(5) It should be a high priority item to make full use of satellite data to extend the spatial coverage of existing data. This will require considerable effort in establishing ground truth and in calibrating satellite observations against surface instrumental data.

(6) In addition to the changing concentrations of CO_2 and the other greenhouse gases, changes in stratospheric aerosol concentration (through changes in volcanic activity and, possibly, due to anthropogenic influences) and changes in solar irradiance are almost certainly important climate forcing factors. Improved monitoring of both of these variables is required.

ACKNOWLEDGEMENTS

Jim Hansen (Goddard Institute for Space Studies) and David Parker and Chris Folland (United Kingdom Meteorological Office) kindly provided unpublished data. A large number of persons provided important feedback on earlier versions of the manuscript. Comments from Eero Holopainen (Department of Meteorology, University of Helsinki), Syukuru Manabe (Geophysical Fluid Dynamics Laboratory), Roland Madden (National Center for Atmospheric Research) and David Parker were particularly valuable. Other useful comments were received from T. Asai, A. L. Berger, H. W. Ellsaesser, H. L. Ferguson, R. M. Gifford, G. S. Golitsyn, J. Goudriaan, J. A. Laurmann, A. S. Monin, O. Preining, G. de Q. Robin, J. W. Tukey, J. D. Woods and Du-zheng Ye. Much of the authors' work reported here was funded by the U.S. Department of Energy, Carbon Dioxide Research Division, under contract numbers DE-AC02-79EV10098 and DE-AC02-81EV10738.

6.7 REFERENCES

Andrews, J. T., and Barry, R. G. (1978). Glacial inception and disintegration during the last glaciation. *Annual Review of Earth and Planetary Sciences*, **6**, 205–228.

Angell, J. K., and Gruza, G. V. (1984). Climate variability as estimated from atmospheric observations. In Houghton, J. T. (Ed.) *The Global Climate*: 25–36. Cambridge University Press.

Angell, J. K., and Korshover, J. (1977). Estimate of the global change in temperature, surface to 100 mb, between 1958 and 1975. *Monthly Weather Review*, **105**, 375–385.

Angell, J. K., and Korshover, J. (1978a). Global temperature variation, surface—100mb: an update into 1977. *Monthly Weather Review*, **106**, 755–770.

Angell, J. K., and Korshover, J. (1978b). Estimate of global temperature variations in the 100-30mb layer between 1958 and 1977. *Monthly Weather Review*, **106**, 1422–1432.

Angell, J. K., and Korshover, J. (1983a). Global temperature variations in the troposphere and stratosphere, 1958–1982. *Monthly Weather Review*, **111**, 901–921.

Angell, J. K., and Korshover, J. (1983b). Comparison of stratospheric warmings following Agung and Chichón. *Monthly Weather Review*, **111**, 2129–2135.

Atkinson, T. C., Briffa, K. R., Coope, G. R., Joachim, M. J., and Perry, D. W. (1985). Climatic calibration of coleopteran data. In Berglund, B. D. (Ed.) *Handbook of Holocene Palaeoecology and Palaeohydrology*, 851–858. John Wiley and Sons Ltd.

Barnett, T. P. (1978). Estimating variability of surface air temperature in the Northern Hemisphere. *Monthly Weather Review*, **106**, 1353–1367.

Barnett, T. P. (1983a). Recent changes in sea level and their possible causes. *Climatic Change*, **5**, 15–38.

Barnett, T. P. (1983b). Long-term changes in dynamic height. *Journal of Geophysical Research*, **88**, 9547–9552.

Barnett, T. P. (1984). Long-term trends in surface temperature over the oceans. *Monthly Weather Review*, **112**, 303–312.

Barnett, T. P. (1985a). On long-term climate changes in observed physical properties of the oceans. In MacCracken, M. C., and Luther, F. M. (Eds). *Detecting the Climatic Effects of Increasing Carbon Dioxide*, U.S. Dept. of Energy, Carbon Dioxide Research Division, Washington D.C.

Barnett, T. P. (1985b). Long-term changes in precipitation patterns. In MacCracken, M. C., and Luther, F. M. (Eds). *Detecting the Climatic Effects of Increasing Carbon Dioxide*, 149–162. U.S. Dept. of Energy, Carbon Dioxide Research Division, Washington D.C.

Barron, E. J., and Washington, W. M. (1984). The role of geographic variables in explaining paleoclimates: Results from Cretaceous climate model sensitivity studies. *Journal of Geophysical Research*, **89**, 1267–1279.

Bell, T. L. (1982). Optimal weighting of data to detect climatic change: application to the carbon dioxide problem. *Journal of Geophysical Research*, **87**, 11161–11170.

Bennett, T., Broecker, W., and Hansen, J., Eds. (1985). *North Atlantic Deep Water Formation*, NASA Conference Publication 2367. NASA Scientific and Technical Information Branch: 66 pages.

Berger, A. (1977). Support for the astronomical theory of climatic change. *Nature*, **268**, 44–45.

Berner, R. A., Lasaga, A. C., and Garrels, R. M. (1983). The carbonate–silicate geochemical cycle and its effect on atmospheric carbon dioxide over the last 100 million years. *American Journal of Science*, **283**, 641–683.

Borzenkova, I. I., Vinnikov, K. Ya., Spirina, L. P., and Stekhnovskii, D. I. (1976). Variation of Northern Hemisphere air temperature from 1881 to 1975. *Meteorologiya i Gidrologiya 1976*, no. 7, 27–35 (in Russian).

Bradley, R. S., and Jones, P. D. (1985). Data bases for isolating the effects of the increasing carbon dioxide concentration. In MacCracken, M. C., and Luther, F. M. (Eds). *Detecting the Climatic Effects of Increasing Carbon Dioxide*, 29–53. U.S. Dept. of Energy, Carbon Dioxide Research Division, Washington D.C.

Brewer, P. G., Broecker, W. S., Jenkins, W. J., Rhines, P. B., Rooth, C. G., Swift, J. H., Takahashi, T., and Williams, R. T. (1983). A climatic freshening of the deep North Atlantic north of 50°N over the past 20 years. *Science*, **222**, 1237–1239.

Brinkmann, W. A. R. (1979). Associations between temperature trends. *Annals of the Association of American Geographers*, **69**, 250–261.

Broecker, W. S., and Takahashi, T. (1984). Is there a tie between atmospheric CO_2 content and ocean circulation? In Hansen, J. E., and Takahashi, T. (Eds.) *Climate Processes and Climate Sensitivity* (Maurice Ewing Series, No. 5): 314–326. American Geophysical Union, Washington, D.C.

Broecker, W. S., Peteet, D. M., and Rind, D. (1985). Does the ocean–atmosphere system have more than one stable mode of operation? *Nature*, **315**, 21–26.

Bryan, K., Komro, F. G., Manabe, S., and Spelman, M. J. (1982). Transient climate response to increasing atmospheric carbon dioxide. *Science*, **215**, 56–58.

Bryan, K., Komro, F. G., and Rooth, C. (1984). The ocean's transient response to global surface temperature anomalies. In Hansen, J. E., and Takahashi, T. (Eds.) *Climate Processes and Climate Sensitivity* (Maurice Ewing Series, No. 5): 29–38. American Geophysical Union, Washington, D.C.

Bryson, R. A. (1974). A perspective on climatic change. *Science*, **184**, 753–760.

Bryson, R. A., and Dittberner, G. J. (1976). A non-equilibrium model of hemispheric mean temperature. *Journal of the Atmospheric Sciences*, **33**, 2094–2106.

Bryson, R. A., and Goodman, B. M. (1980). Volcanic activity and climatic change. *Science*, **207**, 1041–1044.

Budyko, M. I. (1962). Polar ice and climate. *Izvestia Ak. Nauk. SSR*, Ser. Geog. 1962, no. 6, 3–10 (in Russian).

Budyko, M. I. (1969). The effect of solar radiation variations on the climate of the Earth. *Tellus*, **21**, 611–619.

Budyko, M. I. (1982). *The Earth's Climate: Past and Future*. Academic Press, New York: 307 pages.

Budyko, M. I., Vinnikov, K. Ya., Drozdov, O. A., and Efimova, N. A. (1978). The forthcoming climatic change. *Izvestia Ak. Nauk. SSR*, Ser. Geog. 1978, no. 6, 5–20 (in Russian).

Butzer, K. W. (1980). Adaptation to global environmental change. *Professional Geographer*, **32**, 269–278.

Callendar, G. S. (1961). Temperature fluctuations and trends over the Earth. *Quarterly Journal of the Royal Meteorological Society*, **87**, 1–12.

Chen, R. S. (1982). *Combined Land/sea Surface Air Temperature Trends, 1949–1972*. MSc dissertation, Massachusetts Institute of Technology.

CLIMAP Project Members, (1976). The surface of the ice-age Earth. *Science*, **191**, 1131–1144.

CLIMAP Project Members, (1984). The last interglacial ocean. *Quaternary Research*, **21**, 123–224.

Corona, T. J. (1978). *The Interannual Variability of Northern Hemisphere Precipitation*. Environmental Research Paper No. 16, Colorado State University, Fort Collins: 27 pages.

Corona, T. J. (1979). *Further Investigations of the Interannual Variability of Northern Hemisphere Continental Precipitation.* Environmental Research Paper No. 20, Colorado State University, Fort Collins: 20 pages.

Craig, H., and Chou, C. C. (1982). Methane: The record in polar ice cores. *Geophysical Research Letters,* **9**, 1221–1224.

Delmas, R., Ascencio, J.-M., and Legrand, M. (1980). Polar ice evidence that atmospheric CO_2 20,000 yr BP was 50% of present. *Nature,* **284**, 155–157.

Denton, G. H., and Hughes, T. Eds. (1981). *The Last Great Ice Sheets.* Wiley–Interscience. New York.

Diaz, H. F. (1981). Eigenvector analysis of seasonal temperature, precipitation and synoptic scales system frequency over the contiguous United States, Part II: spring, summer, fall and annual. *Monthly Weather Review,* **109**, 1285–1304.

Diaz, H. F., and Fulbright, D. C. (1981). Eigenvector analysis of seasonal temperature, precipitation and synoptic scales system frequency over the contiguous United States, Part I: winter. *Monthly Weather Review,* **109**, 1267–1284.

Dickinson, R. E. (1986). This volume, Chapter 5.

Dorman, C. E., and Bourke, R. H. (1979). Precipitation over the Pacific Ocean, 30° S to 60° N. *Monthly Weather Review,* **107**, 896–910.

Dorman, C. E., and Bourke, R. H. (1981). Precipitation over the Atlantic Ocean, 30° S to 70° N. *Monthly Weather Review,* **109**, 554–563.

Drozdov, O. A. (1966). On the precipitation changes in the Northern Hemisphere when the temperature of the polar basin changes. *Trudy GGO 1966,* no. 198, 3–16 (in Russian).

Drozdov, O. A. (1974). Anthropogenic impact on the hydrological cycle. *Trudy GGO 1974,* no. 316, 83–103 (in Russian).

Eddy, J. A. (1976a). The Maunder Minimum. *Science,* **192**, 1189–1202.

Eddy, J. A. (1976b). Climate and the changing Sun. *Climatic Change,* **1**, 173–190.

Eddy, J. A., Gilliland, R. L., and Hoyt, D. V. (1982). Changes in the solar constant and climatic effects. *Nature,* **300**, 689–693.

Elliott, W. P., and Reed, R. K. (1984). A climatological estimate of precipitation for the world ocean. *Journal of Climate and Applied Meteorology,* **23**, 434–439.

Ellsaesser, H. W., MacCracken, M. C., and Walton, J. J. (1985). Global climatic trends as revealed by the recorded data. *Reviews of Geophysics and Space Physics* (in press).

Epstein, E. S. (1982). Detecting climate change. *Journal of Applied Meteorology,* **21**, 1172–1182.

Fleer, H. (1981). Large-scale tropical rainfall anomalies. *Bonner Meteorologische Abhandlungen* No. 26: 114 pages.

Fletcher, J. O. (1984). Clues from sea-surface records. *Nature,* **310**, p. 630.

Flohn, H. (1977). Climate and energy: a scenario to a 21st century problem. *Climatic Change,* **1**, 5–20.

Flohn, H. (1979). A scenario of possible future climates—natural and man-made. *Proceedings of the World Climate Conference Geneva,* WMO-No. 537, 243–266.

Flohn, H. (1980). *Possible Climatic Consequences of a Man-Made Global Warming,* IIASA publication RR-80-30. International Institute for Applied Systems Analysis, Laxenburg, Austria.

Flohn, H. (1981). *Major Climatic Events Associated with a Prolonged CO_2-Induced Warming,* ORAU/IEA-81-8(M). Institute for Energy Analysis, Oak Ridge Associated Universities, Oak Ridge, 80 pages.

Folland, C. K., and Kates, F. E. (1984). Changes in decadally averaged sea surface temperature over the world, 1861-1980. In Berger, A., Imbrie, J., Hayes, J., Kukla,

G., and Saltzman, B. (Eds.) *Milankovitch and Climate*, Part 2, 721–727. D. Reidel, Dordrecht.

Folland, C. K., Parker, D. E., and Kates, F. E. (1984a). Worldwide marine temperature fluctuations, 1856–1981. *Nature*, **310**, 670–673.

Folland, C. K., Parker, D. E., and Newman, M. (1984b). Worldwide marine temperature variations on the season-to-century time scale. *Proceedings of the Ninth Annual Climate Diagnostics Workshop*, U.S. Dept. of Commerce, NOAA, 70–85.

Gates, W. L. (1976a). Modeling the ice-age climate. *Science*, **191**, 1131–1144.

Gates, W. L. (1976b). The numerical simulation of ice-age climate with a global general circulation model. *Journal of the Atmospheric Sciences*, **33**, 1844–1873.

Gilliland, R. L. (1980). Solar luminosity variations. *Nature*, **286**, 838.

Gilliland, R. L. (1981). Solar radius variations over the past 265 years. *Astrophysical Journal*, **248**, 1144–1155.

Gilliland, R. L. (1982). Solar, volcanic and CO_2 forcing of recent climatic change. *Climatic Change*, **4**, 111–131.

Gilliland, R. L., and Schneider, S. H. (1984). Volcanic, CO_2 and solar forcing of Northern and Southern Hemisphere surface air temperatures. *Nature*, **310**, 38–41.

Groisman, P. Ya. (1979). Algorithm of estimation of linear structural relation between macroclimatic parameters. *Trudy GGI 1979*, no. 257, 76–80 (in Russian).

Groisman, P. Ya. (1981). The empirical estimates of the relationship between the processes of global warming and cooling and the moisture regime over the territory of the USSR. *fIzvestia Ak. Nauk. SSR*, Ser. Geog. 1981, no. 5, 86–95 (in Russian).

Gruza, G. V., and Apasova, Ye. G. (1981). Climatic variability of the Northern Hemisphere precipitation amounts. *Meteorologiya i Gidrologiya, 1981*, No. 5, 5–16 (in Russian).

Hammer, C. U., Clausen, H. B., and Dansgaard, W. (1980). Greenland icesheet evidence of post-glacial volcanism and its climatic impact. *Nature*, **288**, 230–235.

Hansen, J. E., Johnson, D., Lacis, A., Lebedeff, S., Lee, P., Rind, D., and Russell, G. (1981). Climatic impact of increasing atmospheric carbon dioxide. *Science*, **213**, 957–966.

Hansen, J. E., Lacis, A., Rind, D., Russell, G., Stone, P., Fung, I., Ruedy, R., and Lerner, J. (1984). Climate sensitivity: analysis of feedback mechanisms. In Hansen, J. E., and Takahashi, T. (Eds.) *Climate Processes and Climate Sensitivity* (Maurice Ewing Series, No. 5): 130–163. American Geophysical Union, Washington, D.C.

Harvey, L. D. D., and Schneider, S. H. (1985a). Transient climate response to external forcing on 10^0–10^4 year time scales. 1: Experiments with globally averaged, coupled atmosphere and ocean models. *Journal of Geophysical Research*, **90**, 2191–2206.

Harvey, L. D. D., and Schneider, S. H. (1985b). Transient climate response to external forcing on 10^0–10^4 year time scales. 2: Sensitivity experiments with a seasonal, hemispherically averaged, coupled atmosphere, land and ocean energy balance model. *Journal of Geophysical Research*, **90**, 2207–2222.

Hasselmann, K. (1976). Stochastic climate models, part 1, theory. *Tellus*, **28**, 473–485.

Hasselmann, K. (1979). On the signal-to-noise problem in atmospheric response studies. In Shaw, D. B. (Ed.) *Meteorology over the Tropical Oceans*: 251–259. Royal Meteorological Society.

Hays, J., Imbrie, J., and Shackleton, N. (1976). Variations in the Earth's orbit: pacemaker of the ice ages. *Science*, **194**, 1121–1132.

Hoffert, M. I., Callegari, A. J., and Hsieh, C. T. (1980). The role of deep sea heat storage in the secular response to climatic forcing. *Journal of Geophysical Research,* **85,** 6667–6679.

Hoffert, M. I., Gaffin, S., Wang, Z. Y., Hsieh, C. T., and Volk, T. (1985). Interannual climate oscillations in past and future temperature records. Extended Summaries, *Third Conference on Climate Variations and Symposium on Contemporary Climate 1850–2100,* American Meteorological Society, 118–119.

Hoffert, M. I., and Flannery, B. P. (1985). Model projections of time-dependent response to increasing carbon dioxide. In *U.S. Dept. of Energy State of the Art Report on the Climatic Effects of Increasing CO*$_2$ (in press).

Hollin, J. T. (1980). Climate and sea level in isotope stage 5: an East Antarctic ice surge at ~95,000 BP? *Nature,* **283,** 629–633.

Hoyt, D. V. (1979a). An empirical determination of the heating of the Earth by the carbon dioxide greenhouse effect. *Nature,* **282,** 388–390.

Hoyt, D. V. (1979b). Variations in sunspot structure and climate. *Climatic Change,* **2,** 79–92.

Imbrie, J., and Imbrie, J. Z. (1980). Modeling the climatic response to orbital variations. *Science,* **207,** 943–953.

J. S. C. (1984). *Scientific Plan for the World Climate Research Programme.* WCRP Publ. Series No. 2, WMO/TD-No. 6.

Jäger, J., and Kellogg, W. W. (1983). Anomalies in temperature and rainfall during warm Arctic seasons. *Climatic Change,* **5,** 39–60.

Jones, P. D., and Kelly, P. M. (1983). The spatial and temporal characteristics of Northern Hemisphere surface air temperature variations. *Journal of Climatology,* **3,** 243–252.

Jones, P. D., Raper, S. C. B., Bradley, R. S., Diaz, H. F., Kelly, P. M., and Wigley, T. M. L. (1986a). Northern Hemisphere surface air temperature variations 1851–1984. *Journal of Climate and Applied Meteorology,* **25,** 161–179.

Jones, P. D., Raper, S. C. B., and Wigley, T. M. L. (1986b). Southern Hemisphere surface air temperature variations 1851–1984. *Journal of Climate and Applied Meteorology* (in press).

Jones, P. D., Wigley, T. M. L., and Kelly, P. M. (1982). Variations in surface air temperature: Part 1, Northern Hemisphere 1881–1980. *Monthly Weather Review,* **110,** 59–70.

Jones, P. D., Wigley, T. M. L., and Wright, P. B. (1986c). Global temperature variations. *Nature* (in press).

Kellogg, W. W. (1977). Effects of human activities on climate. *WMO Technical Note 156,* WMO No. 486, World Meteorological Organization, Geneva.

Kellogg, W. W. (1978). Global influences of mankind on the climate. In Gribbin, J. (Ed.) *Climatic Change:* 205–227. Cambridge University Press.

Kellogg, W. W., and Bojkov, R. D. Eds. (1982). *Report of JSC/CAS Meeting of Experts on Detection of Possible Climate Change.* WCP-29, World Meteorological Organization, Geneva.

Kellogg, W. W., and Schware, R. (1981). *Climate Change and Society. Consequences of Increasing Atmospheric Carbon Dioxide.* Westview Press, Boulder, Colorado: 178 pages.

Kelly, P. M., Jones, P. D., Sear, C. B., Cherry, B. S. G., and Tavakol, R. K. (1982). Variations in surface air temperatures: Part 2. Arctic regions 1881–1890. *Monthly Weather Review,* **110,** 71–83.

Kerr, R. A. (1984). Carbon dioxide and the control of the ice ages. *Science,* **223,** 1053–1054.

Knox, F., and McElroy, M. B. (1984). Changes in atmospheric CO_2: influence of the marine biota at high latitude. *Journal of Geophysical Research*, **89**, 4629–4637.

Köppen, W. (1873). Über Mehrjährige Perioden der Witterung, insbesondere Über die 11-jährige Periode der Temperatur. (Concerning multi-year climatic periods, in particular 11-year periods of temperature.) *Zeitschrift der Österreichischen Gesellschaft für Meteorologie*, **8**, 241–248; 257–267.

Köppen, W. (1914). Lufttemperaturen, Sonnenflecken und Vulkanausbrüche. (Air temperatures, sunspots, and volcanic eruptions.) *Meteorologische Zeitschrift*, **31**, 305–328.

Kukla, G., Berger, A., Lotti, R., and Brown, J. (1981). Orbital signature of inter-glacials. *Nature*, **290**, 295–300.

Kutzbach, J. E. (1981). Monsoon climate of the early Holocene: climatic experiment using the earth's orbital parameters for 9000 years ago. *Science*, **214**, 59–61.

Kutzbach, J. E., and Otto-Bliesner, B. L. (1982). The sensitivity of the African–Asian monsoonal climate to orbital parameter changes for 9000 years B.P. in a low-resolution general circulation model. *Journal of Atmospheric Science*, **39**, 1177–1188.

Kutzbach, J. E., and Guetter, P. J. (1984). Sensitivity of late-glacial and Holocene climates to the combined effects of orbital parameter changes and lower boundary condition changes: 'snapshot' simulations with a general circulation model for 18000, 9000 and 6000 years ago. *Annals of Glaciology*, **5**, 85–87.

Kutzbach, J. E., and Street-Perrott, F. A. (1985). Milankovitch forcing of fluctuations in the level of tropical lakes from 18 to 0 kyr B.P. *Nature*, **317**, 130–134.

Lacis, A., Hansen, J., Lee, P., Mitchell, T., and Lebedeff, S. (1981). Greenhouse effect of trace gases, 1970–1980. *Geophysical Research Letters*, **8**, 1035–1038.

Lamb, H. H. (1970). Volcanic dust in the atmosphere; with a chronology and assess-ment of its meteorological significance. *Philosophical Transactions, Royal Society, London*, Series A, 266, 425–533. (See also the updates given by Lamb in *Climate Monitor*, **6**, 57–67, 1977 and *Climate Monitor*, **12**, 79–90, 1983.)

Lamb, H. H. (1977). *Climate: Present, Past and Future.* Vol. 2. Methuen.

Leith, C. E. (1984). Global climate research. In Houghton, J. T. (Ed.), *The Global Climate*: 13–24. Cambridge University Press.

Lloyd, C. R. (1984). Pre-Pleistocene paleoclimates: the geological and paleontolog-ical evidence; modeling strategies, boundary conditions, and some preliminary results. In Saltzman, B. (Ed.), *Advances in Geophysics* (Vol. 26): 35–140. Aca-demic Press.

Lough, J. M., Wigley, T. M. L., and Palutikof, J. P. (1983). Climate and climate impact scenarios for Europe in a warmer world. *Journal of Climate and Applied Meteorology*, **22**, 1673–1684.

MacCracken, M. C., and Moses, H. (1982). The first detection of carbon dioxide effects: workshop summary, 8–10 June 1981, Harpers Ferry, W. Va. *Bulletin of the American Meteorological Society*, **63**, 1164–1178.

Madden, R. A. (1976). Estimates of the natural variability of time averaged sea-level pressure. *Monthly Weather Review*, **104**, 942–952.

Madden, R. A., and Ramanathan, V. (1980). Detecting climate change due to in-creasing carbon dioxide. *Science*, **209**, 763–768.

Madden, R. A., and Shea, D. J. (1978). Estimates of the natural variability of time av-eraged temperatures over the United States. *Monthly Weather Review*, **106**, 1695–1703.

Manabe, S., and Broccoli, A. J. (1984). Ice-Age climate and continental ice sheets:

some experiments with a general circulation model. *Annals of Glaciology*, **5**, 100–105.

Manabe, S., and Stouffer, R. J. (1979). A CO_2-climate sensitivity study with a mathematical model of the global climate. *Nature*, **282**, 491–493.

Manabe, S., and Stouffer, R. J. (1980). Sensitivity of a global climate model to an increase of CO_2 concentration in the atmosphere. *Journal of Geophysical Research*, **85**, 5529–5554.

Manabe, S., and Wetherald, R. T. (1980). On the distribution of climate change resulting from an increase in CO_2 content of the atmosphere. *Journal of the Atmospheric Sciences*, **37**, 99–118.

Manabe, S., Wetherald, R. T., and Stouffer, R. J. (1981). Summer dryness due to an increase of atmospheric CO_2 concentration. *Climatic Change*, **3**, 347–386.

Mercer, J. H. (1969). The Allerod oscillation: A European climatic anomaly? *Arctic and Alpine Research*, **1**, 227–234.

Mercer, J. H. (1978). West Antarctic ice sheet and CO_2 greenhouse effect: a threat of disaster. *Nature*, **271**, 321–325.

Milankovitch, M. M. (1941). *Canon of Insolation and the Ice Age Problem*. Königlich Serbische Akademie, Beograd. English translation by the Israel Program for Scientific Translations, published for the United States Department of Commerce and the National Science Foundation, Washington, D.C.

Miles, M. K., and Gildersleeves, P. B. (1977). A statistical study of the likely causative factors in the climatic fluctuations of the last 100 years. *Meteorological Magazine*, **106**, 314–322.

Miles, M. K., and Gildersleeves, P. B. (1978). Volcanic dust and changes in Northern Hemisphere temperature. *Nature*, **271**, 735–736.

Mitchell, J. M. Jr. (1953). On the causes of instrumentally observed temperature trends. *Journal of Meteorology*, **10**, 244–261.

Mitchell, J. M. Jr. (1961). Recent secular changes of global temperature. *Annals of the New York Academy of Science*, **95**, 235–250.

Mitchell, J. M. Jr. (1963). On the world-wide patterns of secular temperature change. *Changes of Climate, Arid Zone Research XX*, UNESCO, Paris, 161–181.

Mitchell, J. M. Jr. (1983). Empirical modeling of effects of solar variability, volcanic events and carbon dioxide on global scale average temperature since AD 1880. *Weather and Climate Responses to Solar Variations*. McCormac, B. M. (Ed.): Colorado Associated University Press, 265–273.

Moses, H., and MacCracken, M. C., Coordinators (1982). Proceedings of the Workshop on First Detection of Carbon Dioxide Effects. *DOE Report CONF-8106214*, U.S. Dept. of Energy Carbon Dioxide Research and Assessment Program, Washington D.C.: 546 pages.

Mosley-Thompson, E., and Thompson, L. G. (1982). Nine centuries of microparticle deposition at the South Pole. *Quaternary Research*, **17**, 1–13.

Namias, J. (1980). Some concomitant regional anomalies associated with hemispherically averaged temperature variations. *Journal of Geophysical Research*, **85**, 1580–1590.

Neftel, A., Oeschger, H., Schwander, J., Stauffer, B., and Zumbrunn, R. (1982). Ice core sample measurements give atmospheric CO_2 content during the past 40,000 years. *Nature*, **295**, 220–223.

Neftel, A., Moor, E., Oeschger, H., and Stauffer, B. (1985). Evidence from polar ice cores for the increase in atmospheric CO_2 in the past two centuries. *Nature*, **315**, 45–47.

Nicholson, S. E. (1980). The nature of rainfall fluctuations in sub-tropical West Africa. *Monthly Weather Review*, **108**, 473–487.

Nicholson, S. E. (1981). Rainfall and atmospheric circulation during drought periods and wetter years in West Africa. *Monthly Weather Review*, **109**, 2191–2208.

North, G. R., Mengel, J. G., and Short, D. A. (1983). Simple energy balance model resolving the seasons and the continents: application to the astronomical theory of the ice ages. *Journal of Geophysical Research*, **88**, 6576–6586.

N.R.C. (Carbon Dioxide Assessment Committee, Board on Atmospheric Sciences and Climate, National Reearch Council) (1983). *Changing Climate*. National Academy Press, Washington D.C.: 496 pages.

Oeschger, H., Beer, J., Siegenthaler, U., and Stauffer, B. (1984). Late glacial climate history from ice cores. In Hansen, J. E., and Takahashi, T. (Eds.) *Climate Processes and Climate Sensitivity* (Maurice Ewing Series, No. 5): 299–306. American Geophysical Union, Washington, D.C.

Oort, A. H. (1978). Adequacy of the rawinsonde network for global circulation studies tested through numerical model output. *Monthly Weather Review*, **106**, 174–195.

Owen, T., Cess, R. D., and Ramanathan, V. (1979). Enhanced CO_2 greenhouse to compensate for reduced solar luminosity on early Earth. *Nature*, **277**, 640–642.

Paltridge, G., and Woodruff, S. (1981). Changes in global surface temperature from 1880 to 1977 derived from historical records of sea surface temperature. *Monthly Weather Review*, **109**, 2427–2434.

Palutikof, J. P., Wigley, T. M. L., and Lough, J. M. (1984). *Seasonal Scenarios for Europe and North America in a High-CO_2, Warmer World.* U.S. Dept. of Energy, Carbon Dioxide Research Division, Technical Report TR012: 70 pages.

Parker, D. E. (1980). Climatic change or analysts' artifice—a study of grid-point upper-air data. *Meteorological Magazine*, **109**, 129–152.

Parker, D. E. (1981). The effects of inadequate sampling and of circulation pattern on real and apparent zonal mean temperature. *Meteorological Magazine*, **110**, 200–204.

Parker, D. E. (1985). The influence of the Southern Oscillation and volcanic eruptions on temperature in the tropical troposphere. *Journal of Climatology*, **5**, 273–282.

Parkinson, J. H., Morrison, L. V., and Stephenson, F. R. (1980). The constancy of the solar diameter over the past 250 years. *Nature*, **288**, 548–551.

Peterson, G. M., Webb, T., Kutzbach, J. E., van der Hammen, T., Wijmstra, T. A., and Street, F. A. (1979). The continental record of environmental conditions at 18,000 yr BP: an initial evaluation. *Quaternary Research*, **12**, 47–82.

Pisias, N. G., and Shackleton, N. J. (1984). Modelling the global climate response to orbital forcing and atmospheric carbon dioxide change. *Nature*, **310**, 757–759.

Pittock, A. B., and Salinger, J. M. (1982). Towards regional scenarios for a CO_2-warmed Earth. *Climatic Change*, **4**, 23–40.

Pivovarova, Z. I. (1977). *Radiation characteristics of Climate in the USSR*. Leningrad, Gidrometeoizdat, (in Russian).

Quenouille, M. H. (1952). *Associated Measurements*. Butterworth Scientific: 242 pages.

Ramanathan, V., Cicerone, R. J., Singh, H. B., and Kiehl, J. T. (1985). Trace gas trends and their potential role in climate change. *Journal of Geophysical Research*, **90**, 5547–5566.

Raper, S. C. B., Wigley, T. M. L., Jones, P. D., Kelly, P. M., Mayes, P. R., and Limbert, D. W. S. (1983). Recent temperature changes in the Arctic and Antarctic. *Nature,* **306,** 458–459.

Raper, S. C. B., Wigley, T. M. L., Mayes, P. R., Jones, P. D., and Salinger, M. J. (1984). Variations in surface air temperatures: Part 3. The Antarctic, 1957–1982. *Monthly Weather Review,* **112,** 1341–1353.

Raynaud, D., and Barnola, J. M. (1985). An Antarctic ice core reveals atmospheric CO_2 variations over the past few centuries. *Nature,* **315,** 309–311.

Reed, R. K. (1979). On the relationship between the amount and frequency of precipitation over the ocean. *Journal of Applied Meteorology,* **18,** 692–696.

Reed, R. K., and Elliott, W. P. (1977). The comparison of oceanic precipitation as measured by gauge and assessed from weather reports. *Journal of Applied Meteorology,* **16,** 983–986.

Reed, R. K., and Elliott, W. P. (1979). New precipitation maps for the North Atlantic and North Pacific Oceans. *Journal of Geophysical Research,* **84,** 7839–7846.

Rind, D., and Lebedeff, S. (1984). *Potential Climatic Impacts of Increasing Atmospheric CO_2 with Emphasis on Water Availability and Hydrology in the United States.* U.S. Environmental Protection Agency Report.

Robin, G. de Q. (1985). This volume, Chapter 7.

Robock, A. (1979). The 'Little Ice Age': Northern Hemisphere average observations and model calculations. *Science,* **206,** 1402–1404.

Rodda, J. C. (1969). Rainfall measurement problems. *IASH, Proceedings of the Bern Association,* **78,** 215–231.

Roemmich, D., and Wunsch, C. (1984). Apparent changes in the climatic state of the deep North Atlantic Ocean. *Nature,* **307,** 447–450.

Roxburgh, I. W. (1980). Long term variations in the solar constant. In *Sun and Climate, CNES/SNRS/DGRST Conference Proceedings:* 261–268.

Ruddiman, W. F., and McIntyre, A. (1981). North Atlantic ocean during the last deglaciation. *Palaeogeography, Palaeoclimatology, Palaeoecology,* **35,** 145–214.

Sarmiento, J. L., and Toggweiler, J. R. (1984). A new model for the role of the oceans in determining atmospheric PCO_2. *Nature,* **308,** 621–624.

Sarnthein, M. (1978). Sand deserts during glacial maximum and climatic optimum. *Nature,* **272,** 43–45.

Schneider, S. H., and Londer, R. (1984). *The Coevolution of Climate and Life.* Sierra Club Books, San Francisco: 563 pages.

Schneider, S. H., and Thompson, S. L. (1981). Atmospheric CO_2 and climate: importance of the transient response. *Journal of Geophysical Research,* **86,** 3135–3147.

Semtner, A. J. (1984). On modeling the seasonal thermodynamic cycle of sea ice in studies of climatic change. *Climatic Change,* **6,** 127–137.

Shackleton, N. J., Hall, M. A., Line, J., and Can Shuxi, (1983). Carbon isotope data in core V19-30 confirms reduced carbon dioxide of the ice-age atmosphere. *Nature,* **306,** 319–322.

Shackleton, N. J., Backman, J., Zimmerman, H., Kent, D.V., Hall, M. A., Roberts, D. G., Schnitker, D., Baldauf, J. G., Desprairies, A., Homrighausen, R., Huddlestun, P., Keene, J. B., Kaltenback, A. J., Krumsiek, K. A. O., Morton, A. C., Murray, J. W., and Westberg-Smith, J. (1984). Oxygen isotope calibration of the onset of ice-rafting and history of glaciation in the North Atlantic region. *Nature,* **307,** 620–623.

Shackleton, N. J., and Opdyke, N. D. (1973). Oxygen isotope and paleomagnetic stratigraphy of equatorial Pacific core V28-238: Oxygen isotope temperatures and ice volumes on a 10^5 year and 10^6 year scale. *Quaternary Research,* **3,** 39–55.

Siegenthaler, U. (1984). 19th century measurements of atmospheric CO_2—a comment. *Climatic Change*, **6**, 409–411.

Siegenthaler, U., Eicher, U., and Oeschger, H. (1984). Lake sediments as continental $\delta^{18}O$ records from the glacial/post-glacial transition. *Annals of Glaciology*, **5**, 149–152.

Siegenthaler, U., and Oeschger, H. (1984). Transient temperature changes due to increasing CO_2 using simple models. *Annals of Glaciology*, **5**, 153–159.

Siegenthaler, U., and Wenk, Th. (1984). Rapid atmospheric CO_2 variations and ocean circulation. *Nature*, **308**, 624–626.

Spelman, M. J., and Manabe, S. (1984). Influence of oceanic heat transport upon the sensitivity of a model climate. *Journal of Geophysical Research*, **89**, 571–586.

Stauffer, B., Hofer, H., Oeschger, H., Schwander, J., and Siegenthaler, U. (1984). Antarctic CO_2 concentration during the last glaciation. *Annals of Glaciology*, **5**, 160–164.

Stefanick, M. (1981). Space and time scales of atmospheric variability. *Journal of the Atmospheric Sciences*, **38**, 988–1002.

Stone, P. H. (1984). Feedbacks between dynamical heat fluxes and temperature structure in the atmosphere. In Hansen, J. E., and Takahashi, T. (Eds.) *Climate Processes and Climate Sensitivity* (Maurice Ewing Series, No. 5): 6–17. American Geophysical Union, Washington, D.C.

Street, F. A., and Grove, A. T. (1979). Global maps of lake-level fluctuations since 30,000 yr BP. *Quaternary Research*, **12**, 83–118.

Swift, J. H. (1984). A recent Θ-S shift in the deep water of the northern North Atlantic. In Hansen, J. E., and Takahashi, T. (Eds.), *Climate Processes and Climate Sensitivity* (Maurice Ewing Series, No. 5): 39–47. American Geophysical Union, Washington, D.C.

Tabony, R. C. (1981). A principal component and spectral analysis of European rainfall. *Journal of Climatology*, **1**, 283–294.

Trenberth, K. E., and Shin, W.-T. K. (1984). Quasi-biennial fluctuations in sea level pressures over the Northern Hemisphere. *Monthly Weather Review*, **112**, 761–777.

Thompson, S. L., and Schneider, S. H. (1982). Carbon dioxide and climate: the important of realistic geography in estimating the transient temperature response. *Science*, **217**, 1031–1033.

Vinnikov, K. Ya., and Groisman, P. Ya. (1979). Empirical model of modern climatic change. *Meteorologiya i Gidrologiya 1979*, no. 3, 26–36 (in Russian).

Vinnikov, K. Ya., and Groisman, P. Ya. (1981). The empirical analysis of CO_2 influence on the modern changes of the mean annual Northern Hemisphere surface air temperature (in Russian). *Meteorologiya i Gidrologiya 1981*, no. 11, 30–43.

Vinnikov, K. Ya., and Groisman, P. Ya. (1982). The empirical study of climate sensitivity. *Isvestiya AS USSR, Atmospheric and Oceanic Physics*, **18(11)**, 1159–1169 (in Russian).

Vinnikov, K. Ya., Gruza, G. V., Zakharov, V. F., Kirillov, A. A., Kovyneva, N. P., and Ran'kova, E. Ya. (1980). Current climatic changes in the Northern Hemisphere. *Meteorologiya i Gidrologiya 1980*, no. 6, 5–17 (in Russian).

Vinnikov, K. Ya., and Kovyneva, N. P. (1983). Global warming: the distribution of climatic change. *Meteorologiya i Gidrologiya 1983*, no. 5, 10–19 (in Russian).

Wallén C. C. (1984). Present century climate fluctuations in the Northern Hemisphere and examples of their impact. *WCP-87*, WMO/TD-No. 9, World Meteorological Organization, Geneva.

Watts, R. G. (1985a). Climatic transients caused by variations in the thermohaline circulation. Extended Summaries, *Third Conference on Climate Variations and Symposium on Contemporary Climate 1850-2100*, American Meteorological Society, 83–84.

Watts, R. G. (1985b). Global climate variation due to fluctuations in the rate of deep water formation. *Journal of Geophysical Research*, **90**, 8067–8070.

Weare. B. C. (1979). Temperature statistics of short-term climatic change. *Monthly Weather Review*, **107**, 172–180.

Webb, T., III and Wigley, T. M. L. (1985). What past climates can tell us about a warmer world. In MacCracken, M. C., and Luther, F. M. (Eds). *The Potential Climatic Effects of Increasing Carbon Dioxide*, 237–257. U.S. Dept. of Energy, Carbon Dioxide Research Division, Washington D.C.

Weller, G., Baker, J. D., Gates, W. L., MacCracken, M. C., Manabe, S., and von der Haar, T. (1983). Detecting and monitoring of CO_2-induced climate changes. In *Changing Climate*. Report of the Carbon Dioxide Assessment Committee, National Research Council: 292–382. National Academy Press, Washington D.C.

Wigley, T. M. L. (1983). The pre-industrial carbon dioxide level. *Climatic Change*, **5**, 315–320.

Wigley, T. M. L. (1985). Carbon dioxide, trace gases and global warming. *Climate Monitor*, **13**, 133–148.

Wigley, T. M. L., Angell, J. K., and Jones, P. D. (1985). Analysis of the temperature record. In MacCracken, M. C., and Luther, F. M. (Eds). *Detecting the Climatic Effects of Increasing Carbon Dioxide*, 55–90. U.S. Dept. of Energy, Carbon Dioxide Research Division, Washington D.C.

Wigley, T. M. L., and Brimblecombe, P. (1981). Carbon dioxide, ammonia and the origin of life. *Nature*, **291**, 213–215.

Wigley, T. M. L., and Jones, P. D. (1981). Detecting CO_2-induced climatic change. *Nature*, **292**, 205–208.

Wigley, T. M. L., and Jones, P. D. (1982). Signal-to-noise ratios for surface air temperature and the detection of CO_2-induced climatic change. In Moses, H., and MacCracken, M. C. (Eds.), *Proceedings of the Workshop on First Detection of Carbon Dioxide Effects*. DOE. Report CONF-8106214, U.S. Dept. of Energy Carbon Dioxide Research and Assessment Program, Washington D.C.: 143–158.

Wigley, T. M. L., Jones, P. D., and Kelly, P. M. (1980). Scenarios for a warm, high-CO_2 world. *Nature*, **283**, 17–21.

Wigley, T. M. L., and Schlesinger, M. E. (1985). Analytical solution for the effect of increasing CO_2 on global mean temperature. *Nature*, **315**, 649–652.

Williams, J. (1980). Anomalies in temperature and rainfall during warm Arctic seasons as a guide to the formulation of climate scenarios. *Climatic Change*, **2**, 249–266.

Williams, L. D., and Wigley, T. M. L. (1983). A comparison of evidence for late Holocene summer temperature variations in the Northern Hemisphere. *Quaternary Research*, **20**, 286–307.

Woillard, G. M. (1978). Grande Pile peat bog: a continuous pollen record for the last 140,000 years. *Quaternary Research*, **9**, 1–21.

Woillard, G. M. (1979). Abrupt end of the last interglacial s.s. in north-east France. *Nature*, **281**, 558–562.

W.M.O. (1982). *WMO Global Ozone Research and Monitoring Project*, Report No. 14 of the Meeting of Experts on Potential Climatic Effects of Ozone and Other Minor Trace Gases, Boulder, Colorado (13–17 Sept. 1982). World Meteorological Organization, Geneva: 35 pages.

W.M.O. (1983). Report of the WMO(CAS) Meeting of Experts on the CO_2 Concentrations from Pre-industrial Times to I.G.Y., *WCP-53.* Bojkov, R. (Ed.), World Meteorological Organization, Geneva.

Yamamoto, R. (1981). Change of global climate during recent 100 years. *Proceedings of the Technical Conference on Climate — Asia and Western Pacific,* 15–20 Dec. 1980, Guangzhou, China. WMO Report No. 578. World Meteorological Organization, Geneva.

Yamamoto, R., and Hoshiai, M. (1979). Recent change of the Northern Hemisphere mean surface air temperature estimated by optimum interpolation. *Monthly Weather Review,* **107**, 1239–1244.

Yamamoto, R., and Hoshiai, M. (1980). Fluctuations of the Northern Hemisphere mean surface air temperature during recent 100 years. *Journal of the Meteorological Society of Japan,* **58**, 187–193.

CHAPTER 7

Changing the Sea Level

Projecting the Rise in Sea Level Caused by Warming of the Atmosphere

G. deQ. ROBIN

7.1 INTRODUCTION

This chapter discusses the effects of atmospheric warming on sea level. We draw on results in other chapters for the possible rise in atmospheric temperatures caused by the increasing atmospheric concentration of CO_2 and other radiatively active gases.

Changes of global temperature affect components of the global hydrological cycle in different ways and with different response times. Higher temperatures increase the amount of water vapour in the atmosphere. Precipitation patterns alter and affect runoff from rivers and glaciers into the sea. Ocean waters expand. Catastrophic collapse of ice sheets has been suggested as another consequence of rising temperatures that might cause a rapid rise of sea level.

Although components of the hydrological cycle have been studied for some time, it is only during the past few years that quantitative attempts have been made to integrate all relevant data. So far, most studies related to rising sea level have been directed at positive components that together could explain the observed rise of sea level. Negative components that could extract water from the oceans for long periods, such as increases in the mass of the Antarctic ice sheet and of the level of groundwater, have received less attention.

Our approach is to interpret observational data and to emphasize correlations and physical models that involve the minimum of assumptions. The more assumptions involved, the more easily can models be tuned to fit existing data but this does not necessarily improve resultant forecasts.

Discussion in Sections 7.3 to 7.5 on components of the hydrological cycle shows so many deficiencies in our knowledge of factors affecting mean global sea level that we make use of a simple linear correlation between changes of mean global temperature (ΔK) and of mean global sea level (ΔSL). Ideally

323

we should use a correlation such as $\Delta SL = a \Delta K + b (\Delta K)^2 + c(\Delta K)^3 + \ldots$, where $a, b, c \ldots$ are constants determined from correlations of existing data. The overriding factor that limits correlation to the linear is that data scatter is too large to determine higher order terms. A further weakness of this approach is extrapolation from global temperature changes of a few tenths of a degree to several degrees. This is only justified if the processes involved are known to be linear over the greater range.

Several processes involving an approximately linear response to temperature help to justify our use of linear correlation. These include the thermal expansion of sea water (except near 0 °C) and the change of height of the equilibrium line on glaciers, that is the line separating the accumulation zone from the ablation zone. The change of saturation water vapour pressure with temperature is approximately linear over a small temperature range (less than 5 °C, say) and is one factor that determines transfer of water mass from oceans to land. Other factors, such as those involving atmospheric and ocean dynamics, are unlikely to involve linear correlations and in some cases we are not sure whether correlations are positive or negative.

Another problem is the time responses of different processes. If these are small compared to the period considered, say 100 years, then the response will be approaching completion within 100 years and the correlation reasonable. If response to K is long compared to the analysis or forecast period, then SL will be given by the product of the rate of change (e.g. mass) and the time. Use of the same period (100 years) of data analysis and of forecasting helps to eliminate the time factor correction for the (unknown) proportional effect of processes with long response times.

Processes with response times short compared with 100 years include warming of upper layers of the ocean, adjustment of small glaciers to climatic change and runoff of water from land. Response times of the order of 100 years are associated with deeper oceanic layers and larger glaciers while melting of continental ice sheets and resultant adjustment of the Earth's crust to changing ice loading continue over some millennia.

Two other factors related to melting glaciers are the non-linear relation between melting rate and temperature and the apparent polar amplification of global temperature change. During summer months when glacier melting occurs, global circulation models of Manabe and Stouffer (1980) and Washington and Meehl (1984) show the polar amplification over Greenland and Antarctica to be small or negligible and barely significant in relation to melting. Climatic data from some polar regions do show a higher variability than those of lower latitudes, but as with models this is likely to be due to winter rather than summer phenomena that do not affect melting. These include the strength of cold surface inversion layers and delays in oceanic freezing.

This report is concerned with changes of mean global sea level (SL). By this we mean changes of mean level in relation to bench marks on the shore recorded by tide gauges operated over long periods. Global or regional changes of sea level refer to the mean value of such changes at many locations over the globe or region. Finally, we should emphasize that damage done by high sea level usually results from storm surges at the time of high tides. We do not attempt to discuss the problems of surges or tides. The former may well be influenced by dynamic factors involved in atmospheric–ocean interactions that are also related to climatic change. These are, however, local and regional problems that require separate study to that of changing global mean sea level.

7.2 OBSERVED CHANGES OF SEA LEVEL AND GLOBAL TEMPERATURE OVER THE PAST CENTURY

7.2.1 Sea Level

Sea level changes at any location are affected by a combination of local, regional and global factors. In addition to the quality and length of records, variations between stations are caused by changes of meteorological, oceanographic and tectonic conditions. To determine global changes of the water mass we need mean values from instruments spaced in a regular network over the world's oceans. Instrument locations where disturbances are likely to be high should be avoided. Mid-oceanic sites rather than continental coastlines may be more stable tectonically. Unfortunately, the distribution of available records is far from meeting the above criteria. Europe and North America provide many of the records, but often are not tectonically ideal. The Southern Hemisphere is poorly represented.

Analyses of global sea level changes are summarised in Table 7.1. There is some variation in the period covered and in the methods used. The earlier estimates and those covering the whole period ≈1880–1980 give values from 10 to 15 cm/century while the more rapid rise since 1920 is shown in analyses of post-1930 data.

The last four studies made special efforts to avoid regional and other bias. Statistically Barnett's (1984) analyses were rigorous, but presentations of Gornitz *et al.* (1982) and Klige (1982) are more convenient for discussion of global changes of mean sea level (*SL*) and their relation to physical processes over the past century. Their global curves are shown in Figure 7.1. The three authors agree that there are large variations between regions, but that over large sections of the world's coastlines, changes of sea level are spatially and temporarily coherent. Barnett does not present a global curve. He considers that the data from 1881 to 1920 indicates a time of little change and that the period 1920 to 1980 was a time of steady increase of *SL*. Klige and Gornitz *et al.* show the most rapid changes of *SL* from around 1925 to about 1955, after which *SL* stayed about the same level until 1975 (Figure 7.1).

Table 7.1　Estimates of mean 'global' sea level increase

Author	Rate (cm/century)	Method
Thorarinsson (1940)	> 5	Cryologic Aspects
Gutenberg (1941)	11 ± 8	–1937 (many stations)
Kuenen (1950)	12–14	–1942 Different Methods
Lisitzin (1958)	11.2 ± 3.6	Sea Level (six stations)
Fairbridge and Krebs (1962)	12	1900–1950 (selected stations)
Emery (1980)	30	1935–1975 (selected stations)
Gornitz et al. (1982)	12	1880–1980 (many stations)
Klige (1982)	15	1900–1975 (many stations)
Barnett (1984)	14.3 ± 1.4	1881–1980 (many stations)
Barnett (1984)	22.7 ± 2.3	1930–1980 (many stations)

Prior to 1920 data were scarce and derived SL changes may be questioned, but after 1950 data were selected from over 500 stations. Barnett considers it inadvisable to fit other than a linear equation to his 1930–80 data, so he does not confirm the 1955–75 levelling off shown by Gornitz et al. Nevertheless, visual inspection of Barnett's six regional curves shows little change in SL between 1955–75 over three regions, two show a marked rise and one a marked fall. Although variable, Klige's SL decreases after 1955 but levels off in the 1960s. We attach some significance to the change of slope shown in SL curves around 1955, since such a change is linked by physical processes with independent data also shown in Figure 7.1. In particular, comparison of slope changes of the SL and temperature curves indicates that processes involving time constants of around one to three decades have had a major influence on sea level changes over the past century.

The analysis of Gornitz et al. (1982) also helps this study by correcting recent SL changes for long-term trends over the past 6000 years, that is since the ice sheets of the last glaciation finally melted. They used ^{14}C dating of measured elevations of past shore line indicators, such as molluscs, corals and brackish water peats to calculate mean trends. Details of the corrections are not given, but in principle the approach supplies the information needed

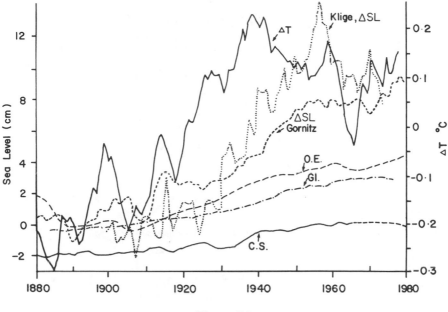

Figure 7.1

ΔK Global mean temperature, 5 year running mean from Hansen *et al.* (1981)
Gornitz Sea level. Global mean—from Gornitz *et al.* (1982)—5 year mean
Klige Sea level. Global mean—from Klige (1982), annual values
O.E. Ocean Expansion to thermocline—from Gornitz *et al.* (1982)—median estimate based on ΔK and constants $\Delta K_{eq} = 2.8$ °C and $K = 1.2$ cm²/sec.
Gl. Glaciers melting. Effect of small glaciers and ice caps melt on sea level.
From Meier (1984)
C.S. Caspian Sea level changes multiplied by 6.6 and expressed in equivalent sea level changes less 2 cm. Data from Micklin (1971) to 1965 and other sources

for this assessment. Figure 7.2 shows the corrected trends for the 10 out of their 14 regions for which corrections were available. Table 7.2 gives numerical details. Column (4) was added to the original table to show the difference between column (2) and column (6), so is not a true mean of actual corrections applied to individual station records. The mean of column (4) of −2 cm/century, is therefore a rough estimate of slow changes due to combined effects of movement of material in the Earth's mantle and of any slow long-term changes in volume of ice sheets. Klige (1982) suggests a figure of around 1 cm/century for the slow rise of *SL* over the past 6000 to 7000 years, half the value shown in Table 7.2, but does not apply this figure as a correction. This is one reason for the smaller change of *SL* by Gornitz *et al.* in Figure 7.1 compared with Klige's data; other possible reasons are the different time spans covered and greater weight given to North Atlantic data by Klige.

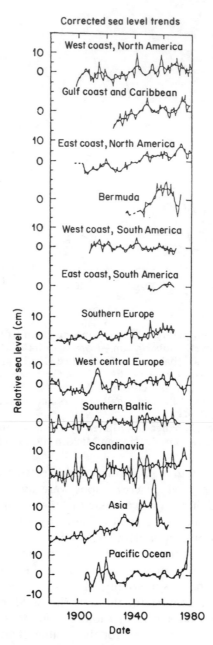

Figure 7.2 Regional mean sea level trends. The heavy lines are 5-year running means. Long-range (6000-year) trends have been subtracted (after Gornitz *et al.*, 1982)

Table 7.2 Sea level trends, 1880 to 1980 including correction for long-term (6000-year) trends (based on Gornitz *et al.*, 1982)

Region	Sea level trend 1830 to 1980				Corrected sea level trend 1880 to 1980		
	(1) Number of stations	(2) Linear trend (cm/100 years)	(3) 95% confidence limit (cm/100 years)	(4) Mean station trend (up to 6000 years) (cm 100 years)	(5) Number of stations	(6) Linear trend (cm/100 years)	(7) 95% confidence limit (cm/100 years)
West coast, North America	16	10	2	−2	1	8	3
Gulf coast and Caribbean	6	23	4	−7	4	16	5
East coast, North America	32	30	2	−15	30	15	2
Bermuda	1	26	16	−6	1	20	16
West coast, South America	8	19	31	−22	2	−3	3
East coast, South America	5	4	11	+12	2	16	11
Africa	2	32	31	–	0	–	–
Southern Europe	15	32	2	−25	7	7	2
West central Europe	7	13	2	−9	5	4	2
Southern Baltic	21	4	2	+1	14	5	2
Scandinavia	47	−37	3	+47	10	10	3
Asia	9	4	3	+18	2	22	4
Australia	9	13	3	–	0	–	–
Pacific Ocean	15	19	3	−13	6	6	4
Global mean	193	12	1	(−2)	86	10	1

The global mean values of uncorrected trends given by Gornitz et al. appear to exclude the two regions for which corrected trend data were not available, while including Scandinavia. If uncorrected trends for all 14 regions are included without weighting by numbers of stations or areas, the global mean is 14 ± 5 cm/century where 5 is the standard error of the mean. The corresponding mean of the corrected trends is 10.5 ± 2.5 cm/century. Avoidance of regional bias by combining adjacent zones in Europe and of the other forms of bias by excluding areas with less than 70 years of data or Bermuda because it has only one record made little difference to the rate of 10.5 ± 2.5 cm/century which is based on data from 75 stations.

7.2.2 Station Trends

Corrections for trends of station elevation in relation to sea level over the past 6000 years have halved the spread of sea level trends between regions in Table 7.2. This justifies the assumption that long-term trends may be extrapolated over the next few centuries to correct for vertical movement of the Earth's surface. Isostatic adjustments and tectonic movement associated mainly with plate crustal boundaries were shown to be the main cause of this vertical movement by Newman et al. (1981). They analysed 3000 dated sea level determinations spanning 12,000 years to provide maps of elevation changes from the present sea level for 1000 year periods, each including 100 to 300 data points. The analysis also revealed a persistent Holocene equatorial bulge which they suggest could be due to changes in the Earth's rotational speed following deglaciation, but its (small?) magnitude is not specified.

Theoretical studies such as Clark et al. (1978) show that the response of world oceans to the disappearance of a major ice sheet will not produce a uniform rise of sea level over the whole globe. A changed gravitational field due to removal of ice plus elastic and isostatic rebound due to removal of ice and increasing ocean loading produce viscous flow of material below the crust. Papers in 'Earth Rheology, Isostasy and Eustasy' Nils-Axel Mörner (ed, 1980) show no consistent agreement over details of viscous flow in the aesthenosphere and mantle, but provide possible explanations for the mean 6000 year trend of around 2 cm/century in Table 7.2 and of variations from that mean due to changes of the shape of the Earth with time.

7.2.3 Sea Level, Salinity and Oceanic Circulation

In addition to expansion of ocean waters with rising temperatures, its density is also strongly dependent on salinity. The surface salinity of the ocean is affected by evaporation and precipitation at the sea surface, by input of fresh water from rivers and melting ice and by salt rejection during sea ice formation. These processes in turn govern convection or the lack thereof in different regions and thus affect the dynamic circulation of the ocean in addition to its response to the global pattern of wind stresses.

The density of the stable mixed layer of 50 to 100 m deep produced by solar heating from around 45° N to 45° S is determined mainly by temperature rather than salinity effects, so we assume that here, the density changes are mainly a function of temperature changes. In high latitudes with lower sea temperatures, salinity changes could have more effect, but where deep convection takes place during winter relative changes of density will be smaller. Around the Arctic Ocean, where river discharge forms a lower density layer 200 m deep at temperatures little above freezing one would expect variations of river discharge with climate to affect sea level by a few centimetres or at most by a few tens of centimetres but its global significance would not be large.

The major effect of salinity changes is likely to be on the circulation of deep ocean waters (see Duplessy and Shackleton, 1985), but knowledge is not sufficient to predict the resultant changes of sea level. Ocean circulation changes driven by wind stresses and oceanic eddies produce surface elevation changes of up to several tens of centimetres that may persist for weeks or months as in the El Niño event of 1982 (Wyrtki, 1985).

Local and regional changes of sea level are smaller and less persistent than global changes. The former include seasonal changes of salinity in estuaries and adjacent regions due to variable discharge from rivers. On the longer time scale, the inverse barometric effect from long-term pressure trends at meteorological stations was investigated by Barnett (1983). The maximum change from 1900–1970 was about 2 mbar, equivalent to a change of relative sea level between stations of 3 cm/century. Major changes of oceanic circulation over long periods, such as the shift of position of the Gulf Stream in the Atlantic since the last glacial maximum will produce relative changes of up to one metre or more while the stream moves across any region.

7.2.4 Temperatures

Similar problems to those of sea level occur in determining mean global trends of temperature due to non-uniform distribution of data and a paucity of records extending back to 1880 in the Southern Hemisphere. These problems are discussed in Chapter 6. Fluctuations of global mean changes of temperature determined by Hansen *et al.* (1981) incorporating Southern Hemisphere data, which are compared with sea level data of Gornitz *et al.* (1982) in Figure 7.1, are very similar to the three Northern Hemisphere analyses of Figure 6.1. The latter show a mean change of around +0.4 °C over the same period as Hansen *et al.* (1981).

7.2.5 Correlation Between Sea Level Changes and Global Air Temperatures

Gornitz *et al.* (1982) tentatively fitted a linear relation between their sea level curve and Hansen *et al.*'s (1981) global temperature trend using

$$\Delta S_t = a\Delta K_{t-t_0} + b \qquad (7.1)$$

where ΔS and ΔK are 5 year means of global sea level and temperature changes respectively and t is time. Parameters a and b were obtained by least squares linear regression and the time lag t_0 was chosen to minimise the variance between sea level curve and temperature. The results were $a = 16$ cm/°C, $b = 0.3$ cm, and $t_0 = 18$ years with a correlation coefficient of 0.8. We can also obtain the coefficient a by equating the average global rise of sea level from 1880 to 1980 of 10.5 cm (corrected for local trends) to the mean rise of 0.4 °C of global air temperature from the same sets of data to give $a = 26$ cm/°C. Barnett's (1984) value of 14.3 cm for sea level rise 1881–1980 and the global temperature rise of 0.5 °C of Chapter 6 would give $a = 29$ cm/°C or the use of 0.4 °C and subtracting 2 cm/century to allow for long-term trends would give $a = 31$ cm/°C. There is no point in using shorter term data after 1930 in this way. There is little net temperature change over this period to match the large sea level rate of change possibly because the time lags involved are then too large. In round figures we therefore suggest limits for a as 16 and 30 cm/century and apply these figures to the estimate of 3.5 ± 2 °C for the response of global temperature to a doubling of CO_2 over the next century. This gives sea level changes using $a = 16$ cm/°C of $\Delta SL = 56 \pm 32$ cm and for $a = 30$ cm/°C of $\Delta SL = 105 \pm 60$ cm which indicates that the sea level rise should be in the range 25–165 cm. We again draw attention to the limitations of this type of linear extrapolation set out in the introduction 7.2.1. It is the only type of estimate that can be made without a full evaluation of the contributions of all physical processes involved. These are now discussed in more detail. Only by their use will we develop a firmer base for forecasting future sea level changes. However, our present state of knowledge is insufficient for this purpose. Linear correlation thus provides a simple forecast that is better than no forecast.

7.3 RESERVOIRS AND EXCHANGE RATES WITHIN THE HYDROLOGICAL CYCLE

7.3.1 Distribution of the Global Water Mass

Figure 7.3 from Woods (1984) based on data in Baumgartner and Reichel (1975) presents a simplified but very useful general picture of the main storages and exchanges within the Earth's hydrological system. Figures for

Table 7.3 Present global water mass distribution

Reservoir	Area (10⁶km²)	Water equivalent (10⁶km³)	% fresh water	Sea level equivalent (mm)	Annual Exchange Volume (km³)	Annual Exchange Sea level equivalent (mm)	Residence time (volume/annual exchange) (years)
Antarctic inland ice	12.0	26.5	70.9	$73.2 \cdot 10^3$	1,800	6	16,200
Greenland ice sheet	1.8	2.46	6.6	$6.8 \cdot 10^3$	500	2	5,400
Other glaciers	0.54	0.12	0.3	330	661	2	180
Permafrost	(≈12)	0.03	0.1	8	0	0	–
Total ice on land	–	29.1	77.8	$80.4 \cdot 10^3$	–	–	–
Lakes, rivers, soil	–	0.225	0.6	$0.62 \cdot 10^3$	37,600	104	6.0
Groundwater	–	8.06	21.6	$22.3 \cdot 10^3$	?	?	0 to 10^4
Atmosphere	504	0.013	<0.1	35	496,000	1,370	9.6 days
Continents } Total freshwater	142	37.33	100	$103.1 \cdot 10^3$	–	–	–
Oceans	362	1,354	–	3,750	–	–	–

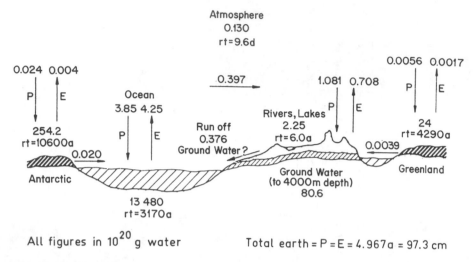

All figures in 10^{20} g water   Total earth = P = E = 4.967a = 97.3 cm

Figure 7.3 The global water balance shown schematically on the basis of data in Baumgartner and Reichel (1975). The residence time of water in the ocean is 3000 years, in the soil one year, and in the atmosphere ten days (after J. D. Woods, 1984)

exchanges are based on available data adjusted to produce global balances between precipitation and evaporation and between runoff to the oceans and the excess of precipitation over evaporation on the continents. Residence times are simply the reservoir mass divided by the exchange rate.

Table 7.3 shows similar details, with data on ice volumes and exchanges from other sources. Also shown in Table 7.3 are areas, volumes instead of masses ($10_7^{20} = 10^5$ km^3) and equivalent sea level thicknesses to the volumes.

7.3.2 Oceans

These contain around 97.3% of the water mass on Earth and provide the major reservoir and the primary source of water via the atmosphere to continents. 86% of global evaporation comes from the oceans and 78% of global precipitation falls directly into the ocean, the remainder returning as runoff of water or ice from the continents. Changes of ocean temperature, radiation and atmospheric circulation will change the rate of precipitation over the continents. The resultant effect on mass distribution between land and sea clearly depends on the reservoir characteristics of other parts of the hydrological cycle as discussed in the introduction (7.1). We regard the ocean as the prime reservoir in which mass changes are evaluated from the response of other reservoirs to changes of climate. We must also consider effects on sea level of thermal expansion of ocean waters caused by climatic changes.

Thermal Expansion

Rising surface temperatures decrease the density of sea water and hence raise sea level. Propagation of temperature changes in the North Atlantic has been discussed by Roemich and Wunsch (1984) and Roemich (1985) using a time series of deep oceanographic observations from 1955 to 1981 at a fixed location off Bermuda. They also compared results of IGY transects of the Atlantic Ocean from the USA to Spain in 1957 that were repeated in 1981. Seasonal temperature variation in the mixed layer about 100 m deep caused variations of around ±8 cm in mean sea level. This layer varies from 70 to 100 m or more in depth between 45° N and 45° S approximately and the thickness response to warming of this layer can be calculated readily. A 3 °C rise of temperature will increase the thickness of a 100 m layer of sea water in low latitudes by around 9 cm.

The next layer extends to a thermocline at around 6 °C with depth of around 1000 m between 45° N and 45° S. This is formed by cooling of surface waters polewards of these latitudes. This water then slides almost horizontally towards lower latitudes beneath the warmer but more saline surface waters. Temperature changes of this layer will be due primarily to changing heat advection from higher latitudes rather than to vertical conduction and diffusion of heat from the surface. Between Bermuda and Spain, the layer was up to 0.6 °C colder in 1981 than in 1957. A mean cooling of around 0.4 °C over the period would thin the layer by around 4 cm (100 m to 1000 m depth). However, in contrast to the above trend, off Bermuda temperatures increased by around 1 °C between 500 and 1000 m depth from 1970 to 1978.

Below 1000 m the Atlantic water warmed by up to 0.2 °C at a fairly uniform rate from 1957 to 1981 down to a depth of 3000 m or more. A 0.2 °C rise would produce an expansion of a 2000 m layer of around 4 cm. This layer is probably fed from Arctic regions and may have been responding to surface warming in northern latitudes in the first half of this century, while the layer above the thermocline may have responded to post-1940 cooling.

The above figures indicate the order of magnitude of expansion effects of layers on sea level. Roemich found that variations of sea level on the east coast of the USA at Charleston were less than at Bermuda some 1000 km offshore, but the trends were similar. This suggests that sea level on continental coastlines follows density changes in the deep ocean, at least partly.

It is clear that oceanographic theory is not yet sufficiently well developed to give an adequate dynamical explanation of existing data on deep sea density distribution, let alone provide forecasts of its response to climatic change. This could involve large changes of circulation and thicknesses of intermediate and deep layers. However, some estimates have been made assuming the pattern of density distribution does not change. Revelle (1983) finds the increase of sea level due to expansion of water above the thermocline to be at least 30 cm over the next century for a global warming averaging about 4.5 °C. He uses a steady state model for the ocean circulation above the thermocline and makes no estimate of temperature changes below this level.

7.3.3 Atmosphere

The mean water vapour content of the atmosphere is equivalent to a layer of water 25 mm thick over the whole Earth, or 35 mm over the ocean surface. If we assume that relative humidities do not change, a rise of equatorial temperature of 3 °C and of double that at higher latitudes would lower sea level by around 7 mm due to additional water vapour in the atmosphere.

7.3.4 Continents

7.3.4.1 Surface Water: Lakes, Rivers, Soil

Residence times of water on land vary widely, depending on the path from the point of precipitation to the ocean including its passage through lakes. However, seasonal variations of river flow indicate that in most areas the average runoff time is a fraction of a year. Over a period of a decade or a century, significant changes of surface water will be due mainly to changed volumes of inland lakes, both natural and man made. Golubev (1983) puts the volume of man-made reservoirs at not less than 5500 km^3, equivalent to subtracting a layer 1.5 cm thick over the ocean surface. Man also extracts around 3500 km^3 per year for use in irrigation and other projects, equivalent to a 1.0 cm ocean layer, but much of this soon rejoins the river systems and oceans by drainage and through evaporation and precipitation. There is, however, no agreement on how conversion of land to agricultural use affects the global water balance.

Overall, man's management of water stocks by storage and depletion of groundwater may lower sea level by around one cm per century, a negligible amount.

In comparison with man's usage, natural runoff from land is an order of magnitude greater while atmospheric transfer is over two orders of magnitude greater (Table 7.3). It is difficult to estimate effect on sea level from natural variability of water transport and storage on continents due to changes of the balance between precipitation, evaporation and runoff, but if global temperatures increase, some changes can be expected. This applies especially to the 25% of continental areas which return no runoff to the sea and to drier areas where runoff is small. Quantitative data from such areas are scarce.

Global variability of precipitation is discussed in some detail in Section 6.2.5. Some coherence in precipitation over regions of 10^6 km^2 or more is found, along with time scale variability (order 10 years) and a few significant long-scale trends (order 100 years). It is suggested that there is no evidence of any overall global mean trend, or if there was, it would almost certainly fail to reach statistical significance. Corresponding data on variations of evaporation are still more limited.

The above lack of data makes variations in the level of the Caspian Sea more significant. This is fed from a watershed of 3×10^6 km^2 comprising 2.3% of all non-glaciated lands. This is the largest basin not returning water to the oceans by river flow and it includes both arid and better-watered temperate lands. We therefore use data from the Caspian Sea to suggest the possible magnitude of variations of continental storage rather than assuming that increases and decreases of precipitation and evaporation patterns over all continents will be in balance and so have no effect on sea level. We multiply the observed volume changes of the Caspian Sea by the square root of the ratio of non-glaciated continental areas to that of the Caspian basin as a suggested estimate of changes of continental storage. The changes are inverted in Figure 7.1 to show the effect on sea level.

We note that Caspian Sea levels changed relatively little from 1880 to 1920, followed by decreasing storage (raising sea level) from 1920 to 1960, then by smaller changes since 1960. This follows a similar pattern to the sea level record.

Since changes in Caspian Sea level correspond closely to the cumulative flow deficit of the Volga River at Volgograd over the same period (Micklin, 1971), they should reflect the precipitation/evaporation balance over the whole basin. Related changes of groundwater storage within the basin are also likely.

7.3.1.2 Groundwater

Although the volume is several times larger than the water equivalent of the Greenland ice sheet, the total volume of groundwater and its residence times are quantitatively the least known part of the hydrological cycle. Nevertheless, some estimates of its importance have been made.

Decreases in groundwater levels in central and western USA during this century are estimated at around 390 km^3 (1.1 cm SL) in Meier (1983). Falls are attributed mainly to extraction by man. Meier suggests global depletion would be four to six times the above figure, that is a rate of 20 to 30 km^3 per year. During this century this would amount to about one half of the volume of water impounded in reservoirs.

Recharging of groundwater reservoirs comes from rainfall not returned to the sea, and so will contribute to extraction of water from the sea. Neither this nor the amount of direct runoff into the sea below sea level is adequately known on a global scale.

7.3.4.3 Smaller Glaciers and Ice Caps

Variations in volume of 25 glaciers measured during 1900 to 1961 were used by Meier (1984) to estimate the contribution to sea level from melting of all small glaciers and ice caps. Antarctica and Greenland were excluded and are discussed later. Many more glaciers have been monitored since 1965, the start of the International Hydrological Decade.

Meier used his long-term data to estimate regional changes of the average mass balance $(\bar{b})$—that is the annual mean deficit (or surplus) expressed in water thickness equivalent over the whole glacier. He then applied this figure to the total area of all glaciers in each region, taking the total mass deficit to be proportional to the annual mass amplitude (a) defined by

$$a = (b_w - b_s)/2 \qquad (7.2)$$

where b_w is the winter balance and b_s the summer balance (normally negative) averaged over each glacier. The correlation coefficient between $\bar{b}$ and a over his 13 regions was -0.55 and the average fraction $\bar{b}/a$ was -0.23 (S.D. 0.12). The errors in the total contribution of all small glaciers and ice caps to sea level (Figure 7.1) are therefore large, but show that melting of all small glaciers and ice caps could contribute from 1/3 to 1/2 of the observed sea level rise during this century. Meier also notes that small glaciers and ice caps were approximately in balance from 1960 to 1975, the effect is approximately in phase with observed sea level changes. Meier's results suggest that the total addition to sea level from these smaller glaciers due to a CO_2-induced temperature rise of 1.5 °C to 5.5 °C over the next century would be from 9 to 31 cm.

Although Meier's extrapolation of the fraction $\bar{b}/a$ over all glaciers of a region is the best available for small temperate glaciers, its use on larger cold glaciers may be questioned. We therefore make an independent estimate of volume changes in terms of the change in height of the equilibrium line ΔH_E with temperature. This line separates the accumulation zone from the ablation zone of glaciers. Its elevation change with temperature approximately follows the dry adiabatic lapse rate, that is $\Delta H_E \approx 100 \Delta K$ with ΔH_E in metres, and ΔK in °C. For comparison, Kuhn's (1981) detailed assessment of factors governing ΔH_E used by Ambach (1985) with field data from West Greenland gave $\Delta H_E/\Delta K = +77$ m/K for constant cloudiness and -4 m per 1/10 cloudiness at constant temperature.

We now make the crude assumption that glacier volumes change in proportion to $2\Delta H_E/H_G$, where H_G is the elevation difference from source to terminus of the glacier and $H_G/2$ is the estimated elevation difference between the source and equilibrium line. The global change of sea level is then given by

$$\Delta SL = \frac{2 \cdot 100 \cdot \Delta K}{\bar{H}_G} \cdot \frac{V_{SG}}{A_o} \qquad (7.3)$$

H_G is the mean elevation span of the world's glaciers (including ice sheets), V_{SG} their total volume in water equivalent (Table 7.3) and A_o the area of the world oceans. $\bar{H}_G = 1320$ m was estimated from the mean elevation span of the 67 glaciers around the world studied during the IGY. For comparison with Meier's extrapolation equation (7.3) gives values of 8 and 28 cm for $\Delta K = +1.5$ °C and $+5.5$ °C respectively.

Our crude figure is likely to overestimate changes because it neglects time constants and uses the dry adiabatic lapse rate rather than Ambach's smaller figure. The neglect of thickness changes with glacier length leads to under-estimation. The assumption that the height range of the accumulation zone is half the height from source to terminus could be in error in either direction and is in any case inferior to Meier's calculations related to total areas and mass balance of both accumulation and ablation zones. While close agreement with Meier's figures is somewhat fortuitous, it helps to support his figures. Furthermore, extrapolation from Meier's figures can be criticised, since melt rates do not bear a linear relation to global temperature change K, or to summer temperatures rather than, say, to $(\Delta K)^3$. However, the assumption of a linear relation between ΔK and ΔH_E as a basis for calculations is reasonable.

In any case it appears likely that both the above figures are an overestimate. If the figure of $V_{SG} = 120,000$ km^3 in Table 7.3 is correct, melting of all small glaciers would add only 33 cm to world sea level, and according to equation (7.3), all small glaciers would disappear if $\Delta K = 6.6$ °C. We need to allow for a realistic spread of glacier elevation ranges to rectify this.

7.3.4.4 Ice Sheets

Although the ice sheets of Greenland and Antarctica contain around 99.5% of land ice shown in Table 7.3, their annual exchange rate amounts to only 78% of glacier ice–ocean exchange, compared with 22% from smaller glaciers. This does not imply that these ice sheets dominate sea level changes due to glacier ice by a factor of 3.5, since once ice is discharged across the flotation line, its subsequent melting history does not change the total oceanic water mass (liquid and frozen) or sea level (minor density effects excepted). Meltwater runoff from Antarctica is low. As no reliable esti-mates are available, we assume it to be 2% of the total discharge, a similar or slightly larger figure than the fraction of the surface area covered by the ablation zone. This suggests a total discharge around 36 km^3 per year of surface meltwater from Antarctica compared to estimates of Greenland melting ranging from 179 to 315 km^3/yr (Table 7.4). The total surface melt-

Table 7.4   Annual mass balance of greenland ice sheet in km^3

	Bader (1961)	Benson (1962) (summary)	Bauer (1967)	Weidick (1984)	Reeh (1985)
Accumulation	+630	+500	+500	+500±100	+487
Melting	−120 to −270	−272	−330	−295±100	−169
Iceberg discharge	−240	−215	−280	−205± 60	−318
Net balance	+270 to +120	+13	−110	0	0

water discharge from these two sources is therefore put at around half all that from small glaciers. The latter figure should be increased by ten per cent to allow for melting of local glaciers on Greenland, not included with these ice sheet estimates or in Meier's figures. These local glaciers are about 5% of the area of the ice sheet and with an ablation area around 10% of that of the ice sheet.

If surface melting of ice sheets has varied with global temperature in the same way as that of smaller glaciers during the past century, we should add 50% to Meier's figures in the previous section to allow for variations of melting of polar ice sheets. This neglects any contributions from basal melting or from variations in the rate of discharge of solid ice across the flotation line due to global changes of temperature.

Penetration of surface temperature changes into cold ice sheets to depths where they could affect flow takes from centuries to millennia or tens of millennia, depending on ice thickness and surface accumulation rates (Robin, 1970; Budd and Young, 1983; Young, 1981; Oerlemans, 1982). For forecasting changes over the next century or two the assumption of constant flow, or of flow changing linearly with time appears satisfactory. Similarly, basal melting which is due to geothermal heat and frictional heat of ice motion but not to surface melting, should not change significantly over our forecast period.

Although we may assume that flow changes are too slow to affect forecasts for the next century, changes in the rate of surface accumulation as well as surface melting resulting from global warming cannot be neglected as they involve immediate mass exchange with the ocean. A 10% increase of accumulation over the Antarctic ice sheet would require extraction of 180 km^3/yr of water from the ocean, lowering sea level by 5 cm/century, while increased melting would have an opposing but probably smaller effect. We therefore discuss polar ice sheets in more detail.

Greenland. Estimates of the mass balance of the Greenland ice sheet over recent decades are shown in Table 7.4. These are affected by poor distribution of data, and do not agree on whether the mass is increasing or decreasing.

Measurements of change of elevation of the ablation zone of central West Greenland from 1948–59 by Bauer showed a general lowering of 0.3 m/year. Seckel (1977) found a mean lowering in the same area of 0.24 m/year from 1959–68 and a thickening in the accumulation zone of around 5 cm per year. Reeh and Gundestrup (1985) interpret data from Dye 3 near the southern dome of the ice sheet as due to thickening of around +3 cm/year with 95% confidence limits of −3 and +9 cm/year. Both changing climate and changing ice flow have been suggested as the reason for central thickening and marginal thinning. Bauer's net balance of −110 km^3/year is a similar proportion of the total mass exchange as Meier's (1984) value of $\bar{b}/a$ for small glaciers and we use this figure later, although the consensus among results suggests a mass balance around zero with an uncertainty of ±100 km^3/year (sea level ±2.8 cm/century).

Antarctica. Since Antarctic inland ice covers an area of one thirtieth of the world oceans, any small imbalance between the total mass accumulation rate of inland ice and the total rate of discharge of ice across the grounding line will have a marked effect on sea level.

Budd and Smith (1985) reviewed available evidence from mass balance studies. Earlier estimates of mass outflow had possible errors of a factor of two and indicated that accumulation over the ice sheet exceeded the outflow of ice to the sea by a factor of up to two. A series of field studies over limited regions using Doppler satellite position fixing has now provided a greatly improved quality of data. Budd and Smith conclude that the total influx of about 2 · 10^3 km^3/year of ice is nearly balanced by the outflow with a discrepancy from 0 to +20%. This corresponds to a sea level *fall* from 0 to 11 cm/century.

Present accumulation rates over East Antarctica vary with location in pro portion to the mean annual air temperature above the surface inversion and suggest that accumulation rates during the 'Last Glacial Maximum' (LGM) were from 30 to 50% lower than at present (Robin, 1977). However, greater winds indicated by increased dust in an ice core at Dome C dated by isotopic events led Lorius *et al.* (1984) to estimate the accumulation was reduced by only 25% during the LGM. More recently, studies of an ice core from Vostok dating back to 160 ka BP by Yiou *et al.* (1985) and by Lorius *et al.* (1985) provide good evidence that during the last ice age precipitation was around half its present value. These figures indicate that a global warming of 3.5 °C should increase Antarctic accumulation by at least 10% and probably by more than 25% above the present level.

7.3.4.5 Global Warming

If global warming does not cause a catastrophic change of flow of the ma-

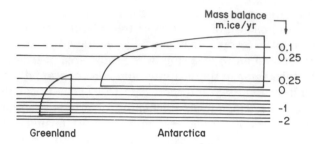

Figure 7.4 Height profiles of ice sheets of Greenland and Antarctica in relation to a simplified mass balance distribution

jor ice sheets during the next century, we may assume that their form and flow does not change significantly in this time. Their effect on sea level can then be estimated in terms of changes of ablation and accumulation over the ice sheets. The problem is summarised schematically in Figure 7.4 from Oerlemans and van der Veen (1984). This portrays profiles of the Greenland and Antarctic ice sheets in relation to a simplified mass balance field. This is highly negative at lower levels in Greenland, rises to a maximum then decreases at greater elevations. To a first approximation a climatic warming implies an upward shift in the mass balance field. This would decrease the mass balance of the Greenland ice sheet but that of Antarctica would increase for a limited warming. The net effect on sea level would depend on which ice sheet is dominant and here estimates vary.

Oerlemans (1982) estimates that a doubling of CO_2 will increase precipitation over Antarctica by 12% and over the next 250 years this would lower sea level by 30 cm. This would be countered by a rise of 20 cm due to melting of Greenland ice according to his use of figures from Ambach (1980). A more detailed estimate is now described.

A warming of 3.5 °C around Greenland would raise the equilibrium line of the ice sheet by around 300 m. This would increase the area of the ablation zone from 16 to 20% of the whole area. If the ablation–elevation relationship remained similar to that at present, ablation near the ice–rock margin in West Greenland would rise from around 3 m/year to perhaps 4.5 m/year, increasing the average rate by 50%. Together these two factors would increase melting by 70 km³/year. The accumulation zone would decrease in area from 84 to 80% of the ice sheet, but if accumulation rates stay constant or increase by 10% (see Figure 7.4) total accumulation would change by −24 or +24 km³/year leaving a net budget decrease of 94 or 46 km³/year respectively.

Over Antarctica, increases of accumulation rate of 10 or 25% due to a 3.5 °C global warming would require 180 or 450 km³/year respectively of water from the ocean while doubling ablation over twice the area would return 108 km³/year to the ocean, leaving net budget increases in Antarctica of 72 and 342 km³/year.

The combined effect of the above changes on Greenland plus Antarctica ranges from -22 to $+296$ km³/year equivalent to a sea level change of $+0.6$ to -8.2 cm/century with the latter figure giving the preferred value.

These very rough estimates indicate that changes of sea level due to global warming affecting Greenland and Antarctica during the next century could be small and cancel each other. However, we may have underestimated changes of ablation or of accumulation associated with global warming on both ice sheets. The net effect of mass balance changes of the two ice sheets on sea level is therefore estimated to be within ± 10 cm for a global warming of 3.5 °C over the next century (see Table 7.5).

The broad conclusion of this section is that while changes in the mass of polar ice sheets during the next century might have a significant effect on sea level, it may not be the dominant effect.

7.3.4.6 Global Cooling

The large volumes of ice extracted from the ocean to form ice sheets of the Northern Hemisphere during the Pleistocene period lowered sea level by between 100 and 200 m. This caused some expansion of Antarctic ice over the deeper continental shelf, followed by retreat during interglacials as sea level rose again. Models, such as those of Denton and Hughes (1981) and Drewry and Robin (1983) suggested that lower sea levels would also cause inland ice to thicken to a varying extent. However, Oerlemans (1982) and Alley and Whillans (1984), when incorporating lower accumulation rates during ice ages into their models, suggest that inland ice elevations may be smaller during ice ages. The latter also included the effect of changing sea level on their ice flow model which acts in the opposite direction by increasing the grounded area during ice ages. It is only recently that Lorius *et al.* (1984) have found evidence that ice in Central Antarctica may have been thinner during the last ice age. The contrast with the Denton and Hughes (1981) model is shown in Figure 7.5. They suggested an increase of Antarctic ice volume during the last ice age around 10×10^6 km³ whereas Figure 7.5 suggests the increase could not have been greater than 3×10^6 km³ and could have been much less.

Recent evidence indicating that ice levels were lower in Central Antarctica during the last ice age includes:

(1) Studies of total gas content of ice cores, equivalent to a recording aneroid barometer. Raynaud and Whillans (1982) and Lorius *et al.* (1984) used this to show that the surface in the region of Byrd Station was around 200 m lower during the last ice age than at present and around Vostok it may have been 100 m lower at that time.

Table 7.5 Changes of water storage affecting the sea level from 1900–1975

Storage type	Klige (1982)		This survey		Source
	km³/yr	SL effect (mm/year)	km³/yr	SL effect (mm/year)	
Lakes	−63	+0.17	−72	+0.20	Caspian SL × 6.6 in Figure 7.1
Underground water	−136	+0.38	−25	+0.07	Meier (1983) mean value due to man's extraction
Antarctic	−315	+0.87	+100	−0.28	Mean from Budd and Smith (1985)
Greenland	−82	+0.23	−110	+0.30	Bauer (in Table 7.4)
Arctic Islands	−12	+0.03	−145	+0.40	Meier (1984)
Mountain Glaciers	−3	+0.01			
Man-made reservoirs	+69	−0.19	+69	−0.19	Golubev (1983)
Total	−542	+1.50	−183	+0.50	
Ocean expansion				+0.38	From Gornitz et al. (1982) $k = 1.2$ cm²/sec.
Observed		+1.50		+1.01	Gornitz et al. (1982)

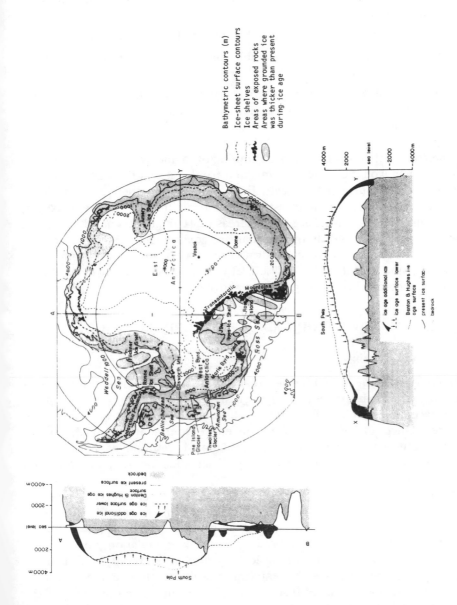

Figure 7.5 Estimated ice-thickness changes over Antarctica since last ice age

(2) Comparison of isotopic $\delta^{18}O$ profiles at Byrd, Vostok and Dome C stations suggests that a similar lowering of surface elevation occurred at all stations, with lowering at Dome C and Vostok being somewhat more than at Byrd.

(3) Analysis of the temperature depth profile at Dome C (Ritz *et al.*, 1982).

(4) Accumulation of meteorites on ablation areas inland of mountains over periods of 10^5 years or more.

(5) Glacial geological evidence of greater flow of ice into McMurdo Sound dry valley during interglacial periods shows that ice inland of mountains was higher at that time (Drewry and Robin, 1983).

One feature that would increase thinning of inland areas while coastal areas thickened would be if ice streams and trunk glaciers through mountains continued to drain inland ice with little change from their present elevation. At present they drain an estimated 80% of the area of the ice sheet, comprising almost all discharge from areas more than 200 km inland of the flotation line. Until the sliding mechanism governing the flow of such ice streams is understood and incorporated in models, the results from other modelling studies cannot be fully effective. Conclusions need to be drawn from the types of field evidence listed above.

7.4 LONG PERIOD AND CATASTROPHIC CHANGES

7.4.1 Climatic Stability

In contrast to the growth and retreat of vast ice sheets over continents in the Northern Hemisphere during the last 1.5 million years, the ice sheets of Antarctica and Greenland appear to have undergone relatively small changes. Once an ice sheet covers an entire continent or island, including the continental shelf, excess accumulation is returned to the ocean in icebergs and iceshelves without melting. Once afloat, melting of ice has no further effect on sea level.

As temperatures increase over such ice sheets, effects will differ between Greenland and Antarctica as seen in Figure 7.4. On Greenland, as the proportion of surface melting to iceberg discharge increases, ice flow will be diverted from the large outlet glaciers to the large ablation zones. When surface melting and the remaining iceberg discharge exceed accumulation, the ice sheet will no longer be stable in relation to climate and it will diminish in size. Our estimate from mass balance sums along the lines of the previous section suggests that this will occur for a temperature rise of 6 °C. Verbitsky (1982) gives a corresponding figure +5 °C. Even if the melting rate increases still further to an excess of 600 km³/year above the accumulation, sea level would rise only 20 cm/century. We use half this figure as our upper limit for possible warming over the next century.

Antarctica

Similar arguments to the above applied to Antarctica suggest that a major climatic retreat of the inland ice of East Antarctica due to surface melting would require a temperature rise of around 10 to 12 °C (Verbitsky, 1982) or around 15 °C (Robin, 1979).

It is useful to relate Figure 7.4 to the development of Antarctic glaciation on the geological time scale. The Antarctic crustal plates drifted to near their present position around 38 Ma BP (Kennett, 1978; Drewry, 1976). Mountain glaciation was followed by development of an ice sheet on East Antarctica some 12 to 14 Ma BP. This ice sheet would have been plotted on Figure 7.4 in a similar position to present-day Greenland. Gradual cooling that started around 22 Ma BP, presumably on a global scale, let this ice sheet expand over West Antarctica by around 4 to 6 Ma BP, at which stage the ice sheet was probably at its maximum volume covering the entire continental shelf. This would take place when the bulk of the accumulation zone of the Antarctic ice sheet lay within the zone of maximum accumulation of Figure 7.4. Further lowering of temperature would be matched by a lowering of the mass balance field on Figure 7.4 and correspond to the formation of the Greenland ice sheet after 3 Ma BP, as shown by studies of marine cores (Berggren, 1972; Shackleton and Opdyke, 1977). This continued global cooling would eventually decrease accumulation on the Antarctic ice sheet and lead to some diminution in its size. It is worth noting that although the global temperature was apparently warmer than today around 4 to 6 Ma BP, melting by warm ocean waters apparently did not prevent the growth of the West Antarctic ice sheet. Kennett's (1978) review suggested that some sea ice was present around Antarctica at 38 Ma BP. He notes ice rafting of debris again at 22 Ma BP which then increased around 12–14 Ma BP, around 4–6 Ma BP and after 1.8 Ma BP during the Quaternary. Microfossil evidence in marine cores indicates that pack ice, if not the source of ice-rafted debris increased along with the debris. These observations suggest that current global circulation models that result in early removal of Antarctic sea ice cover with rising CO_2 should be viewed with caution.

The time constants involved in changing ice levels of Antarctica are important. Rising sea levels after glacial periods led to the disappearance of thicker coastal ice at much the same rate as the ice sheets of the Northern Hemisphere receded, according to both field evidence and model studies by Alley and Whillans (1984). In coastal regions, ice thicknesses that are in equilibrium with the present climate have changed little in several localities since 6000 BP. Further inland time constants must be much longer, probably several tens of millennia. An increase of accumulation of 1 cm/year ($\approx$ 20%) would take 20,000 years to raise the ice level of central regions by 200 m if ice flow was unchanged, and longer if allowance for flow changes is included. Evidence that the ice surface may be rising by this amount 18,000 years after the last glacial maximum, as suggested by Budd and Smith (1985), is consistent with such a long time constant. The central Antarctic ice sheet is therefore unlikely to reach a thickness in equilibrium with the climate during an interglacial period of only 10,000 or 20,000 years.

Although glacial periods tend to be several times longer than interglacials, the same comment applies in relation to lowering of ice levels as a result of lower accumulation. We may therefore expect the amount of lowering to vary with the length and severity of the glacial period.

The glacial period from around 200 ka BP to 135 ka BP preceding the last interglacial was both colder and had a more prolonged cold period than that from around 70 ka BP to 15 ka BP as shown by evidence from glacial deposits in the Northern Hemisphere (Flint, 1971) and suggested by marine core data such as those of Shackleton and Opdyke (1973). Greater lowering of the ice of Central Antarctica before the last interglacial than before the present one, together with the long time constant for recovery would then result in the ice of Central Antarctica being lower during the last interglacial than at present. A difference of 200 to 300 m between these levels during the last and present interglacial would explain one major problem of the last interglacial period (Robin, 1985).

7.4.2 The Last Interglacial

Considerable attention has been given to the possible instability of the West Antarctic ice sheet during the last interglacial for two reasons. The first is evidence of a rise in sea level of 7 m in equatorial regions around 125 ka BP, together with some early marine core data that suggested temperatures may have been some 3 °C warmer at that time. If the sea level rise were worldwide, this would require adding an additional volume of water to the oceans equivalent to the volume of the present Greenland ice sheet or to the volume of water above sea level in West Antarctica today. Because much of the ice rests on rock well below sea level, disappearance of the West Antarctic ice sheet due to instability has received much attention. This is the second reason for interest in this period, so we consider both problems.

The best information on climatic condition during this period is given by 23 well-known research workers in a report 'The last interglacial ocean, CLIMAP', Ruddiman (1984). The main evidence comes from detailed oxygen isotope analyses and biotic census counts in 52 cores across the world ocean. Their summary states 'The first order conclusion about last interglacial climate is that it was extremely similar to the current climate. The total volume of ice was probably equal to or slightly less than today's with any decrease due to calving of West Antarctic ice and atmospheric ablation of Greenland ice'.

Although major changes of ice volumes are deduced from isotopic data from marine cores, the noise level in the $\delta^{18}O$ signal was 50% or more above that of a possible signal corresponding to a worldwide sea level 7 m above the present level during the last interglacial. There is therefore no isotopic evidence of ice sheet retreat. The report states that the strongest evidence for an ice volume difference during the last interglacial is from dated coral reefs.

Effects discussed in Section 7.2.1, such as the persistent equatorial bulge during the Holocene and the non-uniform changes of sea level over the world proposed by Clark *et al.* (1978), might help to explain the 7 m high sea level on tropical coral reefs during the last interglacial without requiring much change from our present ice volume.

Other reasons for doubting that the West Antarctic ice sheet was greatly reduced in volume during the last interglacial are:

(a) There is no evidence from marine cores around Antarctica of any hiatus in sediments due to ice sheet collapse and associated changes of ocean circulation.

(b) The greater extent of the Antarctic ice sheet around 4 to 6 Ma BP when global temperatures were still higher is discussed in Section 7.4.1.

(c) Antarctic bottom water production during the last interglacial was higher than today according to ^{13}C studies of marine cores by Duplessy *et al.* (1984). This suggests that sea surface freezing and the thickness of ice shelves were then no less than at present.

At both Camp Century in Greenland and Byrd station, Antarctica, $\delta^{18}O$ values in near-basal ice indicate surface temperatures were similar or slightly above ($+2$ to $+3$ °C) present temperatures. This could be due to a lowering of surface elevations by 200 to 300 m or to a slightly warmer climate or to a combination of both factors (Robin, 1977). These small changes are incompatible with the collapse of either ice sheet. A third possibility is that $\delta^{18}O$ ratios in ice near bedrock have been changed by isotopic differentiation during the basal melting and refreezing that occurred at both sites.

Results from the first ice core (to 2083 m depth) to span the last interglacial and recover ice from the preceding glacial period are presented in Lorius *et al.* (1985). The $\delta^{18}O$ values at Vostok, where the core was obtained, were around 2‰ higher during the last interglacial than at present. The authors, on the evidence from Lorius *et al.* (1984) discussed in Section 7.3.4.6, assume the surface elevation has undergone little change and interpret the data as indicating the regional climate at the present elevation at Vostok was about 3 °C warmer than now. An alternative interpretation is that since global climate was very similar to the present, the $\delta^{18}O$ change at Vostok was due to a change of surface elevation rather than temperature. The warming of 3 °C would then indicate that the ice surface was then around 300 to 350 m lower. This interpretation fits the suggestion at the end of Section 7.4.1 that the mean surface was 200 to 300 m lower than at present at the end of the last interglacial.

Surface lowering of the above amount over the area of around 8×10^6 km^2 shown in Figure 7.5 would mean that the volume of ice of central Antarctica was some 1.6 to 2.4 $\times$ 10^6 km^3 less than at present. Then if glaciers in Antarctic coastal regions and elsewhere were of similar volume to the present under a very similar climate, the ocean volume would be greater by an equivalent amount. This would provide sufficient water to explain sea level some 5 to 7 m higher during the last interglacial. This hypothesis avoids the difficulty of having to melt large sections of the Greenland and Antarctic ice sheets or having the West Antarctic ice sheet collapse to provide water needed to raise ocean levels by up to 7 m at a time when the climate was similar to the present. Furthermore, the higher sea level during the last interglacial need not be interpreted as an indication of higher temperature (Robin, 1985).

The question of how much the surface elevation and/or the climate around Vostok differed from the present may be answered when total gas content measurements on the ice core become available. This should help to settle the question of whether a catastrophic collapse of the West Antarctic ice sheet took place during the last interglacial period. If the answer is not clear, the controversy is likely to continue since glaciologists' opinions differ widely. Field research on the problem continues.

7.4.3 Dynamic Stability of West Antarctic Ice Sheet

We now consider the problem of stability of marine ice sheets such as West Antarctica against climatic change. It is suggested that ocean warming will greatly increase the rate of bottom melting of ice shelves, such as the Ross and Filchner-Ronne Ice Shelves, and lead to lowering of frictional restraint by grounded ice and ice boundaries, and hence permit rapid deglaciation to take place. Although advances over the past decade have increased our understanding of the problem, a precise answer is not possible because of our lack of understanding of oceanic heat transport as much as that of ice dynamics.

Beneath the Ross and Filchner-Ronne Ice Shelves, it is clear that much melting is due to circulation of continental shelf water at a temperature

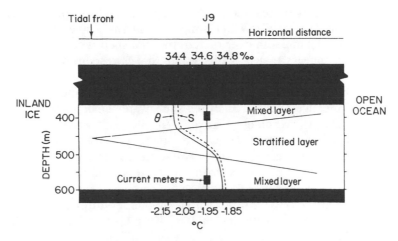

Figure 7.6 The observed temperature (θ) and salinity (S) profiles below the Ross Ice Shelf at the J9 drill camp (from MacAyeal, 1984)

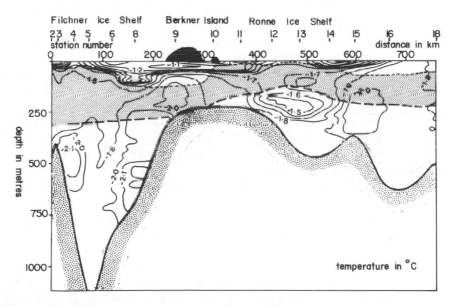

Figure 7.7 Summertime temperature–depth distribution off the front of the Filchner–Ronne Ice Shelves, Antarctica. Also shown is the approximate depth to the base of the ice shelf 20 km south of the ice front (dashed line) and the depth of ice at the ice front (dotted line) estimated from the height of the floating ice cliffs. (Oceanographic data from Gammelsrød and Slotsvik, 1982. Glaciological data from Robin *et al.*, 1983)

of the freezing point of sea water. As salt is rejected during freezing at the ocean surface, the denser water sinks so that the whole column of sea water of around 500 m depth over the continental shelf is cooled to -1.9 °C (Figure 7.6). It appears that it is this water that is responsible for melting ice at still lower temperatures governed by the depth (pressure) of the ice–water interface (-2.2 to -2.8 °C or more) beneath the inner parts of the Ross and Filchner Ice Shelves. Such ice shelf water emerges from beneath the front at -2.0 °C or colder (Figure 7.7.). Seasonal warm water of the top 100 m is mainly at too shallow a depth to penetrate far beneath ice shelves, but a core of warmer water—Antarctic circumpolar deep water—penetrates beneath part of the ice front of the Ross and Ronne Ice Shelves (not Filchner) and contributes to melting beneath some outer parts of these ice shelves (Figure 7.7). The same water, at temperatures up to $+1.0$ °C, dominates melting beneath the ice shelf of King George VI Sound (Potter et al., 1984) and has also been observed on the continental shelf of the Amundsen Sea. This matches the absence of extensive ice shelves where the Pine Island Glacier and Thwaites Glacier discharge into the Amundsen Sea.

The suggestion that rapid collapse of the West Antarctic ice sheet has already started due to discharge of these glaciers into the Amundsen Sea can be countered. McIntyre (1983, 1985) has pointed out that a sharp surface slope around the head of Antarctic ice streams and large glaciers seen on Landsat images at a number of locations indicates the line across which ice transport changes from dominant internal deformation to basal sliding. The line on Landsat images corresponds closely to a sharp increase of bedrock depth in the flow direction seen in radar depth profiles. In deglaciated countries, this corresponds to the steep slope found at the head of many fjords as slopes change abruptly from elevated upland plains. The onset of sliding results from sudden steepening of basal slopes and is stabilized in position where the bottom slope changes. Both Pine Island Glacier (Figure 7.8) (Crabtree and Doake, 1982) and Thwaites Glacier (McIntyre, 1983, 1985) show these features.

Any change of location of the line where rapid sliding starts is unlikely to occur until bottom melting increases sufficiently to bring the ice base above the bedrock scarp. This requires a rise from around -1300 to -1000 m at the flotation line of Pine Island Glacier and -1100 m to 750 m beneath Thwaites Glacier. Crabtree and Doake estimate that a change of thickness of 10% would require an increase of bottom melting of 2 m/year beneath Pine Island Glacier. To decrease thickness at the grounding line to above the scarp level on this basis would require an increase of melting of 3.5 m/year above the present rates. Present melting may already approach the 2 m/year found from salinity and isotopic data beneath the northern part of ice shelf of King George VI Sound by Potter et al. (1984). Once melting raises

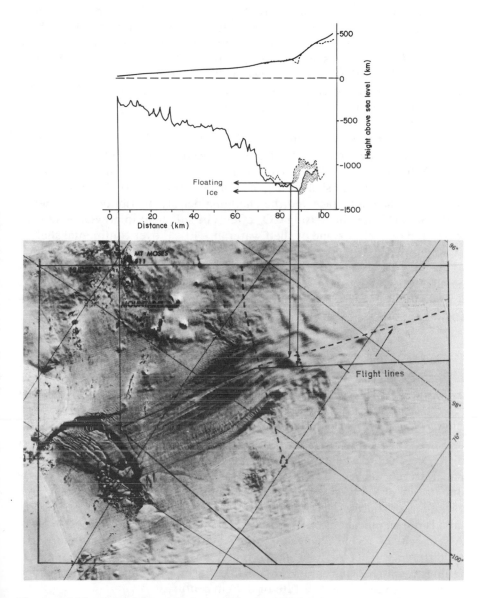

Figure 7.8 Pine Island Glacier, Antarctica. (a) Longitudinal profiles of surface and base obtained by airborne radio echo sounding flight on 6 February (dashed line) and 9 February (solid line), 1981. (b) Landsat image showing flight lines. The grounding line is shown by linked arrows. Here downslope sliding leads to floating ice of decreasing thickness showing stream lines (due to varying thickness across glacier) in contrast to the dimpled appearance of inland ice moving over an irregular bedrock. (Based on Crabtree and Doake, 1982)

the grounding line to the scarp level, rapid sliding of grounded ice could commence. The time taken to affect sea level is then a function of ice sheet dynamics. Bentley (in Revelle, 1983) estimates a minimum of 200 years for the West Antarctic ice sheet to collapse through Pine Island and Thwaites Glaciers. Other estimates for collapse of the West Antarctic ice sheet range from under 100 to many hundred years, but the many estimates available are tentative because of lack of knowledge of the dynamics of sliding of fast ice streams.

Other problems to which an answer is needed include: How much warmer will the Antarctic circumpolar deepwater become due to effects of rising CO_2? What time lag will be involved in this warming? Will circulation of this water beneath ice shelves increase or decrease due to changes of wind stresses on the sea surface and the changing density distribution with depth in the ocean? When may rising CO_2 effects be sufficient to prevent extensive formation of pack- ice and of shelf water at -1.9 °C over continental shelves of Antarctica?

The complexity of effects of changing ice loads on the Earth's crust beneath West Antarctica, of increased accumulation rate accompanying CO_2 warming and of increased melting by oceanic waters is such that a precise answer is not possible at this time.

7.5 COMPARISON

To illustrate lack of knowledge of factors governing global sea level, Table 7.5 compares estimated quantities from this report adjusted to cover the period 1900–1975 with corresponding figures given by Klige (1982).

The emphasis given to ocean water expansion and melting of small glaciers in Figure 7.1 is missing or almost negligible in Klige's paper. The biggest differences come when considering the two largest storage reservoirs. This paper largely neglected natural variations of underground storage, partly due to lack of knowledge and partly because mechanisms causing such large variations of storage are not understood. Agreement over figures for Greenland is not significant as other investigators differ greatly from these figures. Assessment of recent changes of the Antarctic ice mass shows the greatest difference. This paper suggested increased storage equivalent to sea level falling at up to 8.2 cm/century while Klige suggested decreased storage equivalent to a sea level rise at a rate of 8.7 cm/century.

Clearly much research needs to be carried out to obtain a satisfactory knowledge of factors affecting sea level. In this situation use of the simple linear correlation between global sea level and temperature records during the past century to predict future changes appears the only practical basis for a forecast. While such extrapolation is valid for some processes, it is not valid for all.

7.6 CONCLUSIONS

— It is estimated that the prediction of global warming of 3.5 °C ± 2.0 °C due to CO_2 doubling would lead to a sea level rise of 80^{+85}_{-60} cm. This figure is based on the correlation between past changes in sea level and temperature in Section 7.2.

— To predict future changes of sea level at any location, sea level trends over the past few thousand years at that location should be added to the figures above predicted from global warming.

— Over the next century major physical factors affecting sea level due to a postulated global increase of atmospheric temperature of 3.5 °C are likely to be

(1) Thermal expansion of ocean waters could expand the top 100 m of tropical water by 10 cm and the next 900 m by at least 20 cm, and eventually by 50 cm or more. Below that level we would expect cold deep water to continue to flow from polar regions, perhaps at a slightly higher temperature involving a slow expansion of 10 to 20 cm with little effect during the next century. Changes in the depth of the thermocline and hence of the vertical distribution of temperature could, however, produce larger changes than the simple expansion figures given above.

(2) Melting of smaller glaciers and ice caps could produce a rise of around 20 cm ± 12 cm of sea level.

(3) Changes of melting and accumulation of the ice sheets of Greenland and Antarctica will tend to counterbalance each other. Due to uncertainty of opposing trends, the effect on sea level should be within ± 10 cm/century and may well be negative (i.e., contribute to a lower *SL*).

(4) Changes of water storage on land in lakes, rivers, reservoirs and groundwater are very difficult to predict and while unlikely to exceed ± 10 cm/century (Table 7.5) will probably be a fraction of this value.

(5) A catastrophic collapse of the West Antarctic ice sheet is not imminent, but better oceanographic knowledge is required before we can assess whether a global temperature rise of 3.5 °C might start such a collapse by the end of the next century. Even then it is likely to take at least 200 years to raise sea level by another 5 metres.

ACKNOWLEDGEMENTS

Special thanks for provision of relevant material are due to Dr C. S. M. Doake and Mr R. D. Crabtree of the British Antarctic Survey, Dr M. F. Meier of the U.S. Geological Survey, Dr T. M. L. Wigley of the Climatic Research Unit of the University of East Anglia, Dr J. Oerlemans of the

Institute of Meteorology and Oceanography of the University of Utrecht, Dr N. J. Shackleton of the Quaternary Research Laboratory, University of Cambridge, and Dr T. Hushen of the U.S. Academy of Sciences.

Detailed constructive and critical reviews of the first draft of this chapter have been received from Dr M. F. Meier, Dr T. M. L. Wigley, Dr J. Oerlemans, Dr V. Kotlyakov, and Dr B. R. Döös. Further helpful comments from a wider and less specialised review group have come from: Dr R. Bindschadler, Dr A. Berger, Dr H. W. Ellsaesser, Dr H. L. Ferguson, Dr R. M. Gifford, Dr J. Goudriaan, Dr G. McBean, Professor A. S. Monin, and Professor J. D. Woods.

With our lack of knowledge of certain aspects of the hydrological cycle and of the dynamics of the oceans and of polar ice sheets, forecasting of global changes of sea level involves considerable extrapolation and speculation. While comments by referees have resulted in considerable improvement both in the content and in the presentation of this chapter, it has not been possible to incorporate or sometimes to reconcile suggestions from different reviews. The article does not therefore reflect the views of all referees and responsibility for the contents lies with the author.

7.7 REFERENCES

Alley, R. B., and Whillans, I. M. (1984) Response of the East Antarctic ice sheet to sea-level rise, *J. Geophys. Res.*, **89**, C4, 6487–6493.

Ambach, W. (1980) Anstieg der CO_2-Konzentration in der Atmosphäre und Klimaänderung: Mögliche Auswirkungen auf den Grönländischen Eisschild, *Wetter und Leben*, **32**, 135–142.

Ambach W. (1985) Characteristics of the Greenland Ice Sheet for modelling, *J. Glaciol.*, **31**, 3–12.

Bader, H. (1961) The Greenland Ice Sheet, *Cold Regions Science and Engineering Report*, I-B2.

Barnett, T. P. (1983) Recent changes in sea level and their possible causes, *Clim. Change*, **5**, 15–38.

Barnett, T. P. (1984) The estimation of 'global' sea level change: a problem of uniqueness, *J. Geophys. Res.*, **89**, C5, 7980–7988.

Bauer A. (1967) Nouvelle estimation du bilan de masse de l'Inlandsis du Groenland, *Deep Sea Res.*, **14**, 13–17.

Bauer, A. (1968) Mouvement et variation d'altitude de la zone d'ablation ouest, *Meddelelser om Grønland*, **174**, 1–79.

Baumgartner, A, and Reichel, E. (1975) *The World Water Balance*, Amsterdam, Elsevier.

Benson, C. S. (1962) Stratigraphic studies in the snow and firn of the Greenland ice sheet, *SIPRE Res. Rep.* No. 70.

Berggren, W. A. (1972) Late Pliocene–Pleistocene glaciation, in Laughton, A. S., Berggren, W. A., *et al.* (eds) *Initial Reports of the Deep Sea Drilling Projects*, Vol. XII, 953–63, Washington, U.S. Govt. Printing Office.

Budd, W. F., and Young, N. W. (1983) Application of modelling techniques to measured profiles of temperatures and isotopes, in *The Climatic Record in Polar Ice Sheets*, 150–177, Cambridge, Cambridge University Press.

Budd, W. F., and Smith, I. N. (1985) The state of balance of the Antarctic ice sheet—an updated assessment 1984, in *Glaciers, Ice Sheets and Sea Level: Effect of a CO₂-induced climatic change*, Washington, D. C., National Research Council.

Clark, J. A., Farrell, E. E., and Peltier, W. R. (1978) Global changes in postglacial sea level: A numerical calculation, *Quat. Res.*, **9**, 265–287.

Crabtree, R. D., and Doake, C. S. M. (1982) Pine Island Glacier and its drainage basin, *Ann. Glaciol.*, **3**, 65–70.

Denton, G. H., and Hughes, T. J. (1981) *The Last Great Ice Sheets*, New York, Wiley.

Drewry, D. J. (1976) Deep sea drilling from the Glomar Challenger in the Southern Ocean, *Polar Record*, **18**, 47–77.

Drewry, D. J., and Robin, G. deQ. (1983) Form and flow of the Antarctic ice sheet during the last million years, In *The Climatic Record in Polar Ice Sheets*, 28–38, Cambridge, Cambridge University Press.

Duplessy, J-C., Shackleton, N. J., Matthews, R. K., Prelle, W., Ruddiman, W. F., Caralp, M., and Hendy, C. H. (1984) 132C record of Benthic Foraminifera in the Last Interglacial Ocean: Implications for the carbon cycle and the global deep water circulation, *Quat. Res.*, **21**, 225–243.

Duplessy, J-C., and Shackleton, N. J. (1985) Response of global deep-water circulation to Earth's climatic change 135,000–107,000 years ago, *Nature*, **316**, 500–507.

Emery, K. O. (1980) Relative sea levels from tide-gauge records, *Proc. Natl. Acad. Sci., Washington, D.C.*, **77(12)**, 6968–6972.

Fairbridge R., and Krebs, O. (1962) Sea level and the Southern Oscillation, *Geophys. J. R. Astron. Soc.*, **6**, 532–545.

Flint, R. F. (1971) *Glacial and Quaternary Geology*, Wiley, New York.

Gammelsrød, T., and Slotsvik, N. (1982) Hydrographic and current measurements in the Southern Weddel Sea 1979/80, *Polarforschung*. **51(1)**, 101–111.

Golubev, G. N. (1983) Economic activity, water resources and the environment: a challenge for hydrology, *Hydrol. Sci. J.*, **28**, 57–75.

Gornitz, V., Lebedeff, L., and Hansen, J. (1982) Global sea level trend in the past century, *Science*, **215** (4540), 1611–1614.

Gutenberg, B. (1941) Changes in sea level, postglacial uplift and mobility on the earth's interior, *Bull. Geol. Soc. Am.*, **52**, 721–772

Hansen, J., Johnson, D., Lacis, A., Lebedeff, S., Lee, P., Rind, D., and Russell, G. (1981) Climate impact of increasing atmospheric carbon dioxide, *Science*, **213(4511)**, 957–966.

International Association of Scientific Hydrology, *Fluctuations of Glaciers*, (1967) vol 1, *ibid.* (1973), vol 2; *ibid.* (1977), vol 3, Unesco, Paris.

Kennett, J. P. (1978) The development of planktonic biogeography in the Southern Ocean during the Cenozoic, *Mar. Micropaleontol.*, **3**, 301–345.

Klige, R. K. (1982) Oceanic level fluctuations in the history of the earth, in *Sea and Oceanic Level Fluctuations for 15,000 years*, 11–22, Acad. Sc. USSR, Institute of Geography, Moscow, Nauka (in Russian).

Kuenen, Ph. H. (1950) *Marine Geology*, New York, Wiley.

Kuhn, M. (1981) Climate and glaciers, in *Sea Level, Ice and Glaciers*, IAHS Publ., **131**, 3–20.

Lisitzin E. (1958), see Lisitzin (1974) *Sea-level Changes*, Elsevier Oceanography Series, 8, Amsterdam–Oxford–New York, Elsevier Scientific.

Lorius, C., Merlivat, L., Jouzel, J., and Pourchet, M. (1979) A 30,000 year isotope climatic record from Antarctic ice, *Nature*, **280**, 644–648.

Lorius, C., Raynaud, D., Petet, J. R., and Merlivat, L. (1984) Late glacial maximum—Holocene atmospheric and ice thickness changes from Antarctic ice core studies, *Ann. Glaciol.*, **5**, 88–94.

Lorius, C., Jouzel, J. Ritz, C., Merlivat, L., Barkov, N. I., Korotkevich, Y. S., and Kotlyakov, V. M. (1985) A 160,000 year climatic record from Antarctic ice, *Nature*, **316**, 591–596.

MacAyeal, D.R. (1984) Potential effect of CO_2 warming on sub-ice shelf circulation and basal melting, in *Environment of West Antarctica: Potential CO_2 induced changes*, Polar Research Board, National Research Council, Washington, D.C., National Academy Press.

McIntyre, N. F. (1983) *The Topography and Flow of the Antarctic Ice Sheet*, Thesis, University of Cambridge.

McIntyre, N. F. (1985) The dynamics of ice sheet outlets, *J. Glaciol.*, **30**, 108 (in press).

Manabe, S., and Stouffer, R.J. (1980) Sensitivity of a global climate model to an increase of CO_2 concentration in the atmosphere, *J. Geophys. Res.*, **85**, C10, 5529–5554.

Meier, M. F. (1983) Snow and ice in a changing hydrological world, *Hydrol. Sci.*, **28**, 1, 3–22.

Meier. M. F. (1984) Contribution of small glaciers to global sea level *Science*, **226**, 4681, 1418–1421.

Micklin, P. P. (1971) *An Enquiry into the Caspian Sea Problem and Proposals for its Alleviation*: Thesis submitted to University of Washington, USA.

Mörner, N.-A. (ed.) (1980) *Earth Rheology, Isostasy and Eustasy*, New York, Wiley.

Newman, W. S., Marcus, L. F., and Pardi, R. R. (1981) Palaeogeodesy: late Quaternary geoidal configuration as determined by ancient sea levels, *Int. Assoc. Sci. Hydrol.*, Publ. No. 131.

Oerlemans, J. (1982) Response of the Antarctic ice sheet to a climatic warming: a model study, *J. Climatol*, **2**, 1–11.

Oerlemans, J., and van der Veen, C. J. (1984) *Ice Sheets and Climate*, Dordrecht, D. Reidel.

Potter, J. R., Paren, J. G., and Loynes, J. (1984) Glaciological and oceanographic calculations of mass balance and oxygen isotope ratio of a melting ice shelf, *J. Glaciol.*, **30**, 105, 161–170.

Raynaud, D., and Whillans, I. M. (1982) Air content of the Byrd core and past changes in the West Antarctic ice sheet, *Ann. Glaciol.*, **3**, 269–273.

Reeh, N. (1985) Greenland ice sheet mass balance and sea level change, in *Glaciers, Ice Sheets and Sea Level: Effect of a CO_2-induced climatic change*, Washington, D.C., National Research Council.

Reeh, N., and Gundestrup, N. S. (1985) Mass Balance of the Greenland Ice Sheet at Dye 3, *J. Glaciol.*, **31**, 198–200.

Revelle, R. R. (1983) Probable future changes in sea level resulting from increased atmospheric carbon dioxide, *Changing Climate*, Washington, D.C., National Academy Press.

Ritz, C., Lliboutry, L., and Rado, C. (1982) Analysis of a 870 m deep temperature profile at Dome C, *Ann. Glaciol.*, **3**, 284–289.

Robin G. deQ. (1970) Stability of ice sheets as deduced from deep temperature gradients, *Int. Assoc. Sci. Hydrol.*, Publ. No. 86, 141–151.

Robin G. deQ. (1977) Ice cores and climatic change, *Phil. Trans. R. Soc., London*, Series B, **280**, 143–168.

Robin G. DeQ. (1979) Formation, flow, and disintegration of ice shelves, *J. Glaciol.*, **24**, 259–271.

Robin G. deQ. (1985) Contrasts in Vostok core—changes in climate or ice volume, *Nature*, **316**, 578–579.

Robin G. deQ., Doake, C. S. M., Kohnen, H., Crabtree, R. D., Jordan, S. R., and

Möller, D. (1983) Regime of the Filchner-Ronne Ice Shelves, Antarctica, *Nature*, **302**. 582–586.

Robin G. deQ. (1983) (ed.) *The Climatic Record in Polar Ice Sheets*, Cambridge University Press, Cambridge.

Roemich, D. (1985) Sea level and the thermal variability of the ocean, In *Glaciers, Ice Sheets and Sea Level: Effect of a CO_2-induced climatic change*, Washington D. C., National Research Council.

Roemich, D. and Wunsch, C. (1984) Apparent changes in the climatic state of the deep North Atlantic Ocean, *Nature*, **307**, 447–450.

Ruddiman, W. F. (1984) (ed.) The last interglacial ocean — CLIMAP project members (23 listed), *Quat. Res.*, **21**, 123–224.

Seckel, H. (1977) Höhenänderungen im Grönländischen Inlandeis zwischen 1959 and 1968, *Meddelelser om Grönland*, **187**, 4.

Shackleton, N. J. and Opdyke, N. D. (1973) *Oxygen isotope and palaeomagnetic statigraphy of equatorial pacific core, V28-238: Oxygen Isotope Temperature and Ice Volumes on a 10^5 and 10^6 year scale.*

Shackleton, N. J. and Opdyke, N. D. (1977) Oxygen isotope and palaeomagnetic evidence for early northern hemisphere glaciation, *Nature*, **270**, 216–219.

Thorarinsson, S. (1940) Present glacier shrinkage and eustatic change in sea level, *Geogr. Ann.*, **22**, 131–159.

Verbitsky, M. Ya. (1982) Influence of Antarctic and Greenland icesheets on the world oceanic level, in *Sea Level and Oceanic Fluctuations for 15,000 years, 120–124*, Acad. Sci. USSR, Institute of Geography, Moscow, Nauka (in Russian).

Washington, W. M., and Meehl, G. A. (1984) Seasonal cycle experiment on the climate sensitivity due to a doubling of CO_2, with an atmospheric General Circulation Model coupled to a simple mixed-layer Ocean Model, *J. Geophys. Res.*, **89**, No. D6, 9475–9503.

Weidick, A. (1984) *Review of Glacier Changes in West Greenland*, Symposium on climate and Palaeoclimate of lakes, rivers and glaciers, Igles, Austria.

Woods, J. D. (1984) The upper ocean and air–sea interaction in global climate, in Houghton, J. T. (ed.) *The Global Climate*, Cambridge, Cambridge University Press.

Wyrtki, K. (1985) Sea level fluctuations in the Pacific during the 1982–83 El Niño, *Geophys. Res. Lett.*, **12**, 125–128.

Yiou, F., Raisbeck, G. M., Bourles, D., Lorius, C., and Barkov, N. I. (1985) [10]Be in ice at Vostok Antarctica during the last climatic cycle, *Nature*, **316**, 616–617.

Young, N. W. (1981) Responses of ice sheets to environmental changes, in *Sea level, Ice and Climatic Change*, IAHS Publ. No. 131, 331–360.

PART C

The Impacts on Terrestrial Ecosystems

CHAPTER 8

The Effects of Increased CO₂ and Climatic Change on Terrestrial Ecosystems

Global Perspectives, Aims and Issues

R. A. WARRICK, H. H. SHUGART, M. JA. ANTONOVSKY, WITH J. R. TARRANT AND C. J. TUCKER

8.1 TERRESTRIAL ECOSYSTEMS, CLIMATE AND MAN

It is safe to begin with the premise that a climatic change within the range indicated by climate models for a doubling of atmospheric CO_2 could have profound effects on global ecosystems. A cursory, global survey of natural systems reveals an unmistakable correspondence between the broad features of regional climates and the major characteristics of the world's biomes. The transitions from tundra, boreal forests, temperate forests, deserts, grasslands and tropical forests vary systematically with global temperature and precipitation patterns and produce the areal distributions illustrated in Figure 8.1. That climate is, in fact, a primary determinant of the composition and spatial patterns of the major biomes is beyond doubt, although certainly there are feedbacks in the other direction, as through the albedo or hydrologic characteristics of regional plant assemblages and associated soil types. Major changes in the global climate could be expected to alter natural rates of ecosystem change in and between these biomes, particularly within the marginal zones of transition.

If, indeed, the changes in climate that may result from increasing concentrations of greenhouse gases fall in the middle range of estimates provided by climate models (1.5 to 5.5 °C for a CO_2 doubling, as discussed in Chapter 5), we will experience a global climate that is probably at least as warm, or warmer, than at any time within the last 200,000 years. During that time, there were a number of fluctuations of climate, each of which was associated with geographical patterns of terrestrial ecosystems that were markedly—in some cases dramatically—different from those evident today. For example, during the Medieval Warm Epoch (800 to 1200 AD) when average temperatures were perhaps 1 °C warmer than present (at least around the region of the North Atlantic), Canadian boreal forests extended well

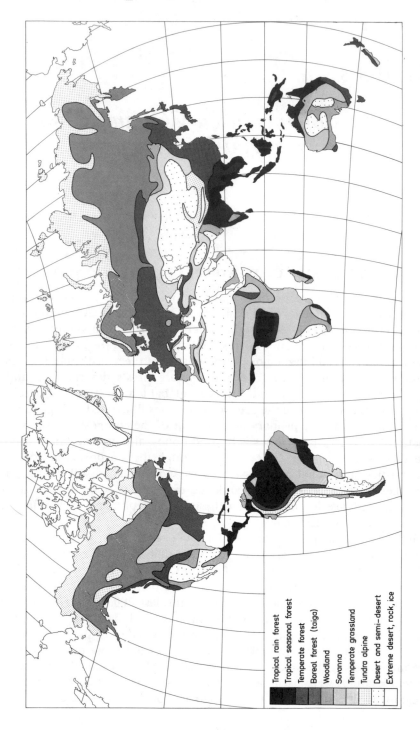

Tropical rain forest
Tropical seasonal forest
Temperate forest
Boreal forest (taiga)
Woodland
Savanna
Temperate grassland
Tundra alpine
Desert and semi-desert
Extreme desert, rock, ice

Figure 8.1 The global distribution of major ecosystem types, based largely on Whittaker (1970) (from Bolin, 1980)

north of the current timberline and cereal cultivation flourished in Iceland and Norway up to 65° N latitude (Lamb, 1977). The warm period of the early Holocene (9000 to 6000 BP) witnessed extraordinarily wetter conditions in the vast subtropical dry zones extending from West Africa to the Indus Valley–Rajasthan area; thriving savannah grasslands existed in large areas that are now unproductive desert (Hare, 1979; Flohn, 1980). The last glacial period (70,000–10,000 years ago) was associated with tundra in Central Europe and with a drastic shrinkage of the tropical rainforest, which quickly expanded with post-glacial warming. The earliest warm peak (Eem) of the last interglacial (around 120,000–80,000 years ago) was perhaps 2–3 °C warmer in the mid- to high-latitudes than today, and broad expanses of deciduous forests extended north into areas now occupied only by non-deciduous species (for reviews, see Flohn, 1980; Gerasimov, 1979). Solely on the evidence of the distant past, the potential for ecosystem change in a warmer future is enormous.

Past changes in climate were also accompanied by variations in the atmospheric concentration of CO_2. However, it is not clear to what degree these CO_2 variations were involved directly in global-scale shifts of terrestrial ecosystems in the past, if at all, notwithstanding the fact that in experimental situations higher CO_2 concentrations have been shown repeatedly to stimulate plant growth and productivity. In this case, the distant past provides few clues for the future.

A fundamental difference between changes in global ecosystems of the past and those of the future is the dominating influence of human intervention in the natural environment. Just as human activity may be responsible for altering the state of the global climate, so, too, are humans capable of manipulating the global biota to a considerable degree. Deforestation by man was a major contributor to past increases in the atmospheric CO_2 concentration, as discussed in Chapter 3; deforestation continues at an alarming rate in many areas of the tropics. In only a matter of centuries, one human activity, agriculture, has added another major terrestrial 'ecosystem', largely at the expense of grasslands and forests. This substitution still continues. Today, cultivated land (excluding pasture and grazing land) occupies about 10% of the world's land surface, an area approximately equivalent to one-third of that occupied by forests (Figure 8.2).

The capacity for human interference may be applied—purposefully or inadvertently—to counteract, retard or accentuate impacts on global biomes that would otherwise occur naturally. For instance, there is strong evidence to suggest that overgrazing and other human activities that are destructive to vegetation have exacerbated the processes of desertification in the subtropical semi-arid lands (Hare, 1983; 1984). Although the future direction of this influence is difficult to foresee, man's role in the process of ecosystem change is so large that any thoughtful assessment of the impacts of

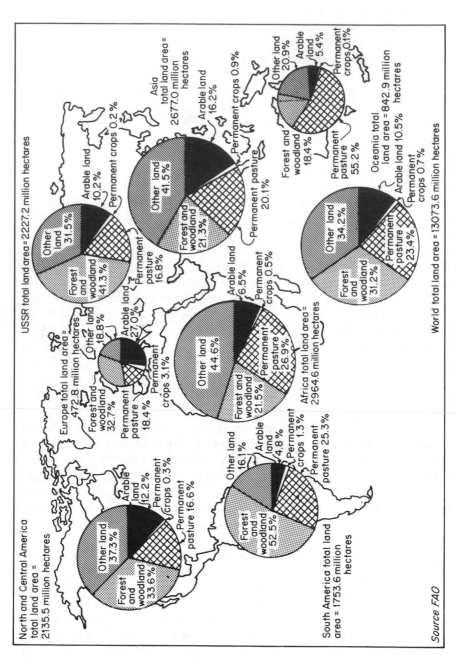

North and Central America
total land area =
2135.5 million hectares

Arable land 12.2%
Permanent crops 0.3%
Permanent pasture 16.6%
Forest and woodland 33.6%
Other land 37.3%

South America total land
area = 1753.6 million
hectares

Other land 16.1%
Arable land 4.8%
Permanent crops 1.3%
Permanent pasture 25.3%
Forest and woodland 52.5%

Europe total land area =
472.8 million hectares

Other land 18.8%
Arable land 27.0%
Permanent crops 3.1%
Permanent pasture 18.4%
Forest and woodland 32.7%

Africa total land area =
2964.6 million hectares

Arable land 6.5%
Permanent crops 0.5%
Permanent pasture 26.9%
Forest and woodland 21.5%
Other land 44.6%

USSR total land area = 2227.2 million hectares

Arable land 10.2%
Permanent crops 0.2%
Permanent pasture 16.8%
Forest and woodland 41.3%
Other land 31.5%

Asia total land area =
2677.0 million hectares

Arable land 16.2%
Permanent crops 0.9%
Permanent pasture 20.1%
Forest and woodland 21.3%
Other land 41.5%

Oceania total
land area = 842.9 million hectares

Arable land 5.4%
Permanent crops 0.1%
Permanent pasture 55.2%
Forest and woodland 18.4%
Other land 20.9%

World total land area = 13073.6 million hectares

Arable land 10.5%
Permanent crops 0.7%
Permanent pasture 23.4%
Forest and woodland 31.2%
Other land 34.2%

Source FAO

Figure 8.2 World land use in 1977 (From Crabbe and Lawson, 1981)

increased atmospheric CO_2, other trace gases and future climatic change must eventually take human management and response into account.

8.2 FOCUS AND AIMS

For this reason, we find it difficult to speak of 'managed' and 'natural' ecosystems in the context of global climatic change. It can be argued that no terrestrial ecosystem is left untouched by the human hand; the ubiquitous nature of the global CO_2 increase is the case in point. It is perhaps more meaningful to refer to a continuum of management, from least managed to most managed. Certainly, agricultural systems fall at the latter end of the spectrum. From a global perspective, forest ecosystems tend to fall toward the other end of the spectrum (although one can cite examples of almost any degree of management) and have important implications for the global carbon cycle as well as economic activity. In Chapters 8–10 we thus focus specifically on *agriculture* and *forest* ecosystems as representative of the management spectrum, hoping to achieve balance while remaining mindful that many more interrelated environmental systems and resources (e.g. the hydrologic cycle) are bound to be affected as well.

In the pages that follow, the human element causes us to tread a difficult path. Our charge is to address the issue of the 'primary' impacts of CO_2 and climatic change on ecosystems—that is, plant response, yield effects, changes in ecosystem composition or areal extent—leaving 'secondary', or social and economic, impacts for subsequent scrutiny. However, especially for agriculture (which, after all, by definition *is* an economic activity) we feel no compunction in occasionally crossing interdisciplinary boundaries in order to incorporate relevant socio-economic factors.

The basic premise of this and subsequent chapters is that at present we do not have available reliable estimates of future climatic change at a regional scale which would allow us to *predict* changes on global ecosystems. Nevertheless, this should not deter the scientific community from examining the *sensitivity* of such systems to changes in climatic variables. This requires an explication of the range of approaches which have been, or could be, taken to do so. Our aim is just that:

- to present the major issues and research questions;
- to outline the methodological approaches by which they could be addressed and to exemplify each through existing studies;
- to summarize what is known concerning the sensitivity of ecosystems to increased CO_2 and climatic change;
- to suggest promising areas of needed research or methodological refinement.

In the remainder of this chapter we provide an overview of the major issues and methodological considerations in assessing impacts in the global context as a prelude to more detailed discussions of agriculture and forest ecosystem in Chapters 9 and 10, respectively. What we cannot provide, unfortunately, are estimates of future changes in ecosystems with a degree of confidence sufficient for informed policy formulation and strategy choice— only possibilities.

8.3 CO_2 AND CLIMATIC CHANGE: INFORMATION FOR IMPACT STUDIES

There are two ways in which ecosystems can possibly be affected by the rising levels of greenhouse gases, and they are not necessarily mutually exclusive. The first is the direct effect of higher ambient CO_2 concentrations on plant growth and development. The direct CO_2 effect has been the subject of numerous research investigations, particularly on agricultural crop species. Understanding of plant response has come about primarily by controlled greenhouse experiments and, to a far lesser extent, by the use of crop-growth simulation models based largely on such experimental results. There are very few data concerning the effects of enhanced CO_2 in actual field conditions. Any direct effects on ecosystems from increasing concentrations of atmospheric CO_2 in this century have yet to be detected convincingly, although some scientists believe that the productivity of natural vegetation and, especially, grain yields has increased and will increase considerably in the future (e.g. Budyko, 1982).

The literature has been extensively reviewed and evaluated (e.g. Lemon, 1983), most recently by the U.S. Department of Energy (1984). Overall, experimental research consistently demonstrates that, directly, CO_2 is potentially beneficial in terms of crop yields and possibly forest productivity. In general, higher ambient CO_2 stimulates greater net photosynthesis—the so-called 'fertilization' effect—and decreases transpiration through a partial stomatal closure, resulting in greater water use efficiency in plants, at least at the microscale. However, the extrapolation from individual plants to dynamic ecosystems is highly tenuous. There are considerable differences between plants regarding their response characteristics. Competition between plants, and the interactions between plants, animals and microbes, are likely to change. For some organisms in terrestrial ecosystems, the net effect of higher CO_2 concentrations in the long-term may be negative, perhaps fatal.

The second way in which ecosystems can be affected by increasing concentrations of greenhouse gases is through changes in climate. Here, too, uncertainty abounds: in relation to the changes in regional climate, to their effects on plants and ecosystems, and to the ways in which enhanced CO_2

will modify the effects of climate on plants. As discussed in Chapter 5, the extent of our knowledge concerning changes in climate can only be expressed confidently in terms of averages at the global scale. Confidence drops precipitously as spatial resolution increases. Currently, GCMs can replicate the major features of the general circulation but are unable to produce realistic simulations of present-day climate at a regional level that is appropriate for impact analysis (at least 200–300 km for agriculture). Furthermore, higher confidence can be placed on GCM predictions of changes in temperature than on changes in precipitation or other climate variables like humidity or radiation. In most cases this situation precludes the possibility of actually predicting the impacts of climatic changes on ecosystems.

8.3.1 Scenarios of Climatic Change

In lieu of predictions, scenarios of changes in regional climate can be used to investigate the sensitivity of ecosystems to climatic variation, and to test and refine the methods of impact analysis. Three approaches have been followed in scenario development (WMO, 1985). The first is to use *GCM simulations* of the effects of CO_2 on climate, despite the lack of realism at the regional scale. The major advantage of GCMs is that, potentially, all required meteorological variables can be generated with true global coverage. Notwithstanding the problems, wide use has already been made of some GCM CO_2 experiments for impact analysis (see Sections 9.4 and 10.4 below).

The use of *past climate data* is another way of generating scenarios. Palaeoclimatic reconstructions of past warm periods (e.g. the Altithermal) have been suggested as analogues for a CO_2 related climatic change (e.g. Kellogg and Schware, 1981), but they, too, suffer from lack of regional detail, among other things. Lately, work has focused on the use of recent changes in climate observed in the instrumental record as analogues for the future (Lough *et al.*, 1983). Although the magnitudes of climatic changes during the past 100 years are considerably less than those suggested by 2 × CO_2 model experiments, they can be used directly as scenarios of climatic conditions *en route* to doubled CO_2. The use of instrumental data in this case means that internal consistency is virtually assured and that scenarios with fine spatial resolution can be developed. This represents a clear advantage over other methods. However, one must remain suspicious of the realism of such instrumentally based scenarios as CO_2 predictions (if such claims are made), since the spatial character of climatic change may not be independent of the method of forcing (see Chapter 3). Moreover, the dependence on observational data means that, for many parts of the world, the length, availability or reliability of records of critical meteorological variables may not be adequate for scenario construction.

The third approach is simply to select *arbitrary changes* in climate variables. For example, a frequently used scenario is a 1 °C temperature rise and a 10% decrease in precipitation, since this has some correspondence to general climatic changes suggested by GCMs for the not-so-distant future, particularly for North America. The approach is recommended by its simplicity and ease by which it can be uniformly applied in comparative analyses. The main disadvantages are that changes in climate are unlikely to be seasonally constant, and that the arbitrary changes may be inconsistent with local meteorological conditions and observed correlations between climate variables.

8.3.2 Applications and Improvements

In impact studies all three kinds of scenario have been employed, but in a somewhat indiscriminate, haphazard fashion. This is not surprising given the diverse nature of impact studies and the research proclivities of impact analysts themselves. The consequence, however, is that existing studies and methodological tools are difficult to compare and evaluate.

Ultimately, the solution to this problem is the improvement of GCMs to the point where detailed, regional climates can be simulated with realism. For analyses of impacts on agriculture and forests in both the high and low latitudes, this requires knowledge of possible changes in precipitation patterns as well as temperature. But this is a long-term goal and vast improvements of this nature are not foreseeable within the next 5–10 years, if ever. In the very long term, it may eventually be possible to develop interactive models capable of describing the interplay between ecosystems and climate. In the meantime, the situation could be improved by the development of sets of regional scenarios of climatic change that could be uniformly applied, thereby increasing the credibility and comparability of climate impact studies.

As models for impact analysis improve, it is becoming increasingly apparent that climate scenarios must provide information on the frequency distributions—the variability and extreme events, as well as the central tendencies—of key meteorological variables. This holds for forest ecosystems as well as agriculture. Short-term occurrences of unusual climatic anomalies during specific stages of the life cycle of forest stands, for example, can have far-reaching, long-term consequences for future forest composition and productivity (Chapter 10). This is certainly the case for agriculture in which changes, for example, in the frequencies of early- or late-season frosts can affect average yield trends, both through the direct impacts on plant growth, and indirectly by, say, the beneficial effects of killing overwintering pests (Chapter 9).

The prospect of developing and applying scenarios of climatic change for impact analyses at the international scale is daunting, to say the least. We must begin by outlining the dimensions of the problem from the global perspective. Let us consider agriculture first.

8.4 GLOBAL AGRICULTURE AND CLIMATIC CONSTRAINTS

The genetic foundation of modern world agriculture is surprisingly narrow. As pointed out by Swaminathan (1979), from the beginning of the Neolithic the domestication of plants and animals has involved a selective filtering of the species (and a reduction in the number of their varieties) upon which mankind depends for agricultural production. On a worldwide basis, there are only 30 crop species whose individual production exceeds 10 million tonnes (Mt) annually. Cereals account for over half of the world's arable land use, and only three crops—wheat, maize and rice—account for 80% of total cereal production (FAO, 1983a). Similarly, only two animal products, beef and pork, make up approximately three-quarters of the world's total animal production.

At local and regional scales, it is unclear whether this 'simplification' increases the vulnerability of agricultural production to climatic change. Some have expressed fears that this indeed may be the case (Swaminathan, 1984) and have argued for a greater effort to introduce species or varieties that are better suited to local environments as, among other things, a hedge against climatic change and variability (NAS, 1972; 1976). On the other hand, new varieties from crop research institutes tend to be more adaptable over a broad range of climatic conditions than the traditional varieties (for rice, say, there are thousands of varieties) that are being replaced and that are often very local in their use. It is possible, therefore, that plant research has actually made it easier to adapt to changing climate (Wortman and Cummings, 1978).

At the global scale, the differences between agricultural regions are immense, which is one reason why total world food production is remarkably stable from one year to the next. The differences pertain not only to environmental and climatic characteristics, but to levels of economic development, technology and human living standards as well. These regional differences are vital to modern patterns of agricultural production, so much of which is based on comparative advantage production and trade. For example, the same major cereal crops are grown and/or consumed in countries as diverse as the USA, Australia, Spain, Western Europe and the USSR; differential regional effects of increased CO_2 and climatic change could tip the balance, with worldwide repercussions. At the broadest scale, the regional differences are most pronounced between the tropics and the temperate regions and their semi-arid and humid zones.

8.4.1 Semi-arid Tropics

Within many areas of the tropics, agricultural production is low and unstable compared to the temperate regions of the world. To a large extent, this is due to the broad problems of underdevelopment, struggling economies and low levels of agricultural technology that are prevalent in the low latitudes. It has been argued often that the potential for increased production is very large given higher (and more expensive) levels of input (e.g. for a recent analysis see Shah and Fischer, 1984). Technology, however, is only one factor among a host of socio-economic and environmental constraints.

The climate of the tropics influences the patterns of agricultural activity and contributes considerably to the persistent agricultural problems. In general, *rainfall* is the climate variable of primary importance in shaping the spatial and temporal variations of agriculture in the tropics. Temperature is secondary but becomes increasingly important further polewards.

This is clearly evident in the semi-arid tropics, a region that contains only 13% of the world's lands, 15% of its people (it includes 48 less developed countries) and a small proportion of its food production; sorghum and millet are the principal crops (Swindale *et al.*, 1981). Here the relationship between rainfall and agriculture is finely tuned: the seasonal cycle of rainfall directly determines the tempo and rhythm of agriculture through its limitation on the length of the growing season (Mattei, 1979). The rainy season *is* the growing season and, over much of the tropical world, is synonymous with the monsoon. Unlike the semi-arid temperate regions, where rainfall is more evenly distributed over the year and where soil moisture reserves can accumulate, the semi-arid tropics receive rain only during concentrated months of the year. The planting date depends critically on the first rains, and delays in the planting date have major effects on end-of-season yields. Thus, even relatively small variations in the amount and timing of rainfall cause high variability in interannual yields.

In many areas, changes in climate could magnify existing problems. Traditional cropping patterns throughout the semi-arid region are finely adjusted to the spatial characteristics of climate (Mattei, 1979). Crop varieties, planting dates and management practices vary markedly along environmental gradients in accordance with climatic expectations. In many cases, even minor alterations in the rainfall regime could disrupt this delicate adjustment and have major repercussions on agricultural productivity. In addition, the inherent environmental variability, along with the limited reserves and access to capital, means that traditional agriculturalists place high priority on minimizing the risks of loss in dry years in lieu of maximizing their gains in the wet years (Swindale *et al.*, 1981). This, in turn, affects long-term average yields and production. Increases in the frequency of dry years could further consume the limited resources of semi-arid farmers of the tropics and have serious impacts on production in countries that already (and increasingly) depend on food imports and suffer from problems of malnourishment. The current plight in Africa of high drought incidence, desertification, high population growth and declining per capita food production may be symptomatic of this process. If, on the other hand, the frequency of dry years decreases in the future, many of the problems could be ameliorated. The direction of precipitation change is highly uncertain.

8.4.2 Humid Tropics

In the humid tropics, rainfall is also a major growth-limiting factor due to its variability and the high potential evapotranspiration. The close dependence of the growing season on rainfall throughout the entire tropics is depicted schematically in Figure 8.3. Food production depends not only on the timely appearance of the monsoon, but also on its strength and reliability throughout the growing season. Indeed, late heavy rains are often just as disruptive as the late arrival of rain, causing flood damage to established crops at a time when it is too late to replant. This applies even to rice which has high water requirements.

Rice is the principal crop of the Asian humid tropics and is the staple food for perhaps 60% of the world's population (van Keulen, 1978). Production of rice in lowland locations is suited to the humid tropics, particularly if well-developed water control and/or irrigation networks are in place. Nevertheless, even lowland rice is susceptible to rainfall variability. Tanaka (1978) has shown a positive relationship between rice yield and rainfall throughout monsoon Asia. The relationship is strongest in areas where water control and irrigation are least developed—India, Burma, Bangladesh, Republic of Korea. The effects of rainfall variability are seen clearly in Sri Lanka where marked differences in rice production occur over short distances between the wetter and drier zones; the Dry Zone is chronically hindered by rainfall fluctuations, which discourage further investment in water control, create financial insecurities and accentuate the impacts of climate (Gooneratne, 1978).

Under optimally irrigated conditions, radiation and temperature influence rice yields through their effects on photosynthesis and respiration rates, respectively (van Keulen, 1978). For instance, in Japan (a temperate region), irrigation has greatly reduced the susceptibility of lowland rice to rainfall variation (Toriyama, 1978). Rather, coolness hampers yields, especially in the northern areas of Japan, a problem which is being solved in part through crop breeding for cool temperature tolerance (Satake, 1978; Nakagawa, 1978).

Expanding use of marginal lands may be increasing vulnerability to climatic change. In most of the developing countries of the low-latitudes, ag-

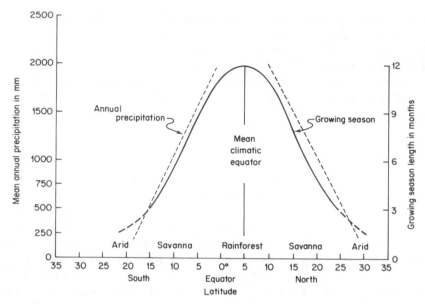

Figure 8.3 A schema of the relationship between mean annual precipitation and growing season length in tropical climates (from Newman, 1977)

gregate food production has managed to keep abreast of population growth over the last few decades. Per capita food production has risen slowly, at about 0.4% per year (Pino *et al.*, 1981), with a few exceptions (like large portions of Africa) where actual declines have occurred. In some areas, yield increases have been responsible for the gains, as in the case of rice production throughout most of south and south-east Asia, or in India where higher yielding varieties of wheat have been adopted. However, in many regions of the developing world, the expansion of total land under cultivation has mainly accounted for production gains. In other areas, a switch to more intensive cultivation practices or crop type has taken place. These areas of new production or rapidly changing agricultural practices often tend to be marginal lands (the best lands are taken first) which are less productive and more susceptible to climatic fluctuations.

For example, Fukui (1979) argues that in South Asia there are three types of agricultural pursuits involving annual crops other than lowland rice that are particularly susceptible to climatic variations. These are currently widespread in the humid tropics and are expanding, and include: (1) shifting cultivation in which traditional fallow cycles are shortened, leading to declines in soil fertility and erosion; (2) continuous cropping (of maize, sorghum, upland rice, cassava, peas, beans, etc., primarily for subsistence); and (3) cultivation of feed crops for export, an activity which has increased enormously in recent years (e.g. maize production in Thailand). These activities are typically relegated to upland areas that are subject to soil deterioration and that are hydrologically marginal when intensively utilized for agricultural production. Population increases, resettlement programmes, urbanization, expanding markets for feed crops and other factors are encouraging greater use of the upland areas. These processes thus may be accelerating environmental degradation and increasing the risks from climatic variability (Fukui, 1979).

Moreover, there is a strong social, political and economic component to the increasing agricultural occupation of marginal lands. The pressures to relocate are felt greatest by the poorer cultural groups in many parts of the developing world, those who stand on the fringe of the growing market economies. The better lands are often converted to cash crop production by those with capital and expertise to do so, while the disadvantaged are forced to accept less than environmentally favourable locations—a 'marginalisation' process (O'Keefe and Wisner, 1975; Susman *et al.*, 1983). Thus, in considering the notion of marginal land, one is forced to consider 'marginalised' people: 'where' and 'who' are inextricably bound.

8.4.3 The Temperate Regions

In this and other respects, the temperate regions of the world differ sharply from the tropics. The mid- and high-latitude zones are the centres of grain production other than rice. North America, Europe, the USSR and China account for over three-quarters of the world's maize and—along with Argentina and Australia in the Southern Hemisphere—wheat production (USDA, 1981). The potential climatic range of most of the important crops grown in the temperate regions is quite large. Wheat and potatoes, for instance, can be grown in any state of the USA (Wittwer, 1980).

In the mid- and high-latitudes, temperature is an important factor in shaping the spatial and temporal variations in agriculture. In contrast to the tropics, the growing season in the temperate zones is generally defined by temperature. Polewards, the geographic extent of crop production is ultimately set by temperature, and the length of the frost-free growing season determines the spatial limits of various agricultural activities. One research issue is whether a warmer climate would allow the geographic extension of cereal production into higher latitudes and elevations, as in expanding the northern limits of wheat and barley production in Canada (Williams, 1975). In areas further equatorward, a warmer climate might present opportunities for an additional crop per year.

But this is not to say that precipitation does not exert a strong, often dominant, force, particularly in the drier, mid-continental locations. In the USA, dry-land wheat growing merges into livestock grazing at its westernmost boundary as a result of low and variable precipitation. In drier areas of the US Great Plains, wheat growing is only feasible with groundwater irrigation, as in the High Plains of Texas. Even in the northern cereal growing regions of the USSR and Canada where average temperatures define the spatial extent of crop production, crop yields within the core region appear to be most sensitive to soil moisture conditions (Stewart, 1985). There, the principal yield response of a change in temperature is often through the effect on evapotranspiration and, hence, soil moisture availability. Thus, in the mid-to high-latitudes, temperature and precipitation have complex and interactive effects on crop production.

The role of climate in the interannual variability of grain production is a major concern. There are some major differences between regions. For instance, the large interannual variability in total wheat production in the Soviet Union is weather-dominated, whereas in the United States the fluctuations are smaller and explained largely by government policy regarding land utilization (Tarrant, 1984). Yield variations at smaller scales are related to climate in complex ways, as attempts at modelling crop–climate relationships have shown (see Chapter 9.4). Large annual fluctuations in region-wide grain yields can result from specific short-term weather anomalies occurring during critical periods of plant growth. For example, in the North American corn (maize) belt, flowering and pollination occur in July, during which time the crop is particularly susceptible to high temperatures. High temperatures can promote premature flowering and the tassels throw their pollen before the silks are formed, with the result of poorly formed ears and low yields. Such was the case in 1980 (McQuigg, 1981). In 1974, the corn yield suffered a large drop from late-season frost, another threat to regional production.

In contrast to the tropics, the large increases in total production during recent decades in the temperate, grain-producing regions have resulted largely from intensification, rather than from expansion of cropped area. (An important exception has been the opening of new lands for the extension of wheat production into Kazakstan and Siberia in the Soviet Union.) Development of new grain varieties, irrigation, increasing applications of fertilizers and pesticides, research and improved management—broadly, 'technology'—have contributed to significant increases in average yields, particularly in the post-war period. In the United States, for example, maize yields have approximately tripled and wheat yields have doubled since 1950. Some countries in Europe have reached average wheat yields that are two and three times those produced in North America (USDA, 1981). For the United States, the government-supported technological advances have created a production capacity that has outstripped demand. A chronic problem has been, and continues to be, one of overproduction and of finding ways of marketing the surpluses. Withholding land—often marginal lands—from production has been one policy response to this problem.

Consequently, the climate-related issues of the higher latitudes, though fundamentally similar to those of the lower latitudes, are expressed slightly differently. There are two sets of issues. First, will technology be able to maintain increases in average yields in the future? Will the rates of increase from technology or possibly higher CO_2 concentrations be sufficient to counterbalance or overcome any adverse impacts on average yields from climatic change? From the long-term perspective, one concern is whether future rates of agricultural change and adaptability will diminish as yields approach the theoretically possible maxima. Some have argued, for instance, that US grain production is reaching a plateau in yield gains (e.g. Thompson, 1975). Others see the yield trends continuing with no anticipated problems in adjusting to climatic change (CIA, 1974; Waggoner, 1983). This issue is further clouded by the uncertainties over continued government support for agriculture in the US and EEC in the face of budget deficits and financial problems. The answers are not clear.

Second, has technology also been effective in reducing the interannual variability of yields resulting from the detrimental year-to-year variations in weather? In other words, is grain production today less vulnerable to the risks of droughts, frosts, freezes, etc., than in the past? And, therefore, do we have less to worry about regarding climatic change and variability in the future? These issues have been subject to considerable debate over the last decade and are still largely unresolved (Schneider and Londer, 1984). Of course, since absolute yields have increased one would expect that the absolute yield variability has also increased. This supposition is borne out by the grain yield statistics in North America (Newman, 1978) and in England and Wales (Dennett, 1980), for instance. The more controversial issue revolves around the relative variability, that is, the departures as a per cent of average (or trend) values. In the context of North America, several studies claim that technology has been triumphant in reducing yield variability (e.g. Newman, 1978; USDA, 1974). Others argue that if one accounts for the fact that from the mid-1950s to the mid-1970s climate was less variable, then grain yields show no decrease in sensitivity to climatic variations and, in fact, the sensitivity may be increasing (NOAA, 1973; Haigh, 1977; Hazell, 1984). The issue so far remains unresolved, principally because of the difficulty in separating the effects of weather and technology.

8.4.4 Global Food Trade

At a global scale, problems of climate and agriculture are superimposed upon a world that is increasingly interconnected through trade, especially cereal trade. From 1960 to 1982, the cereal trade expanded considerably, from about 67 Mt to 221 Mt (FAO, 1961; 1983b). Most of the interaction

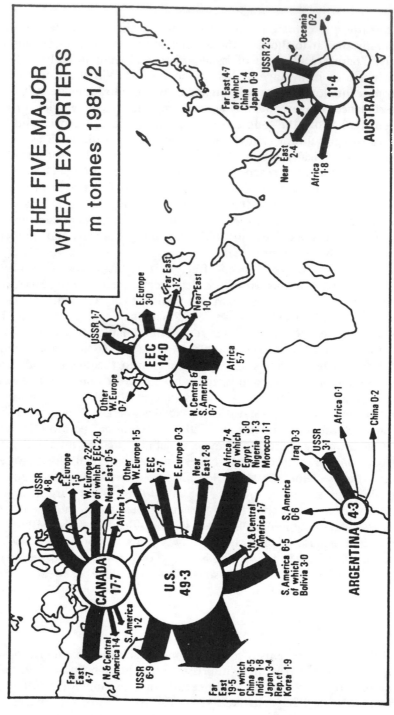

Figure 8.4 Major wheat exporting countries and their markets. Data from the Annual Report of the International Wheat Council (1983) (from Tarrant, 1985)

occurs between the developed countries, those that have the surpluses to sell and those that have the financial ability to buy (Figure 8.4). Wheat dominates the world trade, whereas trade in rice is very small in relation to annual world production.

Imports by developed nations have increased fourfold over these twenty-two years. The importers of the developed countries enter the world market largely to acquire feed grains to accommodate rising domestic demands for higher dietary standards as a result of increasing incomes and policy decisions, rather than population growth. The largest importers of cereals are the Soviet Union and Japan, who accounted for 29% of total imports in 1980 (FAO, 1982). Although the developing nations import less than developed nations, their rate of increase of imports has been slightly higher over the last 20 years. Of the developing countries, it is the middle income countries that account for the larger share of the purchases (Wagstaff, 1982). Grain imports of some OPEC countries have increased sharply following the rise of national incomes during the last decade (Parker, 1978). In contrast, the poorer countries, who, nutritionally, have greatest need for food imports, often simply cannot afford the price.

Exports of cereals originate from only a handful of countries in the mid-latitudes. In 1982, exports from the USA and Canada tallied 127 Mt, or about 58% of the world total, with Argentina, Australia and France contributing a large portion of the remainder (FAO, 1983b). Furthermore, while exportable surpluses have become concentrated in fewer countries, the quantities have increased dramatically. In 1960, North America's share of net grain exports stood at only 39 Mt. Meanwhile, from 1960 to 1982, net imports to Asia increased fourfold and, after entering the world market as a major importer in the early 1970s, the Soviet Union (and Eastern Europe) have increased imports to around 44 Mt in 1982 (FAO, 1983b).

One of the major uncertainties in this interregional reliance is climate. The impairment of grain production in one region of the world is of increasing importance to regions elsewhere. Moreover, the trend toward concentration of the centres of both grain supply *and* demand does not bode well for lower income countries. Of particular concern is the probability of simultaneous crop failures in several major producing or importing countries, as in North America and the Soviet Union. Currently, the probabilities are small (Sakamoto *et al.*, 1980) and production losses in one region are usually offset by gains elsewhere, hence the low interannual fluctuations in aggregate world production. If, however, climatic change in the mid-latitudes proves detrimental to cereal production and the year-to-year risks of shortfalls increase, then the major sources of both supply and demand for grain exports would suffer. The position of the less developed countries with respect to their ability to purchase food could decline drastically under conditions of 'shortage' and higher prices on the world market.

8.4.5 Implications and Issues

In sum, we have made the following broad observations regarding climate and agriculture at the global scale:

- Rainfall is a principal constraint to agriculture in the semi-arid and humid tropics. The effects of increasing atmospheric CO_2 on precipitation patterns and radiation in the tropics (and elsewhere) are highly uncertain.

- Due to climatic, agro-technological and socio-economic factors, yields and food production in many developing countries of the tropics and subtropics tend to be low and unstable. Current interannual variability of yields is an acute and pressing problem. With few food security stocks, any downward turn in production could be critical.

- In the temperate and higher latitudes, temperature and precipitation interactively influence the spatial and temporal variations in agriculture. It is believed that a CO_2-induced warming would be pronounced at these latitudes.

- 'Technology' has steadily increased average grain yields in the developed countries of the temperate latitudes. Will the technological progress continue? Will technology or possibly high CO_2 concentrations compensate for any adverse effects of climatic change? Has technology reduced or increased the interannual variability in crop yields due to weather fluctuations?

- Food production at the margins of crop regions may be particularly sensitive to climatic change. Increasing use of marginal lands for intensive cultivation has contributed to, and may be increasing, the sensitivity of production to climatic change and variability.

- Increased interregional reliance from expanding grain trade has made the climate of one region the concern of another. Both key centres of supply and demand for exportable surpluses are located in the temperate regions. Any adverse effects of climatic change could have disproportionately greater impacts on developing countries of the lower latitudes whose purchasing power cannot always compete during times of scarcity.

How the global systems of cereal production would, or could, respond over time to gradual changes in the regional climate, including CO_2 concentrations, is the key issue. As a starting point, this requires fundamental knowledge of the *sensitivity of crop yields in the core regions* of production, either as it would affect supply in the exporting countries or demand in the importing countries. It also requires knowledge of how regional food production might be altered through possible *changes in cropping patterns at the margins* of production (WMO, 1984). Finally, it requires understanding of the *dynamic response of the agricultural system* as it re-adjusts over time through changes in technological inputs, management practices, pricing mechanisms, government policies, food security stocks and the like. We know there exists considerable flexibility and resilience in agricultural production in the face of climatic change and variability. The question is *how much* and *for whom.*

Against this background we shall address the specific issues and approaches to assessing the impacts of increasing global CO_2 and climatic change in Chapter 9.

8.5 GLOBAL FOREST ECOSYSTEMS AND CLIMATIC CONSTRAINTS

The relatively 'natural' appearance of many forested landscapes can give one the false impression that forest ecosystems are largely unmanaged. In fact, the world's forests are subjected to a wide range of management levels. Almost all forests are managed to some degree for purposes ranging from intensive commercial extraction to extensive resource conservation. Therefore, as in agriculture, the potential impacts of increasing concentrations of atmospheric CO_2 and climatic change must ultimately be examined in the context of human use and manipulation of the natural system.

In the most extreme case, there has been some experimentation with intensively managed biomass plantations in which trees are irrigated, fertilized and harvested in short (ca 2 to 5 years) rotations. This form of cellulose production is the type of forestry that most closely resembles intensive agriculture. In more traditional forestry, forest management involves the regeneration of a commercially valuable tree species by altering sites, planting seedling trees at appropriate spacings, thinning the trees and harvesting the tree crop. In favourable environments, some of these activities are left to natural processes. For example, if a commercial tree species has vigorous regeneration in a given environment, the site preparation or planting management steps can be eliminated, and thinning and harvesting of trees become the only concern. In less intensive forestry, trees are periodically harvested, but the thinning of trees to optimize the forest productivity is omitted. In extensive forms of management, forests are maintained to protect watersheds, wildlife habitats or recreational environments. Even in the most remote forests, there is a degree of management that stems from human intervention with natural processes: for example, reducing the frequency of wildfires in wilderness areas.

Because of this gradient of management intensity, global environmental change could manifest itself in radically different ways. In more intensive forms of forestry, a change in growth and regeneration rates could affect management costs or the techniques used to extract wood products. In the less intensively managed forests, an environmental change might actually change the structure, composition and areal distribution—and consequently the function—of the forest ecosystem.

There are two aspects of the behaviour of forest systems that should be considered in assessing the impacts of environmental change. First, there is a considerable degree of spatial heterogeneity in the potential response of the world's forests to changes in climate, as discussed above. Second, at any given place there is a wide range of temporal scales over which forests will respond dynamically. Unlike the vast majority of agricultural systems, forested systems are dominated by long-lived organisms (trees) that can respond to stress or change at several different time scales. Problems in assessing the response of forest systems with respect to any alteration of environmental conditions are made complex by these multi-level responses (discussed in detail in Chapter 10).

There are three major interrelated issues concerning the impacts of increasing CO_2 and climatic change that, taken together, transcend the spectrum of forest management and combine the considerations of spatial and temporal dynamics: What are the possible impacts on the *areal distribution* of the world's forests? At local scales, what are the potential changes in *forest composition*? What are the possible changes in forest *productivity*? Let us examine each of these issues in turn.

8.5.1 Climatic Change and the Areal (Macroscale) Response of Forests

The clearly observable correspondence between the distributions of global climates and the spatial patterns of vegetation leads one to expect that a change in the former should eventually produce a response in the latter. What climatic factors should be considered in evaluating the potential change in areal distribution of the world's forests?

Some guidance is provided by Holdridge (1947; 1964), who developed a systematic classification of the expected vegetation under differing temperature and moisture conditions (Figure 8.5). The Holdridge diagram is similar to other climate/vegetation mapping systems in that it explicitly recognizes the variables of temperature (expressed in this case as 'biotemperature' which is computed as a heat index for periods during which plants can be photosynthetically active) and moisture (expressed as either rainfall or evapotranspiration). The Holdridge diagram illustrates several relationships that, while perhaps oversimplifying, provide perspective for understanding the response of the global vegetation to climatic change. First, there is a parallel between the latitudinal zonation of the earth (boreal, tropical, etc.) and the zonation of vegetation at different altitudes on mountains (montane, alpine, etc.). Second, the responses to temperature and moisture or precipitation changes depend on relative, rather than absolute, changes. A small

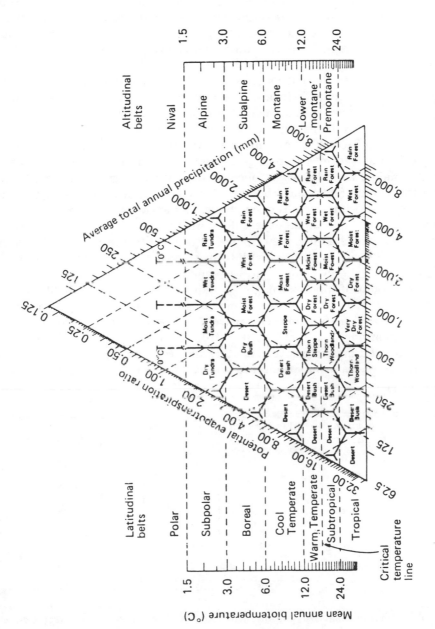

Figure 8.5 The Holdridge Life-Zone Classification System (Holdridge, 1947; 1964)

absolute increase in temperature could be expected to cause a large response
in the ecosystems of the cooler climates of high altitudes or latitudes. Similarly, a small absolute increase in moisture could have a profound effect
in an arid region. To cause a vegetational change of comparable magnitude
in a wet, warm region, the environmental changes would need to be much
greater.

If one plots the present global cover of forest as a function of latitude
(Figure 8.6), the forested regions of the world resolve into two major forest
systems. First, there is a considerable extent of forest in the higher latitudes
that tends to be dominated by evergreen coniferous trees—the circumpolar
boreal forest. Second, at low latitudes a variety of evergreen and raingreen
tropical forests form a second great forest system. The deciduous forests
of the middle latitudes which once covered large areas in Europe, China
and the United States have been reduced greatly in areal extent by land
conversion. For the reasons cited above, of the two great forests now covering the earth, one would expect the boreal forests to be most sensitive to
a warming. To determine the response of tropical forests to climatic change
would necessarily require spatial information on the change in moisture, if
we generalize simply from Figures 8.5 and 8.6.

The amount of time that might be required for the areal distribution of
forests actually to respond to a change in global climate is largely a matter
of conjecture. The time needed for trees to migrate into a region can vary
widely according to the species (e.g. from 25 meters per year in *Fraxinus
ornus* to 2000 meters per year in *Acer* species; Huntley and Birks, 1983).
The ranges of many important tree taxa in both North America and Europe

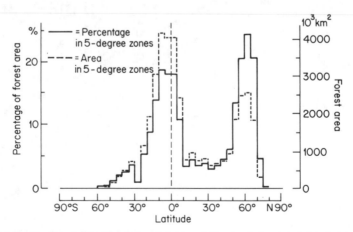

Figure 8.6 Latitudinal distribution of forested land of the earth in 5-degree zones.
Percentages are related to the total area of each zone. The distinct bimodality of the
distribution corresponds with the boreal forests in the higher northern latitudes
and the tropical forests in the equatorial zones (from Baumgartner, 1979)

have been moving since the large alteration of the global vegetation pattern that accompanied the last glaciation and may still be moving.

8.5.2 Local Changes in Forest Composition

Along with the potential areal response of the world's forests (and other vegetation types), there is a smaller scale problem of how forest composition at given locations might change. Results of computer models and of observations of actual forests indicate that, in forest compositional dynamics, the inertia is of the order of the generation of a tree (ca 100 to 200 years). A closer inspection of the processes controlling the structure and composition of a forest reveals that the mechanisms which could be altered by climatic change are numerous and of great potential importance. Changes in climate could be expected to alter differentially the regeneration success and the growth and mortality rates of tree species. The alteration of the competitiveness of the various taxa most likely would be manifested as a change in the forest composition. Changes in moisture conditions (and perhaps thunderstorm frequency) could alter fire probabilities and rates of wildfire spread. Warmer winters could decrease the mortality of overwintering insect pests and thus increase the likelihood or perhaps the intensity of insect outbreaks.

The magnitude of these changes could, in some cases, be quite large, even with relatively small changes in climate. Moreover, these effects are apt to be case specific; it is difficult to generalize across all forests. For instance, a warming in one region could produce increased forest productivity and dominance of a valuable commercial tree species, while the same warming in another region could increase pest populations or wildfire frequency and reduce the extent of commercial forest. Existing computer simulation models for forest phenomena at these time and space scales may prove useful for assessing a number of these local effects and for this reason are reviewed in Chapter 10.

8.5.3 Possible Impacts on Forest Productivity

The immediate responses that one might expect to occur from increased CO_2 or climatic change involve modification of forest productivity. Again, there are numerous differences in productivity and in the factors that could modulate the effects of climatic change from one location to the next. For example, a warming could be expected to do little to increase the productivity of a nutrient-limited forest system. Nonetheless, across a broad range of forests, there are positive relationships (with considerable variability) between the temperature and either the total biomass or the net productivity of forests. Given an adequate supply of water and nutrients, one would expect a global warming generally to enhance the forest productivity. That most of the world's forests, however, are to some degree nutrient- and water-limited with respect to their growth rates is sufficient ground for caution with regard to this generalization.

It may be possible to monitor changes in productivity through the use of remote imagery. For instance, recent research has shown that 1- and 4-km advanced very high resolution radiometer (AVHRR) data from the U.S. National Oceanic and Atmospheric Administration (NOAA) polar-orbiting, sun-synchronous series of operational satellites have immediate utility for repetitive global monitoring of terrestrial vegetation. Daily, the AVHRR sensor records the green leaf dynamics (measured as the photosynthetically active radiation intercepted by terrestrial vegetation) for the entire planetary surface. Daily coverage provides a means to minimize the effects of cloud cover by selection of the most cloud-free data over a given location over several days. Once cloud free data are available at selected time intervals, these data can be used as time series to infer primary production (Goward *et al.*, 1985; Tucker *et al.*, 1985a), to classify continental land cover (Tucker *et al.*, 1985b; Townshend *et al.*, 1985) and to study phytophenological phenomena (Justice *et al.*, 1985).

As an example of the global patterns that can be observed using these satellites, Plate 1 shows the integral of the photosynthetically active radiation absorbed by the terrestrial vegetation for a 12-month period. It represents net primary production (or an index highly correlated with net primary production, after Kumar and Monteith, 1982, and Monteith, 1977). Areas of desert and tundra have integrated values (and hence productivity values) close to zero and are represented by the tan colours (i.e. the Sahara, Arabia, portions of southwest Asia, Namibia, coastal South America, Arctic areas and Antarctica). The orange-tan colours represent areas with low amounts of primary production such as semi-arid grasslands and steppes (i.e. Sahel zone, Patagonia, the interior of Australia, etc. and the more northern areas of the boreal zone and southern areas of the tundra). The purple colours represent areas of highest primary production such as much of South America, Central America, the West coast of the USA, equatorial Africa, southeast Asia, etc.). From the patterns in Plate 1, one can clearly see the general pattern of the earth's vegetation and the location of the forests.

Through the use of remote sensing techniques a potential for global environmental monitoring of productivity now exists. These methodologies also have the potential to increase our understanding of the interactions between the biospheric productivity and the atmospheric systems. For example, latitudinal patterns of primary production inferred from AVHRR data seem to be inversely correlated with annual latitudinal fluctuations in atmospheric CO_2 concentrations (Tucker *et al.*, 1985c). The exact methods or techniques needed to extract quantitative information crucial to many global environmental questions will likely require additional development, but the continued availability of the AVHRR data will allow future comparisons to be made from actual data. Data for such analyses have existed in archive form since July 1981 (with the launch of the NOAA-7) and continued data collection is guaranteed through the mid-1990s.

8.6 SOME FURTHER IMPLICATIONS OF ECOSYSTEM CHANGE: SUMMARY AND CONCLUSIONS

There are no firm grounds for believing that the net effects of increased CO_2 and climatic change will be adverse rather than beneficial. At the extreme, some assessments like the *Global 2000 Report to the President* (U.S. Council on Environmental Quality and Department of State, 1980) see future changes in climate coinciding with deteriorating conditions in agriculture, forests and other resources, and thus paint a very gloomy picture indeed. In contrast, Simon and Kahn (1984) examine the same issues and, in a strongly optimistic tone, reach just the opposite conclusions. In fact, at a global scale the uncertainties that are involved in both sets of analyses are large enough to accommodate both views.

If, in this (and subsequent) chapters, we tend to emphasize the potential negative impacts, it is only because those are the ones which are of most immediate concern to society and which the scientific community should hope to identify and predict. The list of possible adverse consequences of climate-related ecosystem changes is long and speculative, and the following represent a sample.

Conservation. There are many natural parks and reserves that are refuges for rare and endangered plant and animal species. Often these parks occupy a relatively small area in a setting of non-park land. If an environmental change made such parks unsuitable as habitats for these species, it is uncertain whether alternative refuges could be found or whether it would even be possible to transport species to new sites. The risk of widespread extirpation of rare species (particularly those with local distributions) could be high as a result of climatic change.

Forestry. Forestry as a predictive science used in a management context is highly dependent on data or local knowledge of forest response to specific management treatments. Under a sufficently large change in climate, this local knowledge base would have to be used outside its calibration range and the consequences of management actions would be less certain.

Related ecological processes. The global pattern of many of the ecological processes in natural systems could be altered if the climate changed. Insect pests, pathogenic organisms and wildfire frequencies could all change. While the prediction of such changes is highly uncertain, their potential impacts are quite large.

Hydrological systems. The impact of climatic change on regional ecosystems (particularly forests) could alter the hydrological characteristics of watersheds. Decreases in transpiration rates from the direct effects of increased CO_2 on vegetation might increase runoff, for instance, and enhance the effects of precipitation increases or offset the effects of precipitation decreases (Wigley and Jones, 1985). Changes in flooding and river flow rates could have pronounced effects on the rivers themselves, on the ecosystems adjoining the rivers, and on the various human activities that depend on reliable quantity and quality of water.

If, on the other hand, the impacts of increased CO_2, trace gases and climatic change on agriculture, forests and other ecosystems prove, on balance, favourable, all the better.

In summary, this chapter has set the stage for a more detailed examination of the effects of increased CO_2 and climatic change by outlining the major issues and dimensions of the problem in the global context, with an emphasis on agriculture and forest ecosystems. For both agriculture and forests the basic questions are similar: How would crop yields or forest productivity change? How might crop types or forest composition be altered, particularly at the margins of production or at ecological transition zones? How might these effects, integrated over time and space, change global patterns of forests or agricultural production, taking into consideration the interactive natural and human processes that make both systems very dynamic?

In order to derive meaningful, credible answers, it is necessary to interface scenarios of environmental change with research procedures or models capable of testing the sensitivity of the systems. What approaches are available? How have they been used and what questions have been asked of them? What have we learned so far from their specific applications to problems of increased CO_2 concentrations and climatic change? We turn now to consider these questions in the context of agriculture and, in the subsequent chapter, forests.

NOTE ON AUTHORSHIP AND ACKNOWLEDGEMENTS

Primary responsibility for writing the sections concerned with the general ecosystem effects and with agriculture (8.1 to 8.4) was assumed by R. A. Warrick. H. H. Shugart and M. Ja. Antonovsky concentrated on forest ecosystems (Section 8.5). The collaborating authors, J. R. Tarrant and C. J. Tucker, contributed substantially to Sections 8.4 and 8.5.3, respectively. T. M. L. Wigley also wrote portions of Section 8.3, for which we are grateful.

The authors are indebted to the following individuals for their review comments and constructive suggestions: M. A. Ayyad, W. Böhme, B. Bolin, W. Degefu, B. R. Döös, H. W. Ellsaesser, H. L. Ferguson, D. P. Garrity, R. M. Gifford, F. K. Hare, J. Jäger, M. El Kassas, P. M. R. Kiangi, H. H. Lamb, H. E. Landsberg, J. A. Laurmann, A. de Luca Rebello, J. Mattsson, W. J. Maunder, A. S. Monin, O. Preining, A. Rapp, B. R. Strain, M. S. Swaminathan, P. E. Waggoner, G. F. White, T. Woodhead, M. M. Yoshino.

8.7 REFERENCES

Baumgartner, A. (1979) Climatic variability and forestry, in WMO, *Proceedings of the World Climate Conference*, WMO-No. 537, pp. 581–607 Geneva, WMO.

Bolin, B. (1980) *Climatic Changes and their Effects on the Biosphere*, WMO-No. 542, Geneva, WMO.

Budyko, M. I. (1982) *The Earth's Climate: Past and Future*. New York, Academic Press.

Crabbe, D., and Lawson, S. (1981) *The World Food Book*, London, Kogan Page.

Central Intelligence Agency (CIA) (1974) *Potential Implications of Trends in World Population, Food Production and Climate*, OP-401, August.

Dennett, M. D. (1980) Variability of annual wheat yields in England and Wales, *Agricultural Meteorology*, **22**, 109–111.

Flohn, H. (1980) *Possible Climatic Consequences of a Man-Made Global Warming*, Laxenburg, Austria, International Institute for Applied Systems Analysis.

Food and Agriculture Organization (FAO) (1983a) *Production Yearbook*, Rome, FAO.

Food and Agriculture Organization (FAO) (1961, 1982, 1983b) *Trade Yearbook*, Rome, FAO.

Fukui, H. (1979) Climatic variability and agriculture in tropical moist regions, in WMO, *Proceedings of the World Climate Conference*, WMO-No. 537, pp. 426–474. Geneva, WMO.

Gerasimov, I. (1979) Climates of past geological epochs, in WMO, *Proceedings of the World Climate Conference*, WMO-No. 537, pp. 88–111. Geneva, WMO.

Gooneratne, W. (1978) Recent climatic fluctuations and food problems in Sri Lanka, in Takahashi, K., and Yoshino, M. M. (eds) *Climatic Change and Food Production*, pp. 111–123. Tokyo, University of Tokyo Press.

Goward, S. N., Tucker, C. J., and Dye, D. G. (1985) North American vegetation patterns observed with NOAA-7 Advanced Very High Resolution Radiometer, *Vegetacio* (in press).

Haigh, P. (1977) *Separating the Effects of Weather and Management on Crop Production*, Contract report, C.F. Kettering Foundation, Grant No. ST-77-4, Columbia, Missouri.

Hare, F. K. (1979) Climatic variation and variability: empirical evidence from meteorological and other sources, in WMO, *Proceedings of the World Climate Conference*, WMO-No. 537, pp. 51–87. Geneva, WMO.

Hare, F. K. (1983) *Climate and Desertification*, WCP-44, World Climate Programme, Geneva, WMO.

Hare, F. K. (1984) Recent climatic experience in the arid and semi-arid lands, *Desertification Control Bulletin* **10**, May, Nairobi, United Nations Environment Programme.

Hazell, P. B. R. (1984) Sources of increased instability in Indian and U.S. cereal production, *Am. J. of Agric. Econ.*, **66**, no. 3, 302–311.

Holdridge, L. R. (1947) Determination of world plant formations from simple climatic data, *Science*, **105**, 367–368.

Holdridge, L. R. (1964) *Life Zone Ecology*, San José, Costa Rica. Tropical Science Center.

Huntley, B., and Birks, H. J. B. (1983) *An Atlas of Past and Present Pollen Maps for Europe: 0–13000 Years Ago*, Cambridge, Cambridge University Press.

Justice, C. O., Townshend, J. R. G., Holben, B. N., and Tucker, C. J. (1985) Phenology of global vegetation using meteorological satellite data, *Int. J. Remote Sens.* (in press).

Kellogg, W., and Schware, R. (1981) *Climate Change and Society*, Boulder, Colorado, Westview Press.

Kumar, M., and Monteith, J. L. (1982) Remote sensing of plant growth, in Smith, H. (ed), *Plants and the Daylight Spectrum*, London, Pitman.

Lamb, H. H. (1977) *Climate: Present, Past and Future*, Vol. 2, London, Methuen.

Lemon, E. R. (ed) (1983) *CO_2 and Plants: The Response of Plants to Rising Levels of Atmospheric Carbon Dioxide*, AAAS Selection Symposium No. 84, Boulder, Colorado, Westview Press.

Lough, J. M., Wigley, T. M. L., and Palutikof, J. P. (1983) Climate and climatic impact scenarios for Europe in a warmer world, *Journal of Climatology and Applied Meteorology*, **22**, 1673–1684.

McQuigg, J. D. (1981) Climate variability and crop yield in high and low temperate regions, in Bach, W., Pankrath, J., and Schneider, S. H. (eds), *Food–Climate Interactions*, pp. 121–137. Dordrecht, D. Reidel Publishing Company.

Mattei, F. (1979) Climatic variability and agriculture in the semi-arid tropics, in WMO, *Proceedings of the World Climate Conference*, WMO-No. 537, pp. 475–509. Geneva, WMO.

Monteith, J. L. (1977) Climate and the efficiency of crop production in Britain, *Phil. Trans. R. Soc.*, **B-281**, 277–296.

Nakagawa, Y. (1978) Occurrence and prevention of drought injury to crops in western Japan, in Takahashi, K., and Yoshino, M. M. (eds), *Climatic Change and Food Production*, pp. 231–236. Tokyo, University of Tokyo Press.

National Academy of Sciences (NAS) (1972) *Genetic Vulnerability of Major Food Crops*, Washington, D.C., NAS.

National Academy of Sciences (NAS) (1976) *Climate and Food: Climatic Fluctuation and U.S. Agricultural Production*, a Report of the Committee on Climate and Weather Fluctuations and Agricultural Production, National Research Council, Washington, D.C., NAS.

National Oceanic and Atmospheric Adminstration (NOAA) (1973) *The Influence of Weather and Climate on United States Grain Yields: Bumper Crops or Drought*, Washington D.C., U.S. Dept. of Commerce.

Newman, J. E. (1977) Growing seasons as affected by climatic change, in *Proceedings of the Conference on Climate Change and European Agriculture*, Centre for European Agricultural Studies, Wye College, University of London, Ashford, Kent, UK.

Newman, J. E. (1978) Drought impacts on American agricultural productivity, in Rosenberg, N. E. (ed), *North American Droughts*, pp. 59–79. Boulder, Colorado, Westview Press.

O'Keefe, P., and Wisner, B. (1975) African drought: the state of the game, in Richards, P. (ed), *African Environment*, London, International African Institute.

O'Neill, R. V., and DeAngelis, D. L. (1980) Comparative productivity and biomass relations of forest ecosystems, in Rechle, D. E. (ed), *Dynamic Properties of Forest Ecosystems*, Cambridge, Cambridge University Press.

Parker, J. B. (1978) US Farm exports to Saudi Arabia seen doubling this year, *Foreign Agriculture*, **16** (41), 2.

Pino, J. A., Cummings, Jr. R. W., and Toenniessen, G. H. (1981) World food needs and prospects, in Bach, W., Pankrath, J., and Schneider, S. H. (eds), *Food–Climate Interactions*, pp. 47–68. Dordrecht, D. Reidel Publishing Company.

Sakamoto, C., LeDuc, S., Strommen, N., and Steyaert, L. (1980) Climate and global grain yield variability, *Climatic Change*, **2(4)**, 349–361.

Satake, T. (1978) Sterility-type cool injury to rice plants in Japan, in Takahashi, K., and Yoshino, M. M. (eds), *Climatic Change and Food Production*, pp. 245–254. Tokyo, University of Tokyo Press.

Schneider, S. H., and Londer, R. (1984) *Coevolution of Climate and Life*, San Francisco, Sierra Club Books.

Simon, J. L., and Kahn, H. (eds) (1984) *The Resourceful Earth*, Oxford, Basil Blackwell.

Shah, M., and Fischer, G. (1984) People, land, and food production: potentials in the developing world, *Options*, **2**, 1–5.

Stewart, R. B. (1985) The impacts on cereal production, in Part III: The impacts of climatic variations on agriculture in the Canadian Prairies: the case of Saskatchewan, in Parry, M. L., Carter, T. R., and Konijn, N. (eds), *Assessment of Climate Impacts on Agriculture, Vol. 1 in High Latitude Regions*, Dordrecht, D. Reidel Publishing Company (Forthcoming).

Susman, P., O'Keefe, P., and Wisner, B. (1983) Global disasters, a radical interpretation, in Hewitt, K. (ed), *Interpretations of Calamity*, pp. 263–283. Boston, Allen & Unwin Inc.

Swaminathan, M. S. (1979) Global aspects of food production, in WMO, *Proceedings of the World Climate Conference*, WMO-No. 537, pp. 369–405. Geneva, WMO.

Swaminathan, M. S. (1984) Climate and agriculture, in Biswas, A.K. (ed), *Climate and Development*, Natural Resources and the Environment Series Vol. 13, pp. 65–95. Dublin, Tycooly International Publishing Limited.

Swindale, L. D., Virmani, S. M., and Sivakumar, M. V. D. (1981) Climatic variability and crop yields in the semi-arid tropics, in Bach, W., Pankrath, J., and Schneider, S. H. (eds), *Food–Climate Interactions*, pp. 139–166. Dordrecht, D. Reidel Publishing Company.

Tanaka, M. (1978) Synoptic study on the recent climatic change in Monsoon Asia and its influence on agricultural production, in Takahashi, K., and Yoshino, M. M. (Eds), *Climatic Change and Food Production*, pp. 81–100. Tokyo, University of Tokyo Press.

Tarrant, J. R. (1984) The significance of variability in Soviet cereal production, *Trans. Inst. Br. Geogr. N.S.*, **9**, 387–400.

Tarrant, J. R. (1985) A review of international food trade, *Progress in Human Geography*, June (in press).

Thompson, L. M. (1975) Weather variability, climatic change and grain production, *Science*, **188**, 535–541.

Toriyama, K. (1978) Rice breeding for tolerance to climatic injury in Japan, in Takahashi, K., and Yoshino, M. M. (eds), *Climatic Change and Food Production*, pp. 237–243. Tokyo, University of Tokyo Press.

Townshend, J. R. G., Justice, C. O., and Kalb, V. (1985) South American land cover classification (submitted to *Interciencia*).

The Greenhouse Effect, Climatic Change, and Ecosystems

Tucker, C. J., Vanpraet, C. L., Sharmon, M. J., and Von Ittersum, G. (1985a) Satellite remote sensing of total herbaceous biomass production in the Senegalese Sahel: 1980–1984, *Remote Sens. Environ.*, **17** (in press).

Tucker, C. J., Townshend, J. R. G., and Goff, T. E. (1985b) African land cover classification using satellite data, *Science*, **277**, 369–375.

Tucker, C. J., Fung, I. Y., Keeling, C. D., and Gammon, R. H. (1985c) The relationship of global green leaf biomass to atmospheric CO_2 concentrations (submitted to *Nature*).

U.S. Council on Environmental Quality and Department of State (1980) *The Global 2000 Report to the President: Entering the Twenty-First Century*, Vols. I–III, U.S. Government Printing Office, Washington, D.C.

U.S. Department of Agriculture (USDA) (1981) *Agricultural Statistics*, Washington, D.C., Government Printing Office.

U.S. Department of Agriculture (USDA) (1974) The world food situation and prospects to 1985, *Foreign Agriculture Report no. 98*, Washington, D.C., U.S. Government Printing Office.

U.S. Department of Energy (DOE) (1984) Carbon dioxide and climate: summaries of research in FY 1983 and FY 1984, Office of Energy Research, DOE/ER-0202, Springfield, Maryland, National Technical Information Service.

Van Keulen, H. (1978) Simulation of influence of climatic factors on rice production, in Takahashi, K., and Yoshino, M. M. (eds). *Climatic Change and Food Production*, pp. 345–358. Tokyo, University of Tokyo Press.

Waggoner, P. E. (1983) Agriculture and a climate changed by more carbon dioxide, in NRC, *Changing Climate*, Report to the Carbon Dioxide Committee, Board of Atmospheric Sciences and Climate, pp. 383–418. Washington, D.C., National Academy Press.

Wagstaff, H. (1982) Food imports of developing countries, *Food Policy*, **7** (1), 57–65.

Whittaker, R. H. (1970) *Communities and Ecosystems*, New York, Macmillan.

Wigley, T. M. L., and Jones, P. D. (1985) Influences of precipitation changes and direct CO_2 effects on streamflow, *Nature*, **314**, 149–152.

Williams, G. D. V. (1975) Assessment of the impact of some hypothetical climatic changes on cereal production in Western Canada, in *Proceedings of the Conference on World Food Supply in Changing Climate*, Sterling Forest, N.Y., Dec 2–5, 1974.

Wittwer, S. H. (1980) Overview: report of Panel III on the environmental effects on the managed biosphere, in *U.S. Dept. of Energy, Workshop on Environmental and Societal Consequences of a Possible CO_2-induced Climatic Change*, CONF-79044143, pp. 46–48. Washington, D.C., U.S. Department of Energy

World Meteorological Organization (WMO) (1984) *Report of the Study Conference on Sensitivity of Ecosystems and Society to Climate Change*, Villach, Austria, 19–23 September, 1983, WCP-83, Geneva, WMO.

World Meteorological Organization (WMO) (1985) *The Reliability of Crop–Climate Models for Assessing the Impacts of Climatic Change and Variability*, Report of the WMO/UNEP/ICSU-SCOPE Expert Meeting, Geneva, May 19–24, 1984 (in press).

Wortman, S., and Cummings, Jr. R. W. (1978) *To Feed This World: The Challenge and the Strategy*, Baltimore, Johns Hopkins University Press.

CHAPTER 9

CO$_2$, Climatic Change and Agriculture
Assessing the Response of Food Crops to the Direct Effects of Increased CO$_2$ and Climatic Change

R. A. WARRICK AND R. M. GIFFORD, WITH M. L. PARRY

9.1 INTRODUCTION

As discussed in the previous chapter, climate plays a major role in determining the yield levels, the year-to-year variability and the spatial patterns of global agriculture. Increases in atmospheric CO$_2$ concentrations and changes in climate could thus have far-reaching implications for international food production and security. From a global perspective the issues vary from the temperate zones to the tropics, from the core crop regions to the margins of production, and from the developed to the developing countries. The purpose of the present chapter is to gather together what we know (and point out what we do not know) about the effects of CO$_2$ and climatic change, and to elucidate the broad approaches for addressing these issues.

This review is selective in its treatment while attempting to maintain a global perspective. The emphasis is placed on food crops, particularly grains, to the neglect of livestock, grasslands and fibre crops. This is not only because grains account for the vast bulk of global food production and serve to link world regions through international trade, but also because existing models and studies of the agricultural impacts of CO$_2$ and, especially, climatic change have focused largely on these crops. As a whole these studies jump scales to address relevant research questions at different levels of organization ranging from the plant to global food trade; as a consequence, so does this chapter.

The chapter begins by reviewing the evidence for the direct effects of higher CO$_2$ concentrations on individual plant processes, growth and yield. In Section 9.3 we turn to the subject of the agricultural impacts of long-term climatic change. In this section we consider alternative ways of formulating the problem in the light of the complexities created by short-term climatic variability and the response capabilities of agriculture itself. This is followed in Section 9.4 by a discussion of the major approaches to impact assessment and the results of specific analyses. In the concluding section we summarize the main findings and suggest some general directions for future research.

9.2 THE DIRECT EFFECTS OF INCREASED CO_2

In contrast to climatic changes, the rise in atmospheric CO_2 itself will be comparatively smooth and continuous with little year-to-year variability. How will higher CO_2 concentrations affect crop yields in the future in the absence of consideration of climatic change?

Much of the discussion of direct biological effects focuses on the impact of a CO_2 doubling. Unfortunately, there has been no consistency as to the control or base concentration used for purposes of comparison. Although the global average is now 343 ppmv (in 1984), the exact concentration varies with season and latitude (Figure 9.1a), with height above ground and with time of day (Figure 9.1b). Base CO_2 levels used in experiments have varied from 270 to 340 ppmv, and since the response of plant photosynthesis to CO_2 concentration is a saturating one (Figure 9.1c), the reader is cautioned that the quantitative effect of a CO_2 doubling can vary for that reason alone.

The reduction of CO_2 to carbohydrates by photosynthetic carbon fixation accounts for about 90% of the accumulation of plant dry matter. That CO_2 concentration is a limiting factor is shown by the numerous experiments in which higher CO_2 enhances photosynthesis and crop growth. The response has been found to hold even for plants grown under a variety of stressful conditions, despite a frequently repeated generalization that relates to the so-called 'law of limiting factors'. This idea is that when other environmental factors such as water shortage, low light, mineral shortage or excess, and non-optimal temperature limit yield, then higher CO_2 concentration will have little or no effect. Although the generality of that concept has been challenged (Gifford, 1977, 1979a, 1979b, 1980a; Pearcy and Bjorkman *et al.*, 1983) the idea persists (e.g. Kramer, 1981; Liss and Crane, 1983; Tolbert *et al.*, 1983). Indeed, in certain stressful environments the relative photosynthetic response of plants to CO_2 enrichment is actually increased (as is noted below).

Predictions of crop growth and yield are incomplete if based solely on the photosynthetic response at the level of the primary CO_2 fixation mechanism. Other primary and secondary responses (like stomatal conductance and morphological development) and feedbacks interpose between photosynthetic metabolism and crop yield and must be taken into consideration in assessing the effects of higher atmospheric CO_2. Our current understanding of these processes and their effects is reviewed below, building from the un-

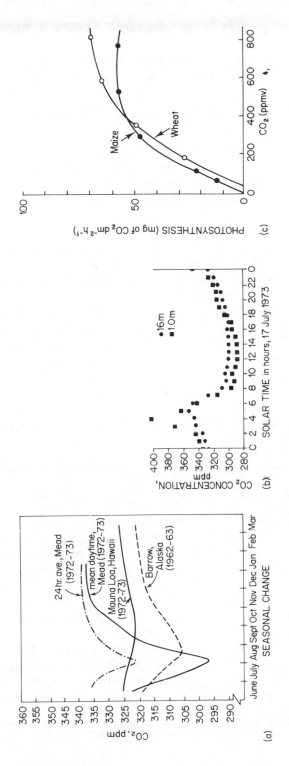

Figure 9.1 Some examples of variations in CO_2 concentrations (a) seasonally, (b) diurnally in an alfalfa crop at Mead, Nebraska, U.S.A., and of (c) net photosynthetic response curves to CO_2 concentration of a C_3 species (wheat) and a C_4 species (maize)

derlying biochemical and physiological responses through to field crop yield, with and without limitations imposed by other environmental restraints.

9.2.1 Biochemistry and Physiology of CO_2 Responses

Plants are grouped photosynthetically into three groups—'C_3', 'C_4' and 'CAM'—according to biochemical distinctions in the mechanism of primary CO_2 fixation. Pineapple is the only representative among commercial crops of CAM plants and would not be expected to respond appreciably to higher CO_2, so it will not be discussed further. At least 95% of the world's biomass is of the C_3 category as are most crop species (Figure 9.2).

The distinctions between the groupings derive largely from the enzymes involved in photosynthetic fixation. Only two, perhaps three, enzymes are known to be of significance in the response of plants to CO_2 enrichment: 'rubisco',[1] 'PEP carboxylase'[2] and perhaps carbonic anhydrase. Rubisco is the primary enzyme for photosynthetic fixation in the C_3 plants and the CO_2 fixation rate per unit leaf area is typically related positively to the amount of this enzyme per unit leaf area. Carbon dioxide itself (together with Mg^{++}) activates rubisco by binding at a non-catalytic site on the enzyme protein (Jensen and Bahr, 1977). Rubisco may not always be fully activated *in vivo* for leaves in normal air (Perchorowicz *et al.*, 1982; Vu *et al.*, 1983), but it is not known whether higher CO_2 concentrations increase the level of activation in the long term (i.e. over the seasonal growth cycle).

Rubisco is not only responsible for catalysing the reduction of the CO_2 to carbohydrates, but, in the presence of light, catalyses a reaction with oxygen. The metabolite produced by the oxygenation releases recently fixed CO_2 during its metabolism—the process of photorespiration. The same catalytic site on rubisco binds both O_2 and CO_2. In C_3 plants the photorespiration rate is high and is determined in part by the relative proportions of CO_2 and O_2 in the leaf. Some of the response of photosynthesis to higher CO_2 concentration is believed to derive directly from the improved competitive advantage of CO_2 molecules over O_2 molecules for the active sites on rubisco. The reduced carbon flow through the photorespiratory cycle leads to less photorespiratory CO_2 loss as well. Even in the absence of competing O_2, however, the fully activated enzyme in leaves at 340 ppmv CO_2 is probably operating only at about half to three-quarters of its substrate-saturated capacity (Edwards and Walker, 1983).

In contrast, the primary carboxylase in C_4 plants is PEP carboxylase which is not competitively inhibited by O_2. Photorespiration is therefore negligible. PEP carboxylase has a higher effective affinity for CO_2 than does rubisco in the absence of O_2, so the enzyme is close to CO_2-saturation at the present

---

[1]   Ribulose 1,5-bisphosphate carboxylase-oxygenase (EC4.1.1.39 or 'rubisco').
[2]   Phosphoenolpyruvate carboxylase (EC 4.1.1.31 or 'PEP carboxylase').

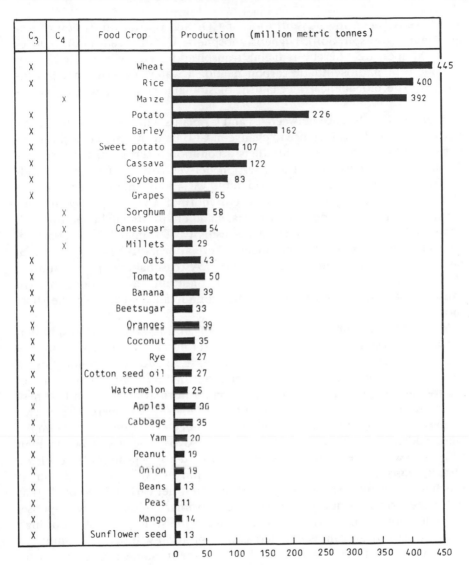

C_3	C_4	Food Crop	Production (million metric tonnes)
X		Wheat	445
X		Rice	400
	X	Maize	392
X		Potato	226
X		Barley	162
X		Sweet potato	107
X		Cassava	122
X		Soybean	83
X		Grapes	65
	X	Sorghum	58
	X	Canesugar	54
	X	Millets	29
X		Oats	43
X		Tomato	50
X		Banana	39
X		Beetsugar	33
X		Oranges	39
X		Coconut	35
X		Rye	27
X		Cotton seed oil	27
X		Watermelon	25
X		Apples	36
X		Cabbage	35
X		Yam	20
X		Peanut	19
X		Onion	19
X		Beans	13
X		Peas	11
X		Mango	14
X		Sunflower seed	13

Figure 9.2 Examples of C_3 and C_4 crops and their global annual production (fresh weight) as reported by *FAO Production Yearbook*, 1980 (adapted from Swaminathan, 1984)

atmospheric CO_2 concentration. Therefore one would not expect a significant enhancement of C_4 crop growth from increased CO_2 in so far as the primary carboxylase properties are concerned.

Leaf surfaces are covered with microscopic pores, or stomata, through which gaseous exchange occurs. The aperture of the pores varies, striking a balance between inwards diffusion of CO_2 and outward diffusion of water vapour (transpiration). Higher atmospheric CO_2 concentration reduces stomatal aperture, thereby reducing transpiration. Hence the efficiency of water use in photosynthetic carbon fixation ('water use efficiency') is increased. The biochemical mechanism of stomatal response to CO_2 is unknown.

There is no difference between C_3 and C_4 plants with respect to the sensitivity of stomatal conductance to change in CO_2 concentration (Morison and Gifford, 1983). This view contradicts a common assertion that C_4 species display greater sensitivity. The assertion derives from leaf chamber studies from which are observed considerable variations among species in the absolute value of stomatal conductance in 'normal' CO_2 for particular combinations of genotype, development stage and environmental conditions. If, however, one focuses on the relative sensitivity of stomatal conductance in control- vs increased-CO_2 experiments, there appears to be little difference between the plant groups (Morison 1985). A reasonable approximation is that for most species and environmental conditions, a CO_2 doubling will cause about a 40% decrease in stomatal conductance, at least in the short term.[3]

Experiments on the effects of high CO_2 concentrations on dark respiration show mixed results. It has been proposed that mitochondrial respiration may increase in plants under high CO_2 in response to sucrose accumulation in leaves (e.g. Tolbert *et al.*, 1983). A mechanism for this is thought to act via the 'alternative pathway of respiration', a normal mechanism that may function to dissipate excess photosynthesized energy (Lambers, 1982). This proposal is consistent with the findings of Hrubec *et al.* (1984), for example, who reported increased respiration rates of soybean leaves grown in high CO_2. However, the converse result was found for wheat (Gifford *et al.*, 1985); plants grown continuously in 590 ppmv CO_2 experienced half as much whole-plant respiration by night (per unit net carbon fixed by day) as did plants grown in normal air. Whether a primary or secondary response, any inhibition of respiration will contribute to the stimulating effect of high CO_2 on net carbon gain while increased respiration will detract from it.

With respect to morphology and development, some species grown in high CO_2 experience greater leaf area expansion and advanced time of flowering (e.g. Hand and Postlethwaite, 1971; Goudriaan and de Ruiter 1983). Although such effects are presumably often a response to improved photosynthate supply, there are also indications of a less direct CO_2 ef-

---

[3]   Morison (1985) plotted conductance at 660 ppmv CO_2 against conductance at 330 ppmv for 80 observations from the literature covering a wide range of species (C_3 and C_4), conditions and methodologies. There was linear correlation through the origin over a 20-fold conductance range, with conductance at 660 ppmv being 0.59 ± 0.04 (99% confidence limit) of conductance at 330 ppmv.

fect. In C_4 species that do not respond photosynthetically to high CO_2, leaf area has been observed to increase. For example, growth analysis of both maize (Imai and Murata, 1978) and itchgrass (Patterson and Flint, 1980) showed that leaf area increased while net dry weight (DW) gain per unit leaf area ('net assimilation rate') was unaffected by CO_2 enrichment to above 600 ppmv. Similarly, with a doubling of normal CO_2, Morison and Gifford (1984b) observed increases in leaf area of the C_4 species *Amaranthus edulis* (15%), *Sorghum bicolor* (29%) and *Zea mays* (40%) grown on declining soil water content. At the same time, the efficiency of conversion of intercepted radiation into dry matter was unchanged by the high CO_2. Thus the increase in growth caused by higher CO_2 in these C_4 species was attributable to greater interception of light because of bigger leaf area, not to increased photosynthesis per unit leaf area. This implies that CO_2 was acting on leaf area development in some way other than via CO_2 effects on photosynthesis rate.

Effects of CO_2 on flowering time are usually minor but not necessarily solely due to change in photosynthate supply. For example, a *slowing* in the rate of flower development in sorghum without any change in dry weight growth (Hesketh and Hellmers, 1973; Marc and Gifford, 1983), seems indicative of some more direct influence of CO_2 on flowering.

9.2.2 Biological Feedbacks

Although leaves of C_3 species photosynthesize faster when transferred to an atmosphere containing higher CO_2 concentration, this initial response may not necessarily persist. Over the growing cycle of the plant, biological feedbacks can come into play acclimating enzymatic activities and leaf photosynthetic rates to the CO_2-enriched environment. But reports on the subject of photosynthetic acclimation offer no consistency as to the direction of change. For example, leaf photosynthetic capacity in high CO_2 concentration has been shown to be higher than (e.g. Bishop and Whittingham, 1968), the same as (e.g. Gifford, 1977), and lower than (e.g. von Caemmerer and Farquhar, 1984) the capacity of plants grown in normal air.

These differences may arise from the interaction of multiple feedback mechanisms. Understanding of photosynthetic acclimation to high CO_2 is hindered by the fact that most reports do not permit separation of effects operating just at the enzymatic and leaf levels, from the longer-term effects emanating from changes in the 'source:sink balance' in the whole plant. Maintenance of enhanced photosynthesis rates and, eventually, yield depend on an adequate sink (or storage organ, like the grain) for the photosynthates. If the growth of sinks does not respond to higher photosynthate supply, then photosynthesis can be depressed—a negative feedback. This occurs, perhaps, by build-up of photosynthetic products in the leaf (Madsen, 1968; Herold,

1980), although the exact mechanism behind this feedback is still unknown (Gifford and Evans, 1981).[4]

Despite limited information, such results suggest that several feedback mechanisms can develop within the CO_2-enriched plant, and that the balance between them varies during plant development and determines the photosynthetic acclimation to higher CO_2. A clearer understanding of this process will depend in part on the ability of further research to separate the influences of these feedback mechanisms on photosynthetic response over time.

An increased rate of senescence (aging) is another possible feedback effect of CO_2 enrichment. Accelerated senescence has been observed in two winter annual species (St. Omer and Horvath, 1983) and in cotton (Chang, 1975), the latter exhibiting concurrent decline in carbonic anhydrase activity in the leaves. Although the observed senescence effect is minor, and is not always detected (e.g. no effect in wheat (Krenzer and Moss, 1975; Gifford, 1977)), it could possibly be pervasive due to increase in ethylene, a natural growth regulator in plants which accelerates senescence. High CO_2 concentrations caused sunflower plants to produce more ethylene, for instance (Dhawan *et al.* 1981). In addition, the CO_2 source for enriching the air might also contain unsuspected traces of ethylene which could promote early senescence. Some Australian sources of CO_2, for example, contained traces of ethylene which were sufficient to hasten senescence of some species (e.g. tomato) but not others (e.g. maize) (Morison and Gifford, 1984a).

If some CO_2-stimulated DW growth were invested in leaf area expansion, a positive feedback effect could be established. Expanded leaf area would allow greater light interception which would promote further DW growth, additional leaf area expansion, and so on, until the leaf canopy becomes dense enough for full interception of incident radiation. Experimental results vary. Soybean and sunflower grown at twice normal CO_2 did not develop more leaf area in some experiments (Carlson and Bazzaz, 1980; Marc and Gifford, 1984), but did in others (Rogers *et al.*, 1984; Morison and Gifford, 1984b). Rice frequently does not increase leaf area appreciably under CO_2 enrichment even though DW growth responds (Yoshida, 1972; Imai and Murata, 1978; Morison and Gifford, 1984b). Conversely, several C_4 species that did not show a response of net CO_2 fixation per unit leaf area or per unit of intercepted radiation, nevertheless responded with an increase in leaf area (as noted above).

The mechanisms involved in CO_2-stimulated leaf area expansion have not been widely investigated. They could be expected to vary, however, since

---

[4]   It is interesting to note, however, that in one experiment in which the sink:source rate in soybean was lowered surgically, enhanced leaf photosynthesis persisted for many weeks at high CO_2 levels (as compared to control-CO_2), despite the inability of the remaining sinks to accept more photosynthetic assimilate (Peet, 1984).

it is known that, depending on the species, the component of leaf area increase under CO_2 enrichment varies between axillary growth (branching; Johnston, 1935), faster rate of leaf emergence (Hofstra and Hesketh, 1975) and development of larger leaves (Goudriaan and de Ruiter, 1983).

9.2.3 Growth and Yield Under Good Environmental Conditions

Under favourable growing conditions, what can we say about the net effect of all the aforementioned responses and feedbacks on plant growth and yield? One attempt to summarize CO_2 enrichment experiments showed mostly positive effects (and a few negative effects) across all groups of C_3 species. Kimball (1983) interpolated 134 observations from the CO_2 enrichment literature published over 64 years to ascertain the average increase of DW growth and yield of a variety of species in response to double 'normal' (330 ppmv) atmospheric CO_2 (Table 9.1). Most of the experiments were conducted under 'good' conditions of nutrient and water supply. For the average of all C_3 species investigated, economic yield increased 26% and immature shoot dry weight increased 40%.

One particularly interesting finding concerns the growth response of small grains. Immature DW (biomass) generally exhibits greater response to high CO_2 than the final economic yields, but this is not so for small grain cereals like wheat. As shown in Table 9.1, the high (36%) increase in grain yields with a CO_2 doubling is nearly twice the increase in biomass of immature crops (20%), a finding that is also supported by work of Goudriaan and de Ruiter (1983). The effects of high CO_2 on wheat seedlings is small (Neales and Nicholls, 1978) compared to the effects once tillering and grain formation occur (Gifford, 1977; Sionit et al., 1981a). This might be a reflection of the powerful influence of CO_2 enrichment before ear emergence on tillering and sink size (ear number) in small-grain cereals (Gifford et al. 1972; Cock and Yoshida, 1973). This result with cereals belies the tentative generalization (Kramer, 1981) that determinate species (i.e. those for which leaf development ceases after flowering, as in cereals) respond less to CO_2 enrichment than do indeterminate species. Given the central role that small grains play in world food production and trade (see Chapter 8), this finding could prove to be of special importance in a CO_2-enriched future.

For C_4 species, the results are mixed. Some growth experiments (Marc and Gifford, 1984; Gifford and Morison, 1985) confirm the biochemically derived expectation of no appreciable growth response of well-watered plants to high CO_2. However, other examples (cited by Kimball, 1983; Morison and Gifford, 1984c) show substantial CO_2 effects on growth and yield. There are two routes whereby this could occur: by some unknown non-photosynthetic effect of high CO_2 on leaf area expansion and hence on light interception (see Section 9.2.1), or via an interaction with water stress, as discussed below.

Table 9.1　Mean predicted growth and yield increases for various groupings of C_3 species for a doubling of atmospheric CO_2 concentration from 330 ppmv to 660 ppmv (adapted from Kimball, 1983). the errors indicated are 95% confidence limits

	Footnote	Immature crops		Mature crops	
		No. of records	% increase of biomass	No. of records	% increase of marketable yield
Fibre crops	1	5	124	2	104
Fruit crops	2	15	40	12	21
Grain crops	3	6	20	15	36
Leaf crops	4	5	37	9	19
Pulses	5	18	43	13	17
Root crops	6	10	49	—	—
C_3 weeds	7	10	34	—	—
Trees	8	14	26	—	—
Av. of all C_3		(83)	40 ± 7	(51)	26 ± 9

Footnotes: The species represented are:
1. cotton (*Gossypium hirsutum*);
2. cucumber (*Cucumis sativus*), eggplant (*Solanum melongena*), okra (*Abelmoschus esculentus*), pepper (*Capsicum annuum*), tomato (*Lycopersicum esculentum*);
3. barley (*Hordeum vulgare*), rice (*Oryza sativa*), sunflower (*Helianthus annuus*), wheat (*Triticum aestivum*);
4. cabbage (*Brassica oleracea*), white clover (*Trifolium repens*), fescue (*Festuca elatior*), lettuce (*Lactuca sativa*), Swiss chard (*Beta vulgaris*);
5. bean (*Phaseolus vulgaris*), pea (*Pisum sativum*), soybean (*Glycine max*);
6. sugar beet (*Beta vulgaris*), radish (*Raphanus lativus*);
7. *Crotalaria spectabilis, Desmodium paniculatum*, jimson weed (*Datura stramonium*), pigweed (*Amaranthus retroflexus*), ragweed (*Ambrosia artemisiifolia*), sicklepod (*Cassia obtusifolia*), velvet leaf (*Abutilon theophasti*);
8. cotton (*Gossypium deltoides*).

9.2.4 Interaction with Growth-limiting Environmental Factors

Under controlled-environment conditions, the percent enhancement of growth owing to high CO_2 concentration has been found to be greater with restricted water supply than with unlimited watering. Since higher CO_2 reduces stomatal conductance (by about 40% for a CO_2 doubling), water use efficiency in the production of dry matter (WUE) increases with CO_2 concentration, even for C_4 species. However, the relative reduction of transpiration rate per unit leaf area is not as great as that for stomatal conductance because, with reduced evaporative cooling, leaf temperature increases, thereby increasing the driving force behind transpiration (viz. the leaf-to-air vapour pressure difference) (Morison and Gifford, 1984c). Thus doubling the CO_2 concentration reduced stomatal conductance of sorghum by 40%, but transpiration rate by only 15% (van Bavel, 1974).

Under growth-limiting water supply, growth of C_3 crops responds to higher CO_2 because of both photosynthetic and stomatal effects (Gifford, 1979a; Morison and Gifford, 1984c), while growth of C_4 species responds because of stomatal effects alone. Thus for both C_3 and C_4 species, the less the availability of water, the greater the percent increase ('relative enhancement') of growth by high CO_2 concentrations.

However, in scaling up from controlled environment to a wide expanse of vegetation in the field, other attenuating phenomena come into play to determine the rate of transpiration. In circumstances where boundary layer conductance is low relative to stomatal conductance (i.e. non-windy conditions), not only is the role of stomatal conductance in controlling transpiration attenuated, but also leaf temperature is higher than under windy conditions and atmospheric humidity close to the crop surface can increase. All these aspects would reduce the impact of CO_2-induced stomatal closure on transpiration. Furthermore, and perhaps even more powerfully, soil water content may be a more important determinant of rate of transpiration on a time scale of days to weeks than is stomatal response to attributes of the aerial environment. For example, for 16 species, the time-course of depletion to exhaustion of stored soil water by individual plants was little affected by twice normal CO_2 because the high CO_2-induced leaf area increases compensated for the reduction in transpiration rate per unit leaf area (Morison and Gifford, 1984b).

Simulation of the CO_2 effect under optimal supply of water also indicated a compensation of decreased leaf transpiration by increased leaf area, resulting in a practically constant transpiration rate per ground area. In other words, the increase in overall water use efficiency approximately equalled the increased growth rate (Goudriaan et al., 1984). In this sense, higher atmospheric CO_2 concentrations may not reduce the frequency of agricultural droughts, as some have claimed. Rather, droughts will still occur but at a higher level of biomass.

Carbon dioxide enrichment increases crop growth and yield at low light intensity which is itself severely growth limiting. The relative enhancement of growth can even be greater than at high light level, as has been found for wheat (MacDowell, 1972; Gifford, 1979a). There are at least two aspects to the mechanism of growth response to CO_2 under photosynthetically limiting light intensities. One is that the quantum yield of leaf photosynthesis close to the light compensation point (i.e. the light intensity at which CO_2 uptake by a leaf is just balanced by respiratory CO_2 release) is CO_2-dependent in C_3 species, but not in C_4 species (Ehleringer and Bjorkman, 1977). High CO_2 increases C_3 species' quantum yield because it suppresses photorespiration. The extent to which the effect on quantum yield manifests itself as plant growth is dependent on the second pertinent aspect—how whole plant (dark) respiration responds. If whole plant respiration is less under high CO_2, then the light compensation point is lowered and some growth is achieved at light intensities that otherwise would prove insufficient for growth to occur. The larger relative enhancement of growth in wheat (reported to show reduced whole-plant respiration in high CO_2) in low light compared to high light intensities might be explained on this basis. For other species such as soybean, which has shown increased respiration under high CO_2, the relative enhancement of growth by high CO_2 appears equal at low and high light (Sionit *et al.*, 1982).

With the prospect of warmer average global temperatures in the future, the response of CO_2-enriched plants under higher temperatures is pertinent. Based on limited information, it appears that in general the positive effect of higher CO_2 in stimulating photosynthesis is increased with higher temperature. However, this effect tends to be counteracted by negative feedback effects over the growth cycle of the plant. For example, for two C_3 species, Berry and Raison (1981) found that the ratio of short-term leaf photosynthesis at 1000 ppmv to that at 330 ppmv CO_2 increased sharply from 1.15 at 15 °C to 3.5 at 50 °C. This is explicable on the basis of the kinetic properties of rubisco and perhaps also on the declining solubility of CO_2 (relative to O_2) with increasing temperature (Jordan and Ogren 1984). However, temperature is important in determining the rate of growth of metabolic sinks (such as developing fruits). Sufficiently high temperatures can adversely affect sink growth (see also Section 9.3.1) and thereby feed back onto leaf photosynthesis and modulate the CO_2 response. This could possibly be one reason why soybeans, grown at supra-optimal temperature (above 30 °C) did not express the large potential CO_2 responsiveness of photosynthesis in enhanced growth (Hofstra and Hesketh, 1975; Hofstra, 1984).

At very low temperatures the inherent capability of sinks to grow is low and not limited by photosynthate supply. Thus even if photosynthesis were highly responsive to CO_2 enrichment at low temperature, it might not have much effect on sink growth that is itself temperature limited. However, high CO_2 can reduce the minimum temperature at which a plant grows and completes its life-cycle. The tropical vegetable okra (*Abelmoschos esculentor*) was unable to complete its life-cycle in normal CO_2 at temperature below 23 (day)/17 °C (night), while plants grown in 1000 ppmv CO_2 at 20/14 °C matured and produced fruit (Sionit *et al.*, 1981b). Thus there was an infinite relative response (i.e. from nothing to something) of fruit yield to high CO_2 at sub-critical temperatures.

Response of plant growth to high CO_2 under nutrient deficiency or surfeit varies with both the nutrient and the species concerned. Low nitrogen supply reduces growth of all species, but with C_3 non-legumes, DW growth of both N-deficient and N-sufficient plants is increased by doubling normal

CO_2 concentration. For instance, the weight of cotton plants almost doubled, irrespective of whether they had received 2 mM or 24 mM nitrate in the nutrient solution (Wong, 1979). Although not as pronounced, perennial ryegrass, wheat and soybean also achieved high per cent increases of dry weight growth from CO_2 enrichment under N-deficiency (Sionit et al., 1981a; Goudriaan and de Ruiter, 1983). The implied improvement in N-use efficiency may emanate from reduced investment in photosynthetic machinery (which has a high N-requirement) per unit of photosynthetic assimilate produced. In nodulated legumes such as soybeans or peas, high CO_2 leads to greater biological nitrogen fixation (Hardy and Havelka, 1974). This effect can be attributed to the production of more nodules on a bigger root system, rather than to greater specific activity of nodules (Phillips et al., 1976; Finn and Brun, 1982). In short, the CO_2 effect is positive under nitrogen stress.

In contrast, Goudriaan and de Ruiter (1983) were unable to show a growth response to CO_2 in phosphorus deficient plants of several species (with the exception of P-deficient bean (*Vicia faba*) plants which were even more responsive to high CO_2 than were plants grown with adequate P).

Potassium is another major nutrient but there is little information on its interaction with atmospheric CO_2. In potato, Goudriaan and de Ruiter (1983) noted a negative effect of increased CO_2, probably associated with higher demand for potassium.

Sodium is an essential element for C_4 photosynthesis. Growth of sodium deficient plants of two C_4 species which do not normally respond to CO_2 enrichment was greatly enhanced by high atmospheric CO_2 (1500 ppmv) (Johnston et al., 1984). While sodium deficiency is uncommon in the field, sodium excess (salinity) is common and causes reduced yield or, at greater excess, toxicity symptoms. Schwarz and Gale (1984) have shown that for diverse species, tolerance of saline conditions is increased by CO_2 enrichment to 2500 ppmv. This effect was ascribed to a shortage of photosynthate in plants suffering salt stress, but it might also be associated with the reduced demand for saline water because of CO_2-reduced transpiration.

9.2.5 Summary

Based on limited experimental results, we can expect a doubling of atmospheric CO_2 concentration from 340 to 680 ppmv to cause a 0 to 10% increase in growth and yield of C_4 crops (such as maize and sugarcane) and a 10 to 50% increase for C_3 crops (such as wheat, soybean and rice), depending on the specific crop and prevailing growing conditions. For C_3 species, the principal source of this response is at the level of the primary carboxylase–oxygenase enzyme, but stomatal, respiratory and morphological responses may also be involved. The latter three effects appear to be the principal sources of growth response for C_4 plants, where response occurs.

Table 9.2   Effects of increased CO_2 on crop response and feedbacks: a tentative compilation

	C_3	C_4
Photosynthesis	+ +	0
Photorespiration	–	NA
Nitrogen fixation (leguminous species)	+	NA
Transpiration	–	–
Dark respiration	M	?
Leaf area development (non-photosynthetic CO_2 response)	+?	+?
Photosynthetic acclimation (at leaf level and via sink:source ratio)	M	M
Senescence	M	0?
Leaf area expansion (via greater photo-synthesis)	0 to +	≈0

+ + = strongly positive

  + = positive

  – = negative

  0 = no effect

NA = not applicable

  M = mixed response (positive or negative)

  ? = not known or uncertain

There are numerous feedbacks operating within the plant that serve both to accentuate and to attenuate the effect of the primary responses. An important positive feedback is the increase in the proportion of incident radiation that is intercepted by leaves because of more rapid leaf expansion in a stand of CO_2 enriched plants. Whereas in C_3 plants this stimulation of leaf expansion is likely to result mainly from the CO_2-stimulated growth itself, in C_4 species circumstantial evidence suggests that it may be a non-photosynthetic effect of higher CO_2. Important negative feedbacks can develop from the build-up of photosynthetic products on the leaves and from changes in the source:sink ratio of the plant. The effects of CO_2 enrichment on the basic biochemical and physiological plant processes and feedbacks are summarized in Table 9.2.

Table 9.3 Relative effects of increased CO_2 on growth and yield: a tentative compilation[1]

	C_3	C_4
Under non-stressed conditions	+ +	0 to +
Under environmental stress:		
Water (deficiency)	+ +	+
Light intensity (low)	+	+
Temperature (high)	+ +	0 to +
Temperature (low)	+	?
Mineral nutrients:	0 to +	0 to +
Nitrogen (deficiency)	+	+
Phosphorus (deficiency)	0?	0?
Potassium (deficiency)	?	?
Sodium (excess)	?	+

[1] Sign of change relative to control CO_2 under similar environmental constraints.

+ + = strongly positive

+ = positive

0 = no effect

? = not known or uncertain

Interactions between the effects of atmospheric CO_2 and other growth-limiting environmental variables on plant growth are complex and not amenable to simple generalization from the so-called 'law of limiting factors'. For example, high CO_2 concentration can reduce the deleterious impacts on growth of water-shortages, low light intensity, temperature extremes or certain mineral deficiencies, notably nitrogen deficiency. On balance, the results of experimental studies indicate that in most conceivable circumstances, the effects of increased CO_2 are beneficial (rarely detrimental) to plant growth and yield, as indicated in Table 9.3.

On the other hand, the scientist's glass-house is not the same as nature's laboratory. Carbon dioxide enrichment also stimulates the growth of weeds, for instance, which compete with crops for available moisture, light, and nutrients in actual field situations. Field studies of CO_2 enrichment have been attempted (e.g. Rogers *et al.*, 1981), but suffer from poorer control of environmental conditions than can be obtained in growth chambers, and have not yet progressed to the stage of studies on competition in mixed plant communities. One important goal of experimental work is to contribute to the development of simulation models (discussed below). In lieu of field studies such models allow one to examine how the biochemical, physiological and environmental factors interact dynamically in the presence of high CO_2 to influence plant growth and yield. Some models have been used in this regard (e.g. Baker and Lambert, 1980; Goudriaan *et al.*, 1984) but greater progress in model development is required before we can place high confidence (at, say, the 5% level) in the results. Such progress is being made.

9.3 PERSPECTIVES ON CLIMATE IMPACTS

How is agriculture affected by changes in climate in the absence of direct CO_2 effects? There are myriad answers: crop varieties are switched, cultivation techniques are modified, plant development is retarded or accelerated, yields become more or less variable, production trends are altered, cropped area expands or contracts, and so on. Ideally, we should like to know the aggregate of these effects, but in the short term this is impracticable. We are forced to be selective.

The selection of research questions depends, in large part, on one's perspective concerning the effects of climate and climatic variation on agriculture. The perspectives vary widely. For instance, Bach *et al.* (1981) expand upon two related but opposing themes: climate as a 'resource' and as a 'hazard'. Riebsame (1985) distinguishes two additional perspectives: climate as 'setting' (the background for agriculture) and climate as 'determinant' (the cause of agricultural patterns and practices). Glantz (1979) identifies yet another dichotomy: CO_2-induced climatic change as an 'event' (a focus on the doubling) and as a 'process' (a focus on the gradual, accumulating environmental change). There is considerable overlap between these perspectives.

Two additional perspectives can be discerned from the rapidly mounting literature on the agricultural effects of climatic change. One view holds that the potential problems (or benefits) for agriculture arise from slow, gradual changes in average climate. The other view portrays the problem as one of slow shifts in climatic risks. The way in which agricultural impacts of climatic change are assessed takes a slightly different twist according to which view predominates. Let us characterize them.

9.3.1 The Slow Change View

The 'slow change' view is implicit in most impact studies. It derives, in part, from the way in which the entire problem of increasing greenhouse gases and climatic change has been analysed. As reflected in this volume, the analysis begins with estimates of past and future emissions of greenhouse gases, followed by estimates of their rates of accumulation in the atmosphere and by predictions of their effects on climate. In general, the changes in climate variables are presented in terms of their central tendencies, based on the equilibrium response of climate models. It is not unexpected to find an extension of this chain of analyses to assess the impacts on agriculture and other ecosystems. The slow changes in average temperature or precipitation predicted in the previous step are quite literally assumed to be the potential problem faced by agriculturalists: a *gradual, long-term, cumulative alteration of climate and, consequently, a slow deterioration (or enhancement) of the growing environment.*

The specific research questions derived from this slow change perspective are cast in a similar mould. For example,

• How would average wheat yields be affected by a 2 °C rise in average annual temperature?

• How might the boundaries of North American corn production slowly recede from its present semi-arid margins with a gradual regional drying and warming?

• In the light of growing global food demands, how might world-regional grain production trends and trade balances be affected by gradual climatic changes?

• How would (or could) agriculturalists perceive and adapt to the slow changes in their environment?

This last question concerning response is particularly vexing from the slow change view. On the one hand, it has been argued that the magnitudes of some estimated changes in climate are so large that they are unprecedented in recorded history and thereby fall outside the realm of human experience. In order to adapt, agriculture may have to devise wholly unique and imaginative strategies for dealing with the effects (Cooper, 1978; 1982). On the other hand, it has also been argued that because the climatic changes will occur in a slow, cumulative fashion, agriculture has plenty of time and can most likely adapt in pace (e.g. Wittwer, 1980). The rate of change should be slow enough to allow farmers to perceive the changes in their growing environment and to switch crops, to adopt more suitable varieties, and to modify their farming practices accordingly. In short, the transition, while formidable, is eased by the luxury of time. These appraisals hinge upon assumptions concerning adaptive capacity and rates of response, to which we shall return shortly.

9.3.2 The Shift-in-risk View

The CO_2 problem as a shift in climatic risks presents an interesting contrast. According to this view, the potential agricultural problems arise mainly from *changes in the frequencies of unusually disruptive (or beneficial) climatic events.* There is no denial that changes in regional climates may occur slowly and gradually. It is argued, however, that the long-term changes in average temperature or precipitation *per se* are of relatively little importance to agriculturalists. Rather, the year-to-year risks from climatic events such as droughts, frosts, or excessive moisture are more important (e.g. Fukui, 1979). The impacts of these relatively infrequent events on crop yields cause financial (or other social, economic or human) stress and play a large role in determining agricultural viability. With a change in climate, it is likely that the frequencies of such events would shift.

In part, this 'shift-in-risk' view is a result of reversing the chain of analyses described earlier, that is, by commencing with agriculture and working toward changes in climate resulting from increasing concentrations of greenhouse gases. The questions are posed: What are the processes or resources (climatic or otherwise) critical to agricultural activities and yields? How might changes in climate affect these processes or resources? Parry and Carter (1984) call this an 'adjoint approach'. This approach, incidentally, creates greater complexity and uncertainty in climate analysis and places the onus on the climate modeller to provide more impact-specific climatological detail at relevant scales of resolution—hence the existing gap between needed and available information for impact studies (WMO, 1984).

The shift-in-risk view leads one to formulate specific research questions rather differently. For example,

• What climatic conditions lead to particularly poor yields of sorghum and millet in the semi-arid sub-Saharan tropics? How would the probabilities of their occurrence shift with a change in climate?

• How would the boundaries of the wheat belt in the drought-prone margins of North America be altered from increased drought risk that may result from a warmer and drier climate?

• What would be the impacts on global grain production from simultaneous poor harvests in North America and the Soviet Union? How might the frequencies of such occurrences change if the climate of the Northern Hemisphere were to change?

In these questions, the emphasis falls on the interannual variability of climate.

The focus on year-to-year events rather than long-term means is partly grounded in assumptions concerning agriculturalists' perception of climatic change. In the absence of scientific information, agriculturalists may encounter considerable difficulty in perceiving and reacting to changes in mean trends. Indeed, atmospheric scientists themselves can only identify a trend in climate if a sufficiently long period of record is available to separate the 'signal' from the 'noise'. Instead, in the face of changing climate, the impacts from the occurrence of particularly unfavourable (or favourable) growing seasons are, in effect, the principal stimuli to which agriculturalists can, and do, react (as in changing crop type or variety, migrating elsewhere or adopting different technologies or cultivation techniques).

Thus, from the shift-in-risk view the environmental cues to which agriculture will respond are not unprecedented (even though the magnitude of the climatic change itself may be unprecedented). The cues are the same year-to-year events that agriculturalists now experience, albeit at different frequencies of occurrence. Therefore, many tactics and strategies already exist for dealing with these familiar climatic risks, although different levels of adoption, changes in farm structure and organization, or possibly new risk strategies may be warranted. This viewpoint is implicit, for example, in the remarks by Clark (1982) who states that '...what *we would* be doing if we were certain about CO_2 predictions is what *we should* be doing anyway to cope with droughts, heat waves, etc.' (p. 3).

9.3.3 Blending the Views: Adaptation and Adjustment

While we have dichotomized the slow change view and the shift-in-risk view for purposes of explication, they are not mutually exclusive. This is evident in many statistical crop impact studies (Section 9.4.1) in which the effects of climatic extremes are reflected in mean yields; the mean and variance are inextricably bound in their long-term effects on yield trends (Mearns *et al.*, 1984).

Furthermore, it has been hypothesized that long-term adaptation to climate results from the aggregate of short-term responses to risk (Parry, 1978, 1985; Whyte, 1981; de Vries, 1980). As agriculturalists strive to achieve the best returns and to build resiliency in the face of interannual climatic variability, the agricultural system becomes best fitted to the most frequently occurring climatic conditions over the long run—that is, those described by measures of central tendency. For example, both the heart and the spatial extent of the Australian wheat growing regions may be largely a reflection of the spatial gradients of drought risk, but may be well described by mean rainfall.

In this respect, the two views just simply may be emphasizing two separate bands of the same response spectrum. We illustrate this notion in Figure 9.3, partly adapted from Fukui (1979). Figure 9.3a displays a hypothetical curve of precipitation, conveniently distributed normally (in reality precipitation is usually described better by non-Gaussian curves). Let us assume that an agricultural system is perfectly centred on the mean, so that any deviation from the mean has negative effects on yields. Superimposed on the distribution in Figure 9.3b are three categories of agricultural response (after Burton *et al.*, 1978; Kates *et al.*, 1985).

First, for frequently occurring amounts of precipitation there exists a set of responses that are labelled *adaptations*. From one year to the next, agri-

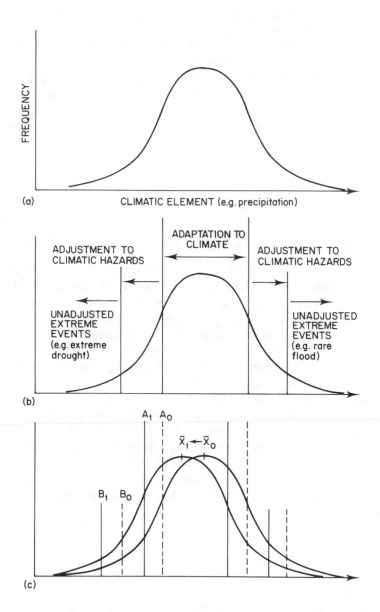

Figure 9.3 Schema of adaptation and adjustment to climate and climatic change (adapted partly from Fukui, 1979). (a) Frequency distribution of a climatic element, upon which are superimposed 'bands' of adaptation and adjustment (b). (c) A change of mean (X_0 to X_1) requires a shift in adaptation (A_0 to A_1) and adjustment (B_0 to B_1) in order to compensate for higher frequencies of dry (hatched) and extreme drought (cross-hatched) events

culturalists expect mildly wet or dry conditions to occur. It is this expectation of weather that, to the agriculturalist, is 'climate' (Hare, 1985). Individuals and organizations accumulate a large mix of cultural, technological or behavioural measures—adaptations—to accommodate this expected variation. They are reflected broadly in the timing of farm operations like planting and harvesting, the migration routes of pastoral nomads, or the spatial patterns of major agricultural systems like livestock grazing or dryland wheat farming. Adaptations evolve over the long term (greater than several generations) and may not be consciously recognized as having any relationship at all to climatic or environmental fluctuation. Adaptations allow agriculture to interact freely with the expected environment without disruption or inhibiting stress. Within this 'band' of adaptation, climate is a resource.

Second, toward the tails of the distribution in Figure 9.3b are precipitation amounts that occur with rarer frequency. These events are not expected from one growing season to the next and are perceived as hazards (droughts or floods) if they exceed the adaptive capacity of the system and 'cause' disruption or loss. To deal with such recurring but unexpected annoyances, individuals and organizations make discrete *adjustments*. Adjustments are consciously adopted to cope with environmental risk, and include such measures as drought resistant wheat varieties, flood levees, emergency irrigation or grain reserves. Despite the adjustments adopted, there is always residual loss or disruption (by definition, otherwise the events would no longer be considered hazards). Every adjustment has its associated costs as well. Balancing cost against residual loss, the level of adjustment at any given time might be considered, *de facto*, society's 'acceptable level of risk'.

Third, at the far tails of the distribution are the very extreme, rare events—for example the 1-in-500 year flood or drought. Few specific adjustments are contemplated, either because they would be too costly, no viable alternatives are perceived, the events are considered too rare, or some combination thereof. Herein lurks the potential for catastrophe: a decade of drought in the North China Plains or five consecutive years of monsoon failure in South Asia.

The point we wish to make is that the degree of vulnerability to climatic change and variability depends on the widths of the 'bands' of adaptation and adjustment, and, therefore, on the differences between climatic resources and climatic hazards. And these bands, far from static, are prone to change over time and space. As Heathcote (1985) notes, what is flooding in one set of circumstances is excess water for irrigation in another.

In this sense, the future impacts on crop yields and production depend on the dynamics of agriculture and society as well as the stimulus of environment. For example, it has been asserted (Burton *et al.*, 1978) that, in many developing countries struggling with transition to modern agriculture, the bands have shrunk rapidly. Traditional adjustments and adaptations have been displaced or discarded, while the more technological or market-oriented mechanisms that are characteristic of the developed world have not, as yet, been satisfactorily adopted. This creates situations of high vulnerability to the vagaries of climate, as evidenced by the high tolls exacted by the occurrence of extreme climatic events (Kates, 1980).

What are the effects of climatic change? In Figure 9.3c we have superimposed a change in climate, a slow change in mean from $\bar{X}_0$ to $\bar{X}_1$—drier conditions—assuming no change in variability. For conditions described by the new mean ($\bar{X}_1$) and the expected deviations immediately around it, the change results in lower crop yields more often than before. But these yield changes fall well within the existing band of adaptation; so, on a year-to-year basis they are not really unexpected or disruptive, and there is no lack of mechanisms to accommodate them. The potential problem is rather to re-centre on the central tendency over the long run. The flexibility afforded by the adaptive capacity of the agricultural system will most likely allow it to calibrate fairly easily, closely in pace with the changing climate, as through gradual alterations of planting dates, planting densities, or allocations of irrigation water—a *fine tuning* of the system.

Greater difficulties are encountered further left on the curve in Figure 9.3c. Climatic events that were once infrequent enough to be considered hazards (i.e. moderate droughts) now occur with troublesome regularity. Agriculturalists may begin to perceive them as part of the expected weather. If agriculturalists wish to maintain the levels of acceptable risk previously attained, the higher probability of loss or disruption associated with these events (represented by the hatched area) becomes intolerable. In effect, agriculture is under-adapted. An expansion of the band of adaptation from A_0 to A_1 will be required. This might be accomplished, for example, through a continuous adoption of stress-tolerant crop varieties, an expansion of farm sizes, or a switch to diversified farm operations that are more suitable to the new climatic conditions—an *alteration* of the system.

At the far left tail of the curve in Figure 9.3c are the extreme droughts to which agriculture is largely unadjusted. By virtue of the climatic change, the occurrence of these events has become more probable. Again, if previous levels of acceptable risk are to be maintained, the band of adjustment must be expanded accordingly, from B_0 to B_1 (the cross-hatched area in Figure 9.3c). But, in many cases, the alternatives may be so severely limited or prohibitively costly, and the impacts so disruptive in terms of crop yields and socio-economic consequences, that the only perceived recourse may be abandonment and migration. The Dust Bowl migrations from the U.S. Great Plains during the 1930s (Worster, 1979), the abandonment of cereal and hay production in Iceland with the Little Ice Age (Ogilvie, 1981), or, perhaps, the present situation in vast areas of drought-stricken Africa are illustrative. The

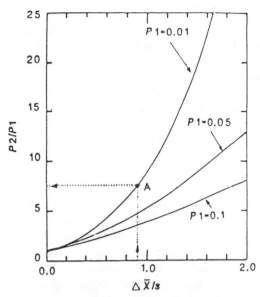

Figure 9.4 The sensitivity of extreme climatic events to changes in the mean, based on normal distribution and constant standard deviation (see text for explanation) (from Wigley, 1985)

long-term effect could be a change in land use and agricultural landscape—a *change* of system.

Of course, on the right—'wet'—side of the curve the frequencies of occurrence have been reduced. For these events it could be argued that agriculture is over-adapted and over-adjusted in relation to apparent levels of acceptable risk.

Three additional points should be emphasized with respect to climatic change and risk. First, the frequency of occurrence of extreme events can be very sensitive to relatively small changes in the mean (Mearns *et al.*, 1984; Wigley, 1985). This relationship is illustrated in Figure 9.4 (from Wigley, 1985). The abscissa shows the change in the mean (ΔX) as a multiple of the standard deviation (S), while the ordinate shows the resulting change in the probability of extreme events with initial probabilities ($P1$) of 0.1, 0.05, and 0.01 (like the previous figure, this diagram is based on an assumed normal distribution and constant standard deviation). For example, if the annual mean precipitation over England and Wales (approximately 920 mm) fell by 100 mm (approximately 0.9 standard deviations—an amount, by the way, projected by some GCMs with a CO_2 doubling), the initial 1-in-100 year drought ($P1 = 0.01$) would become roughly 7.5 times more frequent in any given growing season ($P2/P1 = 7.5$, point A).

Second, as pointed out by Parry (1985; also see Sakamoto *et al.*, 1980), individual farmers and agricultural systems may be especially vulnerable to consecutive years of poor yields, and the probabilities of consecutive occurrences of extreme climatic events could increase dramatically with a change in climate. For instance, while in the previous example the initial 1-in-100 year drought became 7.5 times more frequent, the chances of two consecutive years of drought of this magnitude would increase by over 56 times (assuming independent events). The potential for a catastrophic succession of poor harvests, particularly in areas already sensitive to drought, could escalate rapidly, even if the change in climate itself (as measured by the central tendency) were glacially slow.

Third, it is likely that the agricultural response would not be smooth and gradual. The disruptive climatic events are already infrequent, so considerable time might pass before farmers could perceive that the probabilities were changing. In the absence of credible scientific information, response would come about through direct experience, as through a rash of particularly severe years of unfavourable weather (if, indeed, climatic change is for the worse). In this way, agricultural response is apt to occur in an abrupt, step-like manner as human perception catches up with physical reality. In the meantime, the adverse impacts could be severe.

We have attempted to show that the slow change view and the shift-in-risk view just simply emphasize different aspects of the same problem of climatic change. The climatic effects of increased concentrations of greenhouse gases, although commonly described in terms of long-term, large-scale averages, can be manifested in many ways across a wide range of spatial and temporal scales. In the global context, one danger is that the problem of climatic change may be defined too narrowly. For instance, people who represent the interests of developing countries sometimes claim that the problem of a slow, long-term change in climate is quite secondary to immediate problems of interannual yield variability, and is therefore of limited interest (WMO, 1984). This is unquestionably a valid point from the slow change view. However, from the shift-in-risk perspective the potential agricultural impacts of climatic change could be interpreted as an exacerbation of existing yield variability—a problem which could be felt acutely, abruptly, and possibly in the not-so-distant future.

9.4 THE IMPACTS OF CLIMATIC CHANGE

Four broad approaches to assessing the agricultural impacts of climatic change can be identified. *Crop impact analysis* concentrates directly on estimating the primary effects of environmental variables on crop yields. *Marginal–spatial analysis* examines the possible spatial shifts in cropping patterns (or other characteristics of agriculture) that might result from changes in climate at the margins of production. The third approach, *agricultural sector analysis*, focuses on estimating the range of impacts within and be-

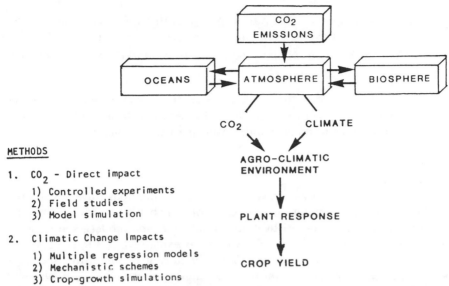

METHODS

1. CO_2 - Direct impact
   1) Controlled experiments
   2) Field studies
   3) Model simulation

2. Climatic Change Impacts
   1) Multiple regression models
   2) Mechanistic schemes
   3) Crop-growth simulations

Figure 9.5 Crop impact analysis. The approach is largely unidirectional and sequential and seeks to estimate the primary, first-order impacts of changes in the growing environment on crop responses and yields as a result of increasing atmospheric concentrations of greenhouse gases

tween agricultural regions, with an emphasis on the positive and negative feedback mechanisms that, in a dynamic fashion, reduce or enhance the primary impacts on crop yields and production. Finally, *historical case studies* ask, What does past experience tell us about the agricultural impacts of climatic change? Let us examine each approach.

9.4.1 Crop Impact Analysis

The first approach is presented schematically in Figure 9.5. Crop impact analyses seem to isolate and to quantify the effects of climate variables (including the direct CO_2 effects, treated separately in Section 9.2) on crop response and yields. In applications to problems of climatic change, such analyses have attempted to estimate the 'before-and-after' yield effects, usually assuming an instantaneous change from one climate state to another. Although frequently unstated, rather constrictive boundary conditions are required, and the results of most crop impact analyses should include the following caveats:

• If no shifts in spatial cropping patterns take place to adapt to changes in regional climate.

• If no changes in perception and managerial response occur.

• If the technologies and cultivation practices that affect crop-climate relationships remain constant over time.

• If no feedbacks to yields and production from market forces or government policies occur.

Of course, these are big 'ifs', and, as we shall see, subsequent approaches (Sections 9.4.2 to 9.4.4) progressively relax these constraints by setting the boundary conditions to include wider aspects of the problem.

Crop-climate models

Most crop impact analyses have relied on three methods for assessing the possible effects of climatic change, each of which has its advantages and drawbacks. In empirical–statistical, multiple *regression models*, some aspect of production—usually commercial yields—is explained by some set of 'independent' climate variables, like monthly values of precipitation and temperature, plus a term to account for any long-term trends in yields that are usually attributed to 'technology'. The constants in the regression equation are determined empirically, and the observations for regression fitting are taken from historical records of agricultural production and climate data. The more explanatory variables included in the regression equation, the larger the number of empirically derived constants. This, in turn, requires a long historical record to provide a sufficient number of observations to derive statistically significant equations and avoid spurious results—a major constraint in many countries where reliable historical records are short. Even where records are sufficiently long, changes in crop varieties, management or technology can alter crop–weather relationships and, in effect, make historical data 'outdated' (Robertson, 1983). This is a serious drawback to using such models to predict the long-term effects of changes in climate on yields.

Regression techniques are not particularly suitable for understanding the interacting physical, biochemical and physiological processes underlying crop growth and yield. They skip the stage marked 'plant response' in Figure 9.5 and attempt a direct link between environmental change and reported yield. This is the 'black-box' criticism frequently levelled at regression models (e.g. see Katz, 1977). Furthermore, differences in crop varieties, management practices and soil conditions are difficult to include as explanatory variables in regression equations (this would also increase the number of constants). Thus regression models tend to be site specific, and it is commonly accepted (but frequently ignored) that they should not be applied outside the region or data range from which they were constructed.

With the advent of computers, it has been possible to construct crop-growth *simulation models* which combine the mathematical equations that describe the physical, chemical and physiological mechanisms and their interaction. Such models focus explicitly on plant processes such as photosynthesis, transpiration and respiration. Data requirements for simulation models are, as a rule, demanding. The simulation time-step can vary from weeks to minutes—hourly is common—and data on radiation, minimum and maximum temperatures, and soil moisture are required at those same time intervals.

One major advantage of simulation models for assessing the impacts of climatic change is their potential 'transportability'. In principle, if the processes of plant growth are described accurately and integrated correctly, the specific region of application should be of little consequence, since the model itself will demonstrate the limiting factors for growth (Baier, 1977). The effects of different management practices or environmental sensitivity can then be examined systematically.

With assumptions about management, soil conditions and planting densities, area-wide yields can be estimated using simulation models. However, Monteith suggests (WMO, 1985) that, despite their complexity and process-orientation, computer simulations have not been conspicuously more successful than simpler models in making predictions of crop yields. In fact, attempts to be comprehensive have sometimes increased the size and complexity of models to the point where confusion eclipses illumination.

Intermediate to the regression and simulation approaches are simpler, deterministic mathematical functions—or *mechanistic schemes* (cf WMO, 1985)—that relate individual climate variables to particular crop growth processes over the stages of plant development. Such schemes are especially useful for analysing the effects of a specific climate variable with respect, say, to its limiting or optimal conditions. However, their simplicity contributes to their principal drawback: the failure to consider the correlation and interaction of elements, the adaptation of plants to stress over the period of growth, and the growth restrictions imposed by nutrient deficiencies, pests or other factors (Monteith, 1981). In short, mechanistic schemes lack comprehensiveness and dynamism—the fundamental rationale for building simulation models. Mechanistic schemes provide the building blocks for process-based simulation models.[5]

The strengths and weaknesses of crop-climate models are summarized in Table 9.4.[6] In general, a common deficiency of all three types of model is the

---

[5] It is instructive to note that mechanistic schemes are the outcome of laboratory and field measurements of plant processes fitted to mathematical functions based on the laws of physics and physical chemistry. As such, they contain statistical summaries of experimental work—and, thus, so do simulation models. Therefore, although we make the distinctions between statistical regression models, mechanistic schemes and crop-growth simulation models, the distinctions are somewhat artificial.

[6] For reviews of crop-climate models, see WMO (1982; 1985), Baier (1977; 1983), Robertson (1983), Biswas (1980), CIAP (1975), Sirotenko (1983), or Nix (1985).

Table 9.4   The (a) uses and (b) criticisms of types of crop-climate models: statistical relations (SR), mechanistic schemes (MS) and crop-growth computer simulations (CS) (after WMO, 1985)

(a) USES

	SR	MS	CS
Summarizing	***	*	
Analysis		**	*
Relative environmental sensitivity		*	***
Prediction (a) interpolation	***	**	**
(b) extrapolation	*	*	*
Development	*	**	***

(usefulness: * = marginal, ** = moderate; *** = substantial; blank = not useful)

(b) CRITICISMS

	SR	MS	CS
Too many 'disposable' constants	+	+	+
Too many disparate sources			+
Too few critical validations	+	+	+
Too site/species specific	+		
Too many physiological forcing functions		+	+
Too comprehensive to comprehend			+
Sinks rather than sources of understanding	+		+

(+ = applicable; blank = not applicable)

lack of rigorous validations. Ultimately, all models should be tested on independent data (not used in model construction or parameter estimation), a criterion which applies to 'process-based' simulation models and mechanistic schemes, as well as to regression models (Robertson, 1983; Haun, 1983). It is likely that many models that are potentially useful for crop impact analysis in various regions of the world have not been adequately validated, although the extent of the situation has yet to be determined (WMO, 1985).

Despite their deficiencies, crop-climate models have been used to examine the possible impacts of climatic change. What have we learned?

Applications and Findings

Crop impact analyses that deal explicitly with the subject of climatic change are surprisingly few in number. Most studies have concentrated on wheat and maize in the mid-latitudes. The major studies of wheat and maize, plus a few pertinent analyses of lesser scope, are listed in Table 9.5. It is evident that a heavy emphasis has been placed on North America and Europe, as reflected in the recent undertakings by the U.S. National Academy of Sciences (NRC, 1983) and by the European Economic Community (Meinl and Bach *et al.*, 1984), and, for the most part, in the decade-old Climate Impact Assessment Program study (CIAP, 1975). The only attempt at systematic, global crop impact analysis was the National Defense University study (NDU, 1980); but in lieu of modelling, the NDU study opted for a consensus of 'expert judgment' regarding climate–yield relationships—a tactic that has drawn heavy criticism (for recent critiques see Stewart and Glantz, 1985; Schneider, 1985. Only the American crop impact analyses are noted in Table 9.5).

All the studies in Table 9.5 implicitly assumed a 'slow change' view of the problem of climatic change. They investigated the changes in average yields accompanying changes in long-term climate. To perturb yields, most of the studies used arbitrary scenarios (see Section 8.3) in which changes in climate were imposed instantaneously and uniformly across seasons; instrumental analogues and GCM scenarios were used in a few cases. In short, the list of crop impact analyses contains a rather diverse mix of methods, with respect both to choice of crop–climate model and to assumptions about climatic change.

Although this diversity should deter all but the most foolhardy from comparing the findings, we have attempted to do so in Figures 9.6a–f. The symbols plotted on the graphs correspond to those noted in Table 9.5 (the 'hollow' symbols represent regression, the 'solid' are simulation and the 'letters' denote mechanistic or other methods). The graphs show the average yield changes expected with specified changes in precipitation and temperature. The general conclusion one can draw from the patterns in Figure 9.6 is that despite the diversity of scenarios and methods—regression, simulation, mechanistic, and even expert judgment, with all their inherent deficiencies—the studies are in basic agreement regarding the expected direction of yield effects in current cultivars of wheat and maize from changes in climate. They are less precise about the relative magnitude of those effects. Two basic observations can be made.

First, warming appears detrimental to yields of wheat and maize in the core crop regions in the mid-latitudes of North America and Europe. (Effects at the cold margins of production or other specific locations might be quite different, as discussed in Section 9.4.2). Keeping in mind the large un-

Table 9.5 Studies of crop yield impacts from climatic change, with particular reference to wheat and maize

Symbol[1]	Study (Specific author)	Region	Crop	Model type	Climate Scenario
□	NRC (Waggoner, 1983)	USA	Wheat	Regression	Arbitrary
		(1) N. Dakota			
		(2) Red River Valley			
		(3) S. Dakota			
		(4) Nebraska			
		(5) Kansas			
		(6) Oklahoma			
		(1) Iowa	Maize		
		(2) Illinois			
		(3) Indiana			
■		N. Dakota	Wheat	Simulation	
△	CEC (Santer, 1984)	European Community	Wheat	Regression	GCM
		(1) Ireland			
		(2) Denmark			
		(3) Netherlands			
		(4) Belgium			
		(5) France			
		(6) W. Germany			
		(7) Italy			

○	CIAP (Ramirez et al., 1975) (Benci et al., 1975)	(1) N. Dakota (2) USA	Wheat Maize	Regression	Arbitrary
	Ritchie (for IMI)	Kansas	Wheat	Simulation	Instrumental
	USDoE (Kanemasu, 1980)	Kansas	Wheat	Simulation	Arbitrary
×	NDU (1980)	USA	Winter wheat Maize	Expert judgement	Arbitrary (expert opinion)
▽	Palutikof et al. (1984)	England and Wales	Wheat	Regression	(A) Instrumental (B) Instrumental
●	Liverman et al. (1985)	Kansas	Maize (irrigated)	Simulation	Arbitrary
M	Monteith (1981)	England	Wheat	Mechanistic & Regression	Arbitrary
G	Goudriaan (WMO, 1985)	General	General	Mechanistic	Arbitrary

[1] Solid symbols = simulations, open symbols = regressions, letters = mechanistic or other methods

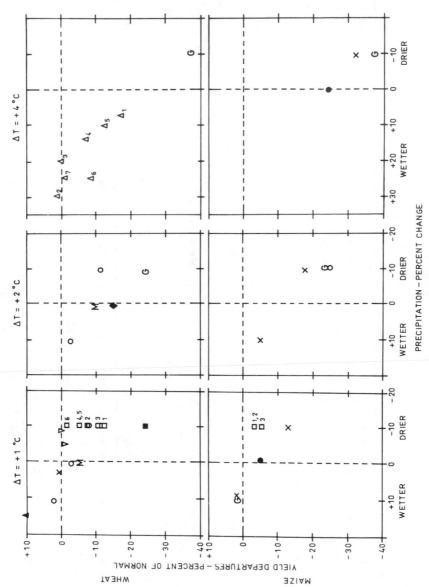

Figure 9.6 Estimates of the impacts on wheat and maize yields of temperature and precipitation changes for given cultivars (for explanation of symbols, see Table 9.5)

certainties in these findings, we may venture a rough guess at the magnitude of impact: with no change in precipitation (or radiation), slight warming ($+1$ °C) might decrease average yields by about $5 \pm 4\%$; a 2 °C might reduce average yields by about $10 \pm 7\%$. To put this into perspective, at current (1983) levels of production, a 10% decrease in wheat and maize yields in North America is equivalent to about 20.5 mt, or 10% of global trade in cereals (FAO, 1983a,b).

Second, reduced amounts of precipitation also decrease yields of wheat and maize in these same core regions. This implies, of course, that both higher precipitation and higher temperatures could have offsetting effects on yields, as indicated by the discernible negative slopes to the data plotted in Figure 9.6. Temperature changes could be expected to have broadly uniform effects on yields over large areas, as compared to precipitation changes whose yield effects would vary over shorter distances depending on local rainfall regimes and soil conditions.

It would be incautious to conclude anything more specific from the results of these studies.

The physical reasons for the reductions in yield with increasing temperature (even in cool regions like Canada or the United Kingdom) are provided by the process-based simulation models and the mechanistic schemes upon which they are based. There are two processes at work. First, higher temperatures are usually associated with higher evapotranspiration and therefore greater moisture stress during critical stages of growth. This effect is likely to be important in regions where inadequate soil moisture is already a characteristic problem.

Second, temperature influences the duration of the growth period of the plant. Yield depends on the rate of photosynthesis (determined principally by available CO_2, light, water and, to a lesser extent, temperature) and the duration of the growth period, which is largely a function of genotype and temperature. Higher temperatures stimulate photosynthesis slightly (which can be beneficial, particularly during emergence and canopy formation) but, to the detriment of yields, accelerate development and shorten the duration of plant growth (an effect that can readily be countered by a switch of cultivar). The latter phenomenon is particularly important during the phenological stages of heading, flowering and ripening in which the economic portion of dry weight yield is formed.[7]

Based on such reasoning, Monteith (1981) simply calculated that, assuming a base temperature of 0 °C for growth and a mean temperature of 15 °C over the growth period, a 1 °C increase in mean temperature would decrease yield by 1/15, or by about 7%. His *a priori* reasoning is in corroboration with the correlation between observed average wheat yields and mean

---

[7] The empirically derived coefficients in the wheat and maize regression models used by Waggoner for the NRC study (1983) are generally consistent with these relationships.

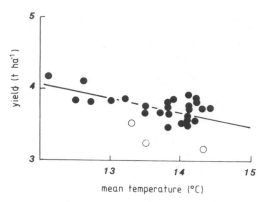

Figure 9.7 Mean yield of wheat in England (1956–1970) as a function of mean temperature for May, June and July for the period 1941–1970. Records for Cornwall, Devon and Surrey (open circles) were omitted from the analysis. Line of best fit has slope of -5% K^{-1} at 14 °C (from Monteith, 1981)

growth-period temperature throughout England (Figure 9.7). In the same way, Goudriaan (WMO, 1985) attempted a rough estimate of the range of possible crop impacts, assuming that temperatures could increase from 1 °C to 4 °C and that precipitation and radiation (from cloudiness) could change 10% in either direction (with relative effects on yields of the same order of magnitude) with a doubling of atmospheric CO_2 concentrations. Ignoring the direct effects of CO_2 and the possibility of change of cultivars, the net influence of climatic change on yields could range from $+13$ to -47%, depending upon whether temperature, radiation and precipitation counteract one another or work in the same direction. (For comparison, we excluded radiation effects and plotted the resultant range, $+3$ to -37%, in Figure 9.4). Monteith's and Goudriaan's rough estimates are not inconsistent with the estimates derived from more complicated techniques.

What do these analyses tell us about the possible 'shifts in risk' as a result of climatic change? Explicitly, very little. The decreases in average yields from higher temperature, for example, might well be more than compensated by a concomitant reduction in the risk of damaging early- or late-season frosts—a troublesome aspect of cereal production in cooler climates (Lough et al., 1983). Drier conditions would not only reduce average harvests, but could also inflate the frequency of harvest 'failures' due to extreme drought—a double burden in semi-arid regions. Implicitly, these effects are reflected in average yield changes generated by regression models, since the regression coefficients capture some of the yield effects of extreme climatic events as they are correlated with average monthly temperatures or precipitation (Mearns et al., 1984). But one cannot glean much regarding their specific yield impact or about changes in their frequency of occurrence.

Several studies have used crop-climate models simply to obtain estimates of yield impacts due to interannual climate variability or anomalous growing seasons. For example, NOAA (1973), TIE (1976) and Warrick (1984) examined impacts on grain production in North America. The latter study, for instance, employed a set of regression models (one of which was essentially the same as that used by Waggoner, 1983, for the NRC study) to investigate the possible impacts on Great Plains wheat yields if the drought years of the 1930s were to recur. It was found that the worst drought year could reduce Plains-wide yields by about 25%. However, while these studies examined yield impacts of anomalous years, they did not make the link to long-term climatic change and thus did not consider possible changes in the frequencies of large yield departures.[8]

An important exception to this last point is the study by Waggoner (1983). Waggoner imposed an arbitrary scenario of climatic change (1 °C warmer, 10% drier) upon an historical climate record (without regard to seasonal or spatial considerations) and, using a crop-growth model, simulated wheat yields (in North Dakota) year by year. He thus derived frequency distributions of yields with and without climatic change, as shown in Figure 9.8. (In contrast, all other studies noted in Table 9.5 simply used mean values of climate variables to generate average yield values.) Waggoner only concerned himself with the change in median yields, but the projected shifts in the frequencies of poor and bumper harvests from his simulation are far more interesting. Under unperturbed climate, the chance of getting extremely low yields, −24% or less below expected median yield, was approximately 1-in-8 years. Under the warmer–drier scenario (and ignoring direct CO_2 effects), the risk of obtaining these same absolute yield levels jumped to about 1-in-2.3 years, a relative increase in risk of over 300%. At the other tail of the distribution, the chances of unusually high yields were drastically reduced Whether North Dakota wheat growers could continue to prosper under such shifts in risk is questionable.

The importance of this study lies not so much in the predicted yield values themselves (there is room for considerable scepticism in this respect), but in the approach. By simulating the frequency distributions of yields, the spectrum of yield effects, including the risks of extreme events, can be examined. In terms of evaluating possible adjustment strategies or policy changes, such analyses are far more useful than simply estimating average yields. In general, the value of crop impact analyses could be enhanced by following similar procedures.

---

[8] Neild *et al.* (1979) did explicitly consider possible shifts in risk, but in terms of climatic events, not yields. One finding of the study was that one type of climatic warming, a lengthening of the series of consecutive days above normal, actually *increased* the incidence of spring freeze damage on early-planted maize in the northern U.S. Corn Belt. The apparent reason is that warmer temperatures promote earlier emergence, thus lengthening the period of exposure to infrequent freezes.

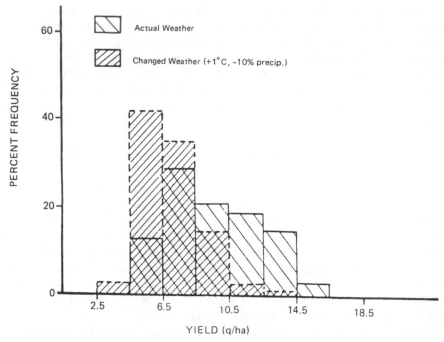

Figure 9.8 North Dakota simulated spring wheat yield (1949–1980) (from Waggoner, 1983)

Global Assessments and Prospects

What can be said about the possible impacts of climatic change on crops other than those grown in the cereal regions of the mid-latitudes? Unfortunately, few studies have been conducted on crops in other parts of the world, particularly in the tropics and sub-tropics. Crop-climate models have been developed world-wide and could be useful in addressing issues related to climatic change. A survey undertaken by WMO (in conjunction with this current CO_2 assessment project) revealed 49 models in 28 countries (of WMO member countries) which potentially could be applied.

It is unrealistic to believe, however, that any crop-climate model can predict the long-term impacts on yields when it is highly certain that large changes in management practices, cultivars or other technologies will take place in the decades to come. To use the models in such a fashion is clearly unwise, and the results of past crop impact analyses must therefore be interpreted critically. Crop-climate models can possibly be more useful as tools for facilitating short-term management and response, rather than for predicting long-term impacts of climatic change. Some potential uses include: analyses of the sensitivity of yields to climate variables; analyses of the interaction of CO_2 and climate variables on yields; determination of the effects of alternative management practices on yields under different climatic conditions; or assessments of crop potentials in given regions.

Further application of crop-climate models for assessing the problems of climatic change should proceed cautiously and should be based on careful model evaluation. The kinds of applications envisaged in climate impact assessment may extend well beyond the original purpose of many models. Careful validations of existing crop-climate models are required to demonstrate their capability for accurately testing the sensitivity of crops to climatic change. Of the 49 models reported in the WMO survey noted above, only 70% were claimed to have been validated. In some cases, model validations that are more rigorous than those already performed by the modellers themselves may even be required (WMO, 1985).

9.4.2 Marginal–Spatial Analysis

Casting a critical eye, the astute observer (e.g. Cooper, 1978) will comment that changes in climate surely will not be unnoticed by agriculturalists. Will not spatial shifts in cropping patterns occur to compensate for climatic changes? And will not such shifts in crop area modulate the impacts on yields and production? If so, it becomes important to identify and estimate the effects of climatic change at the margins of production.

As noted by Parry and Carter (1984), the concept of 'marginality' can be construed in a number of ways—spatial (or environmental), economic and social. There has been a particular research interest in the environmentally marginal areas where the effects of climatic change may be felt acutely. One environmental case is where there exists an apparent *mismatch* of environment and agriculture (coffee growing in the cooler, thermally marginal regions of southern Brazil). In this situation, a slight change in climate could have large, areally widespread effects on 'maladapted' agriculture. *Steep climatic gradients* (temperature gradients along highland slopes in Peru or precipitation gradients across the Sahel) pose another type of environmental marginality in which climatic change could substantively alter the growing environment within short distances and overwhelm even previously well-adapted agricultural activities. Finally, marginality can be viewed as a spatial zone of *transition* between alternative agricultural or other land uses (as between wheat growing and livestock grazing in the semi-arid US Great Plains) as a function of comparative economic advantage vis-a-vis the natural environment.

Marginal–spatial analysis is concerned both with the agricultural impacts in these zones, and with the spatial shifts in the boundaries of, say, crop types, profitability or economic risks that might take place as

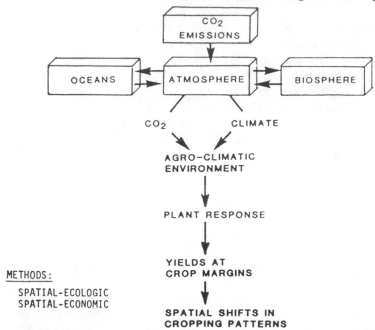

Figure 9.9 Marginal–spatial analysis. This approach is concerned with the effects of climatic change on yields at the margins of production, and with the spatial shifts in crop area or other characteristics of agriculture that might result

a consequence of climatic change. Schematically, this is depicted simply in Figure 9.9. Two specific approaches to the problem can be discerned.

The Spatial–Ecologic Approach

One approach rests heavily on the 'mis-match' connotation of marginality and implicitly draws a strong parallel between natural ecosystems and agriculture. The underlying premise is that the spatial configurations of crop regions are determined largely by the natural environment. It assumes that particular crop types or agricultural systems adapt to climate in the same way that the spatial patterns of tropical rainforests, savannas or mixed temperate forests are influenced by global climate regimes. Any long-term change in climate creates a 'mis-match'—a disequilibrium—which will prompt agriculture to re-adapt. As in natural ecosystems where the spatial manifestation of ecosystem response is most pronounced at the ecotone, agriculture will be most sensitive to climatic change at the margins of the crop region and will expand or contract as average environmental conditions change. Hence the term 'spatial–ecologic' for the approach. This characterization is based on only a handful of studies, most of which are concerned with North American grain production.

For example, in two separate studies, by Newman (1980; 1982) and by Blasing and Solomon (1983), the possible shifts in the U.S. Corn Belt as a result of climatic change were estimated. Notwithstanding some differences in averaging periods, climate data and variable definition, these two studies were essentially alike in approach. The approach required: (1) The identification and quantification of the environmental variables (or indices thereof) that limit the spatial extent of the crop region; (2) The selection of scenarios of climatic change and the modification of meteorological records accordingly; and (3) The calculation of effects on key climatic constraints (e.g. growing season length) and the consequent spatial shift in crop region. Both studies proceeded on the assumption that the Corn Belt is limited in its northern extent by the length of the frost-free growing season and by the thermal requirements for maturation, and in its western extent by inadequate soil moisture. Accumulated growing degree-days (GDDs) — a time–temperature index usually expressed as the number of degrees over a base level for growth (10 °C in this case) per day summed over the growing season — account for these factors in a single index. For this reason, and because the existing northern boundary of the Corn Belt parallels the GDD isolines rather closely, both studies used changes in GDDs from various scenarios of climatic change to approximate the possible geographical shifts in the crop region.

The results of both studies are presented in Figures 9.10 a and b. Newman (1980; 1982) found that a climatic warming would displace the Corn Belt 175 kilometres per degree C in a north-by-northeast direction. Blasing and Solomon (1983) obtained the same results, although the magnitude of the displacement was not quite as large. The latter study also examined the added effects of higher precipitation (indicated by GCM scenarios) and the adoption of earlier planting dates; it was found that, to some degree, both counteract the spatial displacement resulting from higher average temperatures.

Similar spatial–ecologic studies for Canada suggest a potential northerly expansion of small grain production as a result of a lengthening of the growing season that would accompany a climatic warming, barring spatial limitations posed by poor soils or other environmental barriers (Williams, 1975; Williams and Oakes, 1978).

One troublesome aspect of these spatial–ecologic studies, in the absence of a sound physical explanation, is the connection that is made between an environmental index like GDDs and the boundaries of a crop region. A simple spatial correlation between the 1320 GDD isoline and the northern margin of the Corn Belt (as per Newman, 1980) does not by itself infer causal relationship, nor does it constitute firm ground for believing that a shift in the first will necessarily lead to an identical shift in the second. Indeed, taking an historical glimpse of the North American grain belts, one finds considerable movement of both the winter and spring wheat belts (Rosen-

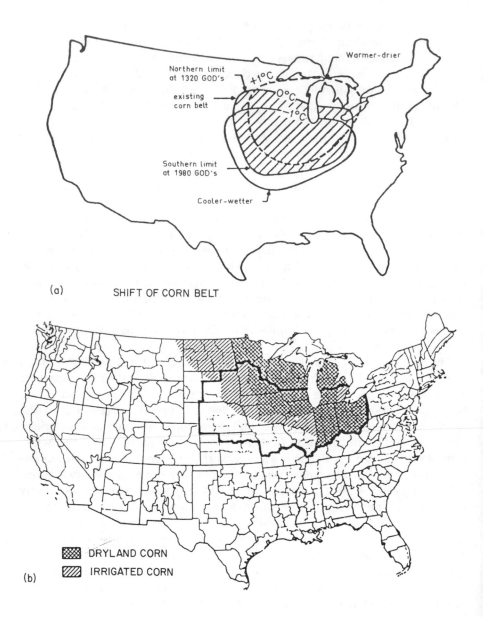

Figure 9.10 Estimations of the impacts of climatic change on the geographical extent of the US Corn Belt using the spatial–ecologic method. (a) Simulated shift based on growing degree days (GDD, in °C) during the frost-free growing season (from Newman, 1980). (b) Shift for 3 °C temperature increase and 8 cm precipitation increase, distributed evenly over the year (Blasing and Solomon, 1983). The solid black line indicates current location of the corn belt

berg, 1982) and the Corn Belt during the last one hundred years—with and without appreciable changes in climate. Furthermore, it should be borne in mind that these studies do not consider crop 'yield' in any detailed fashion. Thus, we are given some approximation of possible spatial displacement of crop area, but little indication of how this might affect area-wide yields and production—which might, in turn, further influence crop production at the margins.

The importance of the spatial–ecologic studies is that they take a first step in recognition of the capacity of agriculture to adapt to climatic change. In addition, the studies explicitly consider the length of the growing season as it shapes crop regions. These two aspects could have considerable effects on changes in crop yields. For example, the finding based on crop-climate models (reported Section 9.4.1) that North American maize yields would decline with climatic warming should be reconsidered in the light of these spatial-ecologic analyses. A concomitant lengthening of the frost-free grow- ing season might well reduce the incidence of frost damage and thereby significantly benefit average yields in the long run. On the other hand, if maize production doggedly pursues GDDs in a north-by-northeast direc- tion, the risks may persist and the yield benefit may be partly or wholly eliminated. The net impact of all these interactions is decidedly unclear.

In general, the major limitation of the spatial–ecologic approach is the implicit assumption (noted above) that managed ecosystems like agriculture will respond slowly to changes in climate in a manner analogous to natu- ral ecosystems. This assumption ascribes a degree of 'climatic determinism' that seems unwarranted. In natural ecosystems, a mix of plant species com- pete with one another for available climate resources—light, moisture—over many seasonal cycles. Slow climatic changes may give the comparative ad- vantage to some plants at the expense of others, and ecosystem boundaries may slowly change as a consequence. But for the vast bulk of the world's food crops, the choice of what seeds to sow, and where, is made afresh with each season. Cropping patterns are subject largely to human decision- making, rather than natural competition. While nature may set the ultimate geographical limits for crop growth, human beings still have plenty of room to manoeuvre.

The Spatial–economic Approach

Another variation of marginal–spatial analysis partly recognizes these crit- icisms and addresses the problem of climatic change from the viewpoint of economic viability. The 'spatial–economic' approach is also concerned with the environmental margins. But unlike the previous approach, the spatial– economic approach inclines toward the 'shift-in-risk' perspective: The focus is on the change in climatic risk and its economic impact on a season-to-

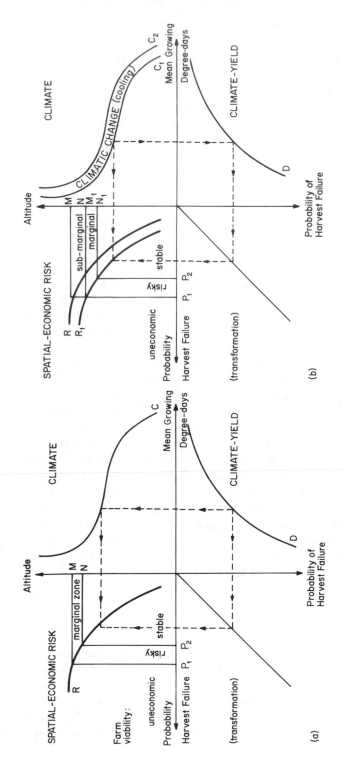

Figure 9.11 Schema of the spatial–economic approach as applied to high altitude cold margins (cf Parry, 1978). (a) The initial probability of harvest failure as a function of elevation (curve R, spatial–economic risk) as derived from altitudinal differences in growing degree-days (curve C) and the probability of harvest failure due to insufficient warmth for crop maturation (curve D). P_1 to P_2 denotes those probabilities for which farming is a risky business, which can be described spatially as the 'marginal zone' (M–N). (b) The effect of a climatic change (cooling, C to C_1) on spatial–economic risk (an increase, R to R_1). The marginal zone shifts downward in elevation (M_1–N_1) while the former marginal zone becomes uneconomic or 'sub-marginal'. With fluctuating climate, recurrent zones of marginal and submarginal land can be estimated and mapped (see Figure 9.12)

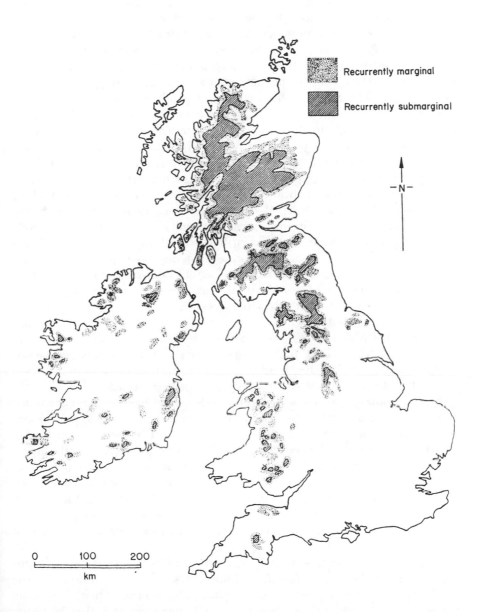

Figure 9.12 Recurrent marginality for oats cultivation in British Isles predicted for 1 °C decrease in mean temperature (from Parry, 1978)

season basis. This is assumed to be the mechanism behind long-term spatial shifts in cropping patterns. Implicit is the notion of risk evaluation and decision-making by farmers. In marginal areas, crop change or farm abandonment could be, and often has been, the choice.

The spatial–economic approach is expressed clearly in the works of Parry (1981, 1985; Parry and Carter, 1984). Parry (1975, 1978) originally investigated the relationships between climate, agriculture and settlement at the margins of oat growing in the uplands of Scotland during the 18th and 19th century. There, the spatial margins of production occur at high elevations with steep temperature gradients. Parry surmised that, because of the steep gradients, climatic cooling (or warming) could substantially increase (or decrease) the year-to-year risks of harvest 'failure' due to lack of sufficient warmth for crop maturation (as measured by GDDs). With slightly lower average temperatures, the risk of harvest failure in consecutive seasons, which is of critical importance to farmers, would be magnified (as discussed in Section 9.3.). If the risks become unacceptably high, the 'marginal zones' (the elevations at which farming is assumed to be just barely profitable given the prevailing risks of harvest failure) would shift to lower elevations as farms are abandoned and settlements recede downslope. Lands that were formerly considered marginal would become 'submarginal' (too risky for farming), and formerly stable, economically viable farms would become marginal. Thus, with a fluctuating climate one would expect to find zones of recurrent submarginal and marginal cropland. Figure 9.11 depicts schematically the methodological procedure by which these spatial shifts are predicted — or 'retrodicted' in this case (Parry, 1981). The results can then be mapped (Figure 9.12). Parry found support for his retrodictions in historical and archaeological evidence of upland settlement patterns.

Recent studies of upland oat-growing in Scotland have used an expanded climatic record to extend the analyses from the 17th century to the present (Parry and Carter, 1985). In this study, the coolest 50-year period (1661–1710) and the warmest 50-year period (1931–80) were compared with respect to estimated changes in the frequencies of harvest failures and the subsequent shifts in the marginal cropping zones (failure frequencies of between l-in-10 and 1-in-50 years). This modest warming (0.7 °C) would lead to an upward shift in elevation of the cropping margins of about 85 metres (equivalent to about one million hectares of land if extrapolated to the whole of Britain).

Similar methods were applied by Z. Uchijima (1978) to examine the latitudinal shifts in GDD isopleths under anomalous weather conditions and the possible effects on rice production in Japan. Uchijima found that, in general, the spatial and temporal variability in GDDs increases at higher latitudes. For changes in temperature expected to occur with a return frequency of 1-in-30 years, the GDD isopleths would shift 150, 200 and 300 km northward or southward from their normal positions in southern, middle and northern parts of Japan, respectively. In the northern region of Hokkaido, thermally marginal for rice growing, a southward shift of this magnitude due to cooling could possibly result in decline in production of about 40%.

A recent, complementary study in Japan by T. Uchijima (1986) considers the possible altitudinal shifts in the limits of rice cultivation in the two northernmost, climatically sensitive districts of Hokkaido and Tohoku. With a GCM scenario of warmer temperatures under doubled atmospheric CO_2 (Hansen *et al.*, 1983), the shifts were estimated to be large—430 to 510 metres upslope.

Discussion

From a global perspective, these studies are valuable in demonstrating how climatic change may create the potential for major spatial shifts in cropping patterns or farming characteristics by influencing decisions at the margins of agricultural regions. However, the caveats ought to be fully recognized.

Two critical assumptions can be identified from the schematic diagrams in Figure 9.11. It is assumed first that the relationship between climate and yield (curve D, which can be equated to technology and management) remains constant over time. The pace of technological change in many parts of the world suggests that the effects of alternative assumptions should be examined. Second, it is assumed that the levels of acceptable and unacceptable risks (P_1 and P_2) that define the zones of marginality and the spatial boundaries of crop production also remain constant over time. Again, in the light of changes in food demand, prices, farm security, government policy, etc., such definitions are bound to change in the future, especially in regions that are experiencing rapid social and economic development.

In addition, a very likely consequence of climatic change would be the spatial readjustment of the particular crop varieties *within* a crop region. For some crops the number of varieties is extremely large, and a broad range of climatic conditions could be accommodated through seed selection. Such genetic diversity, for example, allows wheat to be grown from northern Africa to the high latitudes of Sweden, across a range of temperatures far exceeding those which might be expected from future changes in climate as a result of the greenhouse effect. The possible shifts of boundaries of specific crop varieties have not yet been investigated systematically.

Whether realistic assessments of the global agricultural impacts of climatic change can be made depends on how the rather rigid assumptions of the marginal–spatial analyses can be relaxed, and on how the methods can be adapted to other parts of the world, particularly the tropics and subtropics. Furthermore, there is a need to make better use of existing models, including crop-climate models, in order to specify with greater precision the connections between climate, climatic risks, yields and farming decisions. Finally, there needs to be systematic consideration of the adaptations and adjustments that could, or would, be made in the agricultural sector in the event of changing climate and higher CO_2.

9.4.3 Agricultural Sector Analysis

A group of studies that we label 'agricultural sector analysis' has attempted to make improvements on all these fronts. These studies recognize explicitly that agriculture is one sector embedded in a larger economy, and that the impacts of climatic change will be felt at different levels (e.g. production, income, employment), both within the agricultural sector and between sectors (e.g. manufacturing, services). Estimation of the range of such regional impacts is important not only in its own right, but also because the parts of the sector(s) interact and have capacity for feedback and response. Two specific approaches are evident.

Integrated Regional Impact Analysis

One approach is to use a hierarchy of models, linking them in a sequential fashion to trace the 'cascade' of impacts through the biophysical, economic and social components of a regional system. For example, a combination of crop impact analysis (using crop-climate models) and marginal–spatial analysis allows one to answer two important questions: What is the magnitude of impact? And where does it occur? Outputs from this, in the form of altered average yields or yield probabilities, can be used as inputs to farm simulation models in order to estimate how yield changes might interact with management factors such as fertilizer use to cause changes in farm production and income. These results can then be used as inputs to regional economic input–output models to evaluate the downstream impacts elsewhere in the region (as on grain storage and transport, farm machinery and fertilizer manufacture, retailing or services). Thus there is an attempt to integrate the range of potential impacts and to test the sensitivity of complex regional systems to climatic change and perturbation (see Callaway *et al.*, 1982, for a review of available models). This basic approach was followed, for example, by the U.S. Department of Transportation's Climate Impact Assessment Program (CIAP; Grobecker, 1974), one of the first large-scale climate impact assessments aimed at estimating the possible range of effects resulting from atmospheric ozone depletion (Glantz *et al.*, 1982).

Recent work supported jointly by the International Institute of Applied Systems Analysis (IIASA) and the United Nations Environmental Programme (UNEP) has sought to refine the methods. The IIASA–UNEP Climate Impact Project has developed iterative procedures for testing the effects on agricultural systems of, say, changes in crop cultivars or managerial practices, thus adding a dynamic element that previous studies lacked. The project has pursued eleven case studies, both in high latitude and in semi-arid regions (the results are currently being prepared for publication. See Parry *et al.*, 1986a,b.) The set of semi-arid studies concentrates on short-term climatic variability, particularly drought, with three scenarios comprising a 1-in-10 year event, a single extreme year, and a 'back-to-back' event (consecutive extremes). In the set of high latitude studies the emphasis is placed on medium and long-term climatic changes. Two scenarios are based on the instrumental record and are used to investigate the impacts of a recurrence of an anomalous decade of weather and an extreme weather year. A third scenario based on the gridded outputs from the GISS GCM (Hansen *et al.*, 1983) is used to assess the possible impacts of a CO_2-induced climatic change.

To illustrate, the results of the high-latitude, long-term climatic change analyses are presented in Table 9.6. Some caveats are in order. First, the studies are very recent and not yet finalized, so the findings are unavoidably preliminary. Second, these results should not be regarded as predictions. The high level of uncertainty attached to GCM predictions of changes in regional climate is likely to be magnified as their impacts are traced through the regional agricultural and economic sectors. Rather, the studies should be regarded as experiments designed to evaluate the feasibility of an approach, namely the linking of GCM estimates of climatic change to models of impact.

The specific methods and assumptions vary considerably between the experiments. Most consider changes in both mean air temperature and precipitation, but in some cases precipitation changes are not deemed important (e.g. in Japan where rice is assumed to be fully irrigated). Several experiments assume technological change (e.g. winter rye cultivation in the Leningrad region), while others assume present-day technology (e.g. barley and oats cultivation in Finland). Most of the experiments treat the 2 × CO_2 climate as an equilibrium condition to be compared with the present baseline data, but several experiments introduce a time dimension by assuming a linear trend in climate between the present and future equilibrium conditions. The major assumptions are found in the footnotes to Table 9.6. These differences should dictate against direct comparison across the case studies.

Within each case study, however, the findings illustrate the hierarchy of impacts and linkages in the region. Consider the case of Saskatchewan, Canada. Under the climatic conditions simulated by the GISS model for 2 × CO_2, regionally-averaged mean air temperatures and precipitation for the growing season increased by 3.4 °C and 18%, respectively. Various indices were employed to estimate the possible equilibrium changes in agroclimate and biomass potential. Currently, low temperatures limit the duration of the frost-free season and act as a constraint to crop cultivation in the north of the

Table 9.6 Preliminary results from the IIASA/UNEP Climate Impacts Project, with particular reference to CO_2-induced climatic change in high latitude regions. These estimates describe only one set of scenario experiments. Others not presented here refer to different types of climatic change and to adjusted technology and management. A schema of the approach followed is given in the accompanying notes.[1] Unless otherwise stated, estimates are relative to the 1951–80 climate and to yields simulated for that climate, with management and technology at c. 1980 levels. Direct effects of CO_2 have not been considered (from Parry et al., 1986a)

Study Area	Canada Saskatchewan	Iceland	Finland South	Finland North	Northern U.S.S.R. Leningrad	Cherdyn	Central Zone	Japan
Source for climate scenario	[2]GCM	[2]GCM	[17]South [2]GCM	[18]North [2]GCM	[2]GCM	[2]GCM	Arbitrary Change	[2]GCM
CHANGE IN CLIMATE								
Temperature	[3] +3.4 °C	[10] +3.9 °C	[19] +4.1 °C	[19] +5.0 °C	[25] +4.2 °C	[28] +2.7 °C	[32] +1.0 °C	[41] +3.5 °C
Precipitation	[3] +18%	[10] +15%	[20] +73%	[20] +109%	[25] +52%	[28] +50%	NIL	[42] −5%
Effective temperature sums	[4] +50%		[21] +23%	[21] +37%				
Moisture index	[5] +1 to +13%							
Drought frequency	[6] ×3							
CHANGE IN POTENTIAL								
Cultivable area		[11] +482%						
Cropping limits								[43] +587m [44] +540m
Biomass potential	[7] +1 to +30%							[45] +9%

CHANGE IN AGRICULTURAL PRODUCTIVITY	CROP YIELDS	Spring wheat	[8] −25%	[2] +13%					
		Winter wheat						[33] +28%	
		Barley			[23] −2% [24] −32%	[21] +21% [22] +12%		[34] −5%	
		Oats						[35] −5%	
		Winter rye					[26] −13% [27] +4%		
		Rice					[29] −3% [30] +16% [31] +26%		[46] +8% [47] −7% [48] +2% [49] +9% [50] +26% [51] +5%
		Yield variability							[32] −27%
	LIVESTOCK	Hay		[12] +64%					
		Pasture		[13] +48%				[36] −4%	
		Carrying capacity: pasture rangeland		[14] +242% [15] +64%					
		Carcass weight		[16] +9%					
ECONOMIC CHANGES		Farm income	[9] −26%						
		Farm employment	[9] −3%						
		G.D.P.	[9] −12%						
		Total employment	[9] −2%						
		Additional costs						[37] −22% [38] +5% [39] +5% [40] +4%	
		Food stocks							[53] +1.8MT/yr

NOTES

Data sources: Parry *et al.* (1986a). Names following entries refer to authors of relevant chapters.

1 Schema of approach followed by IIASA UNEP Climate Project:

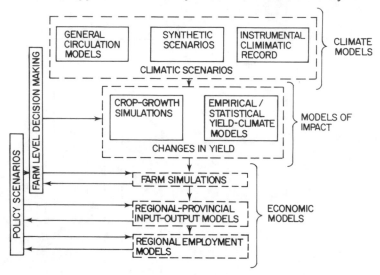

2 Goddard Institute for Space Studies (GISS) General Circulation Model $2 \times CO_2$ experiment; output processed for impact experiments (Bach).

CANADA (Saskatchewan)

3 May-August mean air temperature and precipitation (provincial mean) (Stewart).

4 Effective temperature sum above 5 °C base; approx. provincial average (William and Jones).

5 Precipitation effectiveness index (annual) (Williams and Jones).

6 Percentage months with Palmer Drought Index ≤ -4 (Williams and Jones); baseline = 3.1% (1950–82).

7 Climatic Index of agricultural potential (southern Saskatchewan) (Williams and Jones).

8 Provincial average simulated yields (Stewart).

9 Income, employment and GDP relative to 1980 values (Fautley).

ICELAND

10 Single-representative station (Stykkishólmur)—mean annual air temperature and precipitation (Bergthorsson).

11 Taiga area (hypothetical)—(Bergthorsson).

12 National average hay yield (Bergthorsson).

13 Average experimental pasture yield (Björnsson and Helgadottir).

14 Sheep number on improved grassland. Management as for 1941–9 (Bergthorsson).

15 Sheep number with average carcass weight fixed at 1980–84 level (Bergthorsson).

16 carcass weight with number of sheep fixed at 1965–83 average level (Dyrmundsson and Jonmundsson).

FINLAND

17 Helsinki area (Uusimaa province).

18 Oulu province.

19 Regional mean annual air temperature (Varjo).

20 May–September precipitation (Varjo).

21 Effective temperature sum above 5 °C base; 1971–80 baseline (Varjo).

22 Simulated yields relative to 1959–83 baseline for adapted spring wheat variety with thermal requirements 120 GDD greater than for present variety; southern Finland (Rantanen).

23 Simulation using 1883–1983 baseline climate, relative to 1983 reference yield (Mukula).

24 Simulation using 1959–83 baseline climate, relative to 1983 reference yield (Mukula).

NORTHERN USSR

(25) Mean annual air temperature and precipitation (Iakimets and Pitovranov).

(26) Simulation relative to 1973–80 baseline (Iakimets and Pitovranov).

(27) Simulation for 2035 given transient fertilizer trend and transient climate change for assumed doubling to CO_2 by 2050. Other technology fixed at 1980 levels (Iakimets and Pitovranov).

(28) May–September mean air temperature and precipitation (Sirontenko).

(29) Simulated yield for present variety (Sirontenko).

(30,31) Simulated yield for two new varieties with thermal requirements 50 and 100 GDD greated than for note 29, respectively (Sirotenko).

(32) Transient change from 1980 to 1995 (Kiselev).

(33–6) Estimates using a crop production model relative to 1980 trend yields with technology based on extrapolation to 1995 (Kiselev).

(37) Winter wheat

(38) Barley   Minimized expenditure to maintain production

(39) Oats   at 1980 levels (Kiselev).

(40) Hay

JAPAN

(41) July–August mean air temperature in Hokkaido district. ΔT for Tohoku district = +3.2 °C (T. Uchijima).

(42) June–September precipitation (all Japan) (Z. Uchijima and Seino).

(43) Altitudinal shift for rice (Hokkaido) (T. Uchijima).

(44) Altitudinal shift for rice (Tohoku) (T. Uchijima).

(45) Total net primary production (all Japan) (Z. Uchijima and Seino).

(46) Yield index (Hokkaido) (T. Uchijima).

(47) Simulation for existing early-maturing variety in Hokkaido (Horie).

(48) Simulation for existing early-maturing variety, but for average of annual computations; 1974–83 baseline (Hokkaido) (Horie).

(49) As for note 48, but for new middle-maturing variety (Horie).

(50) As for note 48, but for new late-maturing variety (Horie).

(51) Yield index (Tohoku) (T. Uchijima).

(52) Coefficient of variation for existing early-morning variety relative to 1974–83 baseline (Hokkaido) (Horoe).

(53) Mean annual accumulation of national rice stocks (Tsujii).

Province. Under the scenario of climatic change, the effective temperatures for plant growth increased markedly—a 50% increase in the Effective Temperature Sum (relative to the baseline period, 1951–1980). Thornthwaite's Precipitation Effectiveness Index was used to assess the average moisture situation, which improved with higher precipitation and offset the effects of higher temperature (although the frequency of monthly droughts, as calculated from the Palmer Index, increased).

The calculated effect on biomass potential was positive as well. Using the Climatic Index of Agricultural Potential to assess the combined effect of temperature and precipitation, increases in biomass potential ranged from 1% to 30% in southern Saskatchewan to 74% at Uranium City in the north.

These 'improvements' in agroclimate and biomass potential do not necessarily bode well for spring wheat production in the Province, however. The impact on Saskatchewan spring wheat, which covers much of the southern Province and contributes to about 13–14% of total world wheat trade, was estimated for each crop district using a crop-growth simulation model, and averaged over three broad soil zones. The aggregate effect on total provincial production was a decrease of 25%. Total farm income was estimated to decrease by a similar proportion (relative to 1980 values), as determined from a farm simulation model. When these effects were used as input to regional input–output and employment models, the downstream effects of the climatic change were estimated in terms of change in farm employment (−3%), total provincial employment in all sectors (−2%) and total provincial Gross Domestic Product (−12%). Of course, there are wide confidence intervals attached to all these estimates (and others presented in Table 9.6).

Although most of the estimates are based on the assumption of static agronomic and economic conditions, it is possible to evaluate various options that are available to offset or mitigate the impacts by altering some of these assumptions. In the case of Saskatchewan, for example, options that have been investigated include the substitution of winter wheat for spring wheat and the transfer of marginal crop land to pasture. Experiments such as these serve to generate a new set of impact estimates that can be compared with the initial estimates, and that can help in evaluating appropriate policies of response. This iterative procedure thus explicitly recognizes the potential ability of agricultural systems to respond to environmental change. Some of the estimates in Table 9.6 incorporate adjustments like new crop varieties or changes in fertilizer application.

Agricultural Systems Analysis

Agricultural systems contain many more such feedback mechanisms that, over time, can act to exaggerate or diminish the potential impacts of climatic change on crop yields or spatial cropping patterns. Farmers can change technological inputs, crop varieties or planting decisions, and governments can intervene through price support programmes, export subsidies or disaster payments. The market system itself helps to regulate through forces of supply and demand on price. How do these factors interact to affect agriculture in the face of changes in climate and climatic risk? That is the question addressed explicitly by agricultural systems analyses.

The diagram in Figure 9.13 is a simple illustration of some hypothetical feedbacks that could conceivably influence yields and production under conditions of changing climate. These feedbacks help explain why deteriorating (or improving) climate may not necessarily lead to a concomitant deterioration (or improvement) in yields and production.

One example is the feedback loop between crop response and yields in Figure 9.13: a decline in crop yields due to climatic change may stimulate ge-

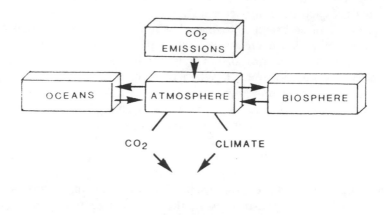

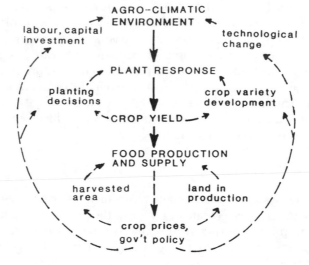

MODELLING METHODS

    Econometric, agricultural sector
    Normative programming
    Agricultural trade (global models)

Figure 9.13 Some feedbacks influencing crop yields and production over time. The *agricultural systems approach* to impact assessment examines the dynamics of agriculture and the mechanisms which can diminish or accentuate the primary yield effects of increased CO_2 and climatic change

neticists to develop (perhaps in the future by biotechnological techniques) a new suite of cultivars, leading to improvements in crop response and raising yields. It has been estimated that, once developed, adoption of a new variety in the USA takes less than a decade (Wittwer, 1980). In many parts of the developing world, the rapidity with which higher yielding grain varieties have been adopted is illustrative of the potential ability of global agriculture to take advantage of future crop research and development.

The central role of price and government policy in the agricultural system is evident. Holding all other factors constant, a decrease in yields would depress production and food supply and could lead to an increase in price. Increased prices have the effect of increasing the amount of land actually harvested (in the US Great Plains, for example, about 8% of planted wheat land, usually the poorer marginal land, remains unharvested, on average, due to economic considerations), thereby raising commercial, area-wide yields (per planted area). Higher price (from any source, including exogenous government price supports) also has the effect of contributing beneficially to farm income and encouraging investments in labour and capital required to produce acceptable yields. The basic premise of agricultural systems analysis is that this economic and political context in which climate and agriculture interact is crucial to understanding the long-term effects of climatic change.

Within this context, one can focus on the problems of climatic change at various scales, from individual farm production to global agriculture and food trade. Callaway *et al.* (1982) describe three types of agricultural sector models that can potentially be used to assess the impacts of climatic change. *Econometric* models relate crop area, yield and production of various crops and regions to explanatory variables (sometimes including weather, but often not) through multivariate regression techniques using historical and cross-sectional data. Similar techniques are used to estimate national demands for export and domestic production. These empirical models attempt to describe the actual behaviour of the system. In contrast, *normative* (or mathematical programming) models, typically based on linear programming techniques, purport to demonstrate how farmers should behave—the 'optimal' behaviour—to satisfy specified economic objectives. These models represent the workings of the production system in physical and technological terms, and specify the flow of resources at various stages of farm production. The farm unit can be representative of any level of regionalization and aggregated up to national scale.

At an international scale, agricultural trade models—or, more generally, *global models*—link up nations or world regions through the mechanisms of world trade in agricultural commodities. Such models can be either normative or econometric. They vary widely in specific method and dynamic qualities, from systems dynamic methods to rather static input–output formulations. A selection of major global models and their characteristics is

Table 9.7 Major global models and their characteristics (adapted from Robinson, 1985)

Model	Authors	Time horizon (yrs)	Method	Geographical aggregation	Aggregation of agricultural sector	Treatment of climate
SARUM	Roberts (1977) SARU (1978)	90	dynamic simulation input–output econometric	3 regions	4 agriculture products, 1 food product, 3 agriculture inputs	generally omitted
MOIRA 1	Linnemann et al. (1979)	45	algorithmic, optimization, econometric	106 nations	1 commodity	repetition of past yield variations
World Integrated Model (WIM)	Mesarovic and Pestel (1974) Hughes (1980)	25–50	dynamic simulation input–output	12 regions (basic) 17 regions (subregional)	5 commodities 3 land types	omitted
International Futures Simulation (IFS)	Hughes (1982)	25+	dynamic simulation	10 regions	2 commodities, 5 land types	represented by yield factor

Latin America World Model	Herrera Scholnik et al. (1976)	100	dynamic optimization	5 regions; 20 regions may exist	livestock crops	omitted
FUGI	Kaya and Onishi (various dates)	10–15	econometric, input–output (dynamic)	14 to 62	4 sector (?)	omitted
USDA Grains, Oils and Livestock (GOL)	Rojko and Schwarz (1976)	10–20	econometric, recursively dynamic	28 regions	up to 14 commodities	omitted
Interactive agricultural model	Enzer et al. (1978)	20	cross impact interactive projection	10 regions	grain as proxy for all foods	stochastic yield prediction
IIASA/FAP	Parikh and Rabar (1981); Fischer and Frohberg (1982)	15–20	linked national models general equilibrium recursively dynamic	23 countries and country groups	9 commodities	omitted

shown in Table 9.7. The time horizons of these models vary from 10 to 200 years, and the geographical aggregation ranges from a single world unit to 106 individual countries. Although in several models yields are varied from year-to-year (based on past yield variability) to represent the influence of weather, none explicitly incorporates climate variables.

However, all the major global models can potentially be used to assess the impacts of rising CO_2 and climatic change by exogenously manipulating the yield components. The major obstacle to doing so is that the models have not been subjected to careful scrutiny with respect to their compatibility and reliability in this problem context (Robinson, 1985; Liverman, 1983). As pointed out by Robinson (1985), global models were not constructed with climate impact assessment in mind, and it is likely that model structure, geographical aggregation and temporal resolution may be inappropriate. (For recent reviews of global models, see OTA, 1982; Meadows *et al.*, 1982; Hughes, 1981; Robinson, 1985; or Callaway *et al.*, 1982.)

Very few explicit climatic 'experiments' have been conducted with global models, despite the recommendations to this effect made by the World Climate Conference over five years ago (WMO, 1979). None could be called 'definitive'. However, two studies, one carried out at NCAR (Liverman, 1983; NCAR, 1984) and the other for the National Defense University study (NDU, 1983), provide examples of the potential use (or misuse) of global models and some insight regarding the possible dynamic responses of the agricultural sector to climatic change.

The study by Liverman (1983) sought to evaluate a single global model, the International Futures Simulation model (IFS; Hughes, 1982) in terms of its reliability for climate impact assessments. The general structure of the IFS model is shown in Figure 9.14 along with the yield-specific relationships within the agricultural sub-component. Like its parent model, the World Integrated Model (Mesarovic and Pestel, 1974), the IFS model does not contain a specific climate component. Instead, climate is represented by a yield factor, a multiplier to yields that can be manipulated as a surrogate for weather. In conducting climate-related sensitivity analyses, Liverman (1983) varied the yield factor in order to examine the direction and magnitude of response of predicted yields, production, exports and imports, crop prices, reserve levels, global starvation and the like. These 'perturbed' runs were compared to 'base' runs with no climatic change (yield factor = 1.0). Such analyses provide clues as to whether a model behaves sensibly.

The IFS model was perturbed with both gradual trend changes and single-year 'pulses'. In the slow trend analysis, the model's 'climate' (the yield factor) was gradually altered beginning in 1985, reaching the maximum change (1.2 or 0.8, depending on the assumed direction of climate impact) in the year 2000. In the pulse analysis, a single year only (1985) receives the maxi-

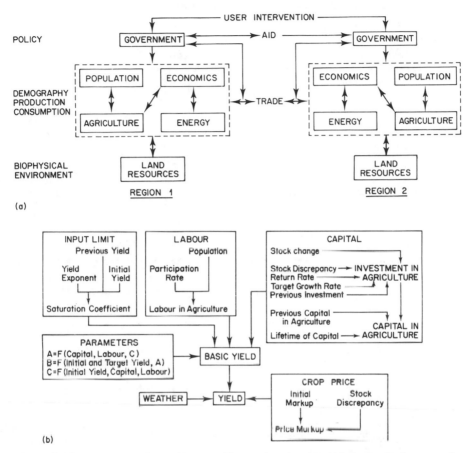

Figure 9.14 The International Futures Simulation model: (A) the basic framework; (B) estimation of crop yield within the agricultural component (from Liverman, 1983, as adapted from Hughes, 1982)

mum change. The analyses were conducted at both regional (e.g. USA) and aggregate global levels (ten regions) and produced two interesting results with respect to yields and production.

First, in the trend analysis the ±20% changes by 'climate' led to predicted yield and production changes in the year 2000 that were noticeably less than 20%. In the USA, for instance, predicted yields increased and decreased by about 16–17% and 14–15%, respectively (Figure 9.15a). This tells us that the model's internal adjustment mechanisms were able to dampen about one-quarter of the adverse yield impact and about one-fifth of the potential gain from climatic change. Similarly, at the global scale total crop production changed by 5–7% in both directions by the year 2000, as shown in

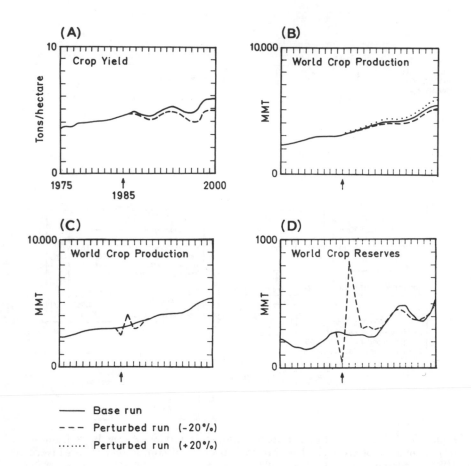

Figure 9.15 Simulated agricultural effects of perturbed 'climate' versus control runs to the year 2000 using the International Futures Simulation Model (adapted from Liverman, 1983). (A) crop yields in the USA with slow trend change in yield factor to −20%; (B) world crop production with slow trend changes in yield factor to ±20%; effect on world crop production (C) and reserves (D) of a single −20% pulse in 1985

Figure 9.15b. Thus, the agricultural system displayed a considerable capacity for absorbing the potential impact of a slow change in climate—by about two-thirds.

Second, single pulses in *either* direction created instabilities in subsequent years, with negative consequences. A sudden jump in yields of +20% flooded crop reserves and depressed prices, which caused less agricultural investment and a rather large drop in production and reserves in the next year. An oscillation was set in motion that took a number of years to disappear. With a -20% pulse, a similar oscillation occurred, but the magnitude of the effects was greater (Figure 9.15c,d). The average change in production over the 15-year period was minor compared to the base run, but the ups and downs generated by the pulse created greater total starvation (the model's indicator of societal impact at the global scale). If, indeed, the importance of climatic change lies in the frequency changes of disruptive climatic events (the 'shift-in-risk' view), this global model would have us believe that the disruptive events are the unusually 'favourable' as well as the unfavourable years for global crop yields.

The study by the National Defense University (NDU, 1983) sought to determine the global agricultural impacts of various scenarios of climatic change, from large cooling to large warming. Yield impacts estimated previously (NDU, 1980) were entered into the USDA Grains–Oilseeds–Livestock (GOL) model to simulate production, trade and other economic effects to the year 2000. The GOL model is an econometric equilibrium model of the agricultural sector with demand, supply and trade components for 28 world regions and 12 agricultural commodities. The model could be described as 'dynamically recursive' in that the estimated values of selected endogenous variables (e.g. price) in one time step are used as exogenous variables in subsequent steps (Callaway *et al.*, 1982). For this reason alone, it would be expected that initial yield impact factors, used to force the model, would differ from the predicted yield and production levels produced by model simulations.

This expectation proved correct. The differences, however, were not consistent from region to region. The global model inflated initial yield impact values in some cases (e.g. USA grains) while deflating them in others (e.g. Australian grains). Not only did magnitudes vary, but the directions of change were even different in several cases (e.g. Argentina). Presumably, the predictions of the model reflect production re-adjustments and changing patterns of comparative trade advantage from changing prices, investment levels and shifts in land use as the model markets clear at each time-step to balance supply and demand. The region by region production estimates for a large warming scenario as per cent deviations from base level projections (no climatic change) are shown in Table 9.8.

Overall, for a large global warming the NDU study projects no change in net global grain production. However, some countries would gain appreciably (e.g. Canada, USSR) and others lose (e.g. USA, Australia, South Asia). The USSR becomes a net exporter, and the USA exports less and Canada more. The study concludes that the global impacts of climatic change on grain production are relatively small compared to the sensitivity of production to variations in assumptions about

454 The Greenhouse Effect, Climatic Change, and Ecosystems

Table 9.8 Simulated global grain production in the year 2000 under a large warming scenario, as a percent of base level projections[1] (adapted from NDU, 1983)

Group/country	% from base level
I. Developed Countries:	−2.4
United States	−3.8
Canada	6.0
European Community	−2.2
Other Western Europe	−2.0
Australia	−3.0
South Africa	−4.1
II. Centrally Planned Countries:	3.1
Eastern Europe	1.1
USSR	6.1
China	0.7
III. Developing Countries	−1.4
Indonesia	0.5
Thailand	−3.6
Other Southeast Asia	−0.0
India	−1.8
Other South Asia	−2.0
High Income North Africa/Middle East	−2.8
Low Income North Africa/Middle East	−2.9
Central America	−2.8
Brazil	0.3
Argentina	2.6
IV. Total Above	0.0
V. Warming Countries' Total[2]	3.3

[1] Large warming scenario (ΔT, ΔP) = 1.4°C, 6% high-mid latitudes; 1.0°C, 2% mid-low latitudes; 0.75°C, 2% sub-tropics.

[2] Countries favourably impacted by warming (Canada, E. Europe, USSR, China).

population growth rates, per capita income, agricultural investment and technological change.

Discussion

One should be cautious about accepting these findings at face value, however. Serious criticisms have been expressed concerning the specific manner in which the NDU study was conducted (e.g. Stewart and Glantz, 1985). More generally, major questions can be raised concerning the reliability of global models for climate impact assessment. Few extensive validations of global models have been performed (Meadows *et al.*, 1982). Global modellers face chronic problems of limited historical data (needed for calibration as well as validation). These are compounded by the occurrence of 'anomalous' years which hinder model validation: for example, sudden policy decisions (large jump in Soviet grain imports in the early 1970s or USA grain embargoes in the 1980s), exogenous sector changes (OPEC and oil prices in the mid-1970s) or unusual combinations of weather events (as in 1972). For these reasons an attempt to validate the IFS model by recalibrating on pre-1970 data and validating on post-1970 data gave poor results (Liverman, 1983). But this does not necessarily mean that the model itself is poor; it may only mean that there was too much 'noise' to determine if it is valid.

This presents a dilemma: '...Is the real world so complex that no simplification (model) can capture its behaviour?' (Robinson, 1985). In the absence of solid validations, belief in the results of models rests largely on faith. Careful attempts at validation and sensitivity analyses are required. One slow, but potentially effective, solution to validation, given limited historical data, is to test model predictions against observed data for each new year. This is being pursued in an effort to validate further IIASA's Food and Agriculture Programme model (Frohberg, 1984).

If we set our sights lower than quantitative prediction of climate impacts, global models may serve us better. If one believes that a model behaves reasonably with regard to the direction and magnitude of change, then in a more qualitative fashion it could be used

• as a pedagogic tool for understanding the interactions of many variables;

• to examine the effects of strategies and policies for responding to climatic change;

• as a framework for organizing what we know and do not know about the behaviour of the system.

In the context of CO_2 and climatic change, little has been made of global models for these purposes.

Even so, agricultural sector analyses tell us, in general, that crop impact analyses and marginal–spatial analyses are only a start, not the answer. In the event of changing climate, the dynamics of the agricultural sector would, to some degree, readjust crop yields and production with the passage of time. It appears likely that in some regions of the world, yields and production may be just as, if not more, sensitive to changes in technology, price or policy as they are to changes in climate, even large ones. Since these factors are largely manipulatable, whereas climate is not, this should give us some confidence in the face of possible climatic change.

9.4.4 Historical Case Studies

An alternative approach to climate impact assessment asks, What impacts actually *did* occur during past climatic changes? Descriptive, historical case studies seek to shed light on possible future effects on agriculture by examining past situations in which nature provided the climatic 'experiment'. The approach is complementary to the previous approaches in that it provides an empirical check on assumptions and model-derived estimates regarding crop yield impacts, spatial shifts in crop margins (other indicators of impact) or the long-term dynamic effects of adjustments and other feedback mechanisms of agricultural systems.

There exist scores of historical case studies that deal with agriculture and climatic change and variability (e.g. see list by Rabb, 1983). Several attempts have been made to pull together the threads of climate and history (e.g. Wigley *et al.*, 1981; Rotberg and Rabb, 1980; Smith and Parry, 1981), but the implications for future changes in climate remain elusive. Even case studies dealing directly with specific slow climatic changes often find it difficult to generalize about climate's impact on agriculture and society. For example, both Le Roy Ladurie's (1972) search through European history for 'times of feast, times of famine' and Post's (1977) investigations of the last great subsistence crises in Europe found the role of climate intermingled with social, political and economic factors and therefore difficult to define.

Others have pointed to the lack of demonstrable impact in particular historical cases as a sign of agriculture's resilience to climatic change. Wittwer (1980), for instance, notes that, in the USA, the state of Indiana experienced a total change of $+2$ °C over this century (with a 0.1 °C per year trend from 1915 to 1945), while agriculture continued to grow and prosper. Similarly, Rosenberg (1982) claimed that, historically, the shifting spatial pattern of wheat varieties with differing climate tolerances in the US Great Plains is evidence of agriculture's potential adaptability to climatic change.

Most historical case studies (and other anecdotal examples), however, skim the peripheries of the CO_2 and climatic change issue and lack scientific rigour. Extraction and comparison of relevant conclusions from the literature are complicated by at least three problems. First, many studies focus on the *distant past* (e.g. the Little Ice Age) when agricultural technologies and socio-economic conditions were radically different from today. Second, many suffer from problems of *time-coincidence* (Parry, 1981); due to lack of control, the cause-and-effect relationship between climatic change and agricultural impact is not clearly established (e.g. the coincidence of drought and economic depression in the US Great Plains during the 1930s). And third, the literature represents an eclectic mix of *different hypotheses* (or lack thereof) and methods which renders individual studies largely incomparable. For these reasons, historical case studies often appear simply idiographic and non-generalizable.

There is much to be gained by designing studies of the impacts of past climatic changes to overcome these problems. One study currently underway, for instance, is systematically examining recent climatic 'changes' (over a decade or so) in a number of climatic divisions in the USA, each with adjacent non-affected areas as control cases (Kates *et al.*, 1984). By examining recent cases of actual changes in climate, it is possible not only to look for evidence of primary biophysical crop and yield effects, but also to determine if, or how, agriculturalists perceived the changes and adjusted to new conditions over time. That a large number of potential case studies exist at this scale in the USA has been demonstrated by Karl and Riebsame (1984) (see Figure 9.16 for example). At a much broader scale, a survey of 20th century climatic fluctuations in the Northern Hemisphere was recently reported by Wallen (1984).

It has also been suggested that studies of cases analogous to climatic change may prove fruitful: for instance, the slow depletion of irrigation water from the Ogallala Aquifer underlying the US Great Plains (Glantz and Ausubel, 1984), or cases of the migration of agriculturalists to regions of unfamiliar climate (Nix, 1985). Rosenberg (1982) suggests spatial 'crop migration histories' as evidence of both climate impact and response at agricultural margins.

Furthermore, if changes in the frequencies of extreme events prove to be important manifestations of climatic change, then research on natural hazards could be particularly relevant (e.g. for global assessments see White, 1974; Burton *et al.*, 1978). Because agricultural (and other) impacts of extreme events are disruptive and visible, the research emphases have been placed on analyses of individual events (e.g. Garcia, 1981, on the international impacts of droughts in 1972); on hazard perception and adjustment (e.g. Heathcote, 1973, and Saarinen, 1966, on Australian and US Great Plains droughts, respectively); on trends in vulnerability (e.g. Warrick, 1980, on US Great Plains droughts, or Kates, 1980, on global trends); or on theories of hazard vulnerability (see Hewitt, 1983, for a range of viewpoints). So far, however, there has been little attempt to link the methods and findings of

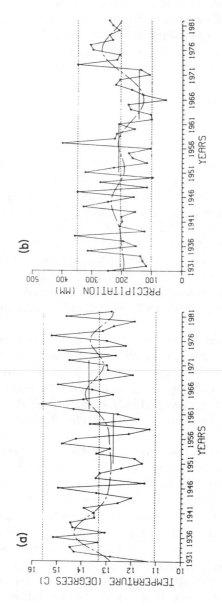

Figure 9.16 Time series plots depicting climatic fluctuations of (a) decreasing temperature and (b) increasing precipitation, from 1943–59 to 1960–76 (from Karl and Riebsame, 1984)

this large body of research to problems of long-term climatic change (as per our discussion in Section 9.3, for instance). The opportunity to learn from our actual experience with climatic change and variability should not be overlooked.

9.5 SUMMARY AND CONCLUSIONS

In this chapter, we have reviewed the possible direct effects of CO_2 on crops, as derived primarily from glass-house experimentation, the various approaches and their applications and findings with regard to assessing the possible effects on agriculture from climatic change. From a global perspective, what can we conclude?

9.5.1 The Possible Impacts of Increased CO_2 and Climatic Change

One approach is to consider crop impact in a static fashion, that is, as if CO_2 doubling or climatic change were instantaneous with no reaction from agriculturalists or the agricultural system. From such 'crop impact analyses' some clear differences exist between the effects of CO_2 and climatic change effects; between the core and margins of crop regions; and between impacts in higher and lower latitudes. With respect to direct CO_2 effects,

• A 'doubling' of ambient CO_2 concentrations has a positive effect on growth and yield of major food and fibre crops. These may range from 10% to 50% for C_3 plants to 0% to 10% for C_4 plants.

• Globally, the potential benefits of CO_2-enhanced yields might well be unevenly distributed because of the differences in where C_3 and C_4 crops are grown. For instance, tropical and sub-tropical regions of Africa and Latin America that are dependent on maize, sorghum or millets (C_4 plants) may be less favoured than rice, wheat, barley or potato (C_3 plants) regions in South Asia, North America or Europe.

• The positive growth and yield response from elevated CO_2 levels is obtained under most environmentally stressful as well as optimal conditions. Thus both the core and the margins of crop regions could benefit from increased CO_2 relative to current yield levels.

• In relative terms (i.e. per cent of control CO_2 levels), the growth and yield response is actually higher under some stressful environmental conditions, like moisture stress. This could possibly have the effect of decreasing interannual yield variability in some cases, as in drought-prone areas.

• In absolute terms, yield response to increased CO_2 concentrations is greatest under good growing conditions, including adequate soil nutrients.

In many developing countries where soil nutrients shortage is a chronic problem, the full benefits of enhanced yields may not be realized, particularly if phosphorus is deficient.

While the rise in global atmospheric CO_2 concentrations will accumulate comparatively smoothly over time and uniformly over space, climatic changes are apt to vary in direction and magnitude from region to region, and to occur against a background of relatively large interannual climatic variability. As yet, the regional patterns of climatic change cannot be forecast reliably. This presents the major obstacle to predicting actual crop yield and production impacts from climatic change. However, the sensitivity of crop yields can be investigated by using scenarios of climatic change—arbitrary, instrumental or GCM-derived. Employing crop-climate models, a number of studies have found that, in the absence of managerial adjustments or direct CO_2 effects,

• For the *core* areas of the North American and European mid-latitude grain regions, the probable effect of an instantaneous increase in average temperatures would be to decrease crop yields. For a 2 °C increase, grain yields might decline from 3% to 17%, as a rough approximation.

• The negative impact of higher temperatures on grain yield derives from associated increases in evapotranspiration, and from accelerated rates of plant development and a shortening of the period of yield formation.

• Increases in precipitation would tend to offset the reductions in grain yield from warming, while decreases in precipitation would accentuate them (even in more humid grain regions like Western Europe).

• The impacts of climatic warming at the margins of production could be less than, greater than (e.g. semi-arid margins) or in the opposite direction from (e.g. at the cold margins) those observed at the core areas of production, depending on local environmental conditions.

• Few systematic studies of the impacts of possible changes in climate have been conducted for the tropics and sub-tropics.

At all latitudes, the potential for severest adverse (or most beneficial) impacts of climatic change on crops may, in fact, be located in the marginal areas, variously defined. Localities with steep environmental gradients, with limiting temperature or precipitation amounts or with economic transition zones may be sensitive to even slight changes in long-term trends, particularly as they may result in shifts in the frequencies of extreme climatic events on a year-to-year basis. But this is to say nothing of the agricultural consequences of, or responses to, potential crop impacts. It may well be that because climatic variability is a feature of those marginal areas, agriculturalists have incorporated a large range of adjustments and adaptations to deal with year-to-year variations. Furthermore, the socio-economic consequences of yield impacts are important only at a local level, whereas yield effects of climatic change in core areas, where the largest proportion of food production takes place, could have major implications at national or international scales. Rarely, however, is the issue of 'impact *for whom*' addressed in crop impact analyses.

The impacts of climatic change in marginal areas of agriculture might well be expected to elicit spatial shifts in crop areas or practices—the concern of the 'marginal–spatial approach' to impact assessment. Such shifts could be considered as either ecological adaptation to, or economic impacts of, climatic change. Research with an ecological orientation indicates that, at least in the mid- and high-latitudes, the potential shifts in crop boundaries based on average crop-climate associations alone could be on the order of hundreds of kilometres per °C change (e.g. rice in Japan, wheat in Canada or maize in the USA). Other marginal–spatial studies interpret and predict possible spatial changes through effects on climatic risks and agricultural decision-making. Still, the potential for spatial re-adjustment looms large.

Spatial readjustment, of course, is only one way in which agriculture could respond to increasing CO_2 and climatic change. In Section 9.3 we characterized the range of responses as 'bands' of adaptation and adjustment, and emphasized that the potential impacts of climatic change and variability are a function of the 'widths' of these bands at any given time. Much of the response capability is internal to agriculture: feedback mechanisms can help to self-regulate the system to environmental change over time. The issue is, how much?

This question is addressed by 'agricultural sector analyses'. One approach is to link models in a sequential fashion to identify, estimate and integrate the 'cascade' of impacts which may occur at the regional scale. At national and international scales, global models that focus on agricultural production, consumption and trade are one means of examining the interactions within the entire system, although they remain largely untested and unused for purposes of climatic impact assessment. Based on limited applications, it can be suggested that

• A large proportion of any potential adverse effects on yields and production as a result of gradual change in climate can be absorbed or avoided through policy and market feedback mechanisms.

• The disruption from single extreme years (which could become more or less frequent with global warming) could cause over-reaction in the system with oscillating impacts in subsequent years. Extremely favourable, as well as unfavourable, production years may cause similar effects.

• The impacts of climatic change on production in one region could be transferred to another over time through the network of global market and trade.

Rather than as findings, it is prudent to advance these points as hypotheses for further analyses.

The general lesson from agricultural sector analyses is that close attention needs to be paid to the *dynamics* of the system. Furthermore, the response of the system depends critically on the assumed *rate* of environmental change. Just as in the case of using GCMs to investigate the climate sensitivity to doubled CO_2, it becomes important to consider the transient response of agricultural systems as climate is changed over time.

In lieu of modelling approaches, 'historical case studies' can draw upon actual experience with climatic change in order to investigate crop impacts and agricultural response. For various reasons, it has been difficult to generalize from existing historical case studies. However, there is ample opportunity to design studies which could provide valuable empirical evidence to balance and complement modelling results.

9.5.2 Further Considerations

As reflected in the structure and content of this review, assessments of the agricultural impacts of increasing atmospheric CO_2 have been neglectful in at least three general ways. First, specific studies of the effects of higher CO_2 concentrations and of climatic change have proceeded quite independently of one another, following different approaches at different scales of analysis. For the most part, the direct effects have been examined by laboratory experiments at the scale of single leaves or plants, starting at the enzymatic level. Studies of the effects of climatic change tend to rely on models of crop yields, production or agricultural activity at regional and global scales. The gulf between the two is wide and largely unsatisfactory. In reality, CO_2 and climate variables like temperature, precipitation and radiation will affect crops simultaneously and interactively, and should be studied accordingly across a range of scales.

Second, agricultural pests and diseases, which exact heavy tolls on crop yields worldwide and which account for a large portion of expended labour and capital in agricultural production, are also likely to be affected by climatic changes (and perhaps CO_2). Nevertheless, they are rarely included in any systematic fashion in CO_2-specific impact studies.

Third, the research emphasis has rested heavily on agricultural crops, particularly grains, in temperate regions. Yet for many parts of the world the possible effects of increased CO_2 and climatic change on grasslands and live-stock production is a primary concern. Again, these effects have received scant attention in CO_2-specific impact studies.

In part, the above issues could be addressed by turning to existing literature in an array of relevant disciplines. But that would require a more exhaustive and discerning review than that attempted by this chapter.

9.5.3 Some Next Steps

The overall conclusion is that we have barely scratched the surface with respect to assessing the possible agricultural impacts of increased CO_2 and climatic change. Problems abound. While hundreds of direct CO_2 experiments have been performed, these are in the safety of the scientist's glasshouse and not subject to the vagaries of nature—weather, pests, disease—nor to the whims of human management practiced in the field. Extrapolations of results from climate-crop models, many of which have not been adequately validated, fail to consider the dynamic response capability of agriculture. Clear distinctions have not been made between the core and the margins of crop regions where entirely different patterns of yield impacts and agricultural response may be experienced. Modellers of global agricultural relations have rarely considered climate, one of the most basic determinants of crop production. We are stuck with patching together scattered studies; the global picture is far from complete. The research community could help in five ways.

First, systematic consideration of the combined effects of CO_2 and climate variables is needed. From the preceding review, it is clear that the effects are interactive, and that simply adding together independently conceived experiments and model results is neglectful of these interactions.

Second, crop-climate models are the principal tool for assessing the yield impacts of climatic change, but from a global perspective the reliability of available models for these purposes is not well established. An international collaborative effort of crop-climate model validations and cross-model comparisons would strengthen the methodological foundation for impact assessment.

Third, in all aspects we know least about the possible agricultural impacts in the tropics and sub-tropics and most about the temperate and high latitudes. The imbalance needs to be redressed.

Fourth, in practically all analyses, scenarios of climatic change are imposed on a crop or agricultural system for which 'impacts' are then estimated. Considerably less thought, *a priori*, has been given to how climate actually influences agricultural decision-making, management and, consequently, yields. Is it the shifts in climatic risks, the slow changes in climatic averages, or some combination thereof which matters most to agriculturalists? Given knowledge of the sensitivities and of the ability or inability of agriculture to cope with change, one can *then* ask how increases in CO_2 and climatic change could make a difference.

Finally, there needs to be a better integration of available methods in impact assessment. Crop-climate models, marginal–spatial methods, and global agricultural models are rarely linked sensibly in order to build a coherent, realistic appraisal of agricultural impacts. Considerable testing and refinement of models is a necessary prelude to such an integrated approach, and the IIASA-UNEP project (discussed above) has taken some tentative, encouraging steps in this direction. This conclusion is consistent with the recent SCOPE review impact assessment and with the philosophy and recommendations of the World Climate Conference (WMO, 1979). In the long run, this is the goal which current climate impact activities should aim to contribute.

NOTE ON AUTHORSHIP AND ACKNOWLEDGEMENTS

Of the two principal authors, R.A. Warrick assumed responsibility for organization of Chapter 9 and for preparation of sections dealing with the possible indirect effects of climatic change (9.1, 9.3 to 9.5). R.M. Gifford wrote Section 9.2, the direct effects of higher CO_2 concentrations on crop plants. The collaborating author, M. L. Parry, contributed to the portion of Section 9.4.3 concerned with integrated regional impact assessments, for which T. Carter prepared Table 9.6. In addition, D. Liverman provided guidance in the preparation of Section 9.4.3, dealing with agricultural systems analysis.

The authors wish to thank the following persons for commenting on earlier drafts: B. Bolin, W. Böhme, B. R. Döös, T. Carter, W. Degefu, H. W. Ellsaesser, H. L. Ferguson, F. K. Hare, P. G. Jarvis, J. Jäger, R. W. Kates, M. R. Kiangi, H. E. Landsberg, D. Liverman, W. J. Maunder, J. E. Newman, O. Preining, C. Sakamoto, B. R. Strain, P. E. Waggoner, M. M. Yoshino.

Many useful ideas and proposals were forthcoming from a small WMO/UNEP/ICSU Expert Meeting on the Reliability of Crop-Climate Models for Assessing the Impacts of Climatic Change and Variability, held in Geneva, 21–25 May, 1984. The authors wish to acknowledge the contribution of the following invited experts at this meeting: J. Goudriaan, T. Hodges, J. L. Monteith, R. D. Stern, E. Ulanova, and T. M. L. Wigley.

9.6 REFERENCES

Akita, S., and Moss D. N. (1973) Photosynthetic responses to CO_2 and light by maize and wheat leaves adjusted for constant stomatal apertures, *Crop. Sci.*, **13**, 234–237.

Bach, W., Pankrath S., and Schneider S. (eds) (1981) *Food–Climate Interactions*, Dordrecht, D. Reidel.

Baier, W. (1977) Crop–weather models and their use in yield assessments, *WMO Tech. Note* No. 151, Geneva, WMO.

Baier, W. (1983) Agroclimatic modelling: an overview, in Cusack, D. (ed.) *Agroclimatic Information for Development: Reviving the Green Revolution*, Boulder, Colorado, Westview Press.

Baker, D. N., and Lambert, J. R. (1980) The analysis of crop responses to entranced atmospheric CO_2 levels, in U.S. Department of Energy, Carbon Dioxide Effects Research and Assessment Program: *Workshop on Environmental and Societal Consequences of a Possible CO_2-induced Climatic Change*, Annapolis, Maryland, April 2–6, 1979, pp. 275–293.

Benci, J. F., Runge, E. C. A., Dale, R. F., Duncan, W. G., Curry R. B., and Schaal, L. A. (1975) Effects of hypothetical climatic change on production and yield of corn, in CIAP Monograph 5, *Impacts of Climatic Change on the Biosphere*, Washington D.C., US Department of Transportation.

Berry, J. A., and Raison, J. K. (1981) Responses of macrophytes to temperature, in Lang, O. L., Nobel, P. S., Osmond, C. B., and Ziegler, K. (eds) *Encyclopedia of Plant Physiology*, Vol. 12A, Berlin, Springer-Verlag.

Bishop, P. M., and Whittingham, C. P. (1968) The photosynthesis of tomato plants in a carbon dioxide enriched atmosphere, *Photosynthetica*, **2**, 31–38.

Biswas, A. K. (1980) Crop–climate models: a review of the state of the art, in Ausubel, J., and Biswas, A. K. (eds) *Climatic Constraints and Human Activities*, IIASA Proceeding Series, pp. 75–92, Oxford, Pergamon Press.

Blasing, T. J., and Solomon, A. M. (1983) *Response of North American Corn Belt to Climatic Warming*, DoE/NBB-004. Prepared for the U.S. Dept. of Energy, Office of Energy Research, Carbon Dioxide Research Revision, Washington, D.C.

Burton, I., Kates, R, and White, G. (1978) *The Environment as Hazard*, New York, Oxford Univ. Press.

Callaway, J. M., Cronin, F. J., Currie, J. W., and Tawil J. (1982) *An Analysis of Methods and Models for Assessing the Direct and Indirect Impacts of CO_2-induced Environmental Changes in the Agricultural Sector of the U.S. Economy*, PNL-4384, Pacific Northwest Laboratory, Battelle Memoril Institute, Richland, Washington.

Carlson, R. W., and Bazzaz, F. A. (1980) in Singh J. J., and Deepak, A. (eds) *Environmental and Climatic Impact of Coal Utilization*, pp. 609–622, New York, Academic Press.

Chang, C. W. (1975) Carbonic anhydrase and senescence in cotton plants, *Plant Physiol.*, **55**, 515–519.

Clark, W. C. (ed.) (1982) *Carbon Dioxide Review· 1982*, New York, Oxford University Press.

Climate Impact Assessment Program (CIAP) (1975) *Impacts of Climatic Change on the Biosphere*, Monograph 5, Washington, D.C., U.S. Department of Transportation.

Cock, J. H., and Yoshida, S. (1973) Changing sink and source relations in rice (*Oryza sativa L.*) using carbon dioxide enrichment in the field, *Soil Sci. Plant Nutr.*, **19**, 229–234.

Cooper, C. F. (1978) What might man-induced climate change mean? *Foreign Affairs*, **56**, 500–520

Cooper, C. F. (1982) Food and fiber in a world of increasing carbon dioxide, in Clark, W. (ed.), *Carbon Dioxide Review: 1982*, pp. 299–319, New York, Oxford University Press.

Dhawan, K. R., Bassi, P. K., and Spencer, M. S. (1981) Effects of carbon dioxide on ethylene production and action in intact sunflower plants, *Plant Physiol.*, **68**, 831–834.

Edwards, G. E., and Walker, D. A. (1983) C_3, C_4: *Mechanisms, and Cellular and Environmental Regulation, of Photosynthesis*, Berkeley, Univ. of California Press.

Ehleringer, J., and Bjorkman. O. (1977) Quantum yields for CO_2 uptake in C_3 and C_4 plants: dependence on temperature, CO_2, and O_2 concentrations, *Plant Physiol.*, **59**, 86–90.

Enzer, S., Drobnick, R., and Alter, S. (1978) *Neither Feast Nor Famine*, Lexington, Massachusetts, Lexington Books.

Finn, G. A., and W. A. Brun (1982) Effect of atmospheric CO_2 enrichment on growth, non-structural carbohydrate content, and root-nodule activity in soybean, *Plant Physiol.*, **69**, 327–31.

Fischer, G., and Frohberg, K. (1982) The basic linked system of the Food and Agriculture Program at IIASA: An overview of the structure of the national models, *Mathematical Modeling*, **3**, 1–22.

FAO (Food and Agriculture Organization) (1983a) *Production Yearbook*, Rome, FAO.

FAO (Food and Agriculture Organization) (1983b) *Trade Yearbook*, Rome, FAO.

Frohberg, K. (1984) Personal communication to R. Warrick.

Fukui, H. (1979) Climatic variability and agriculture in tropical moist regions. In WMO, *Proceedings of the World Climate Conference*, WMO- No. 537, pp. 426–474, Geneva, WMO.

Garcia, R. V. (1981) Drought and Man: The 1972 Case History. Vol 1 of *Nature Pleads Not guilty*, Oxford, Pergamon Press.

Gifford, R. M. (1977) Growth pattern, CO_2 exchange and dry weight distribution in wheat growing under differing photosynthetic environments, *Aust. J. Plant Physiol.*, **4**, 99–110.

Gifford, R. M. (1979a). Growth and yield of CO_2-enriched wheat under water-limited conditions, *Aust. J. Plant Physiol.*, **6**, 367–378.

Gifford, R. M. (1979b) Carbon dioxide and plant growth under water and light stress: implications for balancing the global carbon budget, *Search*, **10**, 316–318.

Gifford, R. M. (1980a) Carbon storage by the biosphere, in Pearman G.I. (ed.), *Carbon Dioxide and Climate: Australian Research*, Canberra, Australian Academy of Science, 167–181.

Gifford, R. M., Bremner P. M., and Jones D.B. (1972) Assessing photosynthetic limitation to grain yield in a field crop, *Aust. J. Agric. Res.*, **4**, 297–307.

Gifford, R. M., and Evans L. T. (1981) Photosynthesis, carbon partitioning and yield, *Annual Rev. Plant Physiol.*, **32**, 485–567.

Gifford, R. M., Lambers H., and Morison J. I. L. (1985) Respiration of crop species under CO_2 enrichment, *Physiol. Plant.*, **63**, 351–356.

Gifford, R. M., and Morison J. I. L. (1985) Photosynthesis, growth and water use of a C_4 grass stand at high CO_2 concentration, *Photosynthesis Res.* (in press).

Glantz, M. (1979) A political view of CO_2, *Nature*, **280**, 189–190.

Glantz, M. H., Robinson, J., and Krenz, M. (1982) Climate-related impact studies: a review of past experience, in Clark, W. C. (ed.) *Carbon Dioxide Review:1982*, pp. 57–93, New York, Oxford University Press.

Glantz, M. H., and Ausubel, J. H. (1984) The Ogallala Aquifer and carbon dioxide: comparison and convergence, *Environ. Conserv.*, **11**, No. 2, 123–131.

Goudriaan, J., and de Ruiter H. E. (1983) Plant response to CO_2 enrichment, at two levels of nitrogen and phosphorus supply 1. Dry matter, leaf area and development. *Neth. J. Agric. Sci.*, **31.**, 157–169.

Goudriaan, J., van Laar, H. H., van Keulen, H., and Louwerse, W. (1984) *Photosynthesis, CO_2 and Plant Production*, NATO Advanced Workshop, Wheat Growth and Modeling, Long Ashton, UK.

Grobecker, A. J. (1974) Research program for assessment of stratospheric pollution, *Acta Astronautica*, **1**, 179–224.

Hand, D. W., and Postlethwaite, J. D. (1971) Response to CO_2 enrichment of capillary watered single truss tomatoes at different plant densities and seasons, *J. Hortic. Sci.*, **46**, 461–470.

Hansen, J., Russell, G., Rind, D., Stone, P., Lacis, A., Lebedeff, S., Ruedy, R., and Travis, L. (1983) Efficient three dimensional global models for climate studies: Models I and II, *Mon. Weather Rev.*, **110**, 609–662.

Hardy, R. W. F., and Havelka, U. D. (1974) Photosynthate as a major factor limiting nitrogen fixation by field-grown legumes with emphasis on soybeans, in Nutman, P. S. (ed.) *Symbiotic Nitrogen Fixation in Plants, International Biological Program* Pub. 7, Cambridge, Cambridge University Press, 421–439.

Hare, F. K. (1985) Climatic variability and change, in Kates, R. W. with Ausubel, J. H., and Berberian, M. (eds) *Climate Impact Assessment: Studies of the Interaction of Climate and Society*, SCOPE 27, pp. 37–68 Chichester, Wiley.

Haun, J. R. (1983) *Mathematical Models in Agrometeorology*, CAgM Report No. 14, Geneva, WMO.

Heathcote, L. (1973) Drought perception, in Lovett, J. V. (ed.) *The Environmental, Economic and Social Significance of Drought*, pp. 17–54, Sydney, Angus and Robertson.

Heathcote, L. (1985) Extreme event analysis, in Kates, R. W. with Ausubel, J. H., and Berberian M. (eds) *Climate Impact Assessment: Studies of the Interaction of Climate and Society*, SCOPE 27, pp. 369–401, Chichester, Wiley.

Herold, A. (1980) Regulation of photosynthesis by sink activity–the missing link, *New Phytol.*, **86**, 131–144.

Herrera, A. O., Scholnik, H. D., et al. (1976) *Catastrophe or New Society: A Latin American World Model*, International Development Research Centre, Ottawa.

Hesketh, J. D., and Hellmers, H. (1973) Floral initiation in four plant species growing in CO_2 enriched air, *Environ. Control. in Biol.*, **11**, 51–53.

Hewitt, K. (ed.) (1983) *Interpretations of Calamity*, Boston, Allen and Unwin Inc.

Hofstra, G. (1984) Response of source–sink relationship in soybean to temperature, *Can. J. Bot.*, **62**, 166–169.

Hofstra, G., and Hesketh, J. D. (1975) The effects of temperature and CO_2 enrichment on photosynthesis in soybean, in Marcelle, R. (ed.), *Environmental and Biological Control of Photosynthesis*, The Hague, Dr W. Junk.

Hrubec, T. C., Robinson, J. M., and Donaldson, R. P. (1984) Effect of CO_2 enrichment on soybean leaf and mitochondrial respiration, *Plant Physiol. Suppl.* **75**, 158.

Hughes, B. B. (1981) *Global Modeling*, Lexington, Lexington Books.

Hughes, B. B. (1982) *International Futures Simulation: User's Manual*, Iowa City, Conduit.

Imai, K., and Murata, Y. (1978) Effect of carbon dioxide concentration on growth and dry matter production of crop plants. III. Relationship between CO_2 concentration and nitrogen nutrition in some D_3- and C_4-species, *Japanese J. Crop Sci.*, **47**, 118–123.

Jensen, R. G., and Bahr, J. T. (1977) Ribulose 1,5-bisphosphate carboxylase-oxygenase, *Annual Rev. Plant Physiol.*, **28**, 379–400.

Johnston, E. S., (1935) Aerial fertilization of wheat plants with carbon dioxide, *Smithsonian Inst. Misc. Collections*, **94**, 15.

Johnston, M., Grof, C. P. L., and Brownell, P. F. (1984) Responses to ambient CO_2 concentrations by sodium-deficient C_4 plants, *Aust. J. Plant Physiol.*, **11**, 137–141.

Jordan, D. B., and Ogren, W. L. (1984) The CO_2/O_2 specificity of ribulose 1,5-bisphosphate carboxylase/oxygenase, *Planta*, **161**, 308–313.

Kanemasu, E. T. (1980) Effects of increased CO_2 and temperature on winter wheat yields, in U.S. Department of Energy, *Carbon Dioxide Effects Research and Assessment Program*: Workshop on Environmental and Societal Consequences of a

Possible CO_2-Induced Climatic Change, Annapolis, Maryland, April 2–6, 1979, pp. 314–318.

Karl, T. R., and Riebsame, W. E. (1984) The identification of 10- and 20-year temperature and precipitation fluctuations in the contiguous United States, *J. Clim. Appl. Meteorol.*, **23**, 950–966.

Kates, R. W. (1980) Climate and society: lessons from recent events, *Weather*, **35**, 17.

Kates, R. W., Changnon, S. A., Jr., Karl, T. R., Riebsame, W., and Easterling, W. E. (1984) *The Climate Impact, Perception, and Adjustment Experiment (CLIMPAX): A Proposal for Collaborative Research.* Climate and Society Research Group, Center for Technology, Environment, and Development, Clark University, Worcester, Massachusetts.

Kates, R. W., with Ausubel, J. H., and Berberian, M. (eds) (1985) *Climate Impact Assessment: Studies of the Interaction of Climate and Society*, SCOPE 27, Chichester, Wiley.

Katz, R. (1977) Assessing the impact of climatic change on food production, *Climatic Change*, **1**, 85–96.

Kimball, B. A. (1983) *Carbon Dioxide and Agricultural Yield: An Assemblage and Analysis of 770 Prior Observations*, WCL Report 14, Water Conservation Laboratory, Agricultural Research Service, Phoenix, Arizona, US Dept. Agriculture.

Kramer, P. J. (1981) Carbon dioxide concentration, photosynthesis, and dry matter production. *BioScience*, **31**, 29–33.

Krenzer, E. G., and Moss, D. N. (1975) Carbon dioxide enrichment effects upon yield and yield components in wheat, *Crop Sci.*, **15**, 71–74.

Lambers, H. (1982) Cyanide-resistant respiration: a non-phosphorylating electron transport pathway acting as an energy overflow, *Physiol. Plant*, **55**, 478–485.

Le Roy Ladurie, E. (1972) *Times of Feast, Times of Famine: A History of Climate Since the Year 1000*, New York, Doubleday.

Linnemann, H., De Hoogh, J., Keyser, M. A., and Van Heemst, H. D. J. (1979) *MOIRA: Model of International Relations in Agriculture*, Amsterdam, North Holland.

Liss, P. S., and Crane, A. J. (1983) *Man-Made Carbon Dioxide and Climate Change: A Review of Scientific Problems*, Norwich, Geo Books.

Liverman, D. M. (1983) *The Use of a Simulation Model in Assessing the Impacts of Climate on the World Food System*, NCAR Cooperative Thesis No. 77, Boulder, Colorado, National Center for Atmospheric Research.

Liverman, D. M., Terjung, W. H., Hayes, J. T. with O'Rourke, P. A., and Todhunter, P. E. (1985) Climatic change and grain corn yields in the North American Great Plains, *Climatic Change* (forthcoming).

Lough, J. M., Wigley, T. M. L., and Palutikof, J. P. (1983) Climate and climatic impact scenarios for Europe in a warmer world, *J. Clim. Appl. Meteorol.*, **22**, 1673–1684.

MacDowell, F. D. H. (1972) Growth of Marquis wheat II. Carbon dioxide dependence. *Can. J. Bot.*, **50**, 883–889.

Madsen, E. (1968) The effect of CO_2-concentration on the accumulation of starch and sugar in tomato leaves, *Physiol. Plant*, **21**, 168–175.

Marc, J., and Gifford, R. M. (1983) Floral initiation in wheat, sunflower and sorghum under carbon dioxide enrichment, *Can. J. Bot.*, **62**, 9–14.

Meadows, D. H., Richardson, W., and Bruckman, G. (1982) *Groping in the Dark: A History of the First Decade of Global Modeling*, New York, Wiley.

Mearns, L. O., Katz, R. W., and Schneider, S. H. (1984) Extreme high temperature events: changes in their probabilities with changes in mean temperature, *J. Clim. Appl. Meteorol.*, **23**, 1601–1613.

Meinl, H., and Bach, W., *et al.* (1984) *Socioeconomic Impacts of Climatic Changes due to a Doubling of Atmospheric CO_2 content*, Commission of the European Communities Contract No. CL1-063-D.

Mesarovic, M., and Pestel, E. (1974) *Mankind at the Turning Point*, New York, E.P. Dutton.

Monteith, J. L. (1981) Climatic variation and the growth of crops, *Q. J. R. Meteorol. Soc.*, **107**, No. 454, 749–774.

Morison, J. I. L. (1985) Intercellular CO_2 concentration and stomatal response to CO_2, in Zeiger, E., Cowan, I. R. and Farquhar, G. D. (eds) *Stomatal Function* (in press).

Morison, J. I. L., and Gifford, R. M. (1983) Stomatal sensitivity to carbon dioxide and humidity: a comparison of two C_3 and two C_4 grass species, *Plant Physiol.*, **71**, 789–796.

Morison, J. I. L., and Gifford, R. M. (1984a) Ethylene contamination of CO_2 cylinders: effects on plant growth in CO_2 enrichment studies, *Plant Physiolog.*, **75**, 275–277.

Morison, J. I. L., and R. M. Gifford (1984b) Plant growth and water use with limited water supply in high CO_2 concentrations. 1. Leaf area, water use and transpiration, *Aust. J. Plant Physiol.*, **11**, 361–374.

Morison, J. I. L., and Gifford, R. M. (1984c) Plant growth and water use with limited water supply in high CO_2 concentration. 2. Plant dry weight, partitioning and water use efficiency, *Aust. J. Plant Physiol.*, **11**, 375–384.

National Center for Atmospheric Research (NCAR) (1984) *Annual Report: Fiscal Year 1983*, Boulder, Colorado, NCAR.

National Defense University (NDU) (1978) *Climate Change to the Year 2000: A Survey of Expert Opinion*, Washington D.C., Fort Lesley, J.McNair.

National Defense University (NDU) (1980) *Crop Yields and Climate Change to the Year 2000*: Vol. 1, Washington D.C., Fort Lesley, J. McNair.

National Defense University (NDU) (1983) *The Global Impacts of Climate Change to the Year 2000*, Washington, D.C., Fort Lesley, J. McNair.

National Oceanic and Atmospheric Administration (NOAA) (1973) *The Influence of Weather and Climate on United States Grain Yields: Bumper Crops or Droughts*, Washington, D.C., U.S. Dept. Commerce.

National Research Council (NRC) (1983) *Changing Climate*. Report of the Carbon Dioxide Committee, Board of Atmospheric Sciences and Climate, Washington D.C., National Academy Press.

Neales, T. F., and Nicholls, A. O. (1978) Growth responses of young wheat plants to a range of ambient CO_2 levels, *Aust. J. Plant Physiol.*, **5**, 45–59.

Neild, R. E., Richman, H. N., and Seeley, M. W. (1979) Impacts of different types of temperature change on the growing season of maize, *Agric. Meteorol.*, **20**, 367–374.

Newman, J. E. (1980) Climate change impacts on the growing season of the North American Corn Belt. *Biometeorology* 7 (part 2), 128–142.

Newman, J. E. (1982) Impacts of a rising atmospheric carbon dioxide level on agricultural growing seasons and crop water use efficiencies. Vol II, part 8 of *Environmental and Societal Consequences of a Possible CO_2-Induced Climate Change*, DOE/EV/10019-8, Washington, D.C., U.S. Dept. Energy.

Nix, H. A. (1985) Agriculture, in Kates, R. W. with Ausubel, J. H., and Berberian, M. (eds) *Climate Impact Assessment: Studies of the Interaction of Climate and Society*, SCOPE 27, pp. 105–130, Chichester, Wiley.

Office of Technology Assessment (OTA) (1982) *Global Models, World Futures and Public Policy: A Critique*, Washington, D.C., OTA.

Ogilvie, A. E. J. (1981) Climate and economy in eighteenth century Iceland, in Smith, C. D., and Parry, M. L. (eds) (1981) *Consequences of Climatic Change*, pp. 54–69, Nottingham, Department of Geography, University of Nottingham.

Palutikof, J., Wigley, T. M. L., and Farmer, G. (1984) The impact of CO_2-induced climatic change on crop yields in England and Wales, *Progress in Biometeorology*, Vol. 3, 320–334.

Parikh, K., and Rabar, F. (eds) (1981) *Food for All in a Sustainable World: The IIASA Food and Agriculture Program*, Laxenburg, Austria, International Institute for Applied Systems Analysis.

Parry, M. L. (1975) Secular climatic change and marginal agriculture., *Trans. of Inst. of Brit. Geog.*, **64**, 1–13.

Parry, M. L. (1978) *Climatic Change, Agriculture and Settlement*, Folkestone, Dawson.

Parry, M. L. (1981) Climatic change and the agricultural frontier: a research strategy, in Wigley, T. M. L., Ingram, M., and Farmer, G. (eds) (1981) *Climate and History: Studies in Past Climate and their Impact on Man*, pp. 319–336, Cambridge, Cambridge University Press.

Parry, M. L. (1985) The impact of climatic variations on agricultural margins, in Kates, R. W. with Ausubel, J. H., and Berberian, M. (eds) *Climate Impact Assessment: Studies of the Interaction of Climate and Society*, SCOPE 27, pp. 351–367, Chichester, Wiley.

Parry, M. L., and Carter, T. (1984) *Assessing Impacts of Climatic Change in Marginal Areas: the Search for Appropriate Methodology*, IIASA Working Paper WP-83-77, Laxenburg, Austria, International Institute for Applied Systems Analysis.

Parry, M. L., and Carter, T. R. (1985) The effect of climatic variations on agricultural risk, *Climatic Change*, **7**, 95–110.

Parry, M. L., Carter, T. R., and Konijn, N. T. (eds) (1986a) *Assessment of Climate Impacts on Agriculture* Vol.I: High Latitude Regions, Dordrecht, D. Reidel (forthcoming).

Parry, M. L., Carter, T. R., and Konijn, N. T. (eds) (1986b) *Assessment of Climate Impacts on Agriculture* Vol.II: Semi-Arid Regions, Dordrecht, D. Reidel (forthcoming).

Patterson, D. T., and Flint, E. P. (1980) Potential effects of global atmospheric CO_2 enrichment on the growth and competitiveness of C_3 and C_4 weed and crop plants, *Weed Sci.*, **28**, 71–75.

Pearcy, R. W., Bjorkman, O., *et al.* (1983) Physiological effects, in Lemon, E. R. (ed.), *CO_2 and Plants: The Response of Plants to Rising Levels of Atmospheric Carbon Dioxide*. AAAS Selected Symposium No. 84, pp. 65-106, Boulder, Colorado, Westview Press.

Peet, M. M. (1984) CO_2 enrichment of soybeans: effects of leaf/pod ratio, *Physiol. Plant*, **60**, 38–42.

Perchorowicz, J. T., Raynes, D. A., and Jensen, R. G. (1982) Measurement and preservation of the *in vivo* activation of ribulose 1,5-biphosphate carboxylase in leaf extracts, *Plant Physiol.*, **69**, 1165–1168.

Phillips, D. A., Newell, K. D., Hassell, S. A., and Felling, C. E. (1976) The effect of CO_2 enrichment on root nodule development and symbiotic N_2 reduction in Pisum sativum, *L. Amer. J. Bot.*, **63**, 356–362.

Post, J. D. (1977) *The Last Great Subsistence Crisis in the Western World*, Baltimore, Maryland, Johns Hopkins Univ. Press.

Rabb, T. K. (1983) Climate–society in history: a research agenda, in Chen, R. J.,

Boulding, E., and Schneider, S. H. (eds) *Social Science Research and Climatic Change*, pp. 62–76, Dordrecht, D. Reidel.

Ramirez. J., Sakamoto, C., and Jensen, R. (1975) Agricultural implications of climatic change, in Climate Impacts Assessment Project (CIAP) (1975) *Impacts of Climatic Change on the Biosphere*. Monograph 5, Washington, D.C., U.S. Department of Transportation.

Riebsame, W. E. (1985) Research in climate–society interaction, in Kates, R. W. with Ausubel, J. H., and Berberian, M. (eds) *Climate Impact Assessment: Studies of the Interaction of Climate and Society*, SCOPE 27, pp. 69–84, Chichester, Wiley.

Roberts, P. C. (1977) *SARUM 76—Global Modelling Project: Research Report No. 19*, UK Department of Environment and Transport, London.

Robertson, G. W. (ed.) (1983) Guidelines on crop-weather models, *WMO World Climate Applications Programme*, WCP-50, Geneva, WMO.

Robinson, J. (1985) Global modeling and simulations, in Kates, R. W. with Ausubel, J. H., and Berberian, M. (eds) *Climate Impact Assessment: Studies of the Interaction of Climate and Society*, SCOPE 27, pp. 469–492, Chichester, Wiley.

Rogers, H. H., Beck, R. D., Bingham, G. F., Curie, J. D., Davis, J. M.. Heck, W. W., Rawlings, J. O., Riordan, A. J, Sionit, N., Smith, J. M., and Thomas. J. F. (1981) *Response of Vegetation to Carbon Dioxide: Field Studies of Plant Responses to Elevated Carbon Dioxide Levels*. Report 005, U.S. Department of Energy, Carbon Dioxide Research Division and U.S. Department of Agriculture, Agricultural Research Service, Washington, D.C.

Rogers, H. H., Sionit, N., Cure. J. D., Smith J. M., and Bingham, G. E. (1984) Influence of elevated carbon dioxide on water relations of soybeans, *Plant Physiol.*, **74**, 233–238.

Rojko, A. S., and Schwartz, M. W. (1976) Modeling the world grains, oilseeds and livestock economy to assess world food prospects, *Agric. Econ. Res.*, **28**, 89–98.

Rosenberg, N. J (1981) The increasing CO_2 concentrations in the atmosphere and its implications on agricultural productivity. I. Effects on photosynthesis, transpiration and water use efficiency, *Climatic Change*, **2**, 387–408.

Rosenberg, N. J. (1982) The increasing CO_2 concentrations in the atmosphere and its implications on agricultural productivity. II. Effect through CO_2 induced climate change, *Climatic Change*, **4**, 239–254.

Rotberg, R. I., and Rabb, T. K. (eds) (1980) History and Climate: Interdisciplinary Explorations. A special issue of the *Journal of Interdisciplinary History*, **10**, 583–858.

Saarinen, T. (1966) *Perception of Drought Hazard on the Great Plains*, Res. Monogr. No. 106, Dept. of Geography, Chicago, Univ. of Chicago.

Sakamoto, C., Leduc, S., Strommen, N., and Steyaert L., (1980) Climate and global grain yield variability. *Climatic Change*, **2** (4), 349–361.

Santer, B. (1984) The impacts of a CO_2-induced climatic change on the agricultural sector of the European Communities, in Meinl, H. and Bach, W., *et al.*, *Socioeconomic Impacts of Climatic Changes Due to a Doubling of Atmospheric CO_2 Content*, pp. 456–642. Commission of the European Communities Contract No. CL1-063-D.

Schneider, S. H. (1985) Science by consensus: the case of the National Defense University study 'Climate Change to the Year 2000'—an editorial, *Climatic Change*, **7**, 153–157.

Schwarz, M., and Gale, J. (1984) Growth response to salinity at high levels of carbon dioxide, *J. Exp. Bot.*, **35**, 193–196.

Sionit, N., Mortensen, D. A., Strain, B. R., and Hellmers, H. (1981a) Growth responses of wheat to CO_2 enrichment at different levels of mineral nutrition, *Agron. J.*, **73**, 1024–1027.

Sionit, N., Strain, B. R., and Beckford, H. A. (1981b) Environmental controls on the growth and yield of okra 1. Effects of temperature and of CO_2 enrichment at cool temperature, *Crop Sci.*, **21**, 885–888.

Sionit., N., Hellmers, H., and Strain, B. R. (1982) Interaction of atmospheric CO_2 enrichment and irradiance on plant growth, *Agron. J.*, **74**, 721–725.

Sirotenko, O. D. (1983) *Development and Application of Dynamic Simulation Models in Agrometeorology*, WMO CAgM Report No. 13, Geneva, WMO.

Smith, C. D,. and Parry, M. L. (eds) (1981) *Consequences of Climatic Change*, Nottingham, Department of Geography, University of Nottingham.

St.Omer, L., and Horvath, S. M. (1983) Elevated carbon dioxide concentration and whole plant senescence, *Ecology*, **64**, 1311–1314.

Stewart, T. R., and Glantz, M. H. (1985) Expert judgment and climate forecasting: a methodological critique of 'climate change to the year 2000', *Climatic Change*, **7**, 159–183

Swaminathan, M. S. (1984) Climate and agriculture, in Biswas, A. K. (ed.) *Climate and Development*. Natural Resources and the Environment Series Vol. 13, pp. 65–95, Dublin, Tycooly International Publishing Ltd.

Systems Analysis Research Unit (SARU) (1978) *SARUM Handbook*, UK Department of Environment and Transport, London.

The Institute of Ecology (TIE) (1976) *Impact of Climatic Fluctuations on Major North American Food Crops*, Dayton, Ohio, Charles F. Kettering Foundation.

TIE (1976), see The Institute of Ecology.

Tolbert, N. E., and Zelitch, I. , *et al.* (1983) Carbon metabolism, in Lemon, E. R. (ed.), *CO_2 and Plants: The Responses of Plants to Rising Levels of Atmospheric Carbon Dioxide*. AAAS Selected Symposium 84, pp. 21–64, Boulder, Colorado, Westview Press.

Uchijima, T. (1986) Variation in climate and growth potential of rice, in Parry, M. L., Carter, T. R., and Konijn, N. T. (eds) *Assessment of Climate Impacts on Agriculture* Vol. 1: High Latitude Regions, Dordrecht, D. Reidel (forthcoming).

Uchijima, Z. (1978) Long-term change and variability of air temperature above 10 °C in relation to crop production, in Takahashi, K., and Yoshino, M. M. (eds) *Climatic Change and Food Production*, pp. 217–229, Tokyo, Univ. of Tokyo Press.

van Bavel, C. H. M. (1974) Antitranspirant action of carbon dioxide on intact sorghum plants, *Crop Sci.*, **14**, 208–212.

Verma, S. B., and Rosenberg, N. J. (1976) Carbon dioxide concentration and flux in a large agricultural region of the Great Plains of North America, *J. Geophys. Res.*, **81**, 399–405.

deVries, J. (1980) Measuring the impact of climate on the economy: separating real from false assumptions, *J. Interdisciplinary Hist. X*, **4**, 599–630.

von Caemmerer, S., and Farquhar, G. D. (1984) Effects of partial defoliation, changes of irradiance during growth, short term water stress and growth at enhanced $p(CO_2)$ on the photosynthetic capacity of leaves of *Phaseolus vulgaris, L. Planta*, **160**, 320–329.

Vu, C. V., Allen, L. H., and Bowes, G. (1983) Effects of light and elevated CO_2 on the ribulose biophosphate carboxylase activity and ribulose bisphosphate level of soybean leaves, *Plant Physiol.*, **73**, 729–734.

Waggoner, P. E. (1983) Agriculture and a climate changed by more carbon dioxide. In NRC, *Changing Climate*. Report of the Carbon Dioxide Committee, Board of Atmospheric Sciences and Climate, pp. 383–418, Washington, D.C., National Academy Press.

Wallen, C. C. (1984) *Present Century Climate Fluctuations in the Northern Hemisphere and Examples of their Impact*, WCP-87, Geneva, WMO.

Warrick, R. A. (1980) Drought in the Great Plains: a case study of research on climate and society in the USA, in Ausubel, J., and Biswas, A. K. (eds) *Climatic Constraints and Human Activities*, pp. 93–123, Oxford, Pergamon Press.

Warrick, R. A. (1984) The possible impacts on wheat production of a recurrence of the 1930s drought in the U.S. Great Plains, *Climatic Change*, **6**, 5–26.

White, G. F. (ed.) (1974) *Natural Hazards: Local, National, Global*, New York, Oxford University Press.

Whyte, I. (1981) Human response to short- and long-term climatic fluctuations: the example of early Scotland, in Smith, C. D., and Parry, M. L. (eds) (1981) *Consequences of Climatic Change*, pp. 17–29, Nottingham, Department of Geography, University of Nottingham.

Wigley, T. M. L. (1985) Impact of extreme events, *Nature*, **316**, 106–107

Wigley, T. M. L., Ingram, M., and Farmer, G. (eds) (1981) *Climate and History: Studies in Past Climate and their Impact on Man*, Cambridge, Cambridge University Press.

Williams, G. D. V. (1975) Assessment of the impact of some hypothetical climatic changes on cereal production in Western Canada, in the *Proceedings of the Conference on World Food Supply in Changing Climate*, Sterling Forest, N.Y., Dec 2–5, 1974.

Williams, G. D. V., and Oakes, W. T. (1978) Climatic resources for maturing barley and wheat in Canada, in Hage, K. D., and Reinelt, E. R. (eds) *Essays on Meteorology and Climatology: In Honour of Richard W. Longley*, Studies in Geog. Mono 3., pp. 367–385, Univ. of Alberta, Edmonton, Alberta, Canada.

Wittwer, S. (980) Overview report of Panel III on Environmental Effects on the Managed Biosphere. In U.S. Dept of Energy, *Carbon Dioxide Effects Research and Assessment Program: Workshop on Environmental and Societal Consequences of a Possible CO₂-induced Climatic Change*, Annapolis, Maryland, April 2–6 1979, pp. 46–48.

Wong, S. C. (1979) Elevated atmospheric partial pressures of CO_2 and plant growth: I. Interactions of nitrogen nutrition and photosynthetic capacity in C_3 and C_4 species, *Oecologia (Berl.)*, 44.

World Meteorological Organization (WMO) (1979) *Proceedings of the World Climate Conference*, Geneva 12–23 February 1979, Geneva, WMO.

World Meteorological Organization (WMO) (1982) *The Effect of Meteorological Factors on Crop Yields and Methods of Forecasting the Yield*, WMO No. 566, Geneva, WMO.

World Meteorological Organization (WMO) (1984) *Report of the Study Conference on Sensitivity of Ecosystems and Society to Climate Change*, Villach, Austria, 19–23 September, 1983. WCP-83, Geneva, WMO.

World Meteorological Organization (WMO) (1985) *The Reliability of Crop-climate Models for Assessing the Impacts of Climatic Change and Variability*. Report of the WMO/UNEP/ICSU-SCOPE expert meeting, Geneva, May 19–24, 1984 (in press).

Worster, D. (1979) *Dust Bowl: the Southern Plains in the 1930s*. Oxford, Oxford Univ. Press.

Yoshida S. (1972) Physiological aspects of grain yield. *Annual Rev. Plant Physiol.*, **23**, 437–464.

CHAPTER 10

CO_2, Climatic Change and Forest Ecosystems

Assessing the Response of Global Forests to the Direct Effects of Increasing CO_2 and Climatic Change

H. H. SHUGART, M. YA. ANTONOVSKY, P. G. JARVIS, AND A. P. SANDFORD

10.1 INTRODUCTION

Potential responses of forests to the direct effects of increasing CO_2 and climatic change should be considered in the context of forest management and the trends in wood supply and demand. The total amount of carbon stored in terrestrial ecosystems has diminished over the past several centuries as a result of anthropogenic actions, especially forest clearance (see Bolin, 1977; Woodwell et al., 1978; Houghton et al., 1983). In the last half of the 1970s, about 2400 million m^3 of wood were consumed annually (FAO, 1982). About half of this wood was used as fuelwood and the other half for industrial purposes (lumber, plywood, chipboard, paper, etc.). For the remainder of this century, FAO (1982) projects an increasing rate of forest-clearance, based on forecasts of rising demands for wood. If this occurs, the present area of forest may be reduced by as much as 20% by the year 2000. Problems of wood scarcity could become critical in regions such as Africa and Asia where fuelwood is the primary source of energy for domestic heating and cooking. Whether increasing CO_2 and climatic change will exacerbate or diminish the problems that loom on the horizon is a fundamental issue which concerns the entire global community.

In this chapter we examine the possible responses of forests to changes in climate and the direct effects of increased atmospheric CO_2 concentrations. Change or stress can induce responses in a wide range of frequencies from a forest ecosystem; these are reviewed in Section 10.2. We review the mechanisms involved in the direct CO_2 effect in Section 10.3. Since prediction of the responses of forests may well involve the use of models, the types of models available are reviewed in Section 10.4. The results of applications of such models and other methods to assess the impacts of climatic change are presented in Section 10.5.

10.2 TIME SCALES AND THE RESPONSE OF FORESTS

Forest ecosystems contain a complex web of interactions among physical, chemical and biological processes. Because of this interactive complexity, direct changes in a given process can be attenuated or amplified, and the responses elicited from a forest will be manifested on many different time scales. In this section, we discuss the responses of forests at various scales of time and space as a necessary preamble to considering the possible effects of increasing CO_2 concentrations and climatic change.

10.2.1 Fast Responses

Several of the important processes in forest ecosystems that have very rapid response times are influenced by CO_2 and climatic variables. Among the most important are those involving the exchanges of water, heat and carbon dioxide between the leaf surfaces and the environment. For example, stomata (microscopic pores on the leaf surface) allow the inward diffusion of carbon dioxide used in photosynthesis and, at the same time, allow the loss of transpired water. An increase in the ambient CO_2 concentration could reduce the opening of the stomata required to allow a given amount of CO_2 to enter the plant and might thus reduce the loss of water from a tree. This could increase the efficiency of water use and raise the productivity of forests in certain circumstances. Changes in the radiation input, temperature or humidity above a forest canopy could also produce almost instantaneous responses in CO_2 uptake and water use by the forest. The rapid, direct responses to CO_2 are considered in more detail in Section 10.3.

The question that arises immediately is, to what degree would these fast responses change the amount of photosynthate and ultimately the amount of production over the course of a year in a forest? Tree growth results from the amount of photosynthate produced and the allocation of this photosynthate within the tree. The growth of trees has been modelled using 'mechanistic' representations of physiological processes, but these models have rarely been used to predict responses over periods longer than a year. The complexity of the problem is illustrated in Figure 10.1, which shows how opposing environmental conditions (high temperature and low precipitation; low temperature and high precipitation) may produce the same response: the production of a narrow annual ring in trees growing in arid sites. It is important to note: (1) that the web of interactions is complex and the details of some of the interactions are not well known, and (2) that the response of the whole plant to a stress may be non-linear across the possible range of that stress. In general, the fast processes that operate in forest ecosystems can only be predicted over the longer term with a considerable degree of uncertainty.

There have been several attempts to develop highly detailed 'mechanistic' models of natural ecosystems including forests. These models are useful as heuristic tools for integrating studies of different ecological processes, but are much less useful for predicting long-term ecosystem behaviour. At the Third Conference on CIAP (Cooper *et al.*, 1974), results were presented on the performance of several of such models (developed during the International Biological Program for different types of ecosystems) when subjected to a change in climate. It was noted that 'The models are not designed to predict long term successional trends or changes in plant or animal composition... The simulation runs are therefore limited to a 5-year period and emphasize changes in primary production and in associated aspects of community structure. It is questionable how far beyond this 5-year period the results can validly be extrapolated.' (Cooper *et al.*, 1974). These cautionary comments still apply to current models of ecosystem processes.

10.2.2 Intermediate Time-scale Responses

Environmental change can affect the growth rates of individual trees and thereby have a cumulative effect in changing the amount of living material in the forest system. However, the relation between the rate of individual tree growth and the rate of forest increase (or yield) is more than a simple additive effect. Relatively low levels of stress on trees can produce large changes in the dynamics and composition of forests. Furthermore, the interactions between the populations of trees and phytophagous insects (and other pests and diseases) are, in many cases, mediated by climate and thus are of importance in assessing the possible effects of climatic change on forests. For example, the oak wilt disease in the USSR appears to be dependent on the decreased ability of the trees to resist leaf-eating insects during drought (Israel *et al.*, 1983).

The prediction of yield from growth has been an important topic in modern forestry (Fries, 1974). Forest yield is a consequence both of the growth of individual trees and of the rates of recruitment and death of trees in the forest stand. For example, Figure 10.2 illustrates the relation between stand biomass and the age of stands of *Picea glauca* forests (Yarie and Van Cleve, 1983). Over the time period indicated by the different ages, the growth increment of the trees was constant (diameter increase = 0.11 cm/year; r^2 of regression = 0.87) but the rate of increase of the total biomass clearly declined as a function of age. In contrast, the rate of biomass change of a single tree, which was enlarging with a constant diameter increment,

(a) Relationships important during the growing season

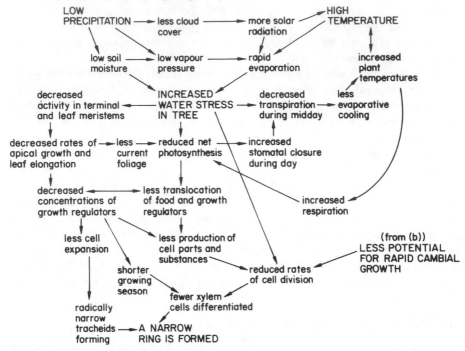

(b) Relationships important prior to the growing season

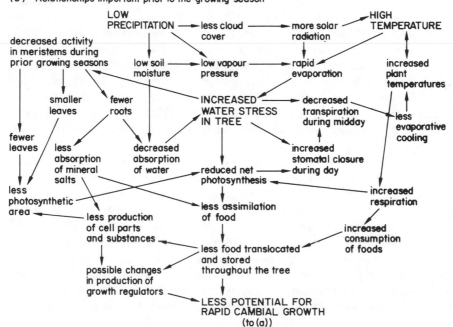

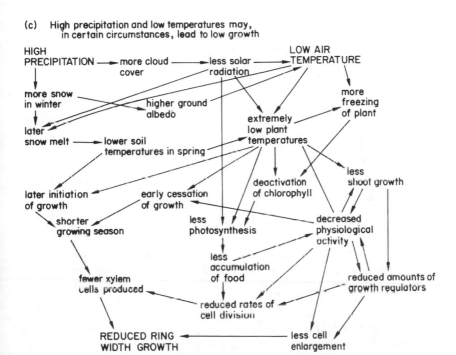

(c) High precipitation and low temperatures may, in certain circumstances, lead to low growth

Figure 10.1 Diagrams representing some of the relationships that cause climate variables to lead to the formation of a narrow ring in trees growing in arid sites. (a) The effect of low precipitation and high temperature during the growing season. (b) The effect of low precipitation and high temperature before the growing season. (c) The effect of high precipitation and low temperature (from Fritts, H. C. 1976)

increased with the size of the tree (because tree biomass is a function of the diameter). Thus, the change in biomass growth rate shown in Figure 10.2 was opposite to that of the individual trees that comprised the forest. Effects such as these prevent direct extrapolation of short-term changes in trees to predict the longer term responses of the forest.

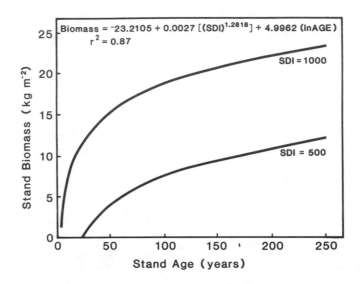

Figure 10.2 The relationship of above ground stand biomass to stand age for white spruce (*Picea glauca*) from the Interior of Alaska (from Yarie and Van Cleve, 1983). A fully stocked stand has a site density index (SDI) of 1000, a stand with one half this density of trees has a SDI = 500

The modern theoretical concept for understanding the intermediate time scale response of a forest is to consider the forested landscape to be a mosaic with each element of the mosaic scaled in relation to the dominant canopy tree (ca 0.1 ha, depending on the size of the trees). This concept was initially developed by Watt (1925, 1947) and has been the topic of a series of papers and books (Raup, 1957; Whittaker and Levin, 1977; Bormann and Likens, 1979 a,b; Shugart, 1984). The concept, in brief, is that the dynamic response of the forest occurs at an areal scale of a large canopy tree and on a time scale that relates to the longevity of the tree, as follows.

'Following the death of a large tree and its fall, a canopy gap forms. The area below this gap becomes the site of increased regeneration and survival of trees. Trees grow, the forest builds, the canopy closes, and the gap disappears. Eventually, the now mature forest in the vicinity of the former gap suffers the mortality of a large tree and a new gap is formed and the cycle is repeated (Shugart 1984).'

The dynamics of a forest are the aggregated dynamics of a large number of such individual gaps. When considered at intermediate time scales (ca 100 years), the pattern of dynamics of a forest can be seen as a cycle of recruitment, death and growth processes; environmental change alters the pattern within the cycle (Figure 10.3). The regeneration phase of the forest cycle and the exact timing of the death of a canopy tree that produces an opportunity for regeneration are highly stochastic processes. This is particularly the case in the regeneration phase when the mortality of small trees is very high. It is in this stochastic part of the cycle that climatic change and variability can have the largest effects in producing change in the forest. For example, a dry, hot year could remove the seedlings of drought-intolerant tree species at the local site of regeneration. It might then be several hundred years (the longevity of a canopy tree) before the eliminated species would again have a reasonable chance to occupy the same area. During the growth, competition and thinning phases of the forest cycle, the dynamics of the forest are much more determinate and the change of the forest in this part of the cycle in response to a climatic change has considerable inertia. In summary, the forest system has great inertia but features short periods when relatively small outside perturbations can greatly alter the eventual path of the system.

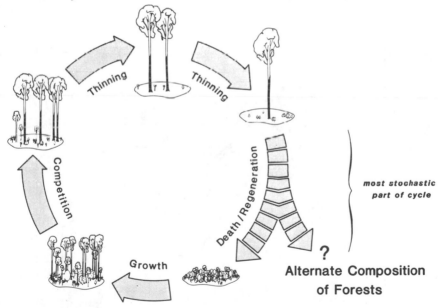

Figure 10.3 Simulation of the long-term regeneration cycle of forests using the BRIND model (Shugart and Noble, 1981) for the Australian Eucalyptus forest. The determinism in the system is great during the growth, competition and thinning phases but more stochastic in death and regeneration phases

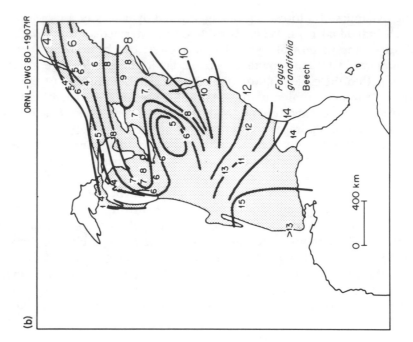

ORNL–DWG 80-19072

Castanea dentata

Chestnut

400 km

(a)

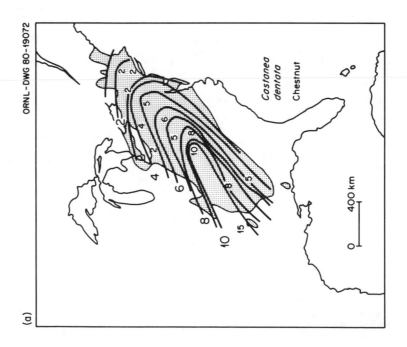

ORNL–DWG 80-19071R

Fagus grandifolia

Beech

400 km

(b)

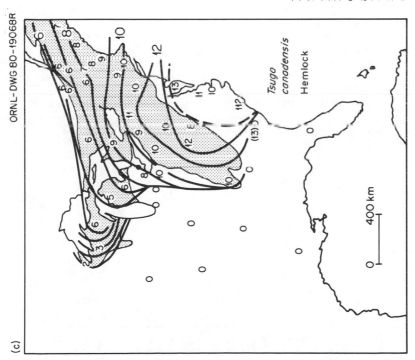

ORNL–DWG 80-19068R

(c)

Tsuga canadensis

Hemlock

400 km

Figure 10.4 Migration maps for three species. The numbers refer to the radio-carbon dates (in thousands of years before present) of the first appearance of the species in a site, as evidenced by an increased pollen abundance or the presence of macrofossils at a site. Isopleths connect points of similar age and represent the leading edge of the expanding population since the Wisconsin glaciation. The dotted area is the modern range of (a) *Castanea dentata*, (b) *Fagus grandifolia*, (c) *Tsuga canadensis*. (From Davis 1981)

10.2.3 Long-term Responses

At the intermediate scale, the successful regeneration of the species suited to particular climatic conditions depends on an adequate supply of seeds. However, if a change in climate was sufficiently large to extend the climate favourable for a species far beyond its present limits, the rate of response to the change would become dependent on the migration rate of the species. Such delays in response have been inferred from data on past climatic changes. The changes in geographical range of different species over time following the last glaciation have been calculated from fossil pollen data. In Figure 10.4, the time required for a species to move between the lines that indicate range boundaries was 1000 years. There is considerable variation in the rates of migration that have been recorded for various taxa (Table 10.1), so the influence of migration in delaying the local response of a forest is potentially large.

Table 10.1 Migration of European fossil pollen taxa (after Huntley and Birks, 1983)

Taxon	Contribution to pollen taxon*	Range of observed migration rates (m/yr)
1 Abies	6	40–300
2 Acer	13	500–1000
3 Alnus	4	500–2000
4 Carpinus betulus	1	50–1000
5 Castanea sativa	1	200–300
6 Corylus type	4	1500
7 Fagus	2	200–300
8 Fraxinus excelsior type	3	200–500
9 Fraxinus ornus	1	25–200
10 Juglans	1	400
11 Larix	2	—
12 Olea europae	1	—
13 Ostrya type	2	—
14 Phillyrea	2	—
15 Picea	2	80–500
16 Pinus	10	1500
17 Pinus (Haploxylon)	3	—
18 Pistacia	3	200–300
19 Quercus (Deciduous)	22	75–500
20 Quercus (Evergreen)	3	—
21 Tilia	4	50–500
22 Ulmus	5	100–1000

* The numbers are the number of native plant species contributing to the pollen taxon. Typically fossil pollen is identified at the genus level.

It is, of course, possible to reduce large, migration-related lag effects by actively moving plant material (seeds and seedlings) to appropriate locations. Suitable propagation techniques to accomplish such a task are better developed for commercially important tree species than for non-commercial species. Such a strategy probably would be prohibitively expensive if pursued on a large scale with native forests. The actual movement of tree species to compensate for a change in climate could most likely be attempted for commercial tree crops only, in a forest management context.

Change in the rate of soil development could also greatly delay the response of a forest to a climatic change. The soil at a given location derives its characteristics from the parent material (the geology at the site), the vegetation and the climate. If both the climate and the vegetation at a given location were to change, there might be considerable delays in the development of the soils and, hence, the forests which one would expect ultimately to develop at the site.

10.2.4 Responses to Climatic Changes in the Past

A very large amount of evidence shows that the world's forests have responded, often dramatically, to changes in climate on various time scales. Even such a small fluctuation as the high-latitude period of warm summers during the 1930s, when mean annual temperatures were typically no more than 1 °C above the average temperatures for previous and subsequent decades, caused a burst of regeneration in boreal forest trees near their polar and altitudinal limits, with measurable advances of the timber line (e g. Hustich, 1958; Morisset and Payette, 1983). The 'Little Ice Age' was a longer period, roughly between 1400 and 1800, during which there were glacial advances in many mid- and high-latitude regions, with recorded or reconstructed mean temperatures in Europe and North America of ca 1–2 °C below present values. The 'Little Ice Age' had substantial effects on forest growth and composition in many northerly areas. For example, in the prairie-forest border region of Minnesota, an increase in precipitation relative to evaporation caused a reduction in the frequency of fires; this, in turn, favoured a forest with more big hardwoods than the oak forest which it replaced. This change took only about a century to accomplish (Grimm, 1983).

Larger climatic changes have taken place since the peak of the last major glaciation ca 20,000 years ago. These changes caused whole biomes to be replaced. The evidence for long-term changes in vegetation comes from analysis of fossil pollen in sediments and other research techniques. Maps based on fossil pollen data show the major patterns of change (e.g. Huntley and Birks, 1983; Peterson, 1983). Some documented changes have been remarkably rapid. For example, the transformation from spruce-dominated to jack pine-dominated forests in eastern Minnesota ca 10,000 years ago took a few hundred years at most, and perhaps less than one hundred years (Wright 1984).

10.3 THE DIRECT IMPACTS OF INCREASING CO_2 CONCENTRATION

10.3.1 The Base of Experimental Evidence

There has been a sufficient number of studies on CO_2 exchange of the dominant tree species and some of the understorey species in northern forests to support the assertion that all the species of major ecological and economic importance are likely to have photosynthesis of the typical C_3 pattern. Therefore, we expect the main effects of rising CO_2 concentrations to result from changes in photosynthesis and stomatal action as has been shown for other C_3 species (see Section 9.2). One can not rule out other direct effects of CO_2 on, say, root and leaf development or CO_2 assimilation in the dark by roots, but little is known about such responses and they are most probably of secondary importance. The likely effects of changes in photosynthesis, stomatal conductance and leaf growth on the growth of trees are indicated in the network diagrams in Figure 10.1.

There is a great dearth of information on the effects of CO_2 concentration on the physiology and growth of particular, relevant forest species. Only a few short-term experiments have been conducted on single leaves, shoots or small plants in assimilation chambers where the response of CO_2 uptake and stomatal action has been characterized in relation to ambient CO_2 concentrations and, in even fewer cases, to mean intercellular space CO_2 concentration. Many measurements have been made on leaves and shoots and sometimes on whole trees using cuvettes in the field, but in these experiments CO_2 has not usually been a controlled variable and higher than normal CO_2 concentrations have not usually been used. There have also been several series of measurements of CO_2 influx from the ambient atmosphere to stands of conifers and broad-leaved species using micrometeorological methods but these have, of course, all been at normal ambient CO_2 concentrations.

A very small number of experiments at elevated CO_2 have been performed on the growth and water use of seedlings and small trees over periods of weeks in controlled environment facilities or in glasshouses, plastic tunnels and open-topped chambers in the field. However, as far as we are aware, there have been no long-term experiments in the field on the response of trees, stands or other parts of forest ecosystems to elevated CO_2 concentrations. To say anything at all about the likely responses of stands or ecosystems to elevated CO_2 concentrations, the only approach available to us is to rely on information concerning short-term responses of processes in leaves and shoots (see Section 10.2.1), and, by using models, to extrapolate up the scales of time and space. This approach offers some hope for the future, but there are formidable problems involved in deriving realistic estimates of forest growth with a 'bottom-up' modelling approach (as discussed in Section 10.3.4).

In the following sections we review the information available concerning responses of CO_2 assimilation, stomatal action and growth of forest species to elevated CO_2 concentrations. We shall begin with CO_2 assimilation and stomatal action at the leaf level and then consider the effects on growth on longer time scales and larger area scales.

10.3.2 Fast, Short-term Responses at the Leaf Level

In short-term experiments, the rate of CO_2 assimilation (A) has been shown to increase more or less linearly over the range of ambient CO_2 concentrations (C_a) from 20 to ca 400 ppmv (e.g. Zelawski, 1967; Brix, 1968; Kriedemann and Canterford, 1971; Ludlow and Jarvis, 1971; Luukkanen and Kozlowski, 1972; Regehr *et al.*, 1975; Gross, 1976; Watson *et al.*, 1978; Masarovicova, 1979; Beadle *et al.*, 1981) in both conifers and broad-leaved species, before starting to level off (Figure 10.5). The onset of significant levelling off is variable, occurring anywhere between 300 and 600 ppmv in different species and may be followed by a sharp decline at ca 1500 ppmv (Koch 1969).

In some conifers, stomatal conductance (g_s) is relatively insensitive to the ambient CO_2 concentration (C_a) compared with the general response of g_s to C_a in other species that use the C_3 photosynthetic pathway (e.g. Beadle *et al.*, 1979; Morison and Jarvis, 1983). This may explain in part why A continues to increase steeply up to high values of C_a in conifers rather than levelling off at lower C_a as in many other C_3 species (Figure 10.6).

We may expect varied responses of A to C_a depending on the capacity of the photosynthetic apparatus to fix CO_2, relative to the rate of supply of CO_2 to the reaction sites. The rate of assimilation depends on two factors. The first is the activity of the photosynthetic enzymes, particularly ribulose-bisphosphate carboxylase, and the enzymes in the carbon reduction cycle that limit the regeneration of the ribulose-bisphosphate acceptor. The second factor is the concentration of CO_2 in the air bathing the mesophyll cells (C_i), since that determines the rate of supply of CO_2 to the reaction sites. C_i is, of course, related to C_a depending on the resistance in the CO_2 transfer pathway between the bulk atmosphere overhead and the intercellular spaces, particularly the resistance of the stomata. These relationships are evident in the curve describing the relationship between A and C_i, the so-called A/C_i curve (Figure 10.7). Through the A/C_i curve we may seek an explanation of the varied responses of A to C_a.

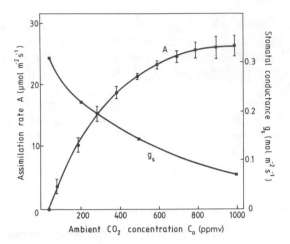

Figure 10.5 The relationship between assimilation rate (A), stomatal conductance for CO_2 (g_s) and ambient CO_2 concentration (C_a) for unstressed *Populus deltoides.* Quantum flux density 2000 μmol m^{-2} s^{-1}; temperature 30 °C; water vapour saturation deficit 1.7 kPa. Vertical bars indicate one standard deviation. Recalculated from Regehr *et al.* (1975)

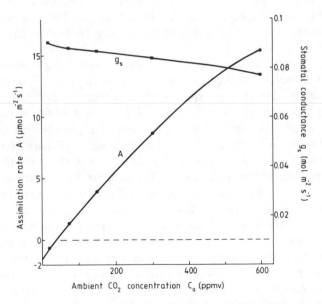

Figure 10.6 The relationship between assimilation rate (A), stomatal conductance for CO_2 (g_s) and ambient CO_2 concentration (C_a) for unstressed *Picea sitchensis.* Quantum flux density 1000 μmol m^{-2} s^{-1}; temperature 20 °C; water vapour saturation deficit 0.6 kPa. Each point is based on at least five measurements. Recalculated from Beadle *et al.* (1979, 1981)

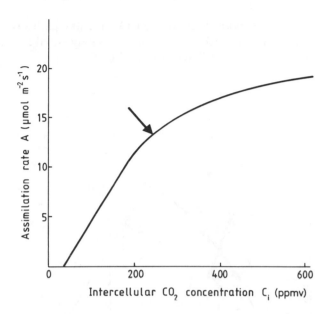

Figure 10.7 The relationship between assimilation rate (A) and mean intercellular space CO_2 concentration (C_i) for unstressed *Picea sitchensis*. Quantum flux density 1100 μmol m^{-2} s^{-1}; temperature 20 °C; water vapour saturation deficit 1.0 kPa. The arrow indicates the value of C_i at the current atmospheric concentration of C_a. The line has been fitted in two sections to a total of 72 data points. Unpublished data of the authors

In Figure 10.8 the point X is the rate of assimilation corresponding to the indicated ambient CO_2 concentration (C_a) and the intercellular space CO_2 concentration (C_i). The line ΓZ represents the 'demand function' of assimilation for CO_2: the line $C_a D$ represents the 'supply function' (Jones, 1973; Raschke, 1979). If the A/C_i curve is approximately linear up to the operating point X, the demand function can be represented by

$$A = g_m(C_i - \Gamma)$$

where g_m is carboxylation efficiency or a mesophyll conductance. (If the relationship is not more or less linear over this range, a more complex treatment is required—see Farquhar and Sharkey, 1982; Jones, 1983). Above the operating point, A becomes a function of the capacity to regenerate ribulose-bisphosphate and reaches an asymptote (A_{max}). The supply function is represented by

$$A = g_c(C_a - C_i)$$

where g_c is the overall conductance of the CO_2 transfer pathway from bulk atmosphere to intercellular spaces and is approximately equal to the stomatal conductance, g_s, for a well-ventilated, isolated leaf.

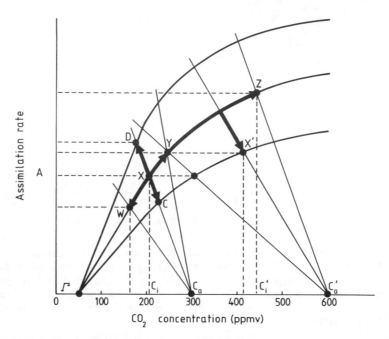

Figure 10.8 A diagram to show the interrelationship between the 'demand function' and the 'supply function' of assimilation. The line ΓZ is a relationship between assimilation rate (A) and mean intercellular space CO_2 concentration (C_i) of the kind shown in Figure 10.9. The other lines show how A may vary depending on changes in the demand function (e.g. $\Gamma X \rightarrow D$ or Γ C), the supply function (e.g. $C_a X \rightarrow C_a W$ or $C_a Y$), or the ambient CO_2 concentration (e.g. $C_a X \rightarrow C'_a Z$) or combinations of these three variables (e.g. $C_a \rightarrow C'_a$ leading to decreases in both the demand and the supply functions but X rising to X')

These functions have the following consequences (assuming for the moment that only one variable changes at a time):

• if the carboxylation efficiency changes, A will change along the line CD; if g_m increases to g'_m, A will increase from X to D.

• if the stomatal conductance changes, A will change along the line WXY; if g_s decreases to g'_s, A will decline from X to W.

• if the ambient CO_2 concentration changes, A will change along the line WXYZ; if C_a increases to C'_a, A will increase from X to Z.

If, for example, there are no changes in g_s or in the demand function, we would expect an increase in C_a to lead to an increase in A along the line XYZ. This is very much the case in *Picea sitchensis* where g_s is largely unaffected by CO_2, so that the response of A is almost linear for C_a between 75 and 600 ppmv (Beadle *et al.*, 1979, 1981). If, however, an increase in C_a were to change the supply function, a smaller increase in A might result. For instance, if the stomata were to close somewhat in response to the increase in C_a, C_i and, hence, A would increase less than in the previous example (e.g. only to Y). In short, it is evident that the increase in A depends on the sensitivity of the stomata to CO_2. Therefore, we would expect the largest, short-term increase in assimilation to occur in plants such as the conifers, which have stomata that are relatively insensitive to CO_2. Smaller, short-term increases in assimilation would be expected to occur in species like *Populus deltoides* that have stomata which are quite sensitive to CO_2.

Other environmental variables may also influence g_s and lead to changes in C_i. For example, in many tree species, both broad-leaved and conifers, the stomata are sensitive to the water vapour saturation deficit of the air (D) (e.g. Appleby and Davies, 1983). As a result, the ratio of internal to ambient CO_2 concentrations (C_i/C_a) decreases with increasing D (Morison and Gifford, 1983). The climatic changes associated with the global increase in CO_2 are very speculative but it seems likely that middle latitudes may become less humid. If an increase in D is correlated with the increase in CO_2, there may be very little change in assimilation rate. Environmental stresses, such as low light (e.g. Morison and Jarvis, 1983), low water (e.g. Beadle *et al.*, 1981; Jones and Fanjul, 1983) or low nutrients (e.g. Wong, 1979) may displace the A/C_i curve downwards, reducing either or both mesophyll conductance (g_m) and A_{max} (see Figure 10.9) and thus reducing the opportunity for A to respond to enhanced CO_2.

It is clear from the preceding discussion that evaluation of the response of assimilation to a change in CO_2 requires knowledge of the A/C_i curves for a range of environmental and physiological conditions. This information is hardly available at the present time for plants grown in controlled environments, let alone for plants grown in the field. One notable exception is a recent experiment in which the CO_2 uptake of a large *Eucalyptus maculata* tree growing on a weighing lysimeter in a forest was measured in relation to a range of CO_2 concentrations by enclosing the tree in a plastic tent (S.C. Wong, personal communication). Assimilation, stomatal conductance and transpiration of the tree responded to CO_2 in much the same way as would a leaf in an assimilation chamber in the laboratory.

All the experiments discussed above investigated short-term responses of assimilation and stomatal conductance in plants that were not given the opportunity to acclimate to higher CO_2 concentrations. There are very few published experiments in which the same responses have been studied in

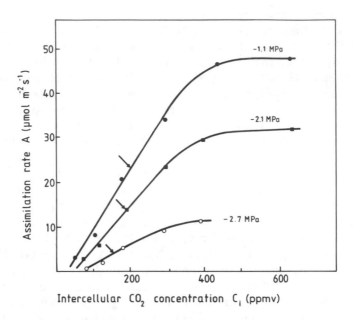

Figure 10.9 The relationship between assimilation rate (A) and mean intercellular space CO_2 concentration (C_i) for leaves of water-stressed apple trees cv James Grieve. The figures by the curves are the leaf water potentials. Quantum flux density 1500–2000 μmol m^{-2} s^{-1}; temperature 15–20 °C; water vapour saturation deficit 0.5–1.0 kPa. The arrows indicate the values of C_i at the current atmospheric concentration of C_a. Each point is the mean of at least four measurements. From Jones and Fanjul (1983)

forest species that have been allowed to grow for extended periods in different CO_2 concentrations (e.g. Rogers *et al.*, 1983). What experiments there are suggest that the response of A and g_s in plants acclimated to elevated CO_2 is qualitatively similar to the responses in plants grown in normal air, i.e. A increases and g_s decreases up to CO_2 concentrations of at least 1000 ppmv.

Studies of the quantitative effects of acclimation are rare. Data for plants other than forest species indicate that acclimation to elevated CO_2 leads to a reduction in certain photosynthetic enzymes (see review by Berry and Downton, 1982) and thus changes the demand function. As a result, A and g_s in air containing 300 ppmv CO_2 become less than they were before a period of acclimation to 600 ppmv (e.g. Wong, 1979). Nonetheless, the relevant observations are that long-term exposure to elevated CO_2 does result in higher A and lower g_s than in plants grown at current normal concentrations.

The combination of the effects of CO_2 on A and g_s may lead to a substantial increase in assimilation per unit of water transpired, the so-called water-use efficiency (i.e. A/E), for plants in enclosures. In *Liquidambar styraciflua*, for example, a doubling of the CO_2 concentration more than doubled the water-use efficiency (Rogers *et al.*, 1983; Tolley and Strain, 1985). Such a response may, of course, vary from species to species depending on the relative sensitivity of A and g_s to CO_2. Whether this response is necessarily likely to occur in field conditions is discussed in Section 10.3.4.

10.3.3 Intermediate Time Scale Effects on the Growth of Trees

There are very few reported experiments in which tree species have been grown for extended periods at elevated CO_2 concentrations. What experiments there are have been confined to the whole or part of one growing season (e.g. Sionit *et al.*, 1985). We are not aware of any published experiments in which woody perennials have been grown at elevated CO_2 concentrations over several growing seasons. This is an important point because quite small differences in relative growth rate, if they persist for several years, lead to large differences in plant size.

The limited evidence available suggests that growth in height, leaf area and dry weight of temperate trees is increased by higher than normal CO_2 concentrations (Yeatman, 1970; Funsch *et al.*, 1970; Krizek *et al.*, 1971; Tinus, 1972; Laiche, 1978; Canham and McCavish, 1981; Lin and Molnar, 1982; Rogers *et al.*, 1983; Tolley and Strain, 1984a and b). However, the results reported so far are extremely variable and in some species no growth response was elicited (Lin and Molnar, 1982). There are too few reliable data from which to generalize about the growth response.

On the basis of results with other species (e.g. Morison and Gifford, 1984a and b) and limited data for tree seedlings (Tolley and Strain, 1984b), we expect the primary consequence of an increase in A to be larger leaf areas that will increase interception of radiation and thus amplify the initial effect of enhanced CO_2. That is not to say that we suppose a direct effect of elevated CO_2 on leaf development; rather, we expect more rapid leaf initiation and lamina expansion as a result both of higher carbohydrate status and possibly higher turgor, as well as greater longevity of leaves. These feedback effects, we suppose, lead to the more rapid attainment and sustention of larger leaf areas and hence to higher growth rates. If leaf growth is inhibited, there may be no growth response to elevated CO_2. For example, ethylene, which inhibits cell division in expanding leaves, when found as a contaminant of artificially synthesized CO_2, completely negated the effect of doubling the ambient CO_2 concentration on plant growth (Morison and Gifford, 1984c). As far as we know, there are no clearly established *initial* effects of CO_2 on leaf initiation and development in herbaceous plants (e.g. Hurd, 1968; Hurd and Thornley, 1974; Sionit *et al.*, 1981; Morison and Gifford, 1984b) or in trees.

There is little relevant information on the effect of elevated CO_2 on partitioning of assimilate. The evidence available suggests that partitioning is largely unaffected (Tinus, 1972; Rogers *et al.*, 1983). More detailed studies on a number of herbaceous species (e.g. Morison and Gifford, 1984b) have demonstrated a consistent decrease in specific leaf area associated with increase in the number and size of cells in leaves, and some changes in leaf weight ratio but not in root to shoot ratio and somewhat similar responses seem to be found in tree seedlings. In a recent study, Thomas and Harvey (1983) found increases in leaf thickness, apparently resulting from the presence of both more and larger cells, in leaves of *Liquidambar stryaciflua* and *Pinus taeda* grown at elevated CO_2 concentrations.

10.3.4 Long-term Impact on Ecosystems

The possible effects of climatic change on the dynamics of species relationships in forest ecosystems are treated in Section 10.4. As far as the direct effects of CO_2 are concerned, we are less able to comment on the effects of an increase in CO_2 concentration on processes of regeneration, competition and species composition because of the lack of data. We know of no experimental studies on relevant species in which the interference between species has been investigated at elevated CO_2 in either field or laboratory. Long-term exposure of undisturbed vegetation to elevated CO_2 is certainly feasible. Unenclosed areas of agricultural crops have been exposed to controlled concentrations of SO_2 throughout an entire growing season in at least two localities and the same could be done with CO_2 in crops, artificial mixtures and natural vegetation, given adequate resources. However, at the present time such projects have not progressed beyond the discussion stage. As is always the case with forest vegetation, the problems are more formidable, both because of the larger scale and the close coupling between forest canopies and the bulk atmosphere.

Recently, two papers have appeared in which annual tree growth of trees in stands in the field has been analysed in relation to the age of the tree over the past 50 or so years and growth was found to be more than expected (LaMarche *et al.*, 1984; Arovaara *et al.*, 1984). It was suggested by the authors that this might be attributed to the increase in CO_2. Such conclusions must be regarded as highly speculative, since growth responds to many environmental variables.

An increase in rate of photosynthesis (and leaf development, if it occurs) would be expected to increase the rate of production of biomass (i.e. the productivity) and the standing crop in forest systems, like plantations, that have not yet achieved a steady state. In forests that have achieved a steady state there may also be increases in the rate of growth of individuals as a result of an increase in CO_2, but this may well not lead to an increase in the biomass of the standing crop. At a steady state the overall, average annual net exchange of CO_2 is zero, although the instantaneous rate of assimilation by some components is very high. This is because the assimilation of CO_2 is balanced by losses of CO_2 in respiration of the living biomass and through mortality and respiratory turnover of leaves, fine roots, branches and entire individuals (Jarvis and Leverenz, 1983). How a rise in ambient CO_2 concentration will affect the checks and balances in this complex system (see Figure 10.1) is very uncertain.

There are two major problems in trying to scale up to the growth and productivity of even comparatively simple stands such as coniferous plantations. First, we lack adequate physiological information to scale up from measurements of physiological processes made over periods of minutes or hours to estimate growth over periods of months or years. We are able to simulate adequately the CO_2 influx to a plantation, in the short term, using data on leaf photosynthetic properties and canopy structure in a model of canopy processes (e.g. Jarvis *et al.*, 1985). However, we still lack sufficient physiological knowledge about processes such as assimilate allocation to make the jump from canopy CO_2 exchange, whether measured or predicted, to growth of individual trees or of the stand. Whilst we can run the same model with higher ambient CO_2 concentrations as input, and predict a new CO_2 influx, we lack sufficient knowledge concerning the feedbacks between CO_2 supply, leaf growth, nutrition and tree growth, to be able to say whether photosynthesis would continue at a higher rate and whether this would lead to a long-term increase in growth of the stand. It is very doubtful that we could, at the present time, even predict the long-term growth responses of young potted trees to elevated CO_2 that have been observed in growth-room experiments, from the short term measurements of CO_2 uptake that have been made on leaves or shoots.

Second, we have scarcely sufficient micrometeorological knowledge about coupling between vegetation and the atmosphere to scale up from a leaf to a stand, let alone to an ecosystem on a regional scale. A leaf in an assimilation chamber or growth room is, by design, tightly coupled to an environment that is imposed upon it. The natural feedback loops between the leaf and the atmosphere are deliberately broken, so that the leaf has little influence over its own environment. In the field, the degree of coupling of vegetation to the atmosphere varies, depending on the area scale and the aerodynamic roughness of the vegetation. As the area scale increases in size, the degree of coupling between vegetation and atmosphere decreases, with the result that transpiration becomes less influenced by a change in stomatal conductance and more strongly dependent on the radiation input. On the regional or global scale the overall rate of transpiration is set by radiation and is not affected by changes in stomatal conductance. Consequently CO_2 cannot be regarded as a global antitranspirant (Jarvis and McNaughton, 1985).

Nonetheless, transpiration from aerodynamically rough vegetation (that is well-coupled to the atmosphere, like a tree canopy) will be reduced by stomatal closure, but at the expense of transpiration from poorly-coupled vegetation which will increase (McNaughton and Jarvis, 1983; Jarvis and McNaughton, 1985). Similarly, we expect CO_2 uptake of well-coupled vegetation such as trees to tend to increase in response to an increase in global CO_2 concentration, depending also on the response of the stomata, whereas the CO_2 uptake of largely decoupled vegetation, such as grasslands and field crops, and probably also the forest understorey, is likely to respond to a lesser extent because of feedback between the local ambient CO_2 concentration and the rate of CO_2 uptake.

Because of these problems in scaling up with respect to both time and space, we regard speculation about the likely effects of rising CO_2 on carbon assimilation and water use by forests as particularly unreliable at present. While empirical experimentation is, in principle, possible, the logistical problems render this impracticable with mature forests. Therefore, there is an urgent need to improve the basis for scaling up from leaf to region and from minutes to years.

10.3.5 Further Research Directions

It is clear from the foregoing survey that a shortage of data at all scales seriously limits an assessment of the likely effects of rising CO_2 on forests. Short-term, fast response data are the most readily obtainable and some progress may be made by scaling up such data with the use of appropriate physiological, canopy and micrometeorological models. For this purpose we need far more extensive characterization of the responses to CO_2 of assimilation, stomatal conductance and transpiration for plants that have been acclimated to high concentrations of CO_2, and for a much wider range of relevant species, than is available at present. However, because of the gaps in our physiological knowledge concerning the linkage between assimilation and transpiration of leaves and the growth and water use of whole plants, it is very necessary that the response of growth and water use to high CO_2 (and other variables such as nutrient and water stress) should be analysed over extended periods of time covering several growing seasons. Thus, we see a need for many more growth experiments on tree species. Finally, we would emphasize that the forest is a functional unit that is likely to respond to high CO_2 in a rather different way to the response of isolated individual leaves or trees. Thus, we identify a need for a much better understanding of the micrometeorological processes that determine the exchanges of CO_2 and water between forests and the atmosphere.

While the use of models offers a means to scale up in both time and space, our present state of knowledge about the processes involved is insufficient to allow this to be done with any real confidence in the results. Consequently, we see a need for the concurrent development of models and empirical studies of the physiological and micrometeorological processes that determine the response to CO_2. Given the paucity of our present knowledge, such empirical studies are needed at each spatial and temporal scale.

10.4 MODELS FOR ASSESSING THE IMPACTS OF CLIMATIC CHANGE

10.4.1 Forest Simulation Models

Determination of the response of a forest to a climatic change involves evaluation of a complex system with many levels of response. One means of attempting to handle this complexity is by the use of quantitative models of forest dynamics as investigative tools. There are several hundred extant models of forest dynamics that simulate the growth of individual trees to determine the temporal response of a forest. These models seem to be most appropriate to 'intermediate time scales' discussed earlier. There have been several reviews of the types and performance of detailed models of forest dynamics (Munro, 1974; Shugart and West, 1980; Shugart, 1984).

The models that have had the greatest success in duplicating the responses of forests over 10- to 40-year time scales have been developed by foresters using empirical calibration of tree growth and form based on large, spatially extensive data sets. Examples of such models include the HUGIN Project (Hägglund, 1981) in Sweden and the STEMS Model of the Lake States in the United States (Belcher *et al.*, 1982). These models are responsive to climatic variables. For example, Hägglund (1981) reports a 10% variation in the yield from plots over a simulated growth period as being attributable to variation in weather. The statistical accuracy of predictions of forest yield from such models is high. These models could, in principle, be used to assess the effects of small changes in climate. However, the models are restricted to conditions within the calibration range of the set of parameters, and the developers of the models are outspoken in their caution against using the models to extrapolate outside this range (Hägglund, 1981; Belcher *et al.*, 1982).

Shugart (1984) categorized several forest simulation models with respect to the structure of the model, and to the age and the diversity of the forest simulated by the model. The models function by simulating the growth (and often death and recruitment) of individual trees in mixed-aged or even-aged and mixed-species or mono-specific forest stands. One important consideration is the form of the competition equations used in a model to simulate the interactions among individual trees spatially (2- and 3-dimensional models), vertically (1-dimensional models called gap models) or nonspatially.

A tabulation of some of these models and a brief indication of their advantages and disadvantages in estimating forest response to climatic change are given in the Appendix. Of these models, only the gap models have been used extensively to simulate the response of forests to climatic change.

10.4.2 Gap Models of Forests

Gap models simulate the diameter increase of each of the trees growing on a small plot (described in Section 10.2 as the gap in the forest cycle). The trees increase in size annually. Mortality of the trees on the plot and the recruitment of new trees are stochastic functions that are applied annually. The models are not constructed to simulate mechanisms on a time scale of less than one year (although monthly temperature and precipitation data are used to compute indices for calculating the regeneration, growth and death of trees). The fundamental approach to developing these models was outlined by Botkin *et al.* (1972) and has been reiterated by several authors since that time. Shugart (1984) provided a detailed explanation of the assumptions in gap models and related the behaviour of such models to modern ecological concepts of forest dynamics. The approach that underlies the development of gap models is to represent the dynamics of the forest by general equations that are parameterized from basic physiology, morphology or forestry. In general, gap models do not require elaborate data sets for parameter estimation, and the data available for the forests in a given locality are often reserved for tests of the models. The models have been used to reconstruct the response of forests to climatic changes associated with the period since the last glaciation (Solomon *et al.*, 1980; Solomon and Shugart, 1984; Shugart, 1984). This application is indicative of the potential of these models to predict the response of forests to future changes in climate.

Since these models can be used to simulate the dynamics of each of the patches comprising a forest mosaic, a two-level model can be used to simulate the dynamics of a region (e.g. Shugart *et al.*, 1973). Such a model functions by using the detailed sub-model to simulate the individual patch dynamics and the higher level sub-model to maintain the inventory of the number of patches that could be ascribed to each of several forest types. Israel *et al.* (1985) have successfully used this approach to simulate the response of production of forest biomass to an increase in temperature.

10.4.3 Dynamic One-dimensional Forest Models

An alternative approach to gap models is to use dynamic, one-dimensional forest models. This approach was followed by Korzukin *et al.* (1986a) to accommodate age distribution in multi-species plant communities, and differs from gap models with respect to averaging procedures. In gap models, most of the dynamic effects derive from the nonlinear equations upon which they are formulated; Monte Carlo procedures are used to generate simulations of gap development by averaging the trajectories to produce the dynamics of plant community behaviour. In the dynamic, one-dimensional model, mathematical averaging is applied to each tree of the community, and then equations are derived to yield an emergent community structure through time. Furthermore, gap models assume constant growth of the taller trees with the growth of shorter trees being reduced by shading, whereas the one-dimensional model does not assume constant growth rates.

Using such a dynamic one-dimensional model, Korzukin *et al.* (1986b) adequately simulated succession in a two-species, post-fire cedar forest in western Siberia. The model described the wave-like, non-stable age dynamics that actually occur in the Siberian cedar forest for 200 years following fires. This type of behaviour is similar to that shown in Figure 10.3 and to that simulated by gap models.

Following the ideas presented by Shugart *et al.* (1973), Antonovsky and Korzukin (1986) developed a three-level (individual-plant community-regional) model that could be used in global–biospheric modelling. This model allows the investigation of the simultaneous effects of climatic factors on forests at each of the system levels. For example, a simulated increase in temperature in the boreal forests of the cold latitudes stimulated the growth of individual trees, increased community biomass (but to a lesser degree than if the various species had not interacted competitively) and increased the frequency of regional fires. The net effect on forest biomass at the regional level can be either positive or negative. As shown by Korzukin *et al.* (1986b) in the case of post-fire succession (mentioned above), predictions of declining total regional biomass are obtained for a 1 °C warming when the difference between the relative acceleration of individual growth (W_m) and the relative increase in fire frequency (W_n) was set to values of $3W_m - W_n < 0$. At values of $3W_m - W_n > 0$, total regional biomass increased with a 1 °C warming. The importance of processes occurring at all levels of forest systems—including the critical role played by fire—are highlighted by this modelling approach.

10.4.4 Evaluation of Models

The use of a model to predict the response of a forest to climatic change necessarily requires extrapolation of one degree or another and for this reason is a procedure that involves some measure of uncertainty regarding the reliability of the estimates. Because it is impossible to conduct long-term experiments on today's forests with respect to climatic change over 10- to 100-year time scales, there will always be a degree of uncertainty in model

Table 10.2 Tests of gap models showing structural and functional responses (after Shugart 1984)

Type of test	Structural response	Functional response
Verification Model can be made to predict known feature of a forest	Is consistent with structure and composition of forests in New Hampshire (JABOWA[2]) Tennessee (FORET[3]), and Puerto Rico (FORICO[4]) and flood-plain of the Mississippi River (FORMIS[5]) Compares to subtropical rain forest of known age (KIAMBRAM[6]) Predicts Arkansas upland forests based on 1859 reconnaissance (FORAR[7])	Predicts forestry yield tables for loblolly pine in Arkansas (FORAR) Predicts relations of forest types in succession in middle altitudinal zone in Australian Alps (BRIND) Predicts response to clear-cutting in Arkansas wetlands (SWAMP) Predicts change in forest types as a function of flood frequency in Arkansas wetlands (SWAMP) and the Mississippi floodplain (FORMIS)
Validation Model independently predicts some known feature of a forest	Predicts frequency of trees of various diameters in rain forests in Puerto Rico (FORICO[4]) and upland in Arkansas (FORAR[7]) Predicts vegetation change in response to elevation in New Hampshire (JABOWA[2]) and Australian Alps (BRIND[8]) Determines effects of hurricanes on diversity of the Puerto Rican Rain Forest (FORICO[4])	Predicts response of Eucalyptus forests to fire (BRIND) Assesses effects of the chestnut blight on forest dynamics in southern Appalachian forest (FORET) Predicts forestry yield tables for alpine ash in New South Wales (BRIND)

Model	Application
predicts the response of the forest to changed conditions	Predicts cha..ges in a 16,000-year pollen chronology from East Tennessee in response to climate change (FORET[3])
	Assesses habitat management schemes for endangered species (FORAR[7]), nongame bird species (FORET[3]) and ducks (SWAMP[9])
	Predicts response of northern hardwood forest to increased levels of CO_2 in atmosphere (JABOWA)
	Predicts response of southern Appalachian hardwood forest to decreased growth due to air pollutions (FORET)
	Predicts response of northern hardwood forest (JABOWA), southern Appalachian forest (FORET) Arkansas uplands forests (FORAR), Arkansas wetlands (SWAMP), and Australian subtropical rain forest (KIAMBRAM) to various timber management schemes.

1) Structural responses describe a forest at one time or on the broad level of species composition only. Functional responses allow for more quantitative detail, examining a forest over a period of time.

2) JABOWA is a model of northern hardwood forest

3) FORET of Tennessee Appalachian hardwood forest

4) FORICO of Puerto Rican Tabonuco montane rain forest

5) FORMIS of Mississippi River floodplain deciduous forest

6) KIAMBRAM of Australian subtropical rain forest

7) FORAR of Arkansas mixed pine-oak forests

8) BRIND of Australian Eucalyptus forests

9) SWAMP of Arkansas wetlands forest

predictions. It is, of course, the difficulty of conducting such experiments on forests that draws one to use models for this purpose in the first place. Careful evaluation of the models can reduce the uncertainty and thus give greater confidence in the estimates of the models.

Among the most stringent evaluation procedures are validations in which a model is tested for its ability to predict values using observations that are independent of those used to develop the model. Generally, gap models have been validated against independent data by using a variety of measurements as tests. For the models to be used to ascertain the sensitivity of forests to future climatic change (Table 10.2) the most important tests involve the ability of the models to simulate vegetational zonation along altitudinal gradients, and to reconstruct changes in the vegetation that have occurred in response to the climatic changes since the last glaciation. The credibility of the models is enhanced by their ability to reproduce the behaviour of a variety of forests on different continents, with different tree species and environmental conditions. Although most gap models have been subjected to a degree of testing, they have not been subjected to a wide range of experimental testing appropriate to the 100-year time scale of the transient responses of forests to changing climate. Consequently, conclusions drawn from the use of these models must be treated with caution.

10.5 STUDIES OF THE RESPONSE OF FORESTS TO CLIMATIC CHANGE

The response of the world's forests to changes in climate can be examined using forest simulation models in conjunction with scenarios of climatic change that are derived, for example, from general circulation models (GCMs; see Chapter 8). The practice of feeding the output of one model into another model should be undertaken with considerable caution, particularly with respect to the potential errors that such procedures can propagate. One should consider the results from such modelling exercises to be indicative of possible responses and look for robust results that appear to be common to several models. Further, any results of this kind should, as far as possible, be corroborated by observations in nature. In the absence of experimental protocols for conducting climatic experiments on forests over centuries, the value of the results of such models must necessarily depend on the internal consistency among the predictions and corroboration through observations.

10.5.1 Global-scale Response of Vegetation

At a global scale, one approach to examining the possible changes in the size and areal extent of the world's forests is to use empirical models of climate and vegetation in a spatial context and to superimpose scenarios of climatic change. Emanuel *et al.* (1985a) used the Holdridge life zone classification (Holdridge, 1947, 1964) to map the distribution of potential vegetation on the Earth's terrestrial surface. The Holdridge classification predicts expected vegetation as a function of a temperature and moisture index. By interpolating monthly temperature and precipitation data from 8000 meteorological stations onto a 0.5 degree latitude by 0.5 degree longitude grid and applying the Holdridge classification scheme to these data, Emanuel *et al.* produced a map of world vegetation. Each of the meteorological records was then altered by a change in the annual average temperature taken from Manabe and Stouffer's (1980) simulation experiment for a CO_2-doubling. The initial procedure was then repeated to obtain a map of the potential vegetation to be expected after the climatic change.

In a subsequent critique of the procedure, Rowntree (1985) noted that the use of mean annual temperatures was less appropriate than the use of seasonally varying temperatures. It was also noted that it would have been more appropriate to use the difference between the $2 \times CO_2$ scenario and the General Circulation Model control run (rather than the difference between the $2 \times CO_2$ scenario and observed data) to derive the magnitude of the temperature changes from which to calculate the effects of climatic change on vegetation. Based on these criticisms, Emanuel *et al.* (1985b) revised the maps of the Holdridge life zones for both the base case (present-day conditions as reflected in the meteorological station data set) and the $2 \times CO_2$ scenario as shown on the endpapers.

At a global scale, the life zone designations of 34% of the 0.5° by 0.5° grid cells were altered. In the higher latitudes, the generally higher temperatures resulted in a 37% decrease in the areal extent of boreal forest and a 32% decrease in the areal extent of tundra (see Table 1 in Emanuel *et al.*, 1985b) Boreal moist forest was replaced by cool temperate steppe and, to a lesser degree, by cool temperate forest and boreal dry bush. Boreal wet forest was replaced by cool temperate forest and boreal moist forest. The boreal forest zone shifted north and replaced about 42% of the 0.5° by 0.5° grid cells designated as 'tundra' in the base case. The northern extent of the tundra was also increased.

Because the temperature changes in the Manabe and Stouffer scenario were smaller toward the equator, there were smaller changes in the tropical life zones. Nevertheless, the areal extents of the subtropical and tropical life zones increased by 8%. The area of subtropical forest life zones decreased by 22%, while the subtropical thorn woodland and subtropical deserts increased by 37% and 26%, respectively.

In the analysis described above, precipitation was left unchanged and thus average evapotranspiration increased. If precipitation were allowed to change, however, a reduction of boreal forest would still result from the higher temperatures, according to the Holdridge life zone classification. Drier conditions would only further decrease the areal extent. Wetter conditions would allow the expansion of boreal forests into areas classified as 'boreal desert' (see the endpapers), but the area of boreal desert is so small that these gains would do little to offset the reduction in boreal forests caused by warmer temperatures. In contrast, the proportions of grasslands (including thorn woodlands and thorn steppe) and deserts would be expected to change considerably under different precipitation regimes. Increased precipitation would have little effect on the area of tropical forests, but decreased precipitation would diminish the area greatly.

Emanuel *et al.* (1985a) identified several sources of uncertainty in these sorts of assessments, including the choice of climate scenario, the choice of mapping algorithm and the relative coarseness of the data grid. Nonetheless, the simulated effects of a warmer climate on the areal extent of the coniferous boreal forests are not inconsistent with the conclusions one might draw from a casual inspection of the position of the boreal forests in relation to key temperature variables. Throughout North America and Eurasia, the northern limit of the boreal forest is delineated by the mean 13 °C isotherm in July (Larsen, 1980). The southern limit of the forest is bounded by the mean 18 °C isotherm in July in regions with favourable moisture conditions (where drier conditions prevail the limit is situated north of this isotherm). Although spatial correlations between climate variables and vegetation do not necessarily establish cause and effect, it is important to note that, with respect to growing season temperatures (indicated by the July isotherms), the boreal forest has a range of only about 5 °C under favourable moisture conditions and less than 5 °C under drier conditions. Thus, increases in average summer surface temperatures of just a few degrees, as projected by GCMs for a CO_2 doubling, might be expected to displace markedly the present boundaries of boreal forests.

Furthermore, there is some evidence that temperature could be a key variable in controlling the internal ecological processes of a boreal forest. For example, in a special issue of the *Canadian Journal of Forest Research* (**13**, No. 5), a set of 20 papers documents a multidisciplinary study of the structure and function of a black spruce (*Picea mariana*) forest in Alaska. The central hypothesis uniting these studies is that 'Black spruce ecosystems, the most widespread vegetation type in the taiga of Alaska, comprise the most nutrient limited, least productive taiga forest type especially in the later stages of succession. These conditions are largely controlled by cold soil temperature.' (Van Cleve and Dyrness, 1983). If, indeed, this hypothesis is substantiated, one might expect that higher soil temperatures as a result of climatic change would increase productivity in boreal forests through their effects on processes which otherwise appear limited by other factors, like nutrients.

10.5.2 Simulation of Response to Climatic Change at a Point Using Gap Models

Gap models have been used to simulate the response of forests in particular locaticns to changes in climate. In one recent study by M. B. Davis and D. B. Botkin (unpublished) various scenarios of climatic change were used to explore the response of the JABOWA model of the northern hardwood forests of the USA. The study described the inertial response of the model to abrupt shifts in climate of varying durations. In general, the study found an approximate 100-year lag in the response of the forest to such changes. For example, an abrupt warming with a duration of 50 years followed by a rapid return to the previous condition, produced a change in forest composition that was not observed until well after the climatic perturbation had terminated. These lag effects result from the time taken for young trees of species favoured by the climatic change to grow to an appreciable size, and are augmented by the delay in the death of trees which are already established but which are less favoured by the new climate. These sorts of lag effects could create difficulties in establishing cause-and-effect when investigating the response of forests to short-term climatic fluctuations: caution should be exercised in associating abrupt changes in vegetation with abrupt, contemporaneous changes in climate. These results are consistent with the modern theory of forest dynamics based on the dynamic mosaic concept that was discussed in Section 10.2.2.

In a series of experiments, the fossil-pollen record at Anderson Pond in the Cumberland Mountain Rim in Tennessee (Delcourt, 1979) was simulated using the FORET model (Shugart and West, 1977) in order to reconstruct the vegetation that occupied the site over the past 16,000 years (Solomon et al., 1980, 1981). During this period, the rich, mixed-hardwood forest that now occupies the site replaced an earlier boreal forest which appears to be much like that near Kenora, Ontario, where the pollen presently being deposited closely resembles that of 16,000 years ago in Tennessee (Davis and Webb, 1975; Webb and McAndrews, 1976). One simulation used climatic data (United States Geological Survey, 1965) from locations where the present-day pollen composition most closely matches the fossil pollen from Anderson Pond. These locations, discussed by Delcourt (1979), are interesting in that, at various times in prehistory, they included sites that were colder as well as warmer and drier than the present climate in central Tennessee. Solomon et al. (1981) also carried out simulations using the past climatic record as estimated by Delcourt (1979) and described additional experiments with the model designed to investigate model sensitivity to variations in the seed sources.

In these experiments the FORET model was able to reproduce the observed patterns of species abundance resulting from variatons in climate over a 16,000 year period. The correlations were highly significant with r^2 values of about 0.50 over virtually the entire simulation. The ability of the model to replicate the exact pattern of a particular taxon was not nearly so good, but was within the uncertainty in the relationships between existing vegetation and present-day pollen deposition. This lack of strong correlation was taken to be as much an indication of the variability that is inherent in the pollen deposition process as of a problem in the model.

One method of testing models involves comparing the vegetation simulated by a model with a fossil-pollen chronology. The potential problem with this procedure is that the climate that is used to drive the model is usually inferred from the fossil-pollen record. Thus, there is necessarily a degree of circularity in the comparison. Such comparisons are not complete tautologies, however. The information simulated by the models is based on the performance of individual species while the less taxonomically detailed fossil-pollen data are mostly reported as genera. Climate reconstructions based on fossil-pollen chronologies provide important information for verifying the ability of a model to predict the effects of climatic change on forests.

10.5.3 Simulation of Response to Climate Variables Varying in Space

Several attempts have been made to use models to reconstruct the spatial variation of a forest system in response to patterns of environmental variables. The ability of a model to estimate the spatial response is a strong indication that the model may be able to estimate temporal response as well. Such tests have been performed with gap models (tabulated along with other tests of the models in Table 10.2) and are discussed briefly in this section. The cases are:

(1) Botkin *et al.* (1972) in their original presentation of the JABOWA model tested the ability of the model to predict the position of the transition zone between a high-altitude coniferous forest and a low-altitude hardwood forest in the Hubbard Brook Forest (Bormann *et al.*, 1970; Bormann and Likens, 1979a, b). The model predicted the transition to occur at an elevation of 750 metres (using a standard lapse rate to adjust temperatures recorded at a base station to changes in altitude and then driving the model with these extrapolated temperatures). This prediction was found to be generally consistent with the pattern actually observed in the states of New Hampshire and Vermont in the USA. The transition was a result of species interactions and the response of individual species to the number of growing degreedays.

(2) Shugart and Noble (1981) tested the ability of the BRIND model to

simulate the response of Eucalyptus forests in the Brindabella Range of the Australian Alps to changes in altitude and to changes in the probability of wildfire. The resultant simulations agreed with patterns observed in the Brindabella range as well as other montane gradients in New South Wales, Victoria and Tasmania.

(3) Kercher and Axelrod (1984) developed a model of the dynamics of the mixed coniferous forest in the San Bernadino Mountains of southern California and tested it along gradients of altitude and with respect to the return frequency of wildfires. The model appears able to simulate the composition and structural responses of the forest to these two variables.

These experiments involving the simulation of the complex responses of forests to environmental gradients that change in time provide some confidence in our ability to use these models to predict the response of a forest system to a change in climate. However, most of the comparisons between predicted and observed data in these tests are largely qualitative. For this reason, the detailed predictions (e.g. the statistical distribution of tree diameters by species) should be considered less reliable than the general patterns predicted by the models.

10.5.4 Simulation of Continental Scale Response Using Gap Models

If gap models can simulate the temporal and the spatial responses of a forest to changes in environmental variables, the next step is to apply such models to predict the response of an entire landscape. Shugart (1984) has discussed the theoretical considerations and the limitations of such applications. Several studies of this sort have been carried out. Important in the context of this chapter is the series of model experiments conducted by Solomon *et al.* (1984) using the FORET model to simulate the response of the eastern half of North America to a CO_2-related change in climate. These results will be discussed in some detail.

Solomon *et al.* (1984) ran the FORET model for 21 locations in eastern North America with 72 species of trees available as seeds at all times. The simulations (10 replicate simulations at each site) were all initiated from a bare plot and were allowed to follow the forest dynamics appropriate to the modern climate with undisturbed conditions for 400 years. After year 400 the climatic conditions were changed linearly to reach the '2 × CO_2' climate in year 500, using a composite scenario of climatic change which strongly resembled that of Manabe and Stouffer (1980; see Solomon *et al.*, 1984, for details). The standard deviations of monthly temperature and precipitation were estimated from instrumental data from a meteorological station near each of the respective sites and were not changed by the shift in climate.

At years 500 to 700 of the simulation, the climate was changed by linear interpolation from the $2 \times CO_2$ climate to a $4 \times CO_2$ climate. The average conditions between years 700 and 1000 were those of the $4 \times CO_2$ climate. In each of the simulations the soil was a 100 cm deep, mesic, silt-loam (17% available water capacity, 35.5% field capacity, 18% wilting point).

The first 400 years of the simulation at each of the points were used to test the response against the observed pattern of species abundance predicted in eastern North America. These tests indicated that the model was able to simulate broad categories of abundance for the 72 species across the 21 sites with about a 90% success rate. The initial 400 years of the simulation also allowed the forest to reach a mature state prior to the imposition of the climatic change. The major effects of the changes in climate that resulted from a doubling and quadrupling of atmospheric CO_2 (see Figure 10.10) were as follows.

- a slower growth of most deciduous tree species throughout much of their geographical range.

- a dieback of many of the dominant trees, particularly in the transition between boreal/deciduous forests.

- an invasion of the southern boreal forest by temperate deciduous trees that was delayed by the presence of the boreal species.

- a shift in the general pattern of forest vegetation similar to the pattern obtained from the Holdridge map experiments (endpapers), with a time lag of up to 300 years.

In addition, Solomon *et al.* (1984) discussed the possible effects of several important ecological processes that were not included in the model. For example, insects and other pathogens, as well as air pollutants, could enhance the mortality simulated by the models. Also, plant migration (and associated lag effects) could have a negative influence on forest productivity.

10.5.5 Discussion of Model Results

The studies described above demonstrate the potential use of simulation models for assessing the effects of a change in climate on forests. For these purposes, it is necessary to identify the appropriate scenarios of climatic change and to standardize the initial starting condition for the various models. Some improvements in the models would be useful. The models that can be used for such simulations still lack several ecological processes that might be important in understanding the response of forests to a change in climate. Some of these additional mechanisms could be incorporated in the

BIOMASS CHANGE SIMULATED
WITH CO₂-INDUCED CLIMATE SHIFTS
IN EASTERN NORTH AMERICA

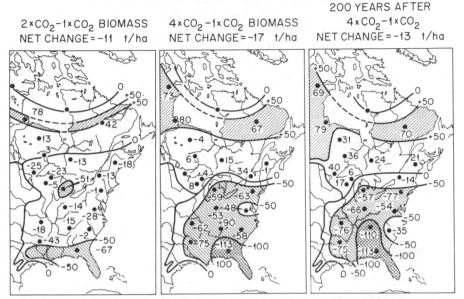

Figure 10.10 Carbon storage dynamics (in megagrams per hectare) simulated at 21 sites in eastern North America. Maps show carbon storage differences between contemporary climate and (A) 2 × CO₂-climate, (B) 4 × CO₂-climate, (C) 200 years after 4 × CO₂-climate stabilizes (from Solomon *et al.*, 1984)

present generation of forest models but several of the additions may well warrant reformulation of the models (particularly those phenomena that include spatial effects). Nevertheless, the potential to perform coarse simulations of the expected responses of forest to a climatic change presently exists.

In general, the results from a limited number of simulations using extant models lead to the tentative conclusion that climatic changes of the order of magnitude that have been predicted by climate models for a doubling of atmospheric CO_2 are sufficient to produce considerable changes in the composition, areal extent and location of the forests of the world. The largest changes are most likely to occur in the temperate and boreal regions. This result seems conservative since several of the phenomena that are not presently included in the models might be expected to amplify these responses. For example, the direct effects of increased concentrations of atmospheric CO_2 on the growth of particular species could reinforce or oppose the changes in forest dynamics and productivity estimated on the basis of a change in temperature alone. There is a large degree of uncertainty involved in these analyses and the results should be treated with caution.

10.6 SUMMARY AND CONCLUSIONS

The forests of the Earth constitute a complex system with many possible responses, both to the direct effects of an increase in atmospheric CO_2 concentration and to the possible changes in climate. These responses may originate from phenomena that operate on very different time and space scales. In general, formidable difficulties are encountered in 'scaling-up' the short-term physiological and biochemical responses of leaves and individual plants to estimate the intermediate and long-term responses of forests. The difficulties arise from the large uncertainties involved in the methods of extrapolation and from the complex interactions that occur at larger scales. The two uncertainties are presently large enough to preclude meaningful discussions of the interactions between CO_2 concentrations and climate except in the most general way.

10.6.1 The Possible Impacts of Increased CO_2 and Climatic Change

With respect to the direct effects of CO_2, these problems of scaling-up are compounded by the lack of experimental evidence for relevant forest species, particularly for plants that have been allowed to acclimate to enhanced CO_2 concentrations over one or more growing cycles. Although higher concentrations of CO_2 have been shown to increase CO_2 assimilation and, consequently, growth rates of individual trees in controlled conditions over the short term, it is highly uncertain whether such effects would be sustained and would lead to increased productivity in actual forest environments over the long term. In uncontrolled environments, the direct CO_2 effects are complicated by micrometeorological differences in the degree of coupling between forests and atmosphere (within as well as between forest systems), and by species competition and interaction. If, indeed, elevated CO_2 concentrations do result in long-term growth enhancement, increases in productivity would be more likely to occur in commercial forests than in mature forests in which the capacities for increased carbon storage are more limited. Direct experimentation at this scale, however, is largely impracticable. In order, therefore, to assess the responses of forest systems to both higher CO_2 concentrations and changes in climate, experimental studies must be augmented by empirical observation and simulation modelling.

With respect to the effects of climatic change, empirical climate–vegetation models and forest simulation models have been used to assess the responses of forests at scales ranging from a single point in a forest to an entire continental system. In general, from the results of a limited number of such studies, together with our understanding of the basic underlying processes, we conclude that:

- Climatic changes of the order of magnitude predicted by climate models for a doubling of atmospheric CO_2 are potentially sufficient to produce substantial intermediate and long-term changes in the composition, size and location of the forests of the world. At continental and regional scales, simulation models indicate considerable spatial heterogeneity in the response of forests.

• The natural forests of the high latitudes in general and the boreal forests in particular, appear sensitive to predicted temperature changes. Warmer conditions could possibly lead to large reductions in the areal extent of boreal forests and a poleward shift in their boundaries. It is at these latitudes that climate models predict the largest warming to occur as a result of increased concentrations of greenhouse gases, with smaller temperature changes in the lower latitudes.

• The forests of the tropical and sub-tropical zones would probably be more sensitive to changes in precipitation than temperature. Because of the high uncertainty regarding future changes in precipitation in the tropics, and because of the present lack of models that can be used to simulate the effects of tropical ecosystems to changes in climate variables, our knowledge of the responses of tropical forests to future climatic changes is meagre.

10.6.2 Some Next Steps

Future research efforts should be encouraged in three general directions. First, a key research problem for forests, as for agriculture, is how changes in climatic variables and CO_2 concentrations will interact to affect the short-term responses of plant processes. Greater understanding will require many more experiments on tree species over one or more growing seasons, with due regard to the micrometeorological effects arising from differences in the coupling between forests and atmosphere.

Second, there is a need to develop monitoring networks to detect shifts in vegetation types and to provide basic data for understanding the processes of ecological change, particularly in tropical regions. Such monitoring could be based on the networks established in the higher latitudes to record fluctuations in timberlines (i.e. marked study plots that are regularly surveyed). Large-scale monitoring to detect changes in regional vegetation patterns or productivity could be pursued with the aid of satellite imagery (see Chapter 8).

Third, further empirical analyses and simulation modelling, including verification of models, are also essential. This is because models are required to extrapolate from the effects of CO_2 and climate variables on short-term plant responses to determine their likely impacts on forest dynamics on time and space scales for which direct experiments are impracticable. Greater emphasis needs to be placed on tropical forest systems in order to achieve a balanced global assessment.

NOTE ON AUTHORSHIP AND ACKNOWLEDGEMENTS

Principal responsibility of writing Sections 10.1, 10.2, 10.4, and 10.5 was undertaken by H. H. Shugart and M. Ja. Antonovsky, Section 10.3 by P. G. Jarvis and A. P. Sandford, and Section 10.6 by all of the principal authors. We are particularly appreciative of the efforts of J. C. Ritchie and I. C. Prentice, who prepared Section 10.2.4. W. R. Emanuel and M. P. Stevenson recomputed global life zone change maps and, along with H. H. Shugart, prepared Section 10.4.1.

The authors also thank several reviewers from the International Meteorological Institute in Stockholm (particularly B. Bolin, B. R. Döös, R. A. Warrick), who read and commented on earlier drafts.

By arrangement with the United Nations Environmental Programme (UNEP), the World Meteorological Organization (WMO) and the International Council of Scientific Unions (ICSU), the International Meteorological Institute in Stockholm convened an expert panel (with expertise in boreal forests) in Stockholm, Sweden on January 7 to 11, 1985. We are particularly pleased to acknowledge the helpful comments of this panel to preparing the final draft of this chapter: B. Bolin, G. Bonan, B. R. Döös, S. Kellomäki, I. C. Prentice, J. C. Ritchie, J. Tucker, K. Van Cleve, R. A. Warrick. Useful comments were also provided by W. Böhme, H. L. Ferguson, R. M. Gifford, F. K. Hare, H. E. Landsberg, J. A. Laurmann, A. de Luca Rebello, A. S. Monin, O. Preining, B. R. Strain, and M. M. Yoshino.

We also wish to acknowledge the support of the United States National Science Foundation (Grant BSR-8510099), the International Meteorological Institute in Stockholm, and the Natural Environment and Climate Monitoring Laboratory of the USSR Goskomgidromet and the USSR Academy of Sciences.

APPENDIX

Classification and characterization of forest simulation models as potential tools for predicting the response of forests to altered climatic conditions.

Mixed-aged, Nonspatial Tree Models

Examples: Hool (1966), Olson and Christofolini (1966), Moser and Hall (1969), Shugart *et al.* (1973), Johnson and Sharpe (1976), Wilkins (1977), Antonovsky and Korzukin (1986a).

Considerations: Parameter estimation generally requires data or insights collected over a long time period. Model output is at regional scales. The models are mathematically tractable and could be coupled with economic models.

Even-aged, Spatial Tree Models

Examples: Newnham (1964), Lee (1967), Mitchell (1969), Lin (1970), Bella (1971), Hatch (1971), Heygi (1974), Lin (1974).

Considerations: Require extremely detailed tree growth data and other model parameters require large data sets. Models are usually for commercial forests only. Models typically do not treat more than a single generation of trees. The model output is very detailed. The output from the models is often in measurements that relate directly to economics.

Even-aged, Nonspatial Tree Models

Examples: Clutter (1963), Curtis (1967), Dress (1970), Goulding (1972), Sullivan and Clutter (1972), Burkhart and Strub (1974), Solomon (1974), Clutter (1974), Elfving (1974), Antonovsky and Korzukin (1985).

Considerations: Require detailed growth data and generally are used only in commercial forests. The models often do not consider tree regeneration and typically are used at time spans of less than a tree generation. Models are computationally fast and can provide data that have direct application in economics models.

Mixed-aged, Spatial Tree Models

Examples: Adlard (1974), Arney (1974), Ek and Monserud (1974), Mitchell (1975).

Considerations: Require very detailed data on a number of ecological processes. The models are computationally slow. The level of detail of the model output is very great. The models have been used in the assessment of the long-term response of forests to pollutant stress and could possibly be used for similar predictions of response to climatic change.

Mixed-species, Nonspatial Tree Models

Examples: Leak (1970), Bosch (1971), Namkoong and Roberts (1974), Forcier (1975), Suzuki and Umemura (1974), Horn (1976), Noble and Slatyer (1978), Waggoner and Stephens (1970).

Considerations: Models are relatively abstract and have been explored for their theoretical content. The level of abstraction in the models can make model testing against forest data problematical.

Mixed-aged, Vertical Tree Models (Gap Models)

Examples: Botkin *et al.* (1972), Shugart and West (1977), Shugart and Noble (1981), Shugart *et al.* (1981), Doyle (1981).

Considerations: The models have complex parameters that are generally inferred from ecological principles. The models predict sufficiently long-term forest responses that model validation can be a problem. The models have been used in applications that involve climatic change.

10.7 REFERENCES

Adlard, P. G. (1974) Development of an empirical competition model for individual trees within a stand, in Fries, J. (ed.) *Growth Models for Tree and Stand Simulation*, Res. Notes 30. Department of Forest Yield Research, Royal College of Forestry, Stockholm, 22–37.

Antonovsky, M. Ya., and Korzukin, M. D. (1985) Hierarchical modelling of vegetation dynamics in biosphere prognosis, in Israel, J. A. (ed.) Integrated global biosphere monotoring, *Proceedings of the Third International Symposium*, Leningrad (in press).

Antonovsky, M. Ya., Korzukin, M. D., and Termikaelian, M. T. (1985) The mathematical modeling of anthropogenic changes in forest ecosystems, in *Proceedings of a Soviet–French Symposium in the USSR*, 91–99.

Antonovsky, M. Ya., and Korzukin, M. D. (1986) Prognostic models of Soviet ecosystem dynamics in background monitoring, in Israel, J. A. (ed.) Ecologically sustainable development of biosphere, *Proceedings of International Symposium, Moscow*, March 1985 (in press).

Appleby, R. F., and Davies, W. J. (1983) A possible evaporation site in the guard cell wall and the influence of leaf structure on the humidity response by stomata of woody plants, *Oecologia*, **56**, 30–40.

Arney, J. D. (1974) An individual tree model for stand simulation in Douglas-fir, in Fries, J. (ed.) *Growth Models for Tree and Stand Simulation*, Res. Notes 30, Department of Forest Yield Research, Royal College of Forestry, Stockholm, 38–40.

Arovaara, H., Hari, P., and Kuusela, K. (1984) Possible effect of changes in atmospheric composition and acid rain on tree growth, *Communicationes Instituti Forestalis Fenniae*, **122**, 15.

Beadle, C. L., Jarvis, P. G., and Neilson, R. E. (1979) Leaf conductance as related to xylem water potential and carbon dioxide concentration in Sitka spruce, *Physiol. Plant.*, **45**, 158–166.

Beadle, C. L., Neilson, R. E., Jarvis, P. G., and Talbot, H. (1981) Photosynthesis as related to xylem water potential and carbon dioxide concentration in Sitka spruce, *Physiol. Plant.*, **52**, 391–400.

Belcher, D. M., Holdaway, M. R., and Brand, G. J. (1982) A description of STEMS—the Stand and Tree Evaluation and Modeling System. *USDA Forest Service General Technical Report NC-79*, North Central Forest Experimental Station, St. Paul, Minn, 18.

Bella, I. E. (1971) *Simulation of Growth, Yield and Management of Aspen*, Ph.D. Thesis, University of British Columbia, Vancouver.

Berry, J. A., and Downton, W. J. S. (1982) Environmental regulation of photosynthesis, in Govindjee (ed.) *Photosynthesis*, Vol II, *Development, Carbon Metabolism, and Plant Productivity*, New York, Academic Press, 263–343.

Bolin, B. (1977) Changes of land biota and their importance to the carbon cycle, *Science*, **196**, 613–615.

Bormann, F. H., Siccama, T. G., Likens, G. E., and Whittaker, R. H. (1970) The Hubbard Brook Ecosystem Study: Composition and dynamics of the tree stratum. *Ecol. Monogr.*, **40**, 377–388.

Bormann, F. H., and Likens, G. E. (1979a) *Pattern and Process in a Forested Ecosystem*, New York, Springer-Verlag, 253.

Bormann, F. H., and Likens, G. E. (1979b) Catastrophic disturbance and the steady state in northern hardwood forests, *Am. Sci.*, **67**, 660–669.

Bosch, C. A. (1971) Redwoods: A population model, *Science*, **162**, 345–349.

Botkin, D. B., Janak, J. F., and Wallis, J. R. (1972) Some ecological consequences of a computer model of forest growth, *J. Ecol.*, **60**, 849–872.

Brix, H., (1968) Influence of light intensity at different temperatures on rate of respiration of Douglas-fir seedlings, *Plant Physiol.*, **43**, 389–393.

Burkhart, H. E., and Strub, M. R. (1974) A model for the simulation of planted loblolly pine stands, in Fries, J. (ed.) *Growth Models for Tree and Stand Simulation*, Res. Notes 30. Department of Forest Yield Research, Royal College of Forestry, Stockholm, 128–135.

Cale, W. G., Jr., O'Neill, R. V., and Shugart, H. H. (1983) Development and application of desirable ecological models, *Ecological Modelling*, **18**, 171–186.

Canham, A. E., and McCavish, W. J. (1981) Some effects of CO_2, daylength and nutrition on the growth of young forest tree plants: I. In the seedling stage, *Forestry*, **54**, 169–182.

Cherkaskin, A. K. (1984) Models of natural and anthropogenic dynamics of forest resources, in *Planning and Prediction of Natural Ecosystems*, Novosibirsk, Nauka, 46–94.

Clutter, J. L. (1963) Compatible growth and yield models for loblolly pine, *For. Sci.*, **9**, 354–371.

Clutter, J. L. (1974) A growth and yield model for *Pinus radiata* in New Zealand, in Fries, J. (ed.) *Growth Models for Tree and Stand Simulation*, Res. Notes 30. Department of Forest Yield Research, Royal College of Forestry, Stockholm, 136–160.

Cooper, C. F., Blasing, T. J., Fritts, H. C., Oak Ridge Systems Ecology Group, Smith, F. E., Parton, W. J., Schreuder, G. F., Sollins, P., Zich, J., and Stoner, W. (1974) Simulation models of the effects of climatic change on natural ecosystems, *Proceedings of the Third Conference on the Climatic Assessment Program* (CIAP), U.S. Dept. of Transportation, Washington, D.C. 550–562.

Curtis, R. O. (1967) A method of estimation of gross yield of Douglas-fir, *For. Sci. Monogr.*, **13**, 1–24.

Davis, M. B. (1981) Quaternary history and the stability of forest communities, in West, D. C., Shugart, H. H., and Botkin D. B. (eds) *Forest Succession: Concepts and Application*, New York, Springer-Verlag, 132–153.

Davis, M. B., and Botkin, D. B. (1983) Sensitivity of fossil pollen records to short term climatic change, (Abstract). *Bull. Ecol. Soc. Am.*, **64**, 156.

Davis, R. B., and Webb, T. (1975) The contemporary distribution of pollen in eastern North America, *Quat. Res.*, **5**, 395–434.

Delcourt, H. R. (1979) Late Quaternary vegetation history of the eastern Highland Rim and adjacent Cumberland Plateau of Tennessee, *Ecol. Monogr.*, **49**, 218–237.

Doyle, T. W. (1981) The role of disturbance in the gap dynamics of a montane rain forest: an application of a tropical forest succession model, in West, D. C., Shugart, H. H., and Botkin, D. B. (eds) *Forest Succession: Concepts and Applications*, New York, Springer-Verlag, 177–184.

Dress, P. E. (1970) *A System for the Stochastic Simulation of Even-aged Forest Stands of Pure Species Composition*, Ph.D. Thesis, Purdue University, Lafayette, Indiana.

Elfving, B. (1974) A model for the description of the structure in unthinned stands of Scots pine, in Fries, J. (ed.) *Growth Models for Tree and Stand Simulation*, Res. Notes 30. Department of Forest Yield Research, Royal College of Forestry, Stockholm, 181–189.

Ek, A. R., and Monserud, R. A. (1974) FOREST: A computer model for simulating the growth and reproduction of mixed forest stands. *Research Report A2635*, College of Agricultural and Life Sciences, University of Wisconsin, Madison, 1 = −14 + 3 appendices.

Emanuel, W. R., Shugart, H. H., and Stevenson, M. P. (1985a) Climate change and the broad-scale distribution of terrestrial ecosystem complexes, *Climatic Change*, **7**, 29–43.

Emanuel, W. R., Shugart, H. H., and Stevenson, M. P. (1985b) Response to comment: Climatic change and the broad-scale distribution of terrestrial ecosystem complexes, *Climatic Change*, **7** (in press)

FAO (1982) *World Forest Products Demand and Supply, 1990 and 2000*, Rome, FAO.

Farquhar, G. D., and Sharkey, T. D. (1982) Stomatal conductance and photosynthesis, *Annual Rev. Plant Physiol.*, **33**, 317–345.

Forcier, L. K. (1975) Reproductive strategies and the co-occurrence of climax tree species, *Science*, **189**, 808–809.

Fries, J. (ed.) (1974) *Growth Models for Tree and Stand Simulation*, Res. Notes 30, Department of Forest Yield Research, Royal College of Forestry, Stockholm.

Fritts, H. C. (1976) *Tree Rings and Climate*, New York, Academic Press, 567.

Funsch, R. W., Mattson, R. H., and Mowry, G. R. (1970) CO_2 supplemented atmosphere increases growth of *Pinus strobus* seedlings. *For. Sci.*, **16**, 459–460.

Goodall, D. W. (1972) Building and testing ecological models, in Jeffers, J. N. R. (ed.) *Mathematical Models in Ecology*, Oxford, Blackwell Scientific Publications, 173–214.

Goulding, C. J. (1972) *Simulation Techniques for a Stochastic Model of the Growth of Douglas-fir*, Ph.D. Thesis, University of British Columbia, Vancouver.

Grimm, E. C. (1983) Chronology and dynamics of vegetation change in the prairie-woodland region of southern Minnesota, U.S.A., *New Phytol.*, **93**, 311–350.

Gross, Von K. (1976) Die Abhängigkeit des Gaswechsels junger Fichtenpflanzen vom Wasserpotential des Wurzelmediums und von der Luftfeuchtigkeit bei unterschiedlichen CO_2-gehalten der Luft, *Forstwissenschaftliches Centralblatt*, **95**, 211–225.

Hägglund, B. (1981) *Forecasting Growth and Yield in Established Forests*. Department of Forest Survey, Swedish University for Agricultural Sciences, Rep. 31 (ISSN 0348-0496), Umea, 145.

Hatch, C. R. (1971) *Simulation of an Even-aged Red Pine Stand in Northern Minnesota*, Ph.D. Thesis, University of Minnesota, St. Paul.

Heygi, F. (1974) A simulation model for managing jack pine stands, in Fries, J. (ed.) *Growth Models for Tree and Stand Simulation*, Res. Notes 30. Department of Forest Yield Research, Royal College of Forestry, Stockholm, 74–87.

Holdridge, L. R. (1947) Determination of world plant formations from simple climatic data, *Science*, **105**, 367–368.

Holdridge, L. R. (1964) *Life Zone Ecology*. Tropical Science Center, San José, Costa Rica.

Hool, J. N. (1966) A dynamic programming—Markov chain approach to forest production control, *For. Sci. Monogr.*, **12**, 1–26.

Horn, H. S. (1976) Succession, in May, R. M. (ed.) *Theoretical Ecology: Principles and Applications*, Oxford, Blackwell Scientific Publications, 187–204.

Houghton, R. A., Hobbie, J. E., Melillo, J. M., Moore, B., Peterson, B. J., Shaver,

G. R., and Woodwell, G. M. (1983). Changes in the carbon content of terrestrial biota and soils between 1860 and 1980: A net release of CO_2 to the atmosphere. *Ecol. Monogr.*, **53**, 235–262.

Huntley, B., and Birks, H. J. B. (1983) *An Atlas of Past and Present Pollen Maps for Europe: 0–13000 Years Ago*, Cambridge, Cambridge University Press.

Hurd, R. G. (1968) Effects of CO_2-enrichment on the growth of young tomato plants in low light. *Ann. Bot.*, **32**, 531–542.

Hurd, R. G., and Thornley, J. (1974) An analysis of the growth of young tomato plants in water culture at different light integrals and CO_2 concentrations, *Ann. Bot.*, **38**, 375–388.

Hustich, I. (1958) On the recent expansion of the Scotch pine in northern Europe, *Fennia*, **82**, 25.

Israel, Yu. A., Filipova, L. M., Insarov, L. M., Semenov, G. E., and Semeniski, F. N. (1983) The background monitoring and analysis of the global change in biotic state, in *Problems of Ecological Monitoring and Ecosystem Modeling*, **IV**, 4–15.

Israel, Yu. A., Antonovski, M. Ya., and Korzukin, M. D. (1985) The prediction of the state of forests in regional background monitoring, *Doklodi ANSSSR* (in press).

Jarvis, P. G., and Leverenz, J. W. (1983) Productivity of temperate, deciduous and evergreen forests, in Lange, O. L., Nobel, P. S., Osmond, C. B., and Ziegler H. (eds) *Encyclopedia of Plant Physiology*, Vol. 12 D, *Physiological Plant Ecology IV*, Berlin, Springer-Verlag, 233–280.

Jarvis, P. G., and McNaughton, K. G. (1985) Stomatal control of transpiration: scaling up from leaf to region, *Adv. Ecol. Res.*, **15**, 1–49.

Jarvis, P. G., Miranda, H., and Muetzelfeldt, R. I. (1985) Modelling canopy exchange of water vapour and carbon dioxide in coniferous forest plantations, in Hutchinson, B. A. (ed.) *The Forest Atmosphere— Interaction*, Dordrecht, Reidel, 521–542.

Johnson, W. C., and Sharpe, D. M. (1976) An analysis of forest dynamics in the northern Georgia Piedmont, *For. Sci.*, **22**, 307–322.

Jones, H. G. (1973) Limiting factors in photosynthesis, *New Phytol.*, **72**, 1089–1094.

Jones, H. G., and Fanjul, L. (1983) Effects of water stress on CO_2 exchange in apple, in Marcelle, R., Clijsters, H, and van Poucke, M. (eds) *Effects of Stress on Photosynthesis*, The Hague, Martinus Nijhoff/Dr W. Junk, 75–93.

Kercher, J. R., and Axelrod, M. C. (1984) A process model of fire ecology and succession in the mixed-conifer forest of the Sierra Nevada, California, *Ecology* (in press).

Koch, W. (1969) Untersuchungen uber die Wirkung von CO_2 auf die Photosynthese einiger Holzgewächse unter Laboratoriumsbedinungen, *Flora*, **158**, 402–428.

Korzukin, M. D., Sedich, K., and Termikaelian, M. T. (1986a) Formulation of a predictive model of the recovery and age dynamics of forests, *Forestry* (in press).

Korzukin, M. D., Sedich, K., and Termikaelian, M. T. (1986b) The application of a prediction model to recover age dynamics of forests to the cedar forests of the central part of the River Ob. *Forestry* (in press).

Kriedemann, P. E. and Canterford, R. L. (1971) The photosynthetic activity of pear leaves, *Aust. J. Biol. Sci.*, **24**, 197–205.

Krizek, D. T., Zimmerman, R. H., Klueter, H. H., and Bailey, W. A. (1971) Growth of crabapple seedlings in controlled environments. Effect of CO_2 levels, and time and duration of CO_2 treatment. *J. Am. Soc. Hortic. Sci.*, **96**, 285–288.

Laiche, A. J. (1978) Effects of refrigeration, CO_2 and photoperiod on the initial and subsequent growth of rooted cuttings of *Ilex cornuta Lindl. et Paxt. cv. Burfordii*, *Plant Propag.*, **24**, 8–10.

LaMarche, V. C., Graybill, D. A., Fritts, H. C., and Rose, M. R. (1984) Increasing atmospheric carbon dioxide: tree ring evidence for growth enhancement in natural vegetation, *Science*, **225**, 1019–1021.

Larsen, J. A. (1980) *The Boreal Ecosystem*, New York, Academic Press, 500 pages.

Leak, W. B. (1970) Successional change in northern hardwoods predicted by birth and death models, *Ecology*, **51**, 794–801.

Lee, Y. (1967) Stand models for lodgepole pine and limits to their application, *For. Chron.*, **43**, 387–388.

Lin, J. Y. (1970) *Growing Space Index and Stand Simulation of Young Western Hemlock in Oregon.* Ph.D. Thesis, Duke University, Durham, North Carolina.

Lin, J. Y. (1974) Stand growth simulation models for Douglas-fir and western hemlock in the Northwestern United States, in Fries, J. (ed.) *Growth Models for Tree and Stand Simulation*, Res. Notes 30. Department of Forest Yield Research, Royal College of Forestry, Stockholm, 102–118.

Lin, W. C., and Molnar, J. M. (1982) Supplementary lightning and CO_2 enrichment for accelerated growth of selected woody ornamental seedlings and rooted cuttings, *Can. J. Plant Sci.*, **62**, 703–707.

Ludlow, M. M., and Jarvis, P. G. (1971) Photosynthesis in Sitka spruce (*Picea sitchensis* (Bong.) Carr.) I. General characteristics, *J. Appl. Ecol.*, **8**, 925–953.

Luukkanen, O., and Kozlowski, T. T. (1972) Gas exchange in six *Populus* clones, *Silvae Genetica*, **21**, 220–229.

Manabe, S., and Stouffer, R. J. (1980) Sensitivity of a global climate model to an increase of CO_2 concentration in the atmosphere, *J. Geophys. Res.*, **85**, 5529–5554.

Mankin, J. B., O'Neill, R. V., Shugart, H. H., and Rust, B. W. (1977) The importance of validation in ecosystem analysis, in Innis, G. S. (ed.) *New Directions in the Analysis of Ecological Systems*. Part I. Simulation Councils Proceedings Series, Vol. 5, Number 1. Simulation Councils of America, LaJolla, California, 63–71.

Masarovicova, E. (1979) Relationships between the CO_2 compensation concentration, the slope of CO_2 curves of net photosynthetic rate and the energy of irradiance, *Biol. Plant.*, **21**, 434–439.

McNaughton, K. G., and Jarvis, P. G. (1983) Predicting effects of vegetation changes on transpiration and evaporation, in Kozlowski, T. T. (ed.) *Water Deficits and Plant Growth* VII, New York, Academic Press, 1–47.

Mitchell, K. J. (1969) Simulation of growth of even-aged models of white spruce, *Yale University School of Forestry Bull.*, **75**, 1–48.

Mitchell, K. J. (1975) Dynamics and simulated yield of Douglas-fir, *For. Sci. Monogr.*, **17**, 1–39.

Morisset, P., and Payette, S. (1983) Tree-Line Ecology, Proceeding of the Northern Quebec Tree Line Conference, *Nordicana*, **47**, 188.

Morison, J. I. L., and Jarvis, P. G. (1983) Direct and indirect effects of light on stomata I., in Scots pine and Sitka spruce, *Plant, Cell and Environment*, **6**, 95–101.

Morison, J. I. L., and Gifford, R. M. (1983) Stomatal sensitivity to carbon dioxide and humidity, *Plant Physiol.*, **71**, 789–796.

Morison, J. I. L., and Gifford, R. M. (1984a) Plant growth and water use with limited water supply in high CO_2 concentrations. I. Leaf area, water use and transpiration, *Aust. J. Plant Physiol.*, **11**, 361–374.

Morison, J. I. L., and Gifford, R. M. (1984b) Plant growth and water use with limited water supply in high CO_2 concentrations. II. Plant dry weight, partitioning and water use efficiency, *Aust. J. Plant Physiol.*, **11**, 375–384.

Morison, J. I. L., and Gifford, R. M. (1984c) Ethylene contamination of CO_2 cylinders, *Plant Physiol.*, **75**, 275–277.

Moser, J. W., and Hall, F. O. (1969) Deriving growth and yield functions for uneven-aged forest stands, *For. Sci.*, **15**, 183–188.

Munro, D. D. (1974) Forest growth models. A prognosis, in Fries, J. (ed.) *Growth Models for Tree and Stand Simulation*, Res. Notes 30, Department of Forest Yield Research, Royal College of Forestry, Stockholm, 7–21.

Namkoong, G., and Roberts, J. H. (1974) Extinction probabilities and the changing age structure of redwood forests, *Am. Nat.*, **108**, 355–368.

Newnham, R. M. (1964) *The Development of a Stand Model for Douglas-fir*, Ph.D. Thesis, University of British Columbia, Vancouver.

Noble, I. R., and Slatyer, R. O. (1978) The effect of disturbance on plant succession, *Proc. Ecol. Soc. Aust.*, **10**, 135–145.

Olson, J. S., and Christofolini, G. (1966) *Model Simulation of Oak Ridge Vegetation Succession*, ORNL/TM-4007, Oak Ridge National Laboratory, Oak Ridge, Tennessee, 106–107.

Peterson, G. M. (1983) *Holocene Vegetation and Climate in the Western USSR*. Ph.D. thesis, University of Wisconsin, Madison, 369.

Raschke, K. (1979) Movements of stomata, in Haupt, W., and Feinleib, M.-E. (eds) *Encyclopedia of Plant Physiology* Vol 7, *Physiology of Movements*, Berlin, Springer-Verlag, 383–441.

Raup, (1957) Vegetational adjustment to the instability of fire. *Proc. Pap. Union Consv. Nature Nat. Resour.*, 36–48.

Regehr, D. L., Bazzaz, F. A., and Boggess, W. R. (1975) Photosynthesis, transpiration and leaf conductance of *Populus deltoides* in relation to flooding and drought, *Photosynthetica*, **9**, 52–61.

Rogers, H. H., Bingham, G. E., Cure, J. D., Smith, J. M., and Surano, K. A. (1983) Responses of selected plant species to elevated carbon dioxide in the field, *J. Environ. Qual.*, **12**, 569–574.

Rowntree, P. R. (1985) Comment on 'Climate change and the broad scale distribution of terrestrial ecosystem complexes' by Emanuel, Shugart, and Stevenson (*Climatic Change*, 7:29–43), *Climatic Change.* 7 (in press).

Shugart, H. H. (1984) *A Theory of Forest Dynamics*, New York, Springer-Verlag, 278.

Shugart, H. H., Crow, T. R., and Hett, J. M. (1973) Forest succession models: a rationale and methodology for modeling forest succession over large regions, *For. Sci.*, **19**, 203–212.

Shugart, H. H., and West, D. C. (1977) Development of an Appalacian deciduous forest model and its application to assessment of the impact of the chestnut blight, *J. Environ. Manage.*, **5**, 161–179.

Shugart, H. H. and West, D. C. (1980) Forest succession models, *BioScience*, **30**, 308–313.

Shugart, H. H., Hopkins, M. S., Burgess, I. P., and Mortlock, A. T. (1981) The development of a succession model for subtropical rain forest and its application to assess the effects of timber harvest at Wiangarree State Forest, New South Wales, *J. Environ. Manage.*, **11**, 243–265.

Shugart, H. H., and Noble, I. R. (1981) A computer model of succession and fire response of the high altitude Eucalyptus forest of the Brindabella Range, Australian Capital Territory, *Aust. J. Ecol.*, **6**, 149–164.

Sionit, N., Strain, B. R., and Hellmers, H. (1981) Effects of different concentrations of atmospheric CO_2 on growth and yield components of wheat, *J. Agric. Sci.*, **97**, 335–339.

Sionit, N., Strain, B. R., Hellmers, H., Riechers, G. H., and Jaeger, C. H. (1985) Long-term atmospheric CO_2 enrichment affects the growth and development of *Liquidambar styraciflua* and *Pinus taeda* seedlings, *Can. J. For. Res.*, **15**, 468–571.

Solomon, A. M., Delcourt, H. R., West, D. C., and Blasing, T. J. (1980) Testing a simulation model for reconstruction of prehistoric forest stand dynamics, *Quaternary Res.*, **14**, 275–293.

Solomon, A. M., West, D. C., and Solomon, J. A. (1981) Simulating the role of climate change and species immigration in forest succession, in West, D. C., Shugart, H. H., and Botkin, D. B. (eds) *Forest Succession: Concepts and Applications*, New York, Springer-Verlag, 154–177.

Solomon, A. M., and Shugart, H. H. (1984) Integrating forest stand simulations with paleoecological records to examine the long-term forest dynamics, in Agren, G. I. (ed.) *State and Change of Forest ecosystems—Indicators in Current Research*, Report 13, Swed. Univ. Agric. Sci., Dept. of Ecology and Environmental Research, Uppsala, Sweden, 333–356.

Solomon, A. M., Tharp, M. L., West, D. C., Taylot, G. E., Webb, J. W., and Trimble, J. L. (1984) *Response of Unmanaged Forests to CO_2-induced Climate Change: Available Information, Initial Tests and Date Requirements.* DOE/NBB-0053. National Technical Information Service, U.S. Dept. Comm., Springfield, Virginia, 93.

Solomon, D. S. (1974) Simulation of the development of natural and silviculturally treated stands of even-aged northern hardwoods, in Fries, J. (ed.) *Growth Models for Tree and Stand Simulation*, Res. Notes 30. Department of Forest Yield Research, Royal College of Forestry, Stockholm, 327–352.

Sullivan, A. D., and Clutter, J. L. (1972) A simultaneous growth and yield model for loblolly pine, *For. Sci.*, **18**, 76–86.

Suzuki, T., and Umemura, T. (1974) Forest transition as a stochastic process. II. in Fries, J. (ed.) *Growth Models for Tree and Stand Simulation*, Res. Notes 30. Department of Forest Yield Research, Royal College of Forestry, Stockholm, 358–379.

Thomas, J. F., and Harvey, C. N. (1983) Leaf anatomy of four species grown under continuous long-term CO_2 enrichment, *Bot. Gaz.*, **144**, 303–309.

Tinus, R. W. (1972) CO_2-enriched atmosphere speeds growth of Ponderosa pine and blue spruce seedlings, *Tree Planters' Notes*, **23**, 12–15.

Tolley, L. C., and Strain, B. R. (1984a) Effects of CO_2 enrichment and water stress on growth on *Liquidambar styraciflua* and *Pinus taeda* seedlings, *Can. J. Bat.*, **62**, 2135–2139.

Tolley, L. C., and Strain, B. R. (1984b) Effects of CO_2 enrichment and water stress on growth of *Liquidambar styraciflua* and *Pinus taeda* seedlings under different irradiance levels, *Can. J. For. Res.*, **14**, 343–350.

Tolley, L. C., and Strain, B. R. (1985) Effects of CO_2 enrichment on growth of *Liquidambar styraciflua* and *Pinus taeda* seedlings grown under different irradiance levels, *Oecologia* (Berlin), **65**, 166–172.

Van Cleve, K., and Dyrness, C. T. (1983) Introduction and overview of a multidisciplinary research project: the structure and function of a black spruce (*Picea mariana*) forest in relation to other fire-affected taiga ecosystems, *Can. J. For. Res.*, **13**, 695–702.

von Caemmerer, S., and Farquhar, G. D. (1981) Some relationships between the biochemistry of photosynthesis and the gas exchange of leaves, *Planta*, **153**, 376–387.

Waggoner, P. E., and Stephens, G. R. (1970) Transition probabilities for a forest, *Nature*, **22**, 1160–1161.

Watson, R. L., Landsberg, J. J., and Thorpe, M. R. (1978) Photosynthetic characteristics of the leaves of 'Golden Delicious' apple trees, *Plant Cell and Environment*,
1, 51–58.

Watt, A. S. (1925) On the ecology of British beech woods with special reference to
their regeneration. II. The development and structure of beech communities on
the Sussex Downs., *J. Ecol.*, **13**, 27–73.

Watt, A. S. (1947) Pattern and process in the plant community, *J. Ecol.*, **35**, 1–22.

Webb, T., III., and McAndrews, J. H. (1976) Corresponding patterns of comtempo-
rary pollen and vegetation in central North America, *Mem. Geol. Soc. Am.*, **145**,
267–299.

Whittaker, R. H., and Levin, S. A. (1977) The role of mosaic phenomena in natural
communities, *Theor. Pop. Biol.*, **12**, 117–139.

Wilkins, C. W. (1977) *A Stochastic Analysis of the Effect of Fire on Remote Vegetation*.
Ph.D. Thesis, University of Adelaide, South Australia.

Wong, S. C. (1979) Elevated atmospheric partial pressure of CO_2 and plant growth.
I. Interactions of nitrogen nutrition and photosynthetic capacity in C_3 and C_4
plants, *Oecologia*, **44**, 68–74.

Woodwell, G. M., Whittaker, R. H., Reiners, W. A., Likens, G. E., Delwiche, C. C.,
and Botkin, D. B. (1978) The biota and the world carbon budget, *Science*, **199**,
141–146.

Wright, H.E. (1984) Sensitivity and response time of natural systems to climatic
change in the Late Quaternary, *Quaternary Science Reviews*, **3**, 91–131.

Yarie, J. and Van Cleve, K. (1983) Biomass productivity of white spruce stands in
interior Alaska, *Can. J. For. Res.*, **13**, 767–772.

Yeatman, C. W. (1970) CO_2 enriched air increased growth of conifer seedlings,
Forestry Chronicle, **46**, 229–230.

Zelawski, W. (1967) A contribution to the question of the CO_2-evolution during
photosynthesis in dependence on light intensity, *Bulletin de l'Academic Polonaise
des Sciences*, **15**, 565–570.

Index

532

Index

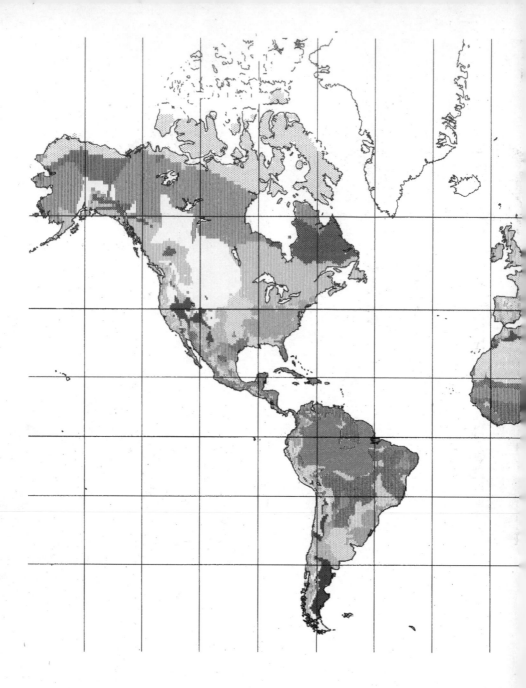

World map of the Holdridge classification with the biotemperature of each of the meteorological stations increased to reflect the climate simulated for a doubling of the concentration of atmospheric CO_2 according to the results generated from a general circulation model as reported by Manabe and Stouffer (1980)